NEXT GENERATION TRANSPORT NETWORKS
Data, Management, and Control Planes

NEXT GENERATION TRANSPORT NETWORKS
Data, Management, and Control Planes

Manohar Naidu Ellanti
Steven Scott Gorshe
Lakshmi G Raman
Wayne D Grover

 Springer

Manohar Naidu Ellanti
Lakshmi G. Raman

Steven Scott Gorshe
Wayne D. Grover

Next Generation Transport Networks
Data, Management, and Control Planes

Library of Congress Cataloging-in-Publication Data

A C.I.P. Catalogue record for this book is available
from the Library of Congress.

ISBN 0-387-24067-5   e-ISBN 0-387-24068-3   Printed on acid-free paper.

Printed in the United States of America.

9 8 7 6 5 4 3 2 1   SPIN 11054764

springeronline.com

Contents

Preface

Are transport networks important? To society? To communications engineering as a field? What about the "fiber glut"? Isn't bandwidth already essentially abundant and free? Despite these and other popular misconceptions, metropolitan, national and international fiber-optic based transport networks are actually one of the engineering marvels of 20th century and have become fundamental infrastructure, crucial to current and future economies and societies. Like many basic civil infrastructures, such as water, roads, power, public health, such engineered systems are almost invisible to the layperson, especially when they work nearly perfectly. But major and unexpectedly severe economic, personal, and societal impacts arise if these systems are removed even temporarily. Like these other basic infrastructures, the fiber optic transport network is now of fundamental importance to our economy, lifestyles, education, entertainment, finance and so on. Advances in computing, wireless, mobility, multi-media, HDTV, the Internet, all come to a halt if it were not for the capabilities of the underlying transport network on which they all ride. The public sometimes asks "What about wireless and cell phones, with them we don't need fiber," but this is based only on technical unawareness that every cell-phone call relies on fiber optic transport for trunking between switches and base stations to complete the calls. Similarly, every DSL and cable modem user of high speed Internet access is also a user of the fiber transport backbone. These "access" technologies, to which we can add phone and bank ATM machines, are best known to us all as users because it is these systems that are "in our face." But all of them rely on a single, ubiquitous, relatively unseen transport network operating behind the scenes.

But there are also important ways in which the transport network differs from the older utility infrastructures mentioned above. Intelligence, survivability, and flexibility is one set of features, but the potential for productivity and wealth impact is an enormous differentiator. In a recent report by the Allen Consulting Group [1], it was estimated that achieving the goal of "true broadband" networking (over 10 Mbps to every home and office) can result in national productivity growth of 10-12%. They comment that this is the fundamental reason why governments should make investment in research networks, ICT, and competitive telecom their number one priority. In their view no other technology, even nanotechnology or bioinformatics is seen as having such a direct and measurable impact on national and global productivity. This means jobs. This means wealth creation. This means a plethora of still unimagined new educational, business, research, recreational, and entertainment possibilities. Transport networks are thus of fundamental importance to a society that wants to reach high and grasp this "prize."

There are many different ways to approach the topic of transport networks. Books typically focus on either the transport (data) plane or the control and management planes, depending the authors' expertise. In order to reduce overhead costs and increase the speed and flexibility of offering new services, however, carriers continually look to automate their operations, administration, maintenance, and provisioning (OAM&P) tasks. The result is an increasingly closer linkage between the transport, management, and control plane technologies. New transport plane technologies are unlikely to be adopted unless they offer OAM&P savings, and new OAM&P tools are unlikely to be adopted unless they can work with the current transport plane infrastructure. One of the objectives of this book is to provide a comprehensive, balanced overview of the transport, management, and control plane technologies. The book is organized with one section devoted to each of these three planes.

Another objective of this book is to provide a useful tutorial and reference information for current and next generation telecommunications network technologies. Since it is not practical for carriers to replace their existing infrastructure, even new equipment and network deployments will need to be compatible with the existing technologies such as SONET/SDH. Both the boom and bust in the telecommunications industry spawned a number of new technologies that are expected to become important in the coming years. Many of these technologies are not covered in existing books. This book provides a detailed tutorial overview of these new technologies, seeking to put them into the proper context with respect to interworking with existing networks and technologies. Examples include the Generic Framing Procedure (GFP), the Link Capacity Adjustment Scheme (LCAS), the

Resilient Packet Ring (RPR) protocol, Automatic Switched Optical Network (ASON), and XML/SOAP (Simple Object Access Protocol) based management technologies. Since TL1 and OSMINE are important for many North American carriers and vendors, a brief discussion of both is included in the management section. Also, discussion of multi-stage switching used in modern digital cross-connects and other switching equipment can be found in Chapter 2 as part of transport technologies. The critical, related topic of network protection is also covered extensively in Chapter 8 to round the book.

What do the four of us especially have to offer on the topic? The authors that Manohar assembled for this book have each been recognized by their peers for their contributions in their respective areas. Each of us has more than 20 years of experience in telecommunications research, product development, and/or standards. Steve and Lakshmi are both recipients of the Committee T1 Alvin Lai Outstanding Achievement Award for their standards contributions. Wayne is an IEEE Fellow for his contributions to survivable and self-organizing broadband transport networks. (More complete biographies of each author appear near the end of the book prior to the Index.) We divided the chapter responsibilities as follows: Manohar – Chapters 2, 7 and 9; Steve – Chapters 1, 3, and 4 (and contributing to 8); Lakshmi – Chapters 5 and 6; and Wayne – Chapter 8.

Due to the multiple authors, the book may have some inconsistencies in style, despite our best efforts to harmonize. If our readers identify any editorial or other types of errors that escaped us, please bring them to our attention - we will update our companion website with corrections to typographical or technical errors as soon as we discover them. This being a very detailed book - we are bound to have some errors and the companion website should be helpful for those looking for corrections. For any questions related to this book, please email Steve Gorshe (sgorshe@ieee.org).

Manohar N Ellanti   Steven S. Gorshe
Fremont, California   Beaverton, Oregon

Lakshmi Raman   Wayne D. Grover
Albany, Oregon   Edmundton, Canada

January 2005

[1] Allen Consulting Group, "True Broadband: Exploring the economic impacts," An Ericsson contribution to public policy debate, September 2003, available online: http://www.ericsson.com.au/broadband/true_broadband.asp

Acknowledgements

Manohar would like to thank his wife, Sumati, and children, Mounika, Vivek for letting him spend time working on this book- numerous versions and revisions of chapters, book proposals and correspondence related to this project. Manohar also likes to thank his co-authors for coming together on this project

Steve would like to primarily thank his wife Bonnie, and boys Alex and Ian for their patience as I worked on the book for many an evening. I would also like to thank Vern Little and Stacy Nichols at PMC-Sierra for their support in allowing me to combine some of my work on the book with my writing of white papers for PMC-Sierra.

The authors would like to express their thanks to the following individuals for their review and input to the material in this book. Thomas Alexander, Tim Armstrong, Richard Cam, Robert Crestani, Huub van Helvoort, Patrice Plante, Winston Mok, Anthony Sack, and Trevor Wilson, Harinder Singh, Rao Lingampalli, Manish Vichare, Mahesh Subramanian, Chiradeep Vittal, Steve Langlois. Manohar would particularly like to mention and thank Jackie Orr in helping with review and contribution to section on OSMINE, a very rarely covered material in other books. Special thanks to Lyndon Ong, who is involved in ITU, OIF and other standards bodies related to control plane, in helping with review of Control Plane chapter and contributing to parts of this chapter. Authors would like to thank Melissa Guasch at Kluwer/Springer for helping with reviews and general correspondence related to this project. Special thanks to Alex Green, the editor at Kluwer/Springer for providing support and encouragement.

Chapter 1

INTRODUCTION TO TRANSPORT NETWORKS

1.1. GENERAL INTRODUCTION

Telecommunications transport networks are the largely unseen infrastructure that provides local, regional, and international connections for voice, data, and even video signals. In fact, most "private" networks are implemented by leased connections through the public transport network infrastructure. Transport networks in telecommunications and data communications networks are changing rapidly with the introduction of new technologies that address the need for new value-added services, high availability, and integration. There has been a considerable amount of effort from equipment vendors and network providers to bridge and unify previously dedicated networks to serve the data and telecommunications market. This effort is reflected in the output of several standards bodies and industry groups and the field trials of new equipment and services.

In the midst of this change, there are two trends that have made transport networks more visible to end-users. The first is the desire for higher bandwidth multi-media enterprise network connections. The enterprise network administrators building the networks have to take into account the various capabilities of the transport networks, including whether they can provide an integration of voice and data services or whether each must be carried on a separate sub-network. The second trend is the increasing deregulation and/or privatization of national telephone carriers. For example, U.S. long distance carriers such as AT&T, Sprint, and MCI can now bypass the local telephone companies, effectively providing direct access for enterprise customers into their transport networks.

Who should care about transport networks, and why? As indicated above, those who construct or administer enterprise networks are often constrained by the capabilities of the public transport networks. Familiarity with transport network technology will allow them to make appropriate decisions regarding WAN connectivity. Service providers working to increase revenues with the introduction of value added services would benefit from understanding how to best utilize the capabilities of the new technologies. There is still a substantial market for transport network equipment, and clearly anyone involved in developing these products needs to be familiar with the existing and emerging technologies. Policy makers should also be familiar with the transport network technologies and their potential impact on policy decisions. Those in academic circles who use or do research related to telecommunications networks also need a thorough understanding of their technology, and the practical constraints on introducing new technologies.

The motivation for this book is to offer, in a single source, information that allows readers with differing requirements to gain a complete picture of the different dimensions of transport networks along with practical perspectives. With the large number of industry standards specifications associated with various aspects of transport networks, it is often difficult to get the big picture view of what will be deployed in the future and how it will facilitate a service provider to meet their business objectives. By bringing together in one place various topics that are spread across multiple specifications, the book enables readers with different goals not only to understand the complete picture but also to access details. The authors also attempt to provide insights on not only what the existing technologies are, but also how they evolved and the constraints and drivers for the future directions. The style of the book is to begin chapters and sections with tutorial background for readers that are new to the subjects. The chapters then move to a more detailed treatment that can be used as a reference for readers more familiar with the subjects. As such, the book is aimed at readers with different levels of prior knowledge, from the student to the network professional.

This introductory chapter begins with a description of transport networks, first from a historical perspective and then from two different taxonomy viewpoints. This discussion also includes a brief introduction to access networks in the context of how they relate to transport networks. With these descriptions in mind, the chapter then summarizes the contents of the remaining chapters of the book. The chapter concludes with a look at some of the current and anticipated future trends in transport networks.

1.2.   WHAT IS A TRANSPORT NETWORK?

In the broadest sense, a transport network can be regarded as the set of facilities and equipment that carry data between the network elements (NEs) that switch or route the customer data into the transport network. These switching NEs use the transport network to carry the customer data to the proper destination, with the transport network being responsible for reliably delivering that data. (Perhaps the simplest definition of a transport network is Simila's Rule "a bit goes in and the bit comes out, no more, no less" regarding the preservation and delivery of the data through the network.[1]). Of course, this definition is somewhat simplistic. As transport networks grow in geographical size and capacity, it becomes increasingly important to have Operations, Administration, Maintenance, and Provisioning (OAM&P) systems associated with the transport networks. Otherwise, it would be impossible to set up and run a transport network with any degree of reliability or cost-effectiveness. As an example, the current lack of adequate OAM&P capabilities has prevented Ethernet from becoming a viable transport network technology except within in networks of very limited scope.

In this section, we examine transport networks from two different viewpoints. In the first approach, transport networks can be broken down according to their geographical or functional scope. The second approach, which provides the outline for the remaining chapters in this book, is to decompose transport networks into the three logical planes; transport, management, and control. To begin the section, however, it is useful to have a brief historical review of the evolution of transport networks to set the discussions in their proper context.

---

[1]   A favorite saying of Ray Simila, former manager of U.S. West's transmission equipment evaluation laboratory who is still very active in the telecommunications field.

Table 1-1. Historical milestone summary of the public telecommunications network

Analog Era	Year	Description	Area
	1876	Telephone Invented	Access
	1878	First Switched Service	Switching
	1879	First Automatic Switch	Switching
	1892	Step-by-step Strowger automatic switch	Switching
	1917	A Carrier (4 voice circuits)	Transmission
	1938	Bell AT&T Crossbar switch	Switching
	1940's	TD carrier microwave (600 voice circuits + video)	Transmission
	1950'	TD carrier microwave	Transmission
Digital Era	1962	T1 digital trunk (24 voice circuits)	Transmission
	1965	1ESS, Computer Controlled Switch	Switching
	1972	Fully Digital Switches, Nortel	Switching
	1976	4ESS Computer Controlled Switch, CCS	Switching, Network Control
	1984	SSN, ISDN	Access, Network Control/Control Plane
Fiber Optic Era	70's-	Kao's paper on the possibility of optical transmission loss of <20db loss per Km, laser, Corning's invention of optical fiber, ubiquitous deployment of fiber, SONET/SDH standardization	Transmission
	1988	First SONET/SDH systems deployed	Transmission
	late 1990s	DWDM systems see significant deployment	Transmission
Policy/Regulation	1984	US Network divided into 160 LATAs	Regulatory
	1996	US Telecom Regulation Act	Regulatory
Mobile era	1985 forward	Introduction of cellular telephone technology made mobile phones practical and attractive.	Access
Internet Era	1975	Vincent Cerf invents TCP/IP protocol	Layer 3 Protocol
	1990's	Netscape, Internet	Applications
Stock Market	Late 1990s	Enthusiasm over the rapidly increasing bandwidth requirements of the Internet cause an unprecedented boom in investment new and existing telephone carriers, and system and component manufacturers.	Market Forces
	2000 -	Bandwidth Glut, Downturn in Telecom and market shakeout.	Market Forces
FR, ATM/QoS	1990s	Deployment of FR/ATM to carry data traffic	Data
OC-192 IP Routers	Late 90's	Juniper's First OC-192 IP router	Data

1.2.1 Historical perspective

The history and milestones of the telecommunications networks is summarized in Table 1-1. The earliest transport network were point-to-point copper cables with separate wires dedicated to each voice channel. (The switching nodes were operators with manual patch panels.) Since this was clearly an unscalable approach, it become important to introduce multiplexing so that multiple voice channels could be carried over the same set of wires. Until the 1960s, when the first digital transmission systems were introduced, the voice channels were transmitted as analog signals. The first multiplexing technology was Frequency Division Multiplexing (FDM) since this was the most appropriate for carrying analog signals. Carrying these analog FDM signals for long distances over copper wires was unattractive since it required running cables through very difficult terrain and required frequent amplification to restore the signal level. Microwave radio transmission provided the answer for many years in the long distance transport networks.

When digital transmission technology was introduced in the early 1960s, it proved to be revolutionary. As those familiar with communications systems are aware, a digital signal can be regenerated such that if there are no bit errors, there will be no degradation to the signal regardless of how many time the signal is regenerated[2]. In contrast, analog signals suffer a decrease in signal-to-noise ratio and some distortion at each point where the signal is amplified. Also, the integrated circuit technology that was also introduced in the 1960s proved much amenable to building low-cost digital circuits than analog circuits. With the introduction of digital transmission technology, the most appropriate multiplexing technology was Time Division Multiplexing (TDM). Digital TDM was used both on copper cable systems and on microwave radio.

In North America, the digital hierarchy was defined by AT&T and referred to as Digital Signal of level n in the hierarchy (DSn). The DS1 signal carried 24 voice channels at a rate of 1.544 Mbit/s, the DS2 multiplexed four DS1s at a rate of 6.312 Mbit/s, and the DS3 multiplexed seven DS2s at a rate of 44.736 Mbit/s. The DSn hierarchy was commonly referred to as the asynchronous hierarchy. In other parts of the world, the hierarchy defined by the CCITT[3] was adopted and referred to as the

---

[2] This, of course, assumes that the signal doesn't accumulate too much jitter at each regenerator.

[3] CCITT stands for International Telegraph and Telephone Consultative Committee. The CCITT changed its name during the 1990s to International Telecommunications Union – Telecommunications Standardization Sector (ITU-T).

plesiochronous digital hierarchy. The PDH signals are the 2.048 Mbit/s signal that carries 30 voice channels, the 8.488 Mbit/s signal that multiplexes four 2.048 Mbit/s signals, the 34.368 Mbit/s signal that multiplexes four 8.488 Mbit/s signals, and the 139.264 Mbit/s signal that multiplexes four 34.368 Mbit/s signals[4].

The only integrated OAM&P capability with analog FDM systems was the inclusion of an orderwire channel in some of the signals. An orderwire is a dedicated, point-to-point voice channel that the crafts people at each of the orderwire channel could use for communications when they configured or maintained that facility or equipment at either end of the channel. One of the characteristics of the DS*n* and PDH hierarchies is that they had very limited integrated OAM&P capability. The DS1 time-shared its framing bit to derive a CRC and its advanced implementations provide a 2 kbit/s OAM&P message channel. The 2.048 Mbit/s channel similarly time-shared its framing byte to derive OAM&P channels. The higher rate signals typically only included some type of error detection bits in the frame. The result was that OAM&P of the network tended to be somewhat labor intensive, and it was difficult to quantify the quality of service being provided to different subscribers (users). In the mid-1980s, it was estimated that over 70% of a telephone companies costs went to OAM&P with less than 10% going to new equipment. Clearly, reducing OAM&P costs was a high priority.

It was also true that providing new services typically took a long time. The long lead times included putting the infrastructure in place to enable those services, but even with the infrastructure in place it would sometimes take weeks or even months before a service could be turned up for a new customer. Long provisioning times and lack of network flexibility thus limited the carriers' ability to bring in new revenue.

Another key historical factor relating to OAM&P was the break-up of the Bell System (AT&T). Now, a typical business connection would involve three different carriers; namely the two local exchange carriers (LEC) and the long distance interexchange carrier (IEC). If a problem existed on a multi-carrier connection, it became extremely important (under the threat of law suits from business customers who were losing revenue) for the carriers to determine whether or not the problem occurred in their network. This situation was a major driver toward more better, more accurate, and faster

---

[4] These PDH signals are often referred to as E1-E4, respectively. The relatively new ITU-T designation for the DS*n* and PDH signals are P11 = DS1, P12 = E1, P21 = DS2, P22 = E2, etc. In this book, the more familiar DS*n* and E*n* designations are typically used. Both the DS*n* and PDH hierarchies have a level higher than the one shown here, but since these saw limited deployment they are omitted.

methods for tracking the performance of connections and facilities, and better fault location and identification. A similar situation occurred in other countries as the government owned carriers were privatized.

Around the same time as the Bell System break-up, fiber optic cables were being deployed in transport networks. Fiber offered huge improvements in capacity and signal quality relative to copper cable systems, and it was often easier to find right-of-way for the cables due to the much smaller size of the optical cables. As the technology improved, it became feasible to begin replacing the microwave radios in the long distance network with optical fibers[5].

Fiber brought some new challenges, but it also offered some critical new opportunities. The early fiber optic systems were built on the existing DSn/PDH multiplexing approach, with each vendor typically using its own proprietary multiplexing frame format for the higher rate signals. Hence, there was little economy of scale and almost no cases where different vendors' equipment could interwork. This meant that at a carrier-to-carrier interface, both carriers would have to agree to a common equipment vendor if they wanted an optical interconnection. The desire for a standard hierarchy for fiber optic signals was one of the primary drivers for the development of the SONET and SDH standards. Since this was a new standard, one of the opportunities was to define a standard that was compatible between North America and the PDH users. As seen in Chapter 3, this objective was largely achieved. The other opportunity derived from the much higher bandwidth capabilities of the optical fiber. With optical transmission, it was now feasible to add a considerable amount of overhead bandwidth that could be used to greatly reduce the cost and improved the capabilities of networks' OAM&P. The combination of the SONET/SDH OAM&P overhead capabilities and the growing availability of computer resources has revolutionized network management and opened the possibility of more automated control of the network.

At this point in history, SONET/SDH forms the backbone of most of the world's transport networks with computer-based network management systems being common. As discussed in the appropriate sections throughout this book, a number of potential future directions are being explored for next generation telecommunications networks. One future direction will be an increasing amount of wavelength-division multiplexing (WDM). WDM is already seeing extensive deployment, and interestingly, is essentially a return to an FDM technology (i.e., a wavelength can be regarded as a carrier frequency). Another direction for transport networks is an increasing capability for efficient, flexible data transport rather than just being

---

[5] Sprint was the first major long distance carrier to convert to an all-fiber long haul network.

optimized for voice traffic. At this time, the focus has been on adding transport capabilities. Some carriers are promoting a migration to carrying and switching all traffic as data traffic rather than using TDM. Multi-Protocol Label Switching (MPLS) is expected to be the core technology in these packet based networks. Voice signals can be packetized and carried as Voice over Internet Protocol (VoIP).

Another important future direction is and an increase in the ability for automated or near-real-time control of the transport network through the introduction of a control plane on top of the management plane as discussed in section 1.2.3. The control plane has the potential to allow much faster and more flexible initiation of new services and modification of services as they are being used. Chapter 9 contains extensive discussion of control plane related aspects.

A historical aspect that can't be overlooked is the increased capability of transport networks to recover from failures in the network. When individual facilities carried relatively few channels, it was not cost-effective to provide a redundant facility or bandwidth to carry the traffic in the event of a failure along the working path. As more channels are multiplexed onto higher-rate signals, it becomes increasingly important not to let a failure disrupt all of these channels. Some of the data traffic that is carried in modern networks is also critical to protect (e.g., communications among air-traffic controllers, and corporate and military data traffic). SONET/SDH integrates some very powerful automatic protection switching capabilities. A number of other protection and restoration options have become feasible due to the increased computing capabilities of network nodes. Chapter 8 is devoted to this topic.

1.2.2   Classification by geography

One traditional approach to classifying transport networks is in relation to their geographic scope. These classifications are illustrated in Figure 1-1. The access network is that portion of the network that connects the end users (subscribers) to the edge switching elements in the network. The metropolitan (metro) transport network is the network that interconnects central offices (COs) within an urban/suburban region. COs within a metro network are typically directly connected to both access networks and core long distance networks. These metro COs are typically owned by the same carrier, and in many cases either allow the carrier to centralize specialized services (e.g., ISDN or Ethernet routing) in just one CO, or to use different COs for back-up redundancy for each other (e.g., to take over switching functions in the event of a failure of the primary CO for that subscriber). The span lengths between metro COs are typically relatively short. The long distance core transport network provides the interconnection between metro networks, smaller community COs, service providers (e.g., Internet), and regional or international gateways.

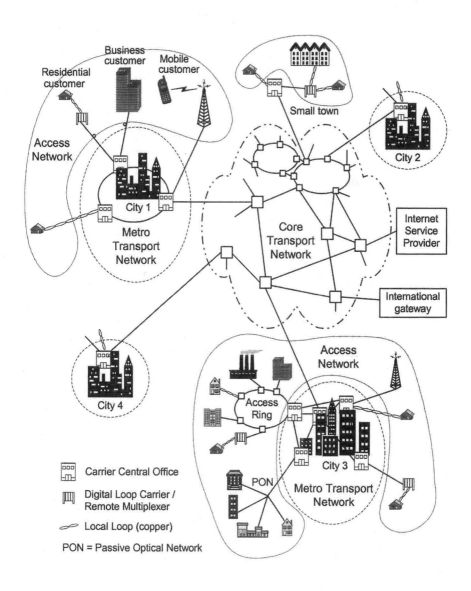

Figure 1-1. Illustration of a telecommunications network

Higher bandwidth technology typically sees its first deployment in the core network since the longer facility lengths necessitate more efficient utilization of the facilities. The technology used in the core networks, however, typically eventually finds its way into the metro network as the cost of the technology decreases and the bandwidth needs of the metro networks increase. From the management, craft training, and equipment inventory perspectives, it is desirable to have as much commonality as possible between the core and metro networks when they exist within the same carrier. LECs typically have both metro networks and core networks to provide interconnection within their region. IECs also typically have both metro and core networks since they often deploy metro networks in order to more efficiently reach their business/corporate subscribers.

Referring again to Figure 1-1, it can be seen that both metro and core transport networks can consist of ring and mesh topologies. Rings have become increasingly popular since they provide inherent route diversity that can be exploited for protection switching. (See City 1 and upper portion of the core network.) Rings have also become increasingly popular in access networks (e.g., City 3). Traffic routing on rings is also more straightforward than in arbitrary mesh networks. Ring topologies are not always convenient, however, due to such constraints as geography or having to use pre-existing right of ways[6]. Arbitrary mesh networks are constructed in order to use convenient cable routings or, in some cases, allow more bandwidth-efficient protection schemes. (See Chapter 8.) Transport networks often consist of a mix of ring and mesh subnetworks, including interconnected rings.

Traditionally, a sharp distinction was drawn between transmission and switching equipment. For the purposes of this book, however, transmission and switching are both considered as part of the transport network. The switches provide the automatic routing of voice (or data) traffic, while the transmission equipment handled the multiplexing and facility connections to carry the traffic between the switches. For example, a voice switch is the equipment to which a subscriber's telephone is connected that does the digit collection when the subscriber dials, and routes the call according to the number that was dialed. Typical transmission equipment includes SONET/SDH terminals. The distinction between transmission and switching has continued to blur over the past 20 years. Transmission networks have increasingly deployed digital crossconnect systems (DCSs) that perform the switching of subscribers' traffic between the various DCS interfaces according to a provisioned route. (See Chapter 2 for a full discussion on switching and crossconnect technology.) DCS-type

---

[6] Of course, a ring can be laid out such that the fibers from different inter-node connections share the same physical right of way or even the same cable. Such rings are called collapsed rings. Collapsed rings don't provide diverse fiber outing in the collapsed portion, and are hence vulnerable to a cable failure (e.g., due to a backhoe) in that region.

crossconnect capability has increasingly been integrated into add-drop multiplexers (ADMs). As illustrated in Figure 1-2, an ADM has two high-speed multiplexed interfaces, each to a different NE. Lower-rate traffic (tributaries) can be added/removed to/from the data transiting the ADM in either direction. ADMs are also typically capable of directly interconnecting two tributaries without that data appearing on one of the high-speed interfaces.[7] A network with switches, DCSs, and ADMs is illustrated in Figure 1-3. Rings nodes are typically ADMs, although they can also be DCSs, while mesh networks are typically constructed of DCSs.

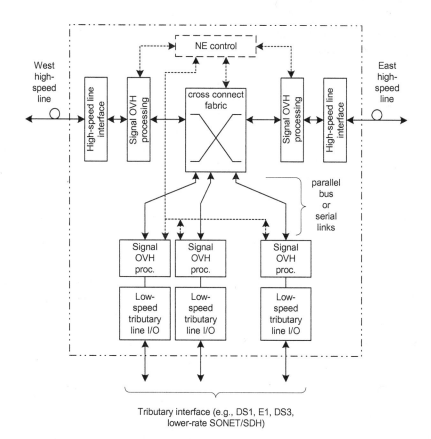

Figure 1-2. Illustration of an Add-Drop Multiplexer (ADM)

---

[7]   This figure illustrates the main general functional blocks in an ADM. Different implementations have used a wide variety of approaches with respect to partitioning these functions among the different printed circuit boards and the interconnections between the functional units.

The distinction between a switch and a DCS/ADM has become a matter of how they are controlled rather than by their switching functions. Switches communicate with each other through a control plane in order to provide the customer-requested connection. (Control planes are introduced in the next section and are discussed in detail in Chapter 9.) DCSs and ADMs rely on provisioning from the network management system (i.e., the management plane that is introduced in the next section and discussed in detail in Chapters 5-7). Switches are very dynamic while crossconnects are relatively static. This distinction is beginning to blur even further, however, as protocols are being developed to allow dynamic re-provisioning of DCSs and ADMs through the control plane. (Again, see Chapter 9.)

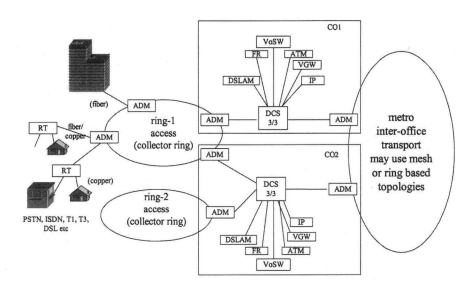

ADM = Add-Drop Mulitplexer

DCS = Digital Cross connect System

DSLAM = Digital Subscriber Loop Access Multiplexer

FR = Frame Relay processing equipment

IP = Internet Protocol processing equpment

ISDN = Integrated Services Digital Network

PSTN = Public Switched Telephone Network

RT = Remote Terminal

VoSW = Voice Switch

VGW = Voice Gateway

Figure 1-3. Illustration transmission and switching equipment

1.2.3   Classification by Logical Layers (planes)

In order to address the complexity associated with transport networks, the well-known methodology of separation into three planes has been introduced. The term logical layers has been used in a couple of contexts and thus it is necessary to understand the difference between them.

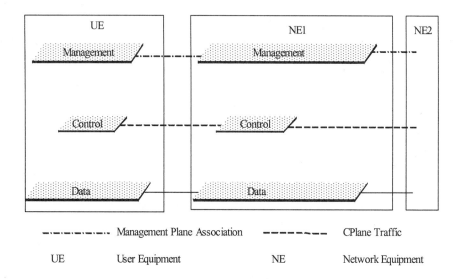

Figure 1-4. Illustration of the three planes[8]

In the first context all the functions associated with providing end-to-end services using the telecommunications networks is divided into three levels of abstraction. These are referred to as data or transport, control, and management planes. Note that in some cases where data and telecommunications merge, new terms such as service plane have emerged and the management plane functions are regarded as part of the control plane. In this book however, the terms are used in the conventional telecommunications sense, as illustrated in Figure 1-4, because of the following logic. A service plane is addressing services offered by the network. To offer a service, the user data is carried in the data plane with the control and management being necessary for enabling and maintaining the service according to the promised quality of service requirements. The need for rapid service introduction has introduced functions such as softswitching to meet the near real time requirements in the control plane compared to the traditional telecommunications practice of setting up a circuit through off-line provisioning, which is part of the management plane. In this context of logical layers, the network elements discussed include data and management plane functions. In some cases they may also include

---

[8] The details of this figure are spread between Chapter 5, 6 and 9 and is presented here as a visual aid. Management and control plane traffic in support of the associations usually flow in the embedded data channels supported by the data plane. However, other options exist, as will be explained in Chapter 9.

control plane features discussed in G.771x series explained in Chapter 9. There are also control plane elements such as signaling control point, signaling switching point network elements that perform only these activities (e.g., setting up a call using an SS 7 network). The management plane functions reside in the network elements and operations support systems discussed below.

The second context in which logical layers is used in the book is in discussing the management architecture in Chapter 5. The management plane functions are separated into different logical layers to provide different views of the management information. Depending on the management function and resources managed, different layers come into play.

1.2.4 Access networks and their relationship to transport networks

Although a full treatment of access networks is beyond the scope of this book, it is worthwhile commenting on several aspects of access networks here that have a bearing on transport networks.

A very large portion of each LEC's capital investment is in the copper wires that connect to the subscribers through the access network. In order to reduce the amount (and length) of access network cable and reduce cable maintenance, carriers increasingly deploy digital loop carrier (DLC) or remote multiplexing (RM) equipment. DLC and RM equipment connect to a number of subscribers over relatively short copper connections and multiplex their signals onto a shared facility (either copper or fiber) back to the CO. A DLC is effectively a remote extension of the voice switch in the CO. RMs include remote digital subscriber line access multiplexers (DSLAMs) that provide DSL service to subscribers. The use of RMs is often necessary in order to reduce the length of the copper connections to the subscribers to a range that will support high-speed data services such as DSL. DLCs and RMs also have the advantage of placing performance monitoring capabilities closer to the customer, which is often very valuable when subscribers use services with a guaranteed service level. The further downstream the network reaches to end subscribers, however, the fewer the number of subscribers that can share that facility or equipment. Due to this reduced sharing, the access facilities and equipment are very cost sensitive.

Access networks typically use transport equipment for connecting to DLCs, RMs and wireless network base stations, and also for dedicated connections to larger business subscribers. For example, many DLCs are connected to SONET multiplexers in the same remote enclosure. Another example is providing DS1, DS3, or SONET interfaces to subscribers from a business campus node with a SONET multiplexer providing the link back to the CO. In many cases, the access networks are deployed with a SONET/SDH ring topology in order to protect against facility or equipment failures. Chapter 8 discusses such protection switching. In a recent development, the IEEE 802.17 Resilient Packet Ring (RPR) technology allows a convenient method for multiple subscribers (typically businesses) to have efficient, flexible data access to SONET/SDH ring networks. RPR is one of the topics covered in Chapter 4. Carriers will typically try to use transport equipment in their access networks whenever the equipment costs allow it, since this commonality can reduce their overall cost of training and equipment inventory. SONET is a prime example of a transport technology that continues to migrate from the transport network to the access network.

One aspect in which the U.S. access network affects the transport network has to do with federal government regulation. In order to encourage competition for local access, the Federal Communications Commission (FCC) requires the incumbent LECs (ILECs) to lease portions of their access network to competitive LECs (CLECs) at a discounted rate. The discounted rate allows the CLEC to provide the same services as the ILEC at a competitive rate. The services, equipment, and facilities that the ILEC must make available to the CLEC are referred to as being unbundled. The nature of the unbundling regulations changes over time. The current regulations include the requirement for the ILEC to provide DSn access to subscribers regardless of whether they are carried over copper cable or are derived from a fiber optic multiplex system (e.g., carried over SONET through the access network). This gives DS1 and DS3 access connections an artificially lower cost for CLECs than equivalent SONET connections[9]. As noted in Chapter 4, this situation can be especially important for the IECs and service providers leasing access to their business subscribers through the ILECs.

Another aspect of the relationship between the access and transport networks is that the access network is often the primary bandwidth bottleneck in the overall network. The amount of traffic carried in the access networks determines the bandwidth needs of the transport networks. For larger business customers, it is usually cost-effective to deploy a dedicated high-speed fiber or copper facility from a CO or a remote fiber node to the business office. The traffic from these connections can often be multiplexed with other traffic in the access network as shown in the City 3 access ring of Figure 1-1. For residential customers, however, it is very expensive to upgrade or replace the copper cables[10] that connect them to the telephone network in order to support broadband service. Meanwhile, there has been

---

[9] ILECs are not required to unbundle their fiber-based interfaces and associated services.

[10] The copper wire connection to a subscriber is typically an unshielded twisted pair of wires. This wire pair is often referred to as a loop (subscriber loop or local loop).

no driving application that motivates subscribers to pay high enough rates to justify the network upgrade. Current technology such as DSL provides adequate speeds for today's applications such as Internet access and telecommuting. So, while this bottleneck is gradually being removed for residential and small business subscribers, it may take many years.

This bottleneck situation has led many in the telecommunications industry to hunt for more cost-effective ways of providing higher bandwidth to residential and small business subscribers. The alternatives fall into four broad categories:

- Higher rates over the existing local subscriber loop
- High-speed data connections through the cable TV network
- Direct fiber connections to the home or near to the home (I.e., FTTx – Fiber to the "x" where x can be a home, curb, business, building, or equipment cabinet.)
- Wireless links

Achieving higher rates over the local loop means deploying remote multiplexers (RMs) much closer to the subscribers in order to keep the loop lengths short enough to allow the higher rates. Very high speed DSL (VDSL) promises rates high enough for video delivery over loops a few hundred to a few thousand feet long. The main cost benefit to this approach is that it re-uses the copper loops. It is still expensive, however, to deploy all the RMs to get close enough to the subscribers. The RMs are connected to the CO through an optical fiber, which means that the each of the RMs needs a local connection for power.

The other existing infrastructure is the cable TV network. Originally, this network was developed for the downstream broadcast of analog video signals. In order to support data services, it is being upgraded to provide an upstream as well as downstream data channels. The connection to the subscriber homes is through a shared coaxial cable with cable modems used by the subscribers for the data connection to the network. Since the coax cable segments are shared by multiple subscribers, the data rate available to each subscriber depends on how many subscribers are using that segment. There may be up to 2000 homes connected to the same segment. The Data Over Cable System Interface Specifications (DOCSIS[TM]) protocol is used to control the medium access for the data. The coax segments are connected to fiber nodes (FNs) that are in turn connected to hub nodes. The hubs are ultimately connected to a head-end for their video signals or to service provider networks for Internet connections. Due to this mix of coax segments and fiber, these networks are commonly revered to as hybrid fiber/coax (HFC) networks. The FNs are analogous to the RMs and the hubs are analogous to either ADMs or DCSs in the telephone network. The main cost advantage of cable modems is the sharing of the coax segments and the inherently higher bandwidth capabilities of coax cable. The cost is similar to the RM/VDSL telephone network approach. Higher rates per subscriber require fewer subscribers per segment, which means more FNs. While this architecture is very good for residential subscriber connections, it is typically not very good for business subscribers who prefer dedicated bandwidth and prefer a non-shared access medium for security reasons. (Also, few businesses have pre-existing connections to the cable TV network.)

The ultimate scenario for telephone and cable TV network providers is to have a fiber optic connection to each subscriber. The fibers can offer virtually unlimited bandwidth to the subscribers. The main cost to fiber connections is the components and circuitry to connect to the fibers. Consequently, passive optical networks (PONs) have been a focus of considerable interest. In a PON, there is a single fiber connection at an optical line terminal (OLT) in the CO, and that fiber branches out to multiple subscriber-end optical network terminal units (ONUs) through a network of passive optical splitters. A time division multiple access (TDMA) protocol is typically used to multiplex data on the fiber tree.

There are two groups actively pursuing PON standards. The first is the Full Services Access Network (FSAN) forum. FSAN has developed a standard referred to as Broadband PON (BPON) that is based on ATM technology[11]. The BPON rates are 155 or 622 Mbit/s downstream (i.e., to the subscriber) and 155 Mbit/s upstream. FSAN recently completed work on a Gigabit PON (GPON) that is based on a GFP-like format. (See Chapter 4 for a full description of GFP. GPON uses modified header information relative to a normal GFP frame and allows fragmentation.) The GPON rates are 2.4 or 1.2 Gbit/s downstream and 155 or 622 Mbit/s upstream. FSAN has brought its work to the ITU-T for publication with G.983 covering BPPON and G.984 covering GPON. The other standards body working on PONS is the IEEE 802.3ah. Not surprisingly, these PONs are based on Ethernet and are referred to as EPONs. EPONs support rates of 10 – 1000 Mbit/s in both the upstream and downstream directions. BPON, GPON, and EPON each have their advantages and disadvantages. While North American carriers have chosen the PONs from FSAN, there is considerable interest in other regions in EPON (e.g., Japan and Korea) where for regulatory competitive reasons carriers need to make a clear distinction between the POTS and data service access networks.

PONs can be very attractive for business customers since they provide a lower cost network for broadband services than point-to-point fibers and

---

[11] An ATM-based PON is commonly referred to as an APON.

they allow the ability to burst data at higher rates (i.e., approach the full PON bandwidth) when the network is lightly loaded. Apart from the cost issues, PONs have one technical issue that has been a major impediment. A major feature of the current telephone network is that the LEC provides power over the local loop to subscribers' phones. Since this telephone company provided power insures that the phones will still operate during power company outages, it is often referred to as lifeline phone service. Clearly this is not possible over an all-fiber network, so the telephone company must either provide an alternative, battery-backed power source, which is very expensive[12], or the subscribers must take responsibility for their own power and maintain their own batteries. Perhaps the growing prevalence of mobile phones will make this a viable option at some point as subscribers become accustomed to idea of being responsible for their own phone batteries. As discussed further below, a number of carriers are counting on this.

The fourth alternative for broadband subscriber access had long been considered the wildcard, but is now looking very promising. That alternative is wireless access. Previously, broadband wireless access required a point-to-point microwave radio link from the subscriber to a base station, with many of these systems requiring a line-of-sight clear path between to two antennas. Here, each subscriber has its own radio channel. One common early system was the Local Multipoint Distribution System (LMDS). These systems were somewhat difficult to engineer and were somewhat costly for many areas. Their main value was for business customers. The Multi-channel Multipoint Distribution Systems (MMDS) addressed some of these shortcomings, but was still expensive for large-scale deployment. The radio technology that appears to be changing the whole situation is IEEE 802.11. With 802.11, multiple subscribers can share a single broadband radio channel. The statistical channel sharing means that subscribers will typically see very high throughput rates. The widespread use of 802.11 technology in LAN applications has driven the cost down to a very attractive point. The IEEE 802.16 technology extends some of the 802.11 concepts to a metropolitan area. The main initial application for 802.16 has been the backhaul of traffic from 802.11 base station sites to central switching sites (e.g., a CO), thus providing metropolitan area network (MAN) connectivity between the 802.11 sites. Some business

---

[12] The cost of maintaining batteries at each subscriber's home makes it a prohibitive option for the LECs. There have been some attempts to solve the problem be having a common power pedestal in the neighborhood that provides a battery-backed power feed to multiple homes. While this is better, it still means a large number of batteries that need routine maintenance and still requires maintaining a copper infrastructure.

subscribers have also started using 802.16 connections as an alternative to wireline DS1 or DS3 service.

Of course, this is a very high-level, somewhat simplified overview of broadband access technology alternatives. The reader may wish to know which of these alternatives will ultimately "win." The answer appears to be that each technology has applications for which it is ideal, and all four will see some degree of deployment. DSL and HFC are already seeing very wide deployment, and 802.11-type wireless access shows tremendous promise. PONs will have the applications for business subscribers, but have seen a much slower growth among residential subscribers. As noted above, there has been a lack of a driving application to justify the cost of upgrading the telephone network local loops to fiber. The application that appears to have the best chance to change this picture is commonly called "triple play" (i.e., voice, data, and video delivery). Cable TV HFC networks clearly have the advantage for video delivery, and the coax cables often give them an advantage for high data rates. Voice services are also now common over these HFC networks, but cable TV HFC networks have typically not provided the same level of reliability that is provided by the telephone network. In many HFC voice implementations (e.g., using Voice over Internet Protocol – VoIP), there is no guarantee of lifeline phone service during power outages. VDSL is required if telephone companies wish to deliver video over the existing subscriber loops, but in order to support the data rates required for High-Definition TV (HDTV), the loop lengths must be very short. Hence, more RMs are required, making VDSL an expensive alternative in many applications. For this reason, some of the telephone companies in Asia and North America are becoming serious about fiber to the home (FTTH) over PON systems. In the U.S., the Federal Communications Commission (FCC) has recently issued a clarification of its rulings to state that an ILEC that deploys a PON system is not required to make it available to CLECs in the same manner as the existing copper systems. Following that FCC ruling, SBC announced in October 2004 that it would begin a very aggressive deployment of PON systems for residential subscribers. Their goal is to provide IP-based triple-play voice, data, and video service. Other U.S. ILECs are expected to follow with similar plans. Separately, the FCC has also mandated that broadcasters begin using digital signals for TV starting in 2007, with a 7-15 window to phase out the use of analog broadcasts. Telephone companies view FTTH as the only viable way to compete with the cable TV companies for triple play services when HDTV (which has a digital format) rolls out. The existence of telephone company FTTH systems will probably spur cable TV companies to also move to an all-fiber network. Wireless access technologies suffer a serious lack of bandwidth for triple play services unless they are coupled with satellite networks for video delivery. Such an 802.11 plus satellite access strategy could be very attractive for many applications.

There are two key points to remember here. First, any broadband access upgrade will place new growth demands on the transport networks. Second, it is services, not technology that will drive access strategies. No one can afford to deploy broadband access infrastructures unless they generate new services that subscribers are willing to pay for.

1.3. SUMMARY OF THE CHAPTER CONTENTS

This introductory chapter introduces transport networks and the various topics that are covered in depth in subsequent chapters. This introduction also touches on other factors such as regulatory policies and access network capabilities that directly or indirectly affect transport networks. The remainder of the book is organized around the concept of the three logical planes that comprise transport networks. The planes, which are described later in this chapter, are the transport, management, and control planes.

1.3.1 Transport Plane (Chapters 2-4 and 8)

This set of chapters cover technologies that carry the data through the network. This is often called the transport or data plane technology.

Chapter 2 discusses fundamentals of digital switching and introduces Crossbar, and multi-Stage Clos techniques used to build SONET/SDH switches. Since scalable switch fabrics are of interest to many, a brief discussion of multi-dimensional switching topologies such as 2-D mesh, hyper cube, and torus is included. Due to the importance of packet switching and the move towards packet-based solutions, a brief discussion of packet switching is also included. Finally, a brief discussion of fault-tolerance of switch fabrics is provided.

After a brief review of some concepts from the legacy asynchronous/plesiochronous networks, Chapter 3 provides an extensive description of the current SONET/SDH networks. SONET/SDH networks form the backbone of the global transport networks. These networks will continue to grow over the next several years due to their excellent track record and the prohibitive cost of replacing them with different technologies. The recent financial difficulties of the telecommunications network providers have made it all the more critical to leverage these SONET/SDH networks. Another factor that secures the ongoing importance of SONET/SDH networks is the flexibility and adaptability that they have shown. The tutorial introduction of Chapter 3 provides a reader unfamiliar with SONET/SDH an understanding of the motivations behind the standards and a basic overview. The details are arranged both by function and by SONET/SDH sub-layer in order to be most readily accessible when the book is used as reference.

Chapter 4 in this division discusses the emerging standards that evolve SONET/SDH networks for efficient data transport and allow more efficient use of optical fibers. The topics of this chapter include virtual concatenation and its associated Link Capacity Adjustment Scheme (LCAS), the Generic Framing Procedure (GFP) that was developed for optimal data transport within SONET/SDH networks, and the IEEE 802.17 Resilient Packet Ring (RPR) that builds on SONET/SDH for access and metropolitan data networks. These technologies will be critical enablers to the growing trend toward providing Ethernet wide area network services over the SONET/SDH infrastructure. Although its extensive deployment will be many years in the future, Chapter 4 also provides a brief overview of the ITU-T G.709 optical transport network (OTN) standard that was developed primarily to carry SONET/SDH client signals in a dense wave-division multiplexed (DWDM) environment.

Chapter 8 is about self-healing in transport networks--in other words protection and restoration processes. This treatment includes a basic discussion of SONET/SDH protection schemes- 1:1, 1+1, SNCP, UPSR, BLSR, RPR, service restoration with LCAS, and the technique of shared backup path protection using GMPLS signaling for protected lightpath services. This chapter goes on to discuss mesh-restoration, a topic that evinces interest of many. The "classic" schemes of span and path restoration are updated with treatments of advanced ideas pertaining to both protection and restoration and the role of both centralized and distributed control. The new service restoration paradigm provided by LCAS is discussed. This approach eliminates the need to pre-allocate network resources to protect services that can tolerate using a reduced bandwidth during the fault condition. The concept of first-failure protection, second failure restoration provides new strategies for ultra-high availability and multi-priority services. Distributed pre-planning offers adaptive self-organized formation of fast-acting protection paths, giving the fastest possible speed but with the adaptability of mesh-restoration. Generalized segment-based protection allows translucent optical networks to be planned in a way that recognized not all nodes may have fault detection and wavelength conversion capabilities. Recent research on Multi-QoP restoration is also covered, showing the way towards a class of mesh-restorable networks with no truly idle "spare" capacity whatsoever. The intriguing new technique of p-cycles, and self-organized formation of p-cycles is also explained. Throughout the chapter insightful and practical perspectives are given on both protection and distributed restoration. For example, why GMPLS "mass reprovisioning" is fundamentally unpredictable, the role of mutual capacity in true path restoration, hybrid networks, SRLG and perspectives on issues such as origins of the "50 ms myth." The treatment is predominantly of a narrative overview and conceptual nature, not a theoretical treatment. References to more in-depth treatments and sources are given throughout.

1.3.2 Management Plane (Chapters 5-7)

Any network, if not properly maintained and managed will yield undesirable performance with the loss of revenue generating customers. The chapters included in this division discuss the management architectures, functions, management information models, and protocols that are necessary to assure a well-maintained network.

Chapter 5 introduces management architecture and discusses various reference points and interfaces, and puts framework and context for rest of this division. The architecture is described in terms of three component architectures, encompassing the functional, physical and information aspects. The functional architectural component includes both the layers of abstraction of management data as well as the management functions performed at each layer defined in various standards M.3010, M.3400, M.3200 and G.7710. The physical architecture introduces the elements of both the transport and management network and communication network that interconnect the physical elements along with a definition of interfaces. The information aspects discussion introduces concepts based on an object oriented information modeling approach to define the management schema required for successful interaction between the managing and managed systems. The key ingredient in network management is the management information and thus it is necessary to keep their semantics independent of the protocols. Defining the management information in a protocol-agnostic manner promotes reusability and minimizes the churn. One method that is popularly used is UML and this chapter contains a brief discussion of this topic. The chapter also includes a brief discussion of the interactions necessary between the control and management planes as some of the traditional management plane functions can be performed by the control plane.

The importance of management information models for successful operation of any network emphasized in Chapter 5 is elaborated with examples in Chapter 6. The chapters include both the requirements in managing SONET/SDH transport networks and an analysis section for each example. The examples also show how the models have been defined both specializing generic technology independent models to SONET/SDH specific requirements as well as those that are relevant only for this technology. Various management functions for managing individual as well as aggregates[13] of network elements (NEs) are included. The approach to modeling presented is based on the current trend in the standards instead of the initial approach taken with the models for SDH in G.774 series for NE management. The chapter uses as examples the models in G.774 series, G.85x series and others available from public forums to show traceability between requirements and the analysis phases. The syntax of the models are not included as they depend on the protocol choices and can be derived without much effort from the analysis phase.

Chapter 7 discusses languages and protocols used in describing and accessing management information respectively. The languages discussed include ASN.1, GDMO, XML, and CORBA IDL. The management information access protocols discussed include SNMP, CORBA IIOP, HTTP, TL1, and CMIP. Augmenting these standards based protocols; this chapter introduces new approaches for developing a new management protocol based on XML/SOAP (Simple Object Access Protocol). XML, used for system-to-system information exchange, is becoming a mainstream technology in many telecom applications. SOAP, discussed in detail in this chapter, is a new transport protocol based on XML. SOAP is widely used by enterprise applications/web services. It is also used as a transport protocol for messages across many messaging interfaces defined in the wireless multimedia messaging standards (MMS). A brief discussion of Bellcore/Telcordia's OSMINE (Operations Systems Modification for the Integration of Network Elements) process is also provided. OSMINE is an important aspect for equipment vendors and carriers, though mainly applicable for ILEC environments.

1.3.3 Control Plane (Chapter 9)

In the recent past, control aspects of transport networks have come under intense discussion. One of the primary reasons for such interest is the operational expense (Op-Ex) associated with operating transport networks using management plane alone. As technologies became available along with computing power in the network nodes, the possibility of automating service creation and network recovery is being looked at as way to save Op-Ex.

Given the importance of control plane and parallels to PSTN, Chapter 9 starts with a discussion of SS7, the first control plane technology introduced into the public network. Concepts such as associated and non-associated

---

[13] The management view of a group of NEs is commonly called the network view. Also, group of NEs and the network they form is sometimes referred to as a subnetwork.

signaling are introduced. Also discussed are inter-carrier signaling interconnection methods used in SS7. Next, the need for the control plane in support of automatic switched optical networks (commonly referred to as ASON) is provided. A detailed discussion of three topics - architecture, routing, and signaling associated with ASON is included. Next, the chapter includes a detailed discussion of PNNI routing as a representative routing protocol while noting that OSPF-TE and IS-IS are two alternative choices for implementing ASON routing. For distributed call and connection signaling, this chapter provides details of RSVP-TE as applied to ASON and GMPLS. The RSVP-TE details are presented in a manner that can serve as a handy reference. A brief discussion of OIF UNI and GMPLS (briefly) is also included.

1.4. WHAT'S HAPPENING WITH TRANSPORT NETWORKS?

This section reviews some of the current status of transport networks and their anticipated future directions. The discussion reviews some of the recent telecommunications history so that the reader can understand the context and rationale behind the status and directions.

1.4.1 Evolution vs. Revolution

The early 1990s saw the first wave of revolution in modern transport networks. SONET/SDH optical transport systems began replacing often-proprietary asynchronous/plesiochronous multiplexing systems that used copper and microwave radio transmission media. SONET/SDH also enabled revolutions in OAM&P through its powerful overhead channels. (See Chapter 3 for the description of SONET/SDH.) Through the latter part of the 1990s, the transport networks saw explosive growth as new fiber and optical component technologies became available and many companies attempted to enter the market to provide long distance bandwidth for the exploding Internet traffic. Unfortunately, there was an over-exuberance that led to an overbuilding of portions of the transport network. In an effort to compete with one another in increasing core network transport capacity, and also in obtaining portions of the wireless access spectrum for the next generation wireless services, carriers exhausted most of their capital. The situation precipitated the severe downturn in the telecom industry in the early 2000s. This situation is starting to turn around as the demand is beginning to catch up to the capacity and portions of the network are in need of upgrade. Carriers are still not in a position to make huge capital expenditures, however. As a result, it is absolutely critical that new technologies be evolutionary; able to build on the existing infrastructure rather than requiring new overlay networks. Another factor that necessitates an evolutionary strategy is the huge investment in network management systems (NMSs). The NMSs not only need upgrade or replacement to support the new features or equipment, the huge craftsperson force has to be trained on using it.

So, what are the drivers and constraints for new equipment and management systems in transport networks? The drivers include the desire to:

- Provide new, revenue-generating services
- Reduce the network OAM&P costs

The constraints include:

- Need to build on (evolve) the existing SONET/SDH infrastructure
- Need to build on existing OAM&P systems and procedures
- Allow a scalable approach to offering ubiquitous service

New revenue is obviously a desirable thing for carriers, but only if it is cost-effective to provide the service. Cost effectiveness is largely determined by this list of constraints.

Reducing OAM&P costs is always important since the operating costs of a transport network are typically much larger than the capital costs. A prime example is SONET/SDH. Its integrated OAM&P capabilities provided a substantial increase in OAM&P capabilities, and allowed automation that substantially lowered the overall OAM&P costs.

For a new service to attract customers, it typically has to be deployed widely enough that a subscriber wanting to use that service will be able to connect to a subscriber at the other end of the network. Since it's obviously not possible to provide a new service ubiquitously, it is desirable to have a scalable solution that can be rolled out based on subscriber demand. One example is the virtual concatenation of existing SONET channels to form a larger channel of a more appropriate size for data transport applications. (Virtual concatenation is discussed in Chapter 4.) One of the key features of virtual concatenation is that it only requires the end-points of the virtually concatenated channel to be aware of its existence, and hence requires no upgrades to the intermediate nodes or their OAM&P procedures. An example of a non-scalable solution for subscriber data transport is leasing "dark" fibers to the subscriber (i.e., fibers that are not currently carrying traffic for the carrier). While dark fibers may be available in certain localized areas, they aren't available across a typical WAN and it would clearly be infeasible to add them for each new subscriber service request.

1.4.2 Looking further into the Future

One of the future themes in the transport network will be convergence. Convergence means being able to transport and manage all types of services with the same transport network. In late 1990s data traffic overtook telephony traffic in terms of public network bandwidth use. The business models associated with this data traffic, however, were not sustainable due to the difference in billing/tariff structures between long distance voice and data traffic. Voice traffic is billed for the use of the circuit for the duration of the connection. Data traffic is bursty and often connectionless, with many of the tariffs structured such that on a bit/second basis data is charged less than voice. This is especially true for Internet connections where high bandwidth channels are leased by the service provider at a bulk rate. Nevertheless, converged network initiatives are beginning to take shape. Depending on regulatory policies, new networks may be engineered to support data first and then engineered to support real time voice traffic. Migration of some voice carrying trunks from the traditional telephone network to new converged networks will not happen overnight. Some of the new technologies need to mature, and new equipment from vendors supporting the new technologies needs to stabilize and mature as well. The primary candidate technologies for the converged network are Multi-Protocol Label Switching (MPLS), and Layer 2 Virtual Private Networks (L2 VPNs), particularly Frame Relay (FR), ATM and Ethernet. Software and protocol-centric solutions will be very important to the emerging data network.

As noted above, the ability to deploy service on a small scale to meet a small set of initial customers and then expand the service as the customer base grows will be very important. One of the key concepts here is that of a sparse network where traffic requiring special treatment is tunneled or concentrated to a select set of nodes that provide those services. This way not all nodes in the network need to be running the new software required to deploy new services. This concept was originally explored by Bellcore for the Integrated Services Digital Network (ISDN) starting in the 1980s. It was believed that updating 10,000 or more CO switches with ISDN-capable call processing and switching software was not an economical and operationally easy thing do. Instead, switches and DLC remote digital terminal (RDT) tunneled the ISDN traffic to a few switches that were upgraded with ISDN capable software. This allowed an LEC to offer ISDN services to its customer with just one or few COs in its region upgraded with ISDN software as opposed to upgrading hundreds of COs. The key aspect was referred to as the virtual CO (VCO). The same virtualization concept will be very important to deploying new data networks and offering existing as well as new data-centric WAN services over such data networks. Over a period of time it will be possible to migrate even voice traffic onto such data networks if reliability and technological problems are solved. The initial demand for data-centric services may not be large enough to install new equipment in every CO or point-of-presence (POP). Transmission capacity is much cheaper today than a decade ago thanks to the decreasing cost of optical components and systems as well as reliable wave-division multiplexing and PON technologies in the metro area. Now services requiring special treatment can be backhauled to few nodes with MPLS/data-centric equipment. Later, as the customer base grows, more nodes can be added to the network.

There are two trends that will have a profound effect on the public switched telephone network (PSTN) if they continue. The one trend is a migration toward carrying voice as well as data in packets, and the other trend is the increasing access rates being offered to subscribers. The increasing access rates were addressed above in section 1.3.4. In addition to the section 1.3.4 applications, however, are the growing importance of Storage Area Networks (SANs). SANs allow customers to periodically back up large databases to remote sites in order to protect against failure or catastrophe at the primary data center. The importance of this remote back-up became particularly apparent in the aftermath of the 9/11 terrorist attacks. In fact, the government is now mandating remote back-ups for some industries.

If voice and data are both carried in packets, then the role of access and edge circuit switches changes or goes away. This is especially true for the large, expensive class 5 switches (e.g., Lucent 5ESS or Nortel DMS) that would be replaced by packet switches. If the digitization can be pushed to the subscriber terminal itself then we have a complete end-to-end digital voice signal with no remaining analog signals in the PSTN. Further, if digitized voice signal can be packetized using an alternative such as VoIP, then we could potentially see the elimination of TDM based circuit switching from PSTN.

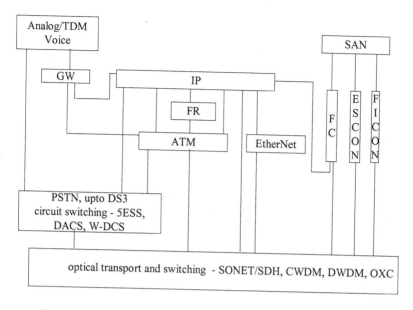

Figure 1-5. Illustration of a circuit-switch-based convergence scenario[14]

There are proponents and opponents of circuit and packet switching. One view is based on circuit switching providing the lowest delay and jitter across the backbone. Circuit switching proponents argue that packet networks can't guarantee low latency and jitter, and therefore some part of carrier network needs to continue to support circuit switching while rest of it can start to migrate to packet switching to support enterprise data connectivity as well as residential broadband. A circuit switched convergence scenario is shown in Figure 1-5. Latency can become an issue at two levels. Once the end-to-end connection latency becomes more that about 20 ms, echo problems caused by imperfect balances in the 2-to-4-wire analog circuit conversion become noticeable. Echo cancellers must be added to the circuit to effectively remove the echo. When the latency gets to the range of hundreds of milliseconds, the delay begins to interfere with speech patterns of the two talkers. Moving the digitization to the subscriber terminal eliminates the source of the echo, which would make latencies of up to 100 ms reasonable in such a fully converged network. Jitter is a bigger problem since the voice packets must compete with variable length (and potentially very large) data packets at each packet switch. The main component of the jitter solution is large buffers at the receiving end of the

---

[14] FICON, ESCON, FC are the storage protocols. The VGW represents a voice gateway – TDM to ATM or TDM to IP.

circuit. However, since increasing the buffer length increases the latency, there are limitations. So, circuit switching still represents a risk-free approach to providing voice quality service. It is anticipated that carriers may start the move towards packet networks and keep voice traffic separate until such time as they feel comfortable with the new packet switching technologies.

Using packet switching instead of routing can provide control for jitter and latency. Packet switching can be used to insure that all the packets associated with a particular flow (e.g., a VoIP connection) always take the same route through the network. Networks that do their switching in this manner are referred to as Connection-Oriented Packet Switched (CO-PS) networks. Traffic engineering, which is available with MPLS as well as other protocols, allows bounds to be placed on jitter and latency by insuring that the path the packets follow has a reasonable number of hops, and has adequate reserved bandwidth and priority for packets carrying jitter and latency-sensitive traffic. Some major carriers see this technology as adequate to begin the deployment of MPLS-based, converged core networks. An illustration of a packet-switched convergence scenario is shown in Figure 1-6.

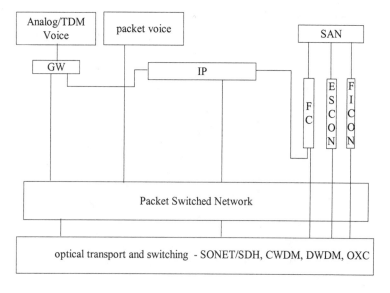

Figure 1-6. Illustration of a packet-switch-based convergence scenario

There is yet another alternative for convergence in the transport network, and that is a move to an all-optical network. In this scenario, each traffic type would be assigned its own wavelength for transmission, and the core network switching would be performed at the wavelength level rather than at the level of the data it contains. There are clearly limits to this approach. Some form of switching and aggregation must be performed at least at the edges of the all-optical network so that the correct information gets mapped into the correct wavelength, and so that enough traffic is aggregated onto a wavelength (through either TDM or packet multiplexing) to be cost effective. It does, however, provide an alternative that may be attractive to long distance providers, and especially to "carriers' carriers" who primarily provide point-to-point connections for other carriers.

Along with the trend toward convergence is a trend toward making better use of network resources for protection and restoration of services. Traditionally, the most typical protection mechanisms have reserved an amount of protection bandwidth that is equal to the working bandwidth that they protect. As discussed in Chapter 8, there has been steady progress on a number of fronts toward greater network efficiency in its use of protection bandwidth.

1.4.3 Advances in Management and Control Planes

The latest advances in the management plane are emerging in both access networks and the multiple technologies present in the networks. Many of the access network management models are addressing the use of ever-expanding Ethernet traffic and ATM in passive broadband optical networks (EPON and BPON). With many different technologies in the network, as noted earlier there is a cost associated with training for different management systems. Industry groups and standards are therefore focusing now at a higher level of abstraction for management of the access, metro and core networks so that the technology specific details are relegated to sub-network management systems, thus reducing both training and interoperability issues. This trend is expected to continue not only because of the different technologies in the network but also because different types of networks (packet technology from internet and circuit technology from telecom networks) are used in conjunction to carry services like voice over IP, multi-media etc.

As has been discussed, the key evolutionary (may be perceived as revolutionary) element is both the separation and merging at the same time of functions across the two planes. In the PSTN the concept of control plane was established almost two decades ago to support 1) automatic call setup (infeasible to be done manually considering the millions of phone calls made on a daily basis) and 2) to optimize resource usage. However, the purpose of introducing a control plane in transport networks (spanning SDH/SONET and OTN technology) stems from a vision with a very similar in end goal to that of the PSTN, and captured and discussed in a framework called Automatic switched optical/transport networks (ASON/ASTN). This is an architecture conceived to offer end-to-end switched connections (electrical or optical) over transport networks spanning one or more carrier networks/domains.   As mentioned earlier in section, ASON/ASTN architecture envisions, using control plane, user-initiated connection setup and tear down akin to telephone call setup and tear down in PSTN supported by PSTN using the SS7 based control plane. While the call volumes today in transport networks may not justify such a bold vision, the transport control plane will be vital for future if on-demand broadband services such as on-demand video, on-demand computing, on-demand storage, and others become viable and popular.   Currently the main driver for control plane introduction in transport networks is operations expense (Op-Ex) savings resulting from automation of service creation. Another lofty goal is speedier and easier interoperability between multiple network domains if they exchange connection setup, status and other relevant information across network-network interfaces (NNI).

Realization of transport control plane vision is made easier, though challenges remain, by increased intelligence and computing power of NEs and new concepts.   See chapter 8 for a discussion of what is called 'self-organizing transport network' in this context.   Availability of transport control plane makes service providers network reconfiguration easy. Such reconfiguration may be needed to meet traffic demand shifts. However, the majority of service provider networks are based on IP (Internet Protocol) and the ability to integrate or bridge the MPLS based control plane with the ASON/ASTN control plane is a challenge though addressed in such user-to-network interfaces (UNI) as OIF UNI1.0 and GMPLS-overlay, both of which are discussed in Chapter 9.   Another possibility, once the control plane is in place, is the ability to construct new software-centric solutions such as virtual CO concept introduced earlier.

1.4.4 Importance of Standards

The role of standards for the transport network has changed during the last 25 years, driven by three major, closely related factors. The first factor was the 1984 breakup of the Bell System in North America (divestiture). Prior to the breakup of the Bell System, AT&T established the de facto standards for North American transport networks by virtue of its near-monopoly status.   In the wake of the break-up, Committee T1 was established under the Exchange Carriers Standards Association (ECSA) and was accredited by ANSI for establishing the U.S. telecommunications network standards. Of the various Committee T1 subcommittees, T1X1 was responsible for carrier-to-carrier interfaces, T1M1 was responsible for various OAM&P definitions, and T1S1 was responsible for switching signaling, including some aspects of ATM. ECSA later changed its name to the Alliance for Telecom Industry Solutions (ATIS) and had just completed a reorganization of its groups that eliminated the Committee T1 layer of the organization and raised several of the former subcommittees to full committee status. Unfortunately, ATIS also chose to change the very well known names of these highly respected groups, in spite of the inevitable confusion that would result. T1X1 became OPTXS and T1M1 became TMOC.

One extremely important consequence of the Bell System breakup was the unleashing of tremendous innovation. While the technology of the Bell System, pioneered by Bell Laboratories, was often second to none in the world, it entered the network according to a carefully controlled corporate business plan. Ideas that failed to win corporate favor were never deployed. Since AT&T bought virtually all its network equipment from its own manufacturing division, there was virtually no opportunity for other equipment vendors to sell equipment into their network. Committee T1 provided a nominally level playing field where different vendors could propose and promote new technology, and individual carriers could request new standards to allow new or better service offerings. SONET (see Chapter 3) was a prime example of multiple equipment vendors working to develop a new signal format and protocol with a great deal of carrier input in spite of initial opposition by AT&T's equipment division (i.e., the remnants of Bell Labs, now known as Lucent). As awesome as Bell Labs record for innovation was, this new environment greatly accelerated the introduction of new technology and often led to much more refined ideas in the final standards.

The second factor in the changing role of standards is the privatization or opening of government-run telephone networks. Previously, it was very common for different countries to run their own telephone networks as part of a national postal, telegraph, and telephone (PTT) administration. It was also common for PTTs to use some of their own proprietary technology in their networks and depend on local companies to build their specialized equipment. Regional standards organizations also existed to establish common signal formats at the interfaces between the national carriers. By far the most important of the regional bodies has been the European Telecommunications Standards Institute (ETSI). Standards for transport network interconnections between different countries falls into the domain of the body that was formerly known as International Telegraph and Telephone Consultative Committee (CCITT), now known as the International Telecommunications Union Telecommunications Standardization Sector (ITU-T). The ITU-T is a branch of the United Nations.

The third major factor is closely tied to the second. With so many different national or regional variations, it was difficult for equipment vendors to achieve the savings that can come from economies of scale. Also, when the networks technology is changing quickly, it is difficult for small, local equipment vendors to provide all the latest technology that their national carriers desire. One of the biggest areas of regional difference was between North America (and those counties such as Japan that adopted the North American formats) and countries that followed ETSI standards. The transport signal formats and even the digital voice encoding techniques were different between these two camps. As discussed in Chapter 3, the optical network standards for SONET and SDH developed in the late 1980s were one of the first major efforts to maximize the alignment of the telecommunications standards on a global basis. Increasingly, carriers have come to consensus that they need global standards in order to drive down the costs of the equipment they purchase, to have ready access to new technologies, and also to reduce the costs of interconnection with other carriers. Equipment vendors, on the other hand, have increasingly seen the value of having their technology or innovations be adopted as part of a standard. While having them standardized opens them up for competing vendors to use, it also insures that there will be a level of industry support that creates a sufficiently large market for their products.

As a result of the growing emphasis on global standards, the ITU-T has increasingly taken on much more importance than the regional standards bodies for new standards. For example, Committee T1 has increasingly chosen to take its new output to the ITU-T for publication rather than publishing a North American version first and then seeking an international version. An example of this conscious decision was the Generic Framing Procedure (see Chapter 4) that was initially developed in T1X1. T1X1 (now OPTX) has increasingly seen its role as the discussion forum and clearinghouse for U.S. transport standards contributions going into the ITU-T Study Group 15 (SG15)[15].

---

[15] With the ITU-T being part of the United Nations, the real voting members of the ITU-T are the national governments. In the U.S., the State Department has recognized different standards bodies as being the authoritative voice for U.S. industry consensus on their area of interest. T1X1 serves that role for the transport network rates, formats, and

A fourth factor that has changed the role of standards has been the growth of industry forums and consortiums. The motivation for forums was the perception (often justified) that the regional standards bodies and the ITU-T moved too slowly and were too narrow in their focus. When a new forum was created, it had the freedom to tailor its scope to its area of interest and to meet as frequently as necessary to make rapid progress. One of the first and most successful of such forums was the ATM Forum. The SONET Industry Forum (SIF) was created to work through various issues associated with deploying SONET networks. Other forums that have had some success include the DSL Forum, the Metro Ethernet Forum, and the Full Services Access Network (FSAN) consortium. Some, like the Optical Industry Forum (OIF) were motivated initially by vendors hoping to promote their technology. OIF ultimately attracted a number of carriers and manufacturers, although its most important output was probably the standardization of chip-to-chip interfaces between telecommunications ICs. The track record of forums relative to the existing bodies is mixed. T1X1, for example, had always been a highly productive and efficient standards body and competed with any forum in terms of quality and timeliness of its output. The ITU-T revamped its procedures to allow it to complete standards in a much more expedient manner. When the telecommunications industry was booming, it was easy to attract participants to a new forum or consortium. During the bust that followed however, companies were forced to prioritize and limit their participation to the bodies that were most important. One result is that some special-topic forums have focused on services or initial implementation agreements that allow services to be deployed prior to the completion of complete standards. Another result is that many forums, such as FSAN, have taken their output as contribution to the ITU-T for publication as an ITU-T standard.

Table 1-2 provides a list of some of the important bodies involved in standards/forum work associated with transport networks, with a brief description of their focus or role.

---

synchronization topics. Hence, contributions targeted at the ITU-T SG15 typically require T1X1 approval before the State Department will authorize them to be submitted to SG15.

Table 1-2. List of some of the important bodies working on standards related to transport networks

Body	Group	Subject area
ITU-T		International Telecommunications Union – Telecommunications Standards Sector
	SG4	Telecommunication management, including TMN
	SG12	End-to-end transmission performance of networks and terminals
	SG13	Multi-protocol and IP-based networks and associated internetworking and OAM issues
	SG15	Optical and other transport networks
ATIS		Alliance for Telecommunications Industry Solutions
	NIIF	Network Interconnection Interoperability Forum – Open forum for issues involving network architecture, management, testing, and operations.
	NIPP	(Formerly T1E1) Network Interface, Power, and Protection Committee – Work includes various UNIs (e.g., SONET or DSL) to the public network
	OPTXS	(Formerly T1X1) Optical Transport and Synchronization Committee – works on network technology for the hierarchical structures and synchronization interfaces for the U.S. telecommunications networks
	PRQC	(Formerly T1A1) Network Performance, Reliability, and Quality of Service Committee
	TMOC	(Formerly T1M1) Telecom Management and Operations Committee – works on OAM&P standards for operations support systems and NE functions.
IEEE 802		Institute for Electrical and Electronics Engineers - *802 is responsible for all LAN and MAN protocols developed by the IEEE*
	802.1	Provider Ethernet bridge, connectivity fault management, Ethernet service OAM
	802.3	10 Gbit/s Ethernet WAN interface, Ethernet OAM, *Ethernet in the First Mile*
	802.17	Resilient Packet Ring (RPR)
ETSI		European Telecommunications Standards Institute
		ETSI's work relating to transport networks includes input to ITU-T, equipment specifications, and various access projects.

Table 1-2 continued

Body	Group	Subject area
Telcordia		Telcordia was originally BellCore, the remnant of AT&T Bell Labs that was formed to serve the research needs of the 7 Regional Bell Operating Companies (RBOCs). Now independently owned by SAIC, it continues to write specifications for equipment and interworking that conform to ITU-T and ATIS standards.
MEF		Metro Ethernet Forum
		MEF work includes service definitions, models and architectures, traffic management, and circuit emulation over networks that use Ethernet.
ATM Forum		The ATM Forum led the development of standards relating to Asynchronous Transmission Mode (ATM) encapsulation and transmission of client signals. Will merge with the MPLS and Frame Relay Alliance.
MPLS and Frame Relay Alliance		The MPLS and Frame Relay Alliance works on applications, deployment, interworking, and interoperability issues for MPLS and Frame Relay networks. Will merge with the ATM Forum.
TM Forum		TeleManagement Forum
		Focuses on "guidance and practical solutions to improve the management and operation of information and communications services."
IETF		Internet Engineering Task Force
		IETF work related to the transport networks includes MPLS-based transport networks, pseudo-wires, and network management.
FSAN		Full Services Access Network consortium
		FSAN works on passive optical network (PON) standards and brings much of its work to ITU-T for final approval and publication.
MSF		Multiservice Switching Forum
		MSF is an association of carriers and equipment vendors that works on implementation agreements for interoperability, and provides input to other standards bodies.

1.5. CONCLUSIONS

The title of the book poses the question whether the introduction of SONET/SDH enhancements as well as the more recent optical transport networks is to be considered as evolutionary or revolutionary. While the reader at the end of reviewing the various chapters may come to their own conclusion, the answer is not a strict yes or no. There are many reasons to consider it to be evolutionary. As an example, there was always the requirement to include management information to indicate when there is a problem in the upstream path of the signal so that no specific action is required by the downstream network elements, thus avoiding an alarm storm. However prior to the introduction of SONET/SDH this information was transferred using different approaches and thus making the interoperability a challenge. The standardization of the overhead bytes alleviated this issue and can be considered to be evolutionary as it is a natural next step once the requirement is established that a standard definition should offer interoperable mid-span meet advantages. On the other hand the equipment vendors, to facilitate better time to market for service offerings by providers, are introducing several disruptive technologies. Many of the network elements are making use of advanced switching, dense computing and storage, and high availability to provide sophisticated algorithms for restoration that can be considered to be revolutionizing traditional equipments in the network. In general, evolutionary technologies are preferred when they provide a path toward a desired carrier goal (e.g., more efficient, flexible, or manageable data transport). Revolutionary technologies need to justify the expense and disruption they bring to the carrier in order to be adopted.

The following chapters address in depth the salient features evolutionary or revolutionary offered by SDH/SONET and OTN networks that will form the backbone of the future networks for the next several years.

REFERENCES

[Bellam]	J. C. Bellamy, *Digital Telephony*, 3rd ed., John Wiley & Sons, New York, 2000.
[BellLab]	Members of Technical Staff of Bell Telephone Laboratories, *Transmission Systems for Communications*, 5th ed., Bell Telephone Laboratories, 1982
[SeRe97]	M. Sexton and A. Reid, *Broadband Networking: ATM, SDH, and SONET*, Artech House, 1997
[KK05]	K. Kazi (editor), *Comprehensive Guide to Optical Networking Standards*, Springer, 2005
[IEEE04]	XML-Based Management of Networks and Services, *IEEE Communications Magazine*, July 2004, pp 56-107

Chapter 2

SWITCHING

2.1. INTRODUCTION

Switches are used in many places in telecommunications networks – class-5 switches at local central offices (CO) to switch voice calls; class-4 switches at tandem central offices to switch inter-office voice calls; digital cross-connects (DCS) at ILEC[16] COs or at IXC POPs to switch intra and inter-office transport signals and so on. Considering the importance and ubiquitous deployment of switching in telecommunications networks, this chapter will try to provide basic information as well as some theoretical results associated with switching.

Switching[17] performed by telecommunications networks is usually called circuit switching where a circuit refers to a voice call or voice call aggregate, such as DS1, DS3, and STS-N. Such aggregate signals are formed by synchronous or asynchronous time division multiplexing (TDM) of lower rate signals as will be explained in Chapter 3. The switches that perform circuit switching are called circuit or TDM switches. They are constructed using single stage crossbar or multi-stage Clos techniques, both of which are explained in this chapter. Switching can be done in space, time, or in combination and will be explained. There can be single or multi-slot (rate) connections and correspondingly single or multi-slot/rate switches (circuit or packet) and all of these concepts are discussed. Also included are

---

[16] ILEC and IXC explained in Chapter 1
[17] Though some discussion here might be applicable, optical switching (switching of light signals) is different.

definitions for strict-sense non-blocking (SSNB), rearrangeably non-blocking (RNB), and wide-sense non-blocking (WSNB. Most importantly, the chapter will provide results from literature on the conditions under which three-stage switching networks are nonblocking for both single-slot and multi-slot connections.

Although the focus of this chapter will be circuit switching, since packet switching is attractive from theoretical standpoint, brief   description of packet switching is also included.   Next, a concept of universal I/O node (UIO) is introduced to draw similarity between transport I/O nodes and processor elements of parallel processing machines. Using the UIO node, some of the interconnection networks used in parallel processing are introduced followed by fault tolerance of the various interconnection networks. Finally, since fault tolerance is an important and mandatory feature of telecom equipment, two techniques used to design fault tolerant circuit switches are explained; redundant switching and fault-tolerant clos (FTC) switching.

2.2. TAXONOMY

Table 2-1 provides classification of packet as well as circuit switched traffic types while Table 2-2 provides classification of connections based on their bandwidth requirements.

Table 2-1. Traffic Types

Traffic Type	Description
Unicast	One input connected to one output only
Multicast	One input connected to more than one output simultaneously
Broadcast	One input connected to all outputs at the same time

Table 2-2. Connection Types

Connection Types	Description	
Circuit Switched	Single Rate	One time slot per connection
	Multi Rate	Multiple Time slots per connection
Packet Switched	CBR	Constant Bit Rate
	ABR (MR, PR)	Available Bit Rate, with some minimum rate and some peak rate

Traffic from video on demand (VoD) or other future applications can be of multicast or broadcast type, though such traffic is not yet common in today's telecommunications networks.

A circuit switched single slot (rate) connection requires only one time slot. The slot can be a 64 Kbps time slot through a switch that switches at 64 Kbps granularity or a 1.5 Mbps (T1) slot a through a switch that switches at T1 granularity. The slot can also be an STS/STM slot associated with SONET/SDH switches. A circuit switched multirate connection requires multiple slots to be connected from input to output through the switch. The switches that are capable of supporting such multi rate connections are called multirate switches and have interesting problems to solve, as will be explained later.

Switches are classified using multiple criteria; Table 2-3 lists, based on blocking property, the various circuit switch types along with definition of packet switch.

Table 2-3. Switch Types based on blocking property

Switch Types	Description	
Circuit Switches	Strict Sense Nonblocking (SSNB)	A switch is said to be strict-sense non-blocking if there is at least one path from a given input to a given output independent of the already existing connections and the path search algorithm
	Wide Sense Nonblocking (WSNB)	A switch is said to be wide-sense non-blocking if it is always possible to setup a path between its idle terminals if, during the operation of the switch, some discipline or heuristic or suitable algorithm is used to assign routes to new calls and suitable repacking of some existing calls is performed after call terminations.
	Rearrangeably Nonblocking (RNB)	A switch is said to be rearrangeably non-blocking if a new call can always be setup, if necessary by rearranging existing connections.
	Blocking	A switch is said to be blocking if a new call from an idle inlet to an idle outlet can't be setup due to unavailability of a path through the switch.
Packet Switches		Switches that use in-band information to determine egress port. Where output port is busy, switches may use scheduling and other complex mechanisms.

Often people use *'switch'*, *'switch fabric','* *matrix,'* and *'switching network'* in the context of switching and such multiplicity of terms needs explanation. Typically, *'switch fabric'* is used to refer to the smallest building block, such as a 32x32 switch matrix, that can be connected to other similar switch matrices to build larger switches. When small switches are connected together in a particular topology or configuration to provide larger switching capacity, such setup is called a '*switching network*'[18]. Generally, all these terms are interchangeable and the reader has to differentiate based on context. . As used in this chapter, the term "*switching network*" should not be confused with telecommunications switching network.

2.3. SPACE DIVISION SWITCHING

This section will discuss techniques used to construct space division switches and applicable to both circuit and packet switching applications.

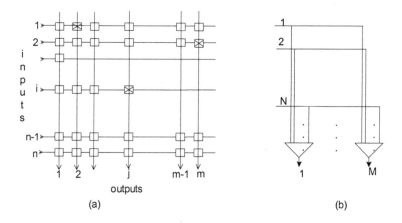

Figure 2-1. Cross-bar switch (a) using cross-points (b) using knock-out selectors at the outlets

2.3.1 Crossbar

The simplest solution to the problem of connecting any input to any output is based on an interconnection network called crossbar and shown in Figure 2-1 (a). First generation automated telephone switching systems were of crossbar type with the cross-points implemented using electro-mechanical relays. With the advent of digital switching systems in 70s', telephone switching systems of today, the cross points are realized using semiconductor gates or memories (or photonic switching devices in an all

---

[18] A virtual switch realized using a set of physical switches in the networks is conceivable. The physical switches involved can be interconnected using any of the interconnection schemes to be presented later.

optical switch). Figure 2-1 (b) shows an alternate way for implementing digital crossbar switches.

The first method is simple and uses cross points using semiconductor gates. The second implementation, at the expense of wiring, brings all inputs to an N: 1 selector and was first used in a packet switch called 'knock-out'[19] [Yu87]. The first method is commonly used to evaluate the performance of crossbar switches. The number of cross-points required represents the cost and is equal to *nxm* (n^2 when *m=n,* and therefore the cost of a cross bar switch is a *quadratic* function of the number of its ports). For switches with small port count, the cost and complexity may not be high. However, to build switches with large port count, such as 512x512 or 1024x1024 port switches[20], the cost of cross points and interconnects becomes too high. But for the cost and complexity, a crossbar switch has the ideal properties of a switch: completely non-blocking and 'any to any' multi-cast capability with a fan-out of *'n'.* It can realize all combinations of unicast input-output combinations[21].

2.3.2 Clos Multi-Stage Switching Network

In 1953, Charles Clos proposed [clos53] a multistage nonblocking switching network to solve the cost and complexity problems associated with crossbar switches used in telecommunications of that time. It was the first proposal with *sub-quadratic* complexity with non-blocking property [Bow77]. Despite many advances in technologies over 50 years, Clos multi-stage switching concept retained its relevance and demonstrated by the fact that many practical switches are based on Clos concept.

Figure 2-2 (a) shows basic representation of a 3-stage NxM asymmetric (symmetric when N=M) Clos switch with N inputs and M outputs while Figure 2-2 (b) shows an alternate representation. The switch has $r_1(=N/n)$ first stage switches, *'k'* second stage switches and, r_2 (=M/m) third stage switches. The notation *C (n, k, m)* will be used to represent a 3-stage Clos switch. Also, without loss of generality, the ensuing discussion will assume *N=M and n=m, i.e.,* the number of inputs is equal to the number of outputs. The Lee graph, shown in Figure2-2 (c), illustrates an interesting point of

---

[19] In a knockout switch, when multiple inputs have packets destined for the same output, the knockout logic at the output, using the N: 1 selector will drop all but one input packet.

[20] Voice Switches used in PSTN can be as big as 100,000x100, 000 ports, albeit 64kbps ports.

[21] The set of supported input-output connection permutations is some times referred to as 'call pattern'. The set of all permissible and realizable input-output combinations may be a subset of theoretical combinations.

Clos architecture- each first stage switch has k ways of reaching a given last stage switch.

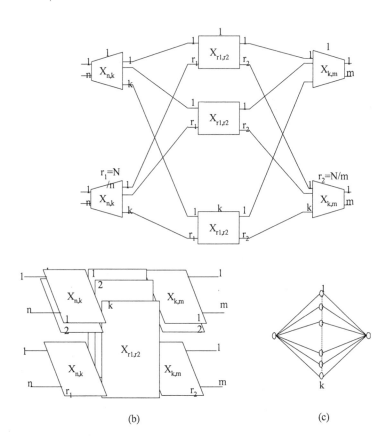

Figure 2-2. Stage Clos Switch (nxkxm) (a) basic view (b) alternate view (c) Lee Graph

As mentioned earlier, for complete connectivity (any to any), normal crossbar switch requires N^2 cross points. In comparision a 3-stage Clos switch, shown in Figure 2-2, needs $2(N/n)\ nk + k(N/n)^2$. If one chooses $n = m = \sqrt{N/2}$ and $k = 2(n-1)$ then the complexity of C (n, k, m) is just under $6N^{3/2}$ [Bow77].

For switches with N=M and n=m, the next section will detail how to choose 'n' and 'k' in order to minimize the total cross point count for a given switch size, N.

2.3.2.1 Strictly Non-Blocking Clos Network

To be strictly non-blocking, the number of middle stage switches, 'k,' in a 3-stage Clos switch must be $\geq 2*n-1$. The result can be understood by considering the worst-case scenario shown in Figure 2-3.

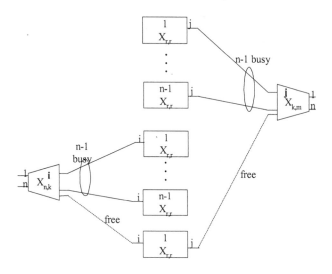

Figure 2-3. Strictly Non-Blocking Clos Network

In this figure, the first stage switch, 'i', is involved in $n-1$ connections thereby using $n-1$ middle stage switches. The third stage switch, 'j', is also involved in another set of $n-1$ connections using separate set of n-1 middle stage switches. Therefore, the total number of occupied middle stage switches k= $2*(n-1)$. Obviously, a new connection, between an idle input on matrix 'i' of 1st stage and an idle output on matrix 'j' of 3rd stage, can be setup if and only if $k \geq 2n-1$. Generalizing this, if $k \geq 2n-1$ then it is always possible to connect any idle input to any idle output regardless of existing connections.

One can derive optimum value of n, the size of the first stage switch, by formulating the total number of cross points, T, in the whole switch in terms of n and then find minimum value of n as shown here:

$$T = N/n * (n*k) + k* (N/n*N/n) + N/n*(k*n) \quad \ldots\ldots\ldots\ldots\ldots(1)$$

Given $k \geq 2n$, for the switch to be completely non-blocking, (1) becomes:

$$T=N*2n + 2N^2/n+ N*2n = 4nN + 2N^2/n \quad \dots\dots\dots\dots\dots\dots\dots(2)$$

The above expression will be minimum when dT/dn is 0:

$$0=4N-2N^2/n^2 \quad \dots\dots\dots\dots\dots\dots\dots\dots\dots\dots\dots\dots\dots\dots\dots\dots(3)$$

Which gives $n = \sqrt{N}/2$.
Example: $N=72$, $n=6$, $k=12$, $r_1= r_2= 12$

Figure 2-4 illustrates 512x512 Clos switch construction ($N=512$, $n=16$, $k=32$)

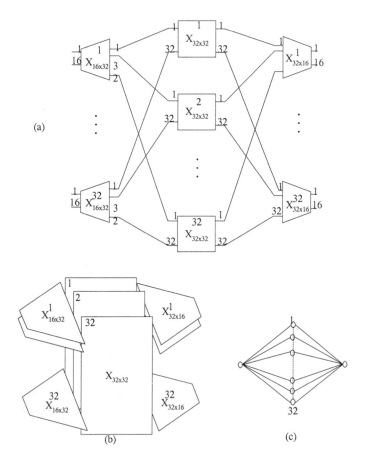

Figure 2-4. Optimal 512x512 Clos 3-stage Switch (a) basic view (b) alternate view (c) Lee Graph

2.3.2.2 **Rearrangeably Non-Blocking Clos Network**

If, k, the number of middle stage switches, equals n, the number of inputs on first stage switches, then the Clos switch is said to have internal blocking. However, by increasing the number of middle stage switches (k>n, but still <2n-1) the switch becomes rearrangeably non-blocking. What this means is that, if a new call can't be setup because all possible paths are already in use then, since k>n and hence there are multiple paths, rearranging existing connections can free up a path that can be used to for the new call. The maximum number of connections that need to be rearranged to make room for the new connection is given by Paul's theorem.

Paul's [Hui90] **Theorem:**

The number of circuits that need to be rearranged is at most min (r1, r3) – 1 , where

r1 is the number of first stage switches and

r3 is the number of third stage switches.

Example:

In a symmetric network r1=r3=N/n. For example, for a 512x512 network n=32, k=16, r1=r2=32. So the maximum number of rearrangements is 31. Since rearrangements also involve   changes in the first and last stage switches, the actual rearrangements is 2*min (r1, r3)-2.

Proof:

> A simple explanation of the result will be provided below. For a formal proof and algorithm, refer to [Hui90].

Without loss of generality assume r1=r3=r.

Assume one wants to establish a connection from a free input on a first stage switch, *'i'*, to a free output on a last stage switch, *'j'*.

As shown in Figure 2-5(a), there will be a free link from *'i'* to some second stage switch *'c'*. Similarly, there will be a free link from *'j'* to some second stage switch *'d'*. If *'c'='d'*, then there is no blocking and therefore there is no further work to be done. Otherwise, switch is in blocking state and can be unblocked by rearranging some existing connections on *'c'* and *'d'*.

As an example, Figure 2-5 (a) shows a switch in blocked state before rearrangement. Shown in this figure is an existing call, P1. This call can be moved so that it is routed via *'d'* thereby freeing up a link from 'c' to 'j'. Since we have a free link from *'i'* to *'c'* and the just freed link from *'c'* to *'j'*, a new connection can be made. In this case, the number of rearrangements is one. Figure 2-5 (b) illustrates the situation after rearrangement.

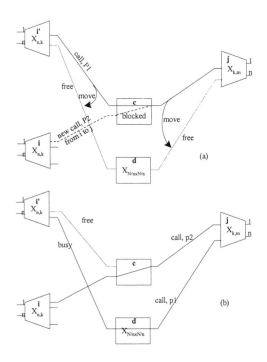

Figure 2-5. Clos Network (k>=n) in (a) before rearrangement (b) after rearranging

The process of rearranging is intuitive, though at first it looks complex. From *'i'*, there has to be one second stage switch, *'c'*, to which *'i'* has a free link. Also there has to be atleast one second stage switch, *'d'*, to which *'j'* has a free link Both *'c'* and *'d'* have at most $r-1$ existing connections, since each one has a free input and output. Now all we need to do is rearrange existing connections on these two switches so that we can make the new connection. Let us start with 'c' and pick an existing path on 'c'. Let us call this P_x. It can be moved from, *'c'* to *'d'* since that will make *'j'* connectable via *'c'*. However, we need to see if P_x conflicts with any existing connection, P_y, on 'd'. If yes, we first need to move P_y onto 'c' and so on. Since there can be at most $r-1$ $(r1=r3=r)$ connections on the second stage switch, 'c' or *'d'*, we see that at maximum we may need to rearrange $r-1$, or in a general case min $(r1, r3)-1$ connections on 'c' and 'd'.

Figure 2-6 illustrates the case for 2-circuit rearrangement.

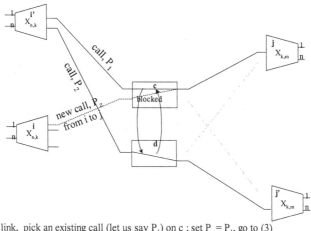

(1) to free c-j link, pick an existing call (let us say P_1) on c ; set $P_x = P_1$, go to (3)
(2) Pick next existing call and assign it to P_x
(3) move P_x to d, however link i-d is already carrying call P_2; set $P_y = P_2$
(4) move P_y to c using i'-c link used by P_x, releasing i-d link. Now move P_x to d using i'-d link used by P_y
(5) Is i-c link , and c-j link free? If yes, stop; else , go back to (2)

Figure 2-6. Rearrangement iteration

Just rearranging around 'c' and 'd' can free the switch from the blocking state. The looping arrows show the iteration around these two switches. This algorithm is also called 'looping'. A formal treatment of rearrangement in Clos networks can be found in [Hui90]. The next section describes general routing issues in Clos networks.

2.3.2.3 Unicast Clos Routing

Routing across a 3-stage Clos network is needed when one or more calls need to be setup or when one or more existing calls need to be rerouted due to some fault. For the first case, one can compute required switch configuration offline or even on-line but the calculation speed is not important. For the second case, re-routing or rearrangement of one or more existing calls because of faults, efficiency is important and is a well-studied topic with different algorithms and techniques proposed to speed up. But for the speed, the basic objective of routing is to find, for a given a call matrix, optimal paths though a multi-stage Clos switching network. Once paths are computed, first, middle and last stage switches are configured. The discussion of Clos routing here will be limited to RNB Clos switching networks. Even with reduced cross point count, a multi-stage SSNB Clos switch is practically large to build for configurations such as 256x256 or 512x512 or higher. By relaxing the strict non-blocking condition, it is possible to design switch such that the overall cost represented by links as well as the number of switching matrices is minimized. With RNB, if a path required for a new call is in use by some other call then, the switch is said to be in blocking condition. The blocking condition, as explained in the previous section, can be removed by rearranging one or more existing connections. In this context, there are two common techniques used for routing or rearrangement- *looping* and *matrix decomposition*. The former was already explained and therefore the next section will discuss the later.

2.3.2.4 Clos Rearrangement using Matrix Decomposition Algorithm

Consider a 9x9 Clos 3-stage switch with n=3, k=3, m=3. Total number of input-output connection permutations will be 16! Let us take one connection permutation p_i that has the following input-output combination shown below with first row representing input and second row representing output.

$$p_i = \begin{pmatrix} Input\# & 1 & 2 & 3 & | & 4 & 5 & 6 & | & 7 & 8 & 9 \\ Output\# & 1 & 5 & 9 & | & 4 & 2 & 6 & | & 7 & 8 & 3 \end{pmatrix}$$

Given a permutation matrix, one can define a connection matrix H_{nxm} whose $(i, j)^{th}$ entry contains the number of connections going from input switch i to output switch j. The entry also represents the number of middle stage switches used to switch all connections from input switch i to output switch j. The following shows H_{3x3} for the above connection permutation:

$$H_{3x3} = \begin{bmatrix} 1 & 1 & 0 \\ 1 & 2 & 0 \\ 1 & 0 & 2 \end{bmatrix}$$

As proposed in [Car93], the above connection matrix can be decomposed into r matrices, each of which can be realized by a single middle-stage switch and an example is shown here:

$$H_{3x3} = \begin{bmatrix} 1 & 1 & 0 \\ 1 & 2 & 0 \\ 1 & 0 & 2 \end{bmatrix} = \begin{bmatrix} 1 & 0 & 0 \\ 0 & 1 & 0 \\ 0 & 0 & 1 \end{bmatrix} + \begin{bmatrix} 0 & 1 & 0 \\ 1 & 0 & 0 \\ 0 & 0 & 1 \end{bmatrix} + \begin{bmatrix} 0 & 0 & 1 \\ 0 & 1 & 0 \\ 1 & 0 & 0 \end{bmatrix}$$

The split matrices provide settings for the middle stage switches. The algorithm provides steps that can be used to split or partition the original connection matrix so that entry at any (i, j) is 1 or 0.

2.3.2.5 Multicast Clos Routing

Theorem:

For a non-blocking switch with n inputs and n outputs, the number of uni-cast connection permutations is n!

Proof:

Consider a 4x4 switch. We have two sets here: input set $\{i_1, i_2, i_3, i_4\}$, and output set $\{o_1, o_2, o_3, o_4\}$. If we choose i_1 then, we can connect it to any outputs and we have four choices. For each of these choices, we can pick another input, let us say, i_2, which can be connected to remaining three outputs and we have 3 choices. Continuing this logic, we can see that we have 4*3*2*1 = 24 combinations. Generalizing for nxn, we have (n-1)*(n-2)*.... *1 = n!

Obviously not all of the combinations are active at the same time in a uni-cast switch. We could not connect i_1 at the same time to $o_1, o_2, o_3,$ and o_4. In a uni-cast switch the total number of connections at any one time =min (n, m).

For a multi-cast switch, the total number of connections (calls) at any one time is less than that for a uni-cast switch. To find the number of permutations, one needs to find the degree of multi-cast or fan out f. For instance, if input port i_1 is multi-casting to all outputs then the remaining inputs, i_2 to i_4, can't be connected to any output, since there are no more idle outputs. Therefore, the total number of connections is one. On the other hand, if the input port i_1 is multi-casting to fraction of total outputs (let us say m/2) outputs then other inputs, $i_2... i_4$, can be connected to the remaining idle outputs in many different ways.

Theorem:

For a non-blocking Clos switch, with n inputs and n outputs, the number of middle stage switches required for a fan out of f:

$$k >= \frac{n + \sqrt{(4f+1)n^2 - 4fn}}{2}$$

For proof, reader can refer to [Dal03].

ATM and multicast switching were active subjects during mid 90's. It was anticipated that distribution of advanced digital services, such as Digital

TV, VoD, using residential fiber networks would take off in a big way creating demand for multicast services. However, due to lack of applications in the early 90's, as well as emergence of alternative broadband distribution networks, such as satellite dish networks, multicast took back seat. However, with the success of DSL and Cable Modem, residential broadband distribution network initiatives are back in vogue albeit with a different name – 'triple-play'. Therefore, there will be renewed interest in multi-cast switching.

In current SONET/SDH networks multicast is used in 1+1 application, though some may not consider it as multicast. In 1+1 protection (c.f., Chapter 8), an input signal is bi-cast onto two output ports. Such bi-cast is also called bridging. Drop and continue used in ring interworking is another application that uses bi-cast.

2.3.3 Recursive Construction

Each of the switches used in the first, second and third stage can be constructed again using the 3-stage factorization method. When a middle stage switch in the basic 3-stage is constructed using another 3-stage network, the result is a 5-stage switch as shown in Figure 2-7.

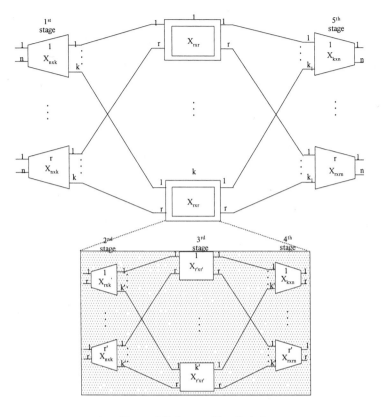

Figure 2-7. 5-stage Clos Switch

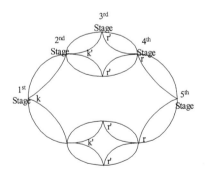

Figure 2-8. Lee Graph for the 5-stage Clos switch in Figure 2-7

2.3.4 Benes Network

When n=m=2 and N is a power of 2, then recursive decomposition of NxN Clos switch   yields a switch that has 2(log$_2$N)-1 stages and is called a

Benes network [Ben65]. The smallest switch will be a 2x2 element. Figure2-9 illustrates a general Benes network along with examples constructions for 4x4 and 8x8 switching networks.

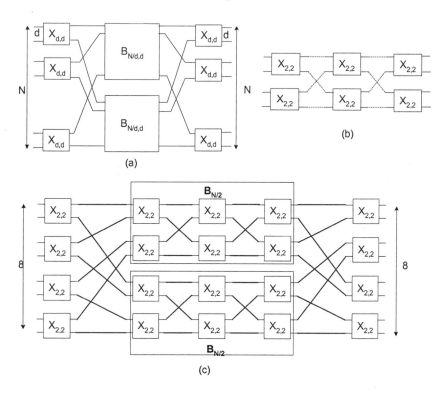

Figure 2-9. Benes Network (a) Generic Construction (b) 4x4 (c) 8x8 (is a 5-stage Clos construction)

2.3.5 Folded Switches

Space switches (single or multi-stage) discussed thus far are two-sided; one side for inputs and one side for outputs. Some applications may require one or more switch outputs to be folded or looped to the input side. Such applications may use static or dynamic folding as shown in Figure 2-10 (a) and (b).

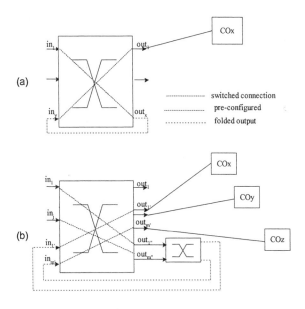

Figure 2-10. Folding Methods (a) Static or hard-wired folding (b) Dynamic/flexible folding

Folded switches were used in the initial days[22] of telephony at remote switching units (RSU). At each RSU, some outputs would be connected back to pre-assigned inputs, which, in turn would be permanently connected to outgoing trunks to the local exchange. When a subscriber, connected to an RSU, makes a call to a remote subscriber, the RSU switches the local subscriber to the folded output. Since the folded output is connected permanently to a trunk facility going to the local exchange, the local CO immediately processes the call without the RSU initiating the time-consuming trunk seizing[23] procedure. It is an old application that may not be used in today's telephone network any more. In the following paragraphs two new applications of folding/looping will be discussed.

A telephone connection involves an input channel and an output channel for each direction of transmission. Given this, if the path through a switch is

---

[22] If broadband networks are built in the future, folded switch application might be used again. It is interesting to note that traffic-engineering extensions being studied in IETF actually mimic the above concept: an IP packet arriving at an input is sent into a Label Switched Path. The label switched path represents a folded output. In SONET switching also, it is possible to use folded switches if one were to off-load VT switching from the main switch onto a smaller switch. All incoming STS-1s that need VT switching can then be switched to a given set of outputs that are connected to a VT switch.

[23] Trunk seizing is used on analog trunk lines such as in R1 DC signaling where on/off is indicated by line seizing which is a slow operation.

uni-directional, then, to support bi-directional connection between terminals connected at ports i and j, one needs to make two paths: inlet i to outlet j, and inlet j to outlet i. On the other hand, with switches that support bi-directional transmission only one path is needed. To understand the implications of this on multi-stage networks, consider two terminals, A and B, connected to a bi-directional switch as shown in (a);

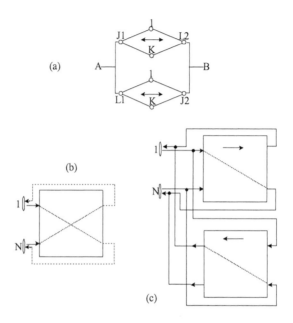

Figure 2-11. (a) Bower's Representation of a bi-directional multi-stage switch; (b) 2-wire and (c) 4-wire folding approaches to get bi-directional switch from uni-directional switches

A is connected at an inlet of a first stage switch matrix J1, and at the like-numbered outlet of last stage switch matrix L1. Similarly, B is connected to first stage switch J2, and last stage switch L2. A telephone call between these two terminals can be setup by finding a path from J1→L2 or from L1→J2 (i.e. path from J1→L2 and L1→J2 are identical). Bowers [Bow77] calls switching networks with such *"permutative-pairing"* (J1→L2=L1→J2) as *"folded switching networks"*. He deduces that a 3-stage folded Clos switching network would need just 'n' (i.e. k=n) middle stage switches to be completely nonblocking as opposed to k=2n-1 in the non-folded case. One can intuitively expect such result based on the fact that out of N terminals connected to a switch, at most N/2 terminals can make calls to the other N/2. Bowers provides detailed analysis of *folded Clos* multi-stage switching networks for single-linkage (v=1) as well as multi-linkage (v>1). Linkage refers to the number of parallel links between switches in consecutive stages. Clos networks with v=1 are the classical type and are called *perfect Clos* networks while those with v>1 are called *imperfect Clos* networks.

A digital crosspoint, by its very nature, is uni-directional and hence switching matrices based on digital crosspoints switch channels in just one direction only. However, one can construct bi-directional digital switches using two approaches (2-wire and 4-wire) as shown in Figure 2-11 (b) and (c). Such bi-directional switches are called *one-sided* or *triangular*.

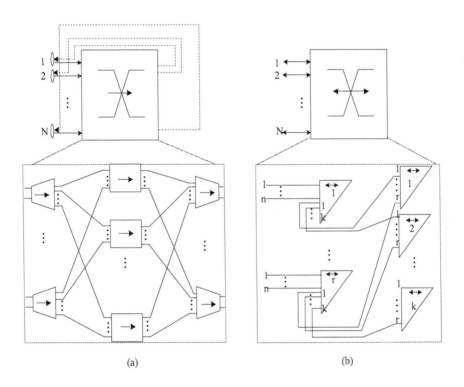

(a)
(b)

Figure 2-12. one-sided/folded multi-stage switch construction (a) using looping (b) using one-sided bi-directional elementary switch matrices

As shown in Figure 2-11 (a), the 2-wire looping concept can be used with two-sided multi-stage switches as well to yield a one-sided/folded multi-stage network. Alternatively, one-sided bi-directional/triangular elementary switch matrices (i.e. as building blocks) can be used to constructed one-sided multi-stage networks as illustrated in Figure 2-11 (b). Each triangular switch matrix can do local switching[24] among its ports without using the next stage. Reader will notice that this construction has only 2-stages.

With the 2-wire looping technique applied to a two-sided multi-stage network, all terminals connect to only one side. As a result the last stage can be folded into first stage. For instance, a 3-stage Clos network can be folded around the middle stage by merging the first, and third stages into one yielding a *one-sided/folded Clos* switch with two stages as illustrated in Figure 2-13 (a).

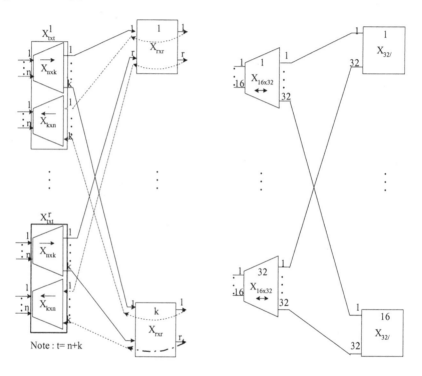

Figure 2-13.(a) Folded Clos Switch (based on looping) construction (b) Example

Each of the shaded boxes in the first stage can be replaced by equivalent one-sided/triangular matrix, and Figure 2-13 (b) illustrates the resulting construction. However, in a 3-stage digital switch with looping or folding, Bower's *"permutative-pairing"* is not possible.

From the standpoint of analysis for non-blocking conditions, one-sided 2-stage Clos networks are identical to their equivalent two-sided 3-stage

---

[24] Also called *hair pinning*. In ADMs, *hair pinning* allows cross-connecting one tributary port to another. In multi-stage switches, hair pinning allows cross connecting at the outer stages without using inner stage thus reducing the probability of rearrangement or blocking.

networks. Table 2-4, from [Jaj03], summarizes theoretical results for folded/one-sided multi-stage Clos networks constructed using triangular elementary digital switch matrices or looping.

Table 2-4. Nonblocking Conditions for one-sided digital Clos Networks

Type of Network	Class	Condition
Triangular	Strict Sense non blocking	1) $k \geq 2n-1$, $r>3$
		2) $k \geq \lfloor 3n/2 \rfloor$, $r=3$
		3) $k \geq n$, $r=2$
	Rearrangeable	$k \geq \lfloor 3n/2 \rfloor$, $r \geq 3$
With Loops	Strict Sense non blocking	1) $k \geq n$, $r \geq 3$
		2) $k \geq n$, $r \leq 3$, n is odd
		3) $k \geq n-1$, $r \leq 3$, n is even
	Rearrangeable	$k \geq \lfloor n/2 \rfloor$

2.4. TIME DIVISION SWITCHING

The previous section focused on space switching. This section will discuss concepts associated with switching time division multiplexed (TDM) traffic arriving at the ports of a switch.

2.4.1 Single-Rate Time Switching

A Time switch (T-Switch) can switch contents of any time slot of one TDM highway, such as SONET/SDH OC-F (F=3,12,48,192, etc), to a time slot of another TDM highway. Such switching is called time division switching. A T-switch that operates on a single TDM input/output highway pair is called a *1x1* T-switch or *'Time-Slot Interchanger (TSI)'*[25].

A 1x1 T-switch/TSI can be realized in two ways. The first one is based on *'all space switch'* approach and is shown in Figure 2-14.

Each input TDM highway is demultiplexed into individual elementary channels, such as STS-1s, and presented to a central space switch. After switching a multiplexer combines the outputs of the central space switch to construct the outgoing TDM highway. The multiplex number, F, determines the size of the central space. For instance, with OC-48 (F=48) TDM highway, the size of the central space switch is 48x48. The operation of the

---

[25] Time Slot Assigner (TSA) is also common and is used to assign tributaries to fixed slots on the line side of ADMs.

overall switch using this approach is simple; each call is interpreted as requiring switching from input time slot, $'t_x'$, to output time slot $'t_y'$. Internally, the call is treated as a call on the space switch requiring switching from input port, P_x, to output port, P_y. The duration of call could be few minutes, as in the case of a voice call, or   days, weeks, months or even years. Longer duration is typical of non-voice call applications such as dialup-data connections, DS1, DS3, or OC-N leased private line services. Whatever may be the duration, the switch settings required for connecting a particular input to a particular output are not changed for the duration of the call.

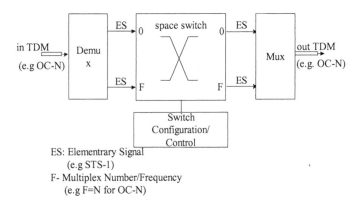

Figure 2-14. T Switch (TSI)- All space switch approach

The second approach to realize TSI, and the most common one, is the memory-based scheme shown in Figure 2-15.

The incoming data from each time slot is written sequentially into data memory (DM). On the outgoing side, DM is read, not sequentially as on the input side, but as per the order stored in the connection memory (CM). By reading CM sequentially 'F' times in a 125-usec period, F locations from DM are read.   The DM is dual-ported and its   speed is such that   F writes and F reads can be performed in one 125-usec window. This type of TSI is called output controlled. One can also construct input controlled TSIs. Both techniques can be found in [Grin83]. To use memories with slower read/write times, odd/even frame buffer techniques are discussed in [Grin83] [Jaj96].

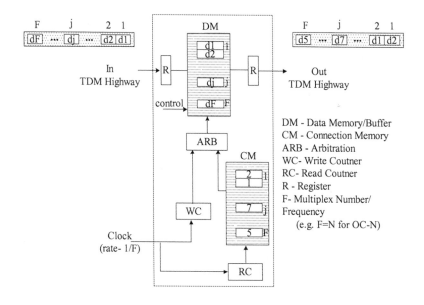

Figure 2-15. T Switch (TSI)- Memory approach

One can equate TSI, since it can interchange any input time slot with any output time slot, to a space switch with '*F*' inlets and '*F*'outles. Because of this, the following equivalence between TSI and space switching can be stated:

$$FxF \text{ space switch } \equiv F\text{-Timeslot TSI}$$

To build T-switches with rxr input/output TDM highways, there are two approaches. The first one is 'an all space switch' approach and is similar to the scheme outlined for the 1x1 T-switch except that the space switch is now larger (F.rXr.F). The second approach is based on sharing, in time domain (F times in a 125-usec period), of a space switch as shown in Figure 2-16.

The size of the space switch is rxr vs. F.rXr.F with the earlier approach. The time-shared central space switch is commonly called time multiplexed space (TMS) switch. The whole construction, consisting of TSI in the outer stages and time-shared space switch in the middle, is called *Time-Space-Time (TST[26])* switch and is used in many practical switching systems. The TST technique uses the fact that if TDM highways are of the same speed and

---

[26] TST is a generic one – actual implementations may use T-S-S-S-S-T (EWSD), T-S-S-S-S-S-T (4 ESS), T-S-T (5 ESS)

are frame synchronized (which is the case in SONTE/SDH networks[27]) then all the highways present F (for the moment assume $F_{ent}=F_{int}=F$) time slots in each 125-usec period with duration of each slot being 125 usec/F. For the duration of a slot, the space switch patches the r inlets with the r outlets as per connection map for that slot. Since there are F slots, the switch needs to store F connection maps.

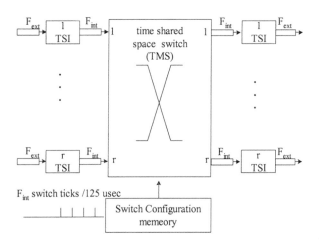

Figure 2-16. Time Shared Space Switch (TST) approach

As one might have noticed, the TMS switches in space domain only and correspondingly its switching function can be expressed as: <port-x, slot-a> to <port-x, slot-a>. If one requires switching from <port-x, slot-a> to <port-x, slot-b>, then either the input or output stage TSI blocks must perform time slot interchange between 'a' and 'b'.

Figure 2-17 (a) shows parameters used to characterize TST; all input TDM highways have F_{ext} time slots, all internal TDM highways have F_{int} time slots. This figure also captures time and space paths from ingress to egress; f_i time paths from a given outer stage switch to a given middle stage switch and, k space paths from each outer stage switch to the middle stage. Next, (b) illustrates the space equivalent of TST with each TSI replaced by a space switch ($F_{ext}xF_{int}$) and the TMS replaced by a set of F_{ext} parallel space switches of rxr capacity each.

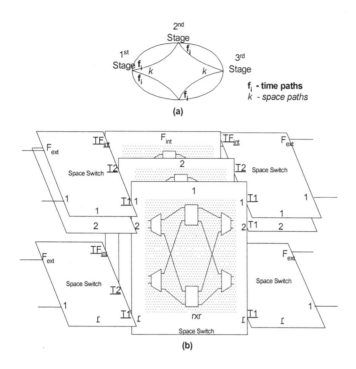

Figure 2-17. (a) TST switch parameter model and (b) space switch equivalent

From this equivalent figure, it is easy to see that F_{int} must be at least twice F_{ext} for the TST to be non-blocking. What this means is that internal TDM rate (and hence the speed at which the space switch is shared in time domain) must be at least 2x F_{ext} for the TST switch to be non-blocking [Weber92]. For more on TST and its use in practical switching systems used in PSTN reader can consult [Bellamy].

In practical SONET/SDH transport switches, an I/O module (IM, can be one or multiple PCBs) collects traffic from m (m≥1) tributary ports and multiplexes them onto a high-speed TDM signal whose aggregate rate is ≥ the sum of the signal rates at the trib ports. The notation (mxn) will be used with IM where 'm' is the number of tributary interfaces and 'n,' the number of high-speed interfaces towards the central switch fabric. The high-speed TDM signals are transmitted to a central switch fabric (a TMS or a TST) using either standard or proprietary signaling. Optical Interworking Forum (OIF) has standardized various electrical interfaces and Figure 2-18 illustrates points of applicability of some of them. They are discussed further in §7.8.

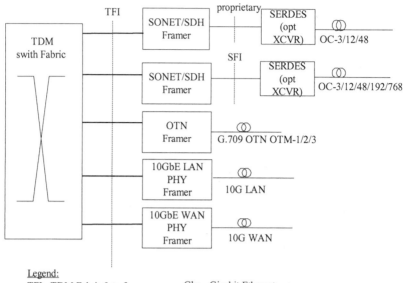

Figure 2-18. Emerging TDM Fabric to I/O interface standards

Also, currently software programmable framers are available that can operate at various speeds (OC-3, 12, 48,192) thus adding flexibility to design and reducing IM inventory cost. The XCVR module converts optical to electrical and vice versa and is usually a pluggable module based on industry standard form factors (MSA GBIC, SFP etc - more on this later).

An mxn IM that terminates 'm' tributary I/O ports may support local switching between any of its I/O ports (also called hair-pinning, to be explained later) without using central switch fabric. If traffic arriving on the I/O ports of an ingress IM is destined for a port on an egress IM then the traffic is sent by the ingress IM to a central TDM fabric.

A 1x1 IM can and may need to perform TSI. Note that TSI is defined as a function that interchanges time slots between two highways of the same rate. A more generic term used to denote the function of switching a time slot of an input highway onto a time slot of an output highway is 'cross-connect'.

The complexity of a TST switch is associated with TSI function in the first and last stages, as well as additional hardware and software required for reconfiguration of space switch in time domain. For instance, a TST for switching time slots between OC-48 highways (i.e. $F_{int}=48$) must reconfigure the central space switch every 125 $\mu sec/48 \cong 2.5$ μsec.

Advances in chip technologies enable one to implement the central space switch and all outer stage TSI blocks in one chip or on a single PCB. Such a chip or PCB is referred to as digital switching matrix (DSM[28]) [Cha79]. They are available in such capacities as 8x8, 64x64, 256x256 (E1/T1 ports) or even higher capacities and are used in practical switching systems used in PSTN. For SONET/SDH switching applications, switching ICs in such capacities as 32x32, 72x72 (OC-48) are available at the time of writing of this book.

As in the case of space switches, outputs of two-sided DSMs/TSTs can be looped back to the input side to construct bi-directional DSM/TSTs (such DSM will be referred to as one-sided/triangular DSM).

Larger time switches can be constructed using smaller TST/DSM matrices as basic building blocks using multi-stage techniques discussed for space switches. Figure 2-19 illustrates general multi-stage time switch construction. Also shown is an example realization for a 256x256 (Pulse Code Modulation-PCM E1 ports) time switch using 8x8 elementary DSMs [Jaj83]. The switch is capable of switching 64kbps voice calls.

---

[28] Also called Digital Symmetrical Matrices. However, it is not necessary that they be symmetrical.

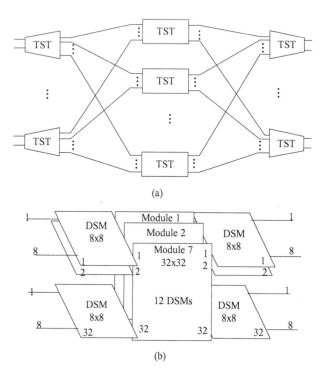

(a)

(b)

Figure 2-19. multi-stage Time Switch (a) Clos construction (single linkage) (b) 256x256
TDM circuit switch   using elementary DSM/TST matrices

Table 2-5. Jajszczyk's non-blocking conditions for uni-rate circuit switches composed of DSM (n, f_{ext})

Switching Network Type	Non blocking condition for 3-stage	Maximum channel capacity (time slots)
Two-Sided	$k \geq 2 \left\lfloor \dfrac{n f_{ext} - 1}{v f_{int}} \right\rfloor + 1$	$N_{3,maz} = \left\lfloor \dfrac{n+1}{2} \right\rfloor fn$
		$N_{5,maz} = \left(\left\lfloor \dfrac{n+1}{2} \right\rfloor \right)^2 fn$
		$N_{s,maz} = \left(\left\lfloor \dfrac{n+1}{2} \right\rfloor \right)^{\frac{s-1}{2}} fn$
One-Sided	$k \geq 2 \left\lfloor \dfrac{n f_{ext} - 1}{v f_{int}} \right\rfloor + 1$, r>3	
	$k \geq \left\lfloor \dfrac{3 n f_{ext} - 2}{2 v f_{int}} \right\rfloor + 1$, r=3	
	$k \geq \left\lfloor \dfrac{n f_{ext} - 1}{v f_{int}} \right\rfloor + 1$, r=2	

Though mulit-stage time switches have both time and space dimension, they can be analyzed for non-blocking conditions using an equivalent space switch with multiple links, as many as the multiplex number (F_{int}), between consecutive stages. For both one-sided and two-sided switching networks composed of DSMs, non-blocking conditions are provided by Jajszczyk [Jaj83] and summarized in Table 2-5; v is the number of links between matrices of adjacent stages, f_{ext} is the multiplex number of each link, and f_{ext} the multiplex number of the TDM I/O ports.

The crosspoint measure as a cost (used with space switches) can't be used with multi-stage circuit switches constructed using elementary DSMs. Instead, one needs to measure the cost interms of the total number of DSMs, and interconnects or links required for connecting switches of consecutive stages as Table 2-6 illustrates (assuming v=1, $f_{int}=f_{ext}=f$).

DSMs may not come in all sizes thereby forcing designers to use only one type of DSM. Therefore, one may need to find maximum number of channels that can be supported by a 3, 5 or s-stage time switch constructed using single type of elementary symmetrical DSM (n,f) where n is the number of inlet/outlet TDM links, and f the multiplex number of the TDM links. Table 2-5 provides maximum channels for 3, 5 and general s-stage switching networks composed of DSMs.

Table 2-6. DSM count for 3-stage switching networks

Overall Switch Capacity	Elementary DSM Capacity	Total # of DSMs	DSM count for 1st / 2nd / 3rd stages	Total Interconnects
2048x2048	64x64	192	64/64/64	8192
1024x1024	64x64	96	32/32/32	2048
882x882	64x64	84	42.5/42/42.5	3698
512x512	32x32	96	32/32/32	2048

To give a practical example, a 1728x1728 OC-48 switch[29] is reported in [Ann01] using single-linkage 3-stage Clos construction and given by C $(n=36, k=72, m=36)$ with $r_1=r_2=48$, $f_{ext}=f_{int}=48$. The asymmetry of the first and third stage switching matrices requires use of DSM's of different size. Naturally, the reported construction has (48) first stage DSMs of 36x72, (72) second stage DSMs of 48x48 and, (48) third stage DSMs of 72x36 for a total of 168 DSMs. Alternatively, one may use single type of DSM across all stages. However, in such a case some ports of DSMs will be unused and therefore one needs to find solution to the problem of left-out ports. The solution to this is to fold the first and third stages as explained earlier and reduce the switching network to a 2-stage construction. In such configuration, first stage switches will need to be of $n+kXn+k$ (or at least of $3nx3n$, since $k>=2n$). Using n=24, the above 3-stage construction yields a folded 2-stage Clos switch with 120 symmetric/square DSMs of 72x72.

Note, though, that the formula for $N_{3,max}$ from Table 2-5 can't be used with a 2-stage folded construction. This is because a 2-stage folded switch still is a 3-stage switch. For instance, the 1728x1728 2-stage folded switch explained above is in fact a 3-stage switch given by C $(n=24, k=72, m=24)$ with $r_1=r_2=72$, $f_{ext}=f_{int}=48$ except that the first and last stage switch are combined into one IC.

Besides the cost of the switch measured in DSM count, there is also cost associated with managing interconnections between various stages, and scalability and upgrade cost (albeit not quantifiable) stemming from large number of interconnections. There are two options to reduce interconnections:

The first option is to avoid multiple stages by designing larger switching capacity in one DSM (realized on a single VLSI chip or a PCB). The internal realization of such large DSM may be based on multi-stage concept.

The second option, useful when one can't avoid but design multi-stage systems, is to use techniques such as optical wavelength division multiplexing (WDM) proposed in [Oki03]. For instance, with the 512x512

---

[29] 4Tbps one-direction/8Tbps bi-directional

switching network shown in Figure 2-4, if all first stage switches can be realized on a single PCB then, the links from this PCB to the PCB housing the middle stage switches can be collapsed into 32 DWDM links, with each link carrying 16 wavelengths. In general, technological advances will continue to shape cost optimization of multi-stage switch constructions. As per Jajszczyk [Jaj03], approaches to optimization of cost of switching network structures have changed over years, taking into account different measures of cost (number of crosspoints, number of DSMs/switch matrix LSIs, etc), or various constraints imposed by different technologies (e.g., the number of waveguide crossover in the case of some types of photonic switches [Vid94]).

2.4.2 Multi-Rate Time Switching (Generalized Circuit Switching)

The above discussion on TST assumes single-slot connections. However, transport and multi-service switches must be capable of switching multi-rate (slot) connections. For instance, a SONET/SDH switch must be capable of supporting any combination of STS-3/3c, STS-12/12c, STS-48/48c connections. Before discussing issues with multi-slot connections, Figure 2-20 shows generalized multi-linkage (v=1 or higher), 3-stage switching network composed of DSMs.

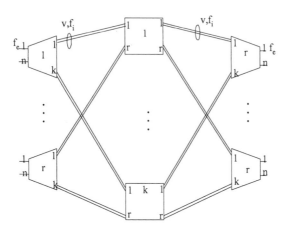

Figure 2-20. Multirate, multi-stage, multi-linkage TDM circuit switch basic structure

The outer stages consists of DSMs(n:f_{ext}, k.v:f_{int})[30] with 'n' TDM external multiplex links and k.v internal TDM multiplex links (to center stage). The middle stage consists of DSMs(k.v:f_{int} ,k.v:f_{int}) with 'v' TDM multiplex links to each matrix in the two outer stages.

There are two approaches to support multi-slot connections. In the first approach, "b" slots of a b-slot connection are switched *independently*. The slots are put together at the egress port. With this switching approach, SONET switches based on STS-1 granularity may need special techniques[31] to switch a contiguous concatenated signal, such as STS-12c. The order in which the channels of a multichannel call appear at the input must be preserved through the switch and is called time slot sequence integrity (TSSI) [Jap88]. However, the independent switching approach may not guarantee TSSI unless special techniques are used [Jap88],[Jaz96]. Also the way in which idle slots are assigned to new call can have an impact on TSSI and nonblocking conditions. Nonblocking conditions for multi-slot switches using this approach are the same as for single-slot switches and already given in Table 2-5.

The second approach involves letting b slots of a b-slot connection occupy the same multiplex link and is called the *parallel* switching approach.   There are three methods of assigning b-slots to a multiplex link[]: random, periodic, contiguous (or bursty). The random assignment doesn't impose any restriction and the b-slots can be assigned to idle slots of a multiplex link in any way. In periodic assignment, the selected idle time slots have equal spacing between them. In contiguous or bursty assignment, the b-slots are assigned to b continuous idle slots of a multiplex link. As per Jajszczyk, the last two methods, though better from TSSI perspective, result in higher probability of blocking. One should note that TSSI will not be a problem if slot assignment is not restricted to frame boundaries. Schemes to preserve TSSI without impacting blocking probability are described in [Jap88].   Kabaciński provides definitions for SSNB and WSNB for multi-rate switches; a multi-stage switching network is non-blocking for multi-slot connections in the strict sense if it is possible to setup a new multi-slot connection   independently of the path search algorithm and the time-slot assignment; if the switching network is nonblocking when special path search or time slot assignment is used then the switching network is said to be nonblocking in the wide sense.

---

[30] A slightly different notation than used earlier, is used here to include port count, and TDM multiplex number of both sides of a DSM. In case of one sided DSM with 'n' I/O ports, and k.v ports to internal stages, the new notation will mean DSM(n,f,k.v,f) with f_{ext}=f_{int}=f.

[31] Such as virtual concatenation or arbitrary concatenation (see chapter 3)

For both SSNB and WSNB, Kabaciński, based on Jajszczyk results for uni-rate 3-stage switching networks composed of DSMs, provides nonblocking conditions and listed in Table 2-7, for the *parallel* switching approach.

With v=1, (the most common case), and $f_{ext}=f_{int}=f$, the WSNB result reduces to $k \geq 2 \left\lfloor \frac{nf-b}{f-b+1} \right\rfloor$. The result is arrived by considering the worst case shown in Figure 2-21.

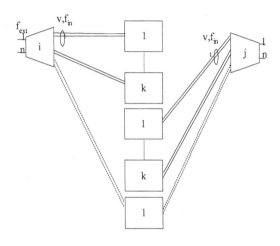

Figure 2-21. Multirate, multi-stage, multi-linkage TDM circuit switch basic structure

A first stage DSM that can accommodate a new b-slot call has at most nf-b time slots that are in use on the links connecting it to switch modules in the second stage. Since an internal link cannot be used by a b-slot call if more than f-b+1 slots of that link are in use, $\left\lfloor \frac{nf-b}{f-b+1} \right\rfloor$ links from the first stage switch matrix to the second stage are busy. Using similar logic, $\left\lfloor \frac{nf-b}{f-b+1} \right\rfloor$ links are busy from center stage to the last stage switch matrix. To accommodate the new b-slot call from the above first stage matrix to the above last stage switch matrix, one more middle stage switch is needed and this completes the proof. The SSNB result is derived by constraining that calls use no more than B time slots, and using worst case time slot assignment (contiguous assignment), and using similar reasoning as given for the WSNB case. Also, each internal link is assumed to have worst-case fragmentation with b-1 contiguous idle slots between used slots. The number of channels a link can carry with b-1 contiguous idle slots between each used slot is f/b. The number of links required to support nf-b channel is therefore $\left\lfloor \dfrac{nf - b}{f/b} \right\rfloor$. Special cases (r<4) are explained in Kabaciński's paper.

Table 2-7. non-blocking conditions for multi-rate connections through a multi-stage DSM switching network

Type of Network	SSNB $\quad$ Note: $c = \dfrac{f_{int}}{b}$	WSNB
One Sided	$k \geq \max\limits_{1 \leq b \leq B} 2\left\lfloor \dfrac{nf_{ext} - b}{vc} \right\rfloor + 1$ when r>3 $k \geq \max\limits_{1 \leq b \leq B} 2\left\lfloor \dfrac{3nf_{ext} - 2b}{vc} \right\rfloor + 1$ when r=3 $k \geq \max\limits_{1 \leq b \leq B} 2\left\lfloor \dfrac{nf_{ext} - 2b}{vc} \right\rfloor + 1$ when r=2	$k \geq 2\left\lfloor \dfrac{nf_{ext} - B}{v(f_{int} - B + 1)} \right\rfloor + 1$ when r≥4 $k \geq \left\lfloor \dfrac{3nf_{ext} - B}{2v(f_{int} - B + 1)} \right\rfloor + 1$ when r=3 $k \geq \left\lfloor \dfrac{nf_{ext} - B}{v(f_{int} - B + 1)} \right\rfloor + 1$ when r=2
Two Sided	$k \geq \max\limits_{1 \leq b \leq B} 2\left\lfloor \dfrac{nf_{ext} - b}{vc} \right\rfloor + 1$	$k \geq 2\left\lfloor \dfrac{nf_{ext} - B}{v(f_{int} - B + 1)} \right\rfloor + 1$

Melen and Tuner [Mel03] generalized the multi-slot circuit-switching problem and their result for non-blocking conditions for multi-rate multi-stage packet switching networks is given by $k \geq 2\left\lfloor \dfrac{n - B}{S - B} - 1 \right\rfloor$; where S is the speedup of the internal links compared to external links given by S= f_{int}/f_{ext}, and B is the maximum virtual circuit rate as a fraction of the bandwidth of the inlet port bandwidth.

Melen and Tuner also note that the number of middle stage switches depends on the speedup factor, S. In the ideal case S=n and only one middle stage switch is needed. However, as will be reemphasized in the next section, it may not be practical to speed up internal links by a factor of 'n'. For instance, assume a packet DSM (n,f_{ext},k,f_{int}) with n=16, f_{ext}= 2.5Gbps. With multi-stage packet switch composed of these DSMs, to speedup internal links by a factor of n (=16) means f_{int}=40Gbps, a technology that is still emerging. So it is reasonable to expect S<n and possibly S=1. Melen and Turner note that with packet switching systems having B=S=1, to

support b-rate virtual circuits with $0 \leq b \leq B$, the number of middle stage switch matrices is infinity i.e. Clos construction with strictly non-blocking property is not possible. Therefore it is better to switch smaller bandwidth circuits using a separate switch fabric dimensioned properly for that rate than use one fabric with large number of middle stage switches. For instance, with DCS that needs to support both SONET/SDH switching and VT- Melen and Turner's conclusions would mean building two fabrics - one for VT switching and one for SONET/SDH switching - each one is dimensioned to their traffic needs. As explained earlier - all SONET/SDH signals that contain VTs can be switched into a adjunct VT fabric where the VTs are switched from one SONET/SDH container to another and the output of VT fabric can be looped back to the main SONET/SDH switch as explained in the §2.3.5

2.5. PACKET SWITCHING

Previous sections focused on space and time division switching concepts and techniques. Space division switching concepts are applicable to realizing packet switches also as will become apparent in this section. However, circuit switching is very specific to switching calls embedded in TDM streams. With packet switching – the each stream carries packets, with each packet identifying the exit port as opposed to TDM streams. This embedded information simplifies switch control though asynchronous nature of packet arrivals introduces new problem to be solved. Though, TDM/circuit switching is the dominant form of switching used in PSTN and transport networks today, enterprise and service provider networks are completely based on cell/packet switching (or routing) using such technologies as X.25, ATM, FR, IP, MPLS. Particularly, FR and ATM based switching networks have been in place since early 90's. Currently IP/MPLS based packet transport and switching is receiving attention as part of an effort to move and merge circuit switching and packet switching/routing.

Independent of the technology, application and network architectures, all packet switches have a single function – switch packets, from input to output, based on information embedded in them. This is different from circuit switches, which don't use or depend on the contents of time slots to determine outgoing time slot or port. As a result of this difference, as well as relative simplicity that is possible with packet switches (since lot of complex control circuitry including synchronization is not required), different type of packet switches have been published. Some of them differ in approach and some in structure. For instance, Banyan, and Batcher–Banyan packet switches use multi-stage switching approach but differ in structure. On the other hand, there are packet switches based on shared-memory approach, entirely different from the ones based on multi-stage switching approach. In general, the switching or routing function performed by a packet switch can be expressed in three different ways as listed in Table 2-8.

Table 2-8. Packet Switching Functional Description

Switching Function	Description
<port-x, header-h1, port-y, header-h2>	Switch packets at ingress port-x, with header-h1 on to port-y. The packets as they leave the switch will have new header-h2.
<*, header-h1, port-y, header-h2>	Switch packets at any ingress port with header-h1 on to port-x. The packets as they leave the switch will have new header-h2.
<*, header-h1, port-y, *>	Switch packets at all ingress ports with header-h1 on to port-y. The packets as they leave will be unmodified.

The switching function, as well as any internal space switch settings, must be configured into the switch by a central controller, much like in a circuit switch. Packet switches that don't need such explicit configuration of their internal switching network   provide an interesting behaviour called *self-routing.* In such type of switches, incoming packets are processed at the ingress port module (IPM).   Each IPM, using information embedded in packets, determines output port module (OPM) for each packet. Next, each packet is tagged   with internal address of OPM. In the case of ATM (FR) switches, each incoming cell ((cells, frames) frame)'s VCI/VPI (DCLI) is replaced with new values. In the case of MPLS switches, standard label processing rules are applied that can result in swapping an existing label with a new one or adding (push) of a new label.   . The modified packets are presented to the switch, which routes them to their respective OPM.

There can be two modes of operation of packet switches – *asynchronous* or *synchronous*:

In *asynchronous* mode, packet switch operates as follows: each input port/module with packets destined to output port, y, will queue them in a queue designated for y. Each output port, as long as its queue is not empty, will keep reading its queue and transmitting the read packets   out of that port.

In *synchronous* mode, packet switch operates by switching inputs to outputs at discrete times spaced by 't', the switching interval. This principle of operation is similar to that of TMS in a circuit switch. Also, packets that arrive at the input ports, before the switching time, will have to wait until the switch is ready to switch. The worst case waiting time will be t.

If average packet arrival rate at input ports is 1/t, then the switch is said to have a relative speed, L, of one unit.   Table 2-9 provides some examples.

Packets that arrive at the input ports, before the switching time, will have to wait until the switch is ready to switch. The worst case waiting time will be t. If packets received at input modules are larger than what the switch can transfer then such packets must be broken into smaller ones. Each output module must be capable of reassembling the original packet in such case. For instance, IP/MPLS switches based on ATM cell switching fabric can only switch 53-byte cells from input to output. Therefore, each IPM will have to break arriving IP/MPLS packets into 53-byte cells. Similarly, each OPM must be capable of reassembling original IP/MPLS packet from the 53-byte cells. In the rest of the discussion packet and cell switching will be used interchangeably.

Table 2-9. (a) Packet Switch Speed (b) Buffering Techniques (c) Throughput for Input buffering vs. speed up

Packet Arrival Rate	Switching Interval	Switch Speed, L
1/100 msec	100 msec	1
1/100 msec	50 msec	2
λ	T	λ/t

(a)

Buffering Technique
Output Buffering
Input Buffering
Input and Output Buffering
Input, Output Buffering with Virtual Output Queuing

(b)

Speed L	Throughput
1	0.5858
2	0.8845
3	0.9755
4	0.9956
5	0.9993
N	1

(c)

With packet switches operating in synchronous mode there can be contention. For instance, at switching time t1, more than one input port can have packet destined for the same output port. Since the switch can transfer only one packet to an output port during any switch slot such situation represents contention. Table 2-9 lists some approaches to reducing or eliminating contention.

The first technique is based on speeding up the switch by a factor of K ($1 \leq K \leq N$). However, with speedup, OPMs can receive potentially K packets during a single switching slot. Therefore, each OPM needs to have a buffer of at least K.

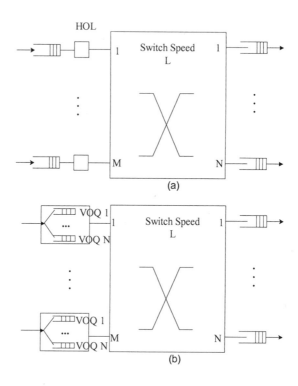

Figure 2-22. Packet Switch (a) System Model (b) Packet – Virtual Output Queuing

In input buffering scheme, IPMs resolve contention using some arbitration scheme. Those that loose arbitration either drop or buffer the packets. Such buffered packets can be presented to the switch at a later switching slot. However, the size of the input buffer has been subject of extensive research. Besides, this technique has an inherent problem. To explain this, consider input port, I, with some packets already in its buffer. Further, assume that, just before switching slot, Tx, port, I, receives a new packet, P_{ij}, destined for output port J. Also assume that no other input port has packet destined to port, J. Now during switching slot, Tx, the switch can switch Pij from I to J. However, since Pij is behind many other previously arrived packets at I, the packet can't be presented to the switch. Therefore the switch can't switch Pij even though output port J is idle and therefore the situation represents a wasted opportunity or underutilization of the switch. This situation is called *Head of Line Blocking* (HOL). Obviously, for an NxN packet switch, if L=N then HOL blocking doesn't exist – because even if all N IPMs have packets destined to the same output port, the switch will transfer them freeing buffers at the IPMs for the next arriving packets.. However, speeding up the switch by N times may not be practical in all cases   and hence HOL will exist. Therefore, to avoid HOL without necessarily increasing the switch speed, a modified input queuing technique called *virtual output queuing* is used and illustrated in Figure 2-22 (b).

Input buffered switch is said to have a throughput of one if 1) all input ports present a packet each to the switch in a given switching slot and 2) the packets are destined to different output ports and 3) the switch switches all of them. On the other hand, if HOL blocking occurs then only N' (N'<N) input ports can present packets to the switch. The throughput of the switch will then be N'/N or less than one. By speeding up the switch, HOL blocking can be avoided thus increasing switch throughput. Table 2-9 (c) provides switch throughput vs. speedup [Yuji92] [Karol87].

The third technique basically combines the two buffering strategies. All the three techniques are applicable to multi-stage switches as well and can be applied at any stage. Figure 2-22 (a) illustrates generalized packet switch with input and output buffers and Figure 2-23 shows general architecture of IP/MPLS packet switch.

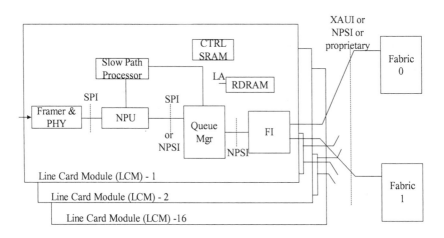

LA - Look Aside Interface   NPU - Network Processor Unit
SPI - OIF System Packet Interface   FI - Fabric Interface
NPSI - Network processor Streaming Interface
XAUI - 10 gigabit Attachment Unit Inteface

Figure 2-23. IP/MPLS Packet Switches and standard interfaces

A typical packet switch consists of a fast path, a slow path, a scheduler, and a fabric (two for redundancy). The switch fabric of a packet switch can be a simple crossbar or a multi-stage Clos network. Recently switches with capacity in the range of 100 Gbps and higher have been reported based on shared memory. The fast path usually refers to movement of IP/MPLS packets from ingress port to egress port through the switch fabric without involvement of software in the slow path. The software in the slow path can include processing of IP packets in terms of finding egress port by route lookup, error processing, reassembly etc. Flow control is done using either in-band or out-of-band control mechanism between switch fabric (the scheduler) and line card modules. The figure also shows various interfaces that are being standardized by many forums, such as OIF, for chip-to-chip or backplane interfaces.

2.5.1   Universal I/O Node (UIO)

Circuit switches perform a simple task – for each call, connect (patch) the involved input and output ports for the duration of the call. The switches are oblivious to the bits and bytes going across. Naturally, one can conceive of a universal switch fabric that switches variety of services – Ethernet, FR, ATM, and SONET/SDH etc. In such a scheme, one can conceive of a universal I/O (UIO) node (terminating single or multiple I/O signals) that performs protocol specific processing of incoming I/O signals before presenting traffic to the switch. For instance, a UIO node terminating SONET/SDH signal needs to perform pointer processing (PP), a SONET/SDH specific function. Similarly, a UIO node terminating GE signal needs to perform Ethernet frame processing specific to Gigabit Ethernet protocol. Also, each UIO node, if terminating multiple I/Os, can route or switch traffic among these I/Os without using the central switch. For instance, UIO node terminating 16 OC-48 signals can perform switching of traffic among these ports without using the central switch (such capability was introduced as hair-pinning earlier). Internally an UIO node may use DSM discussed earlier.

The problem of switching traffic between two ports can now be translated to that of switching traffic between two UIO nodes and there are two alternatives to perform such switching; distributed and centralized.

In the distributed switching option, a set of UIO nodes are directly connected to each other using simple or multi-dimensional topological structures (i.e. topology). The number of neighbors, and hence the number of links, a UIO node will have in a particular dimension depends on the interconnection structure For instance, in 2-D tori each UIO node will have two neighbors each in x and y-dimension. Similarly, in 3-D tori, each node will have two neighbors each in the x, y, and z direction.

In the centralized switching option, a set of UIO nodes are connected using a dedicated space switch in between. The space switch can be based on single or multi-stage Clos concepts discussed earlier. The UIO nodes together with the central space switch can still be considered a distributed switch since UIO nodes have the capability to route packets between them. When a switch has UIO ports the difference between centralized and distributed switching is subtle and we can, without loss of generality, assume that such switches are distributed in nature.

Figure 2-24 illustrates SONET/SDH version of UIO node with SONET/SDH specific pointer processor (PP) logic. Chapter 3 discusses SONET/SDH protocol and frame formats.

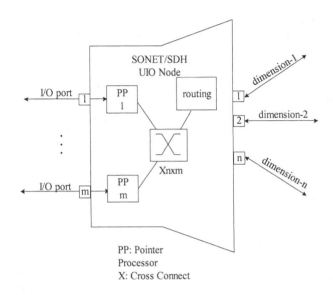

Figure 2-24. SONET/SDH UIO node in a multi-dimensional interconnection network

For now, if traffic on a local port of a SONET/SDH UIO node is destined for a   port on the same UIO node then, that UIO node can switch that traffic locally. Otherwise, the source UIO node needs to send the traffic into the switch fabric using one of its outgoing links. Since there is no destination information embedded in the SONET/SDH frame, the UIO node will have to construct a new frame containing the received TDM timeslot data[32] and embed the destination information. The embedded information can and should identify the destination UIO node and any additional information that is useful for the destination UIO node to determine the final exit port. The routing algorithm running on the UIO nodes will take care of getting the frame to the destination. The destination UIO node can extract the TDM data and insert it into appropriate outgoing SONET/SDH line signal after pointer

---

[32] SONET/SDH frames are TDM in nature but essentially; they are cell-based streams much like ATM.

processing. One should note that in the case of SONET/SDH TDM switching, traffic in each time slot, once it enters a switch (a switch consisting of one or more UIO nodes), must leave the switch in less than 25 usec[33].

Given this view of a UIO node, in the rest of the discussion on interconnection networks, the reader can visualize a UIO node as equivalent to a processor element in a parallel processing machine.

2.6. INTERCONNECTION NETWORKS

Whether interconnecting multiple processors in a SIMD (Single Instruction/Multiple Data) parallel processing machine or connecting multiple processors to shared memory in a MIMD (Multiple Instruction/Multiple Data) machine, all posed challenges that needed solutions and research. Therefore, the parallel processing research community worked on solving such problems and produced a number of interconnection networks, the general view of which is shown in Figure 2-25.

Interconnection networks are classified using *direct* and *indirect* terminology [Pad90]. A *direct network* is one in which processor elements (PE) or UIO nodes are interconnected using point-to-point links. The interconnection between the PEs is *static,* meaning a given PE is connected to some other set of PEs in a given topology in a fixed way. On the other hand, an *indirect network* is one in which the PEs or UIO nodes connect to a reconfigurable switch (cross-bar or multi-stage Clos). The interconnection between the PEs is *dynamic.*

---

[33] Bellcore GR-496-CORE specifies that the through delay for a non-terminated STS payload (including one going through a cross-connect) should not exceed 25 usec. The limit applies to payload that is added/dropped at a node. The basis for such requirement is not clear though one can infer that a voice call going through a number of switches is not expected to have delay of more than 30 msec beyond which expensive echo-cancellers must be installed. It is possible that the budget of 30 msec is used to arrive at the 20-25 usec delay for each individual switch.

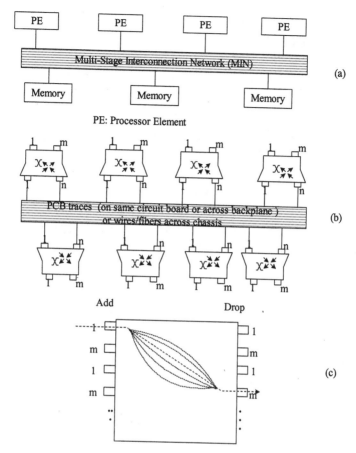

Figure 2-25. (a) Multi-process interconnection network (MIN) system model (b) Equivalence of Multi Processor and UIO Node Based Interconnection Networks (c) alternate paths in MIN between an add and drop port

Classification[34] of interconnection networks is shown in Figure 2-26. However, many of the interconnection networks listed in the figure such as Omega, Baseline etc are not discussed in this chapter.

An indirect network, though the interconnection itself is static, can exhibit *dynamic* behaviour if each PE or UIO node can provide routing and switching function. For instance, consider two processor elements, PE_x and PE_y, that are not directly connected but each one connected to a third

---

[34] Adapted from Peyravi Lecture notes

processor element, PE_z. Both, PE_x and PE_y, can use routing and switching services of PE_z to send messages to each other. PE_z is said to be performing message/packet routing. As reader may be a familiar, nodes that switch or route traffic by processing packets would fit packet switch classification as well. We can generalize this - for a given *static* interconnection topology, a PE can find the best way to send messages to some other PE using shortest path algorithm or some other means. All PEs (UIO nodes) are cooperatively switching/routing traffic between them and such collective behaviour represents distributed switching.

Interconnection network, if *dynamic*, might need to be configured by an outside control entity to connect given input-output permutation. A more interesting case occurs when the interconnection network can configure itself or can route/switch traffic from ingress to egress without external control or configuration. Such interconnection networks, that route or switch packets from input to output without having to be setup by external control, are called self-routing networks [Vid94]. Terms such as cut-through, wormhole, deflection, hot potato, etc are some of the routing techniques employed by self-routing networks. Also, the *self-routing* algorithms and techniques are particular to the interconnection topology.

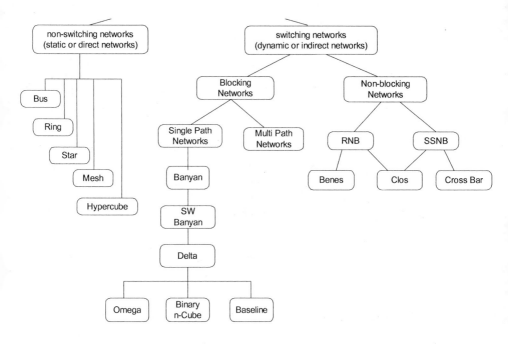

Figure 2-26. Interconnection Network Classification (Pyravi)

What makes *self-routing* networks interesting is their self-organizing behaviour eliminating the need for processor elements i.e. the end users to find optimal paths. There are two things here – a pre configured interconnection network and packet routing over such pre-configured switching network. The job of PEs or UIO nodes in this scheme is to inject traffic into such network with some proprietary header indicating desired destination PE or UIO node address. The self-routing network then routes/switches that packet to the destination.

The equivalence of a multi processor interconnection network (MIN) and SONET/SDH UIO based circuit switch is shown in Figure 2-25. In this scheme, the SONET/SDH UIO node, when a service, such as STS-1, is ready, can send the STS-1 frame into the MIN as a message with some header. The frame can be sent as a whole or in pieces in which case the destination must perform reassembly. The MIN will route the message to appropriate destination UIO node where the service inside the message is extracted and inserted into an outgoing SONET/SDH link by the PP associated with the outgoing link. Though message/packet switching in a multi-processor interconnection network is similar to that in a UIO based interconnection network, there are some differences. A packet belonging to a circuit must be exited with in 125 usec[33]. Packets belonging to a circuit must observe strict sequence. If MIN supports multiple paths then two consecutive packets of a circuit might arrive at the output port out-of-sequence. It is the responsibility of the destination UIO node to re-order them. Except for these restrictions, packet and circuit switching are equivalent.

By now it should be clear that distributed switching is nothing but a set of PEs or UIO nodes that are directly interconnected. The structure of such interconnection network is the focus of the following paragraphs, particularly the multi-dimensional structures. Multi-dimensional [Abb] network is an abstract concept rooted in graph theory. Unlike points and lines in a 3-D space, in an n-dimensional space the entities of lower dimensions become points and lines for higher dimensions as illustrated in Figure 2-27

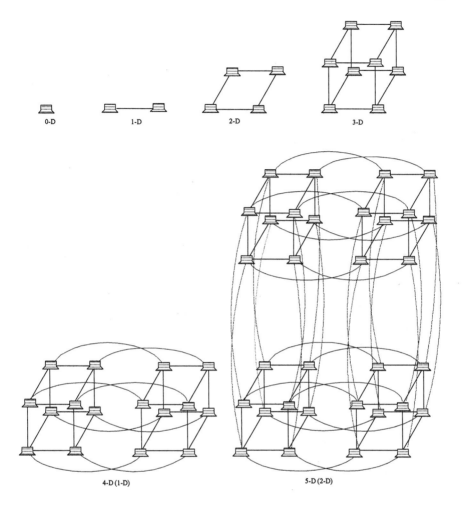

Figure 2-27. n-dimensional construction using lower dimension primitives

If the above or similar multi-dimensional distributed construction is used to build a large TDM switch then the links connecting the nodes can be based on OIF's TFI-5 or some proprietary signaling. If all the nodes can be accommodated in one chassis then the links can be electrical and supported by the chassis backplane. On the other hand, if the construction requires multi-chassis system then nodes in different chassis may be interconnected using optical links - based on TFI-5 or proprietary signaling. Not only that, the links from one chassis to another can be combined using passive optical multiplex technology (DWDM) to avoid fiber/cable crowding though such constructions must be evaluated for cost effectiveness as well as ease of operation interms of upgrade.

2.6.1 2-D Mesh

A distributed switch fabric built using 2-D mesh topology consists of nodes and links arranged in a 2-D grid as shown in Figure 2-28

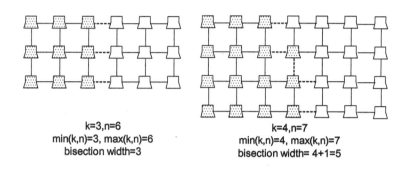

k=3,n=6	k=4,n=7
min(k,n)=3, max(k,n)=6	min(k,n)=4, max(k,n)=7
bisection width=3	bisection width= 4+1=5

Figure 2-28. 2-D Mesh Topology for Distributed Switch Fabric

The nodes can be visualized as UIO nodes in a transport switch or process elements in a parallel processing machine. The mesh is similar to a crossbar except that the cross point is not just connecting input to output but rather a traffic routing entity that either terminates the traffic or routes the traffic. There are only two outgoing links and the decision to choose which outgoing link to use will be based on routing algorithm. A generalized notation called k-ary-n-cube is also commonly used to label such interconnection network topology. A 2-D mesh with 3-rows and 6 columns (3-ary-6-cube) is shown in Figure 2-28 (b). The figure also shows *bisection width* and the term will be explained later.

2.6.2 2-D Torus

A 2-D torus is derived from 2-D mesh with by connecting edge nodes with wrap-around links as shown in Figure 2-29 (a) and (b).

Wrap around links are usually along the same axis. However, 2-D torus constructions using twisted wrap around links also exist. In 2-D torus, every node is connected to four other nodes (degree =4). Each node in a 2-D torus can represent a PE or UIO node. At each node, traffic can be injected and sent on the torus to any other node where the traffic is dropped. The example torus has width (W)=4 and height (H)=4. By breaking four vertical short links and four wrap-around links, for a total eight links, the torus can be bisected horizontally. Similarly, by braking four horizontal short links plus four wrap-around links the torus can be bisected vertically. Obviously, connecting SONET/SDH UIO nodes using torodial network requires PCB traces on a circuit board or on system backplane or wires/fibers across chassis. Such interconnections, particularly inter-chassis connections, can add to complexity as well as cost. Nevertheless, 2-D torus is a viable interconnection structure for distributed circuit switching.

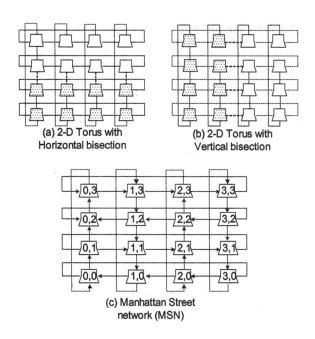

(a) 2-D Torus with Horizontal bisection

(b) 2-D Torus with Vertical bisection

(c) Manhattan Street network (MSN)

Figure 2-29. 2-D Torus and Manhattan Street Network

A special type of 2-D torodial switching network that has elicited interest in the research community is the Manhattan Switching Network (MSN) [Max85]. As illustrated in Figure 2-29 (c), MSN is similar to a 2-D mesh except that every alternate row is in opposite direction. It is similar to how many cities have one-way traffic setup in every alternate street. Each wrap around link is the equivalent of imaginary by-pass between the start and end of a street.

2.6.3 Hyper Cube

Figure 2-29 illustrates a binary hypercube, also called 2-ary-N-Cube. In hyper cube, a node at each corner is connected to n-other nodes along the edges of the cube. Each node is assigned a binary address $(b_n...b_0)$ and addresses of nodes that are adjacent differ only in 1-bit.

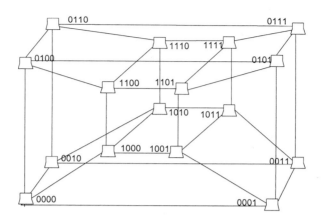

Figure 2-30. Binary HyperCube

Each node in a hypercube can be a UIO node. Traffic arriving at the tributary ports of UIO node is injected into the hypercube, towards a destination UIO node. Intermediate cube nodes, using their embedded routing algorithms, route traffic towards its destination. The routing algorithms try to balance traffic on each outgoing link versus reducing the hop-length.   For more on routing techniques in hypercube switching networks, reader can refer to [Ted90]. The hypercube as a whole exhibits self-organizing behaviour, assisted by routing algorithms.

2.6.4   Cube Connected Cycles (CCC)

The Cube-connected cycles were first proposed by Preparata and Vuilemin in 1981 to solve layout problems in VLSI [Prep81]. Since then, the topology has been extensively studied and is considered a practical substitute for hypercube.

A CCC is constructed from an n-dimensional hypercube by replacing each node in the hypercube with a cycle of n-nodes as shown in Figure   2-30.

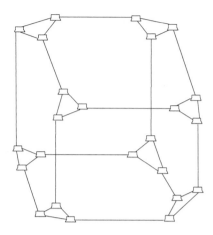

Figure 2-31. Cube Connected Cycle

2.6.5 Parameterization of Interconnection Networks

In the previous section, we discussed some direct interconnection topologies/networks. The nodes in these interconnection networks are intelligent and participate in switching or routing of transit traffic besides adding/dropping traffic. In general, direct networks can be characterized by the following commonly used metrics:

Table 2-10. Metrics for Comparing Interconnection Network

Metric	Description
Degree (D)	The number of links each node has towards its neighbors
Diameter (D)	The longest path length from an input (node) to an output (node) through the switch
Edge Connectivity (C_e)	Minimum number of nodes that need to be removed to disconnect the network.
Node Connectivity (C_n)	Minimum number of links that need to be removed to disconnect the network. Also reflects the number of node disjoint paths between any pair of nodes
Fault Diameter (FD)	Given node connectivity, C_n, fault diameter is defined as the diameter of the network with at most C_n -1 nodes
Bisection Width (BW)	The number of edges to be removed to separate the network into two equal parts
Cost	Number of links or number of links + number of stages + number of switching elements

The significance of the above parameters requires some explanation. Switch fabrics (both OEO[35] and OOO) are compared in terms of fault tolerance, cross talk, latency etc. A large bisection width (BW) reflects that a large number of interconnection links have to fail to make the switch fabric disconnected into two parts. Hence, large BW indicates higher fault tolerance of the switch fabric. A larger diameter indicates that traffic incident at an input port has to travel across multiple inner switch nodes before it reaches destination. Hence, diameter reflects latency as well as cross talk[36]. In the case of synchronous transport switches, such as used for SONET/SDH the latency has to be limited to less than 125 µsec[37].

Table 2-11[38] lists some metrics for some of the interconnection networks though we didn't discuss all of them.

Table 2-11. Commonly used metrics for popular interconnection networks

	Nodes	Degree	Diameter	Bisection Width	Links (cost)
2-D Mesh	N	4,2	$2(\sqrt{N}-1)$	$2\sqrt{N}$	$2(N-\sqrt{N})$
2-D Torus	N	4	$\sqrt{N}$	$2\sqrt{N}$	2N
3-D Torus	2^{2n}	$O(N^{1/3})$		2×2^{2n}	
Ring	N	2	$\frac{N}{2}$	2	N
3-D Cube, n=3	2^n	3	n	n+1	4n
Hypercube	2^n	n	n	N/2	$N\frac{\log N}{2}$
Benes	N	4	$\log_2 N$	~n	

[35] OEO means, ingress I/Os port perform optical to electrical conversion, the egress I/O ports perform electrical to optical conversion, and the central switch fabric switches electrical signals from input to output. In contrast, in OOO type of switches the I/O ports don't perform OE or EO conversion and the central switch fabric is based on all optical switching using optical switching elements.

[36] In optical switching, a 2x2 Directional Couplers (DCs) takes two signals and switches them to two outputs In an optical switch constructed using 2x2 DCs as building blocks, longer series path can add cross talk.

[37] In the case of SONET/SDH protection, an input K1/K2 byte must, in a pass through mode, be switched from input to output port in the next frame and is the reason for the latency to be less than 125 usec.

[38] Collected from presentations of Mihir Mishra, Lehre and others

2.7. CENTRALIZED AND DISTRIBUTED SWITCHING

When all I/O signals are brought to a central place for switching, we call it centralized switching. There is no strict definition or classification of non-centralized switching although terms such as hierarchical switching, distributed mesh switching or multi-dimensional switching are often referred to as distributed switching. Centralized switch itself can be implemented using a multi-dimensional distributed switch fabric and realized on multiple PCBs. In the following discussion, we will use physical sense of centralized switching rather than the theoretical sense- if I/O signals are brought to one place for switching we call it centralized. Any other scheme would be distributed switching without loss of generality.

Investigation and research into alternatives to centralized switching for large switches was required due to physical as well as high-speed backplane transmission limitations of bringing dozens of I/O signals to a central switch. However, some technological advances, mainly from 1995 onwards are making centralized switching as viable solution for small (<64x64) and medium (<256x256) size switches. There are broadly four advances worth mentioning:

First, advances in IC technologies had great impact on the evolution of switching systems. In 1996, .35 micron ASIC technology was mainstream with .18 micron being the case in 2000. In addition, nano-meter technology is expected to be available by 2009. With these advances, it is possible to put millions of gates plus lots of static RAM into a single ASIC enabling one to build large crossbar or multi-stage switches in a single IC.

Second, parallel to the developments in IC semiconductor technologies are advances in packaging. In early 90's 300 pins was considered a large IC package. Now, at the time of writing of this book, ASICs with 1000[39] pin packages are becoming practical. With such large pin count on a single IC, it is possible to connect many input/output ports to a centralized switch implemented in highly integrated silicon thus requiring fewer components and hence providing cost benefits to the design of overall switch.

Third, PCB and backplane edge connector technology has also advanced allowing cards to terminate many more signals than hardware designs of early '90s. Also standardized back planes, such as AdvancedTCA, StarFabric, cPCI, cPSB[40], etc, are available for switching applications. Some of them support centralized and distributed topologies. Optical back planes are also a future possibility. PCBs with as many as 16 layers are now

---

[39] Ball Grid Array (BGA) packages, 0.25, 0.18, 0.13u CMOS process with 90 nm expected soon.
[40] cPCI – compact PCI, cPSB – compact Packet Switch backplane

possible compared to about six in the early 90's. Plus, special connectors that offer good transmission characteristics such as HM-Zd are being used for Gbps back planes. Exotic PCB laminate materials such as Nelco are being considered too. However, manufacturing process using new and exotic materials may not be mature enough to give cost and yield benefits. Therefore, silicon innovations, such as PAM-4/10G NRZ SerDes, to extend the life of legacy FR-4 backplane are getting attention. Such silicon solutions can ease the problem of upgrade of existing 2.5G-10G/slot switching systems to 10G-40G/slot.

Fourth, the speed of I/O is increasing gradually with 2Mbit serial links between chips being mainstream in '94 to Gbps serial links being common in 2000. Table 2-12 lists some of the serial I/O interfaces (some of them are parallel, but commonly referred to as serial I/O).

Table 2-12. Serial I/O Standards

Serial I/O	App	Bus width/rate/Signaling	Comment
SPI-3	2.5G	8 or 32/104Mbps	Framer←→System
SPI-4 Phase 1	10G-12G	64/200Mbps/HSTL	' '
SPI-4 Phase 2	10G	16/622Mbps/LVDS	' '
SFI-4 Phase 1	10G	16/622Mbps/LVDS	Framers←→SerDes
SFI-4 Phase 2	10G-12G	4/2.5G/CML	
SPI-5/SFI-5	40-50G	16/2.5Gbps/CML	
TFI-5	40-50G	16/2.5-3.125Gbps/CML	Framer←→Fabric
PAM-4	6-10G	1/6-10G/PAM-4	NRZ[41] signaling also reported to achieve 10G rate over FR-4 back planes
XAUI	10G	4/3.25Gbps/CML	10GE Applications, 8B/10B, 10Gbps payload rate, Chip-to-chip or backplane, 20" on FR-4 PCBs
InfiniBand IBTA	10G	4/2.5Gbps	Server/Data Center Applications

All MSA transponders (GBIC, SFP, XENPAK, XFP, XPAK) offer one of the above common electrical interfaces. Not only are I/O speeds

---

[41] See www.uxpi.org, a consortium supporting 10G backplane standard based on NRZ.

increasing, the SerDes required to drive signals onto back planes are becoming available as CMOS ASIC macros for system/sub-system vendors to integrate into their custom ASICs. Such integration can eliminate lot of discrete components. Besides advances in chip-to-chip and back plane electrical interfaces in terms of standardization and technologies such PAM-4, on-board as well as card-to-card optical interconnection technologies are being researched that might be viable in future [IEEE03]. Also, OIF is standardizing very short reach (VSR) optical interfaces that could be used for chassis-to-chassis interconnection in multi-stage switches. Components such as Vertical Cavity Emitting Laser (VCSEL) s may also play a role in large switches of future networks.

A number of these above factors are favoring centralized switches over distributed switches, at least for sizes upto 64x64 (small) or 256x256 (medium). However, there is a limit to advances in each of the above physical component technologies; semiconductor sub-micron/nano-meter technology; IC pin count; edge connector pin density; I/O transceiver speed; availability of CMOS technology at higher speeds. For larger switch designs of future and next generation[42], networks one would need to explore distributed switch fabrics as alternatives for centralized switches. Again, as we mentioned earlier, some of the above technological advances do enable one to construct higher capacity switches using centralized Clos multi-stage concept without recourse to distributed switching techniques. However, distributed switching has inherent benefits in terms of generality and simplicity in control and configuration and can be consider as ideal:

- To build switches with large port count (>512x512), particularly with the need to avoid fiber/cable crowding around single chassis systems[43].
- To provide better fault tolerance
- To use self-routing available in distributed switch fabrics thus eliminating the need to perform configuration of matrices in multi-stage Clos switching networks or rearrangement complexity in RNB clos networks

---

[42] The largest traffic nodes today don't seem to be passing or switching traffic of more than 500Gbps - If ever a CO or POP needs to pass or switch 1Tbps+ with the need to access any or all of that bandwidth then the case for switches with large port count will be more justified.

[43] Particularly where bandwidth on each port is a fraction of the signal rate of the internal links. For instance, OC-12 ports on a switch with 10G or 40G internal links between stages. A switch with 1.2Tbps total switching capacity can have as high as 2000 such ports - which is not practical to accommodate around one chassis.

- To support switching/routing of multiple services types in single equipment with service specific termination and processing logic confined to I/O cards using UIO node concept explained earlier.
- To integrate network processor technology at the I/O nodes, which can provide support for QoS, policy based routing/switching and active network and other advanced services.

Practical considerations and pragmatic approaches to building switches will prevail and one can't conclude with certainty that one approach is better than the other.

2.8. FAULT TOLERANCE OF SWITCHING NETWORKS

As mentioned in the introduction, fault coverage of telecom switches is very important. Any equipment that can't switch an input to output because of internal faults is certainly something that will impact its acceptability by carrier customers. The same is true if a fault can cause routing to fail for many input-output pairs. System designers must pay special attention in the choice of architecture as well as in fault models of the components used to build switches. Carriers, the buyers, must seek answers to questions that relate to fault recovery and switch properties, such as blocking, under fault conditions. Interested readers can explore more on fault tolerance in multi-stage interconnection networks in [Ada87]. The following definitions can be used to characterize fault coverage of multi-stage interconnection networks.

- Fault Model
- Full-access
- Fault Tolerance Size
- Full recovery
- Fault Tolerance Size

Fault Model identifies faults that can happen in the switch due to fault-prone components, such as ASICs and interconnection links (used to connect output of one stage to the input of another). When any such component or link fails the tolerance criterion can be used to reassess switch properties - blocking, nodal degree, bisection width etc.

Under fault conditions, a switch meeting **full-access** criterion provides any input to access any output. It is not necessary that path diversity, be same before and after a fault. All it means is that after a given fault switch continues to be non-blocking and any input can be connected to any output.

Fault Tolerant Size is a measure of the number of faults, F, a switch can tolerate without breaking the fault-tolerance criterion. When F=1, the switch is said to be single fault-tolerant. A switch with F>1 is said to be *F-robust*. Generally, switches with higher nodal degree tend to have higher fault tolerance as they have many more alternate paths. Tori, 2-D Mesh, 3-D Mesh, HyperCube have good fault tolerance.

Clos multi-stage networks also exhibit higher fault tolerance due to the fact that there are multiple links between outer and inner stage switches there by providing resiliency against link or middle stage switch failures. While k=2n gives SSNB characteristics, to get more resiliency one needs to increase k to provide better fault tolerance and maintain full-access under all load conditions.

2.8.1   Fault Tolerant Switch Design Techniques

Faults can be a result of internal failures, such as stuck-at semiconductor faults, or a result of faulty links connecting switches in a multistage switch. Faults can also be specific to topology and technology used to construct switches. There are generally two techniques used to provide fault tolerance in circuit switches. One is based on total switch duplication or system level redundancy and the other is based on providing extra paths in the switching network.

2.8.1.1 Redundant Switch Design

Typically, system vendors design switches that are card redundant while theoretical studies focus on functional block fault-tolerance. Theoretical models don't necessarily take into account practical equipment and manufacturing faults. Component or system aging faults, such as those associated with heat that affects soldering or some other physical or electrical properties of components - can only be detected after a switch is in operation for some time and such faults can be categorized as equipment faults as well. Another type of aging fault is the one associated with software – such as memory leak or algorithmic errors that show up after aging. In general, independent of how they occur, a simple and effective scheme to provide resiliency against faults is to duplicate the entire switch as shown in Figure 2-32 for a 32x32 switching capacity.

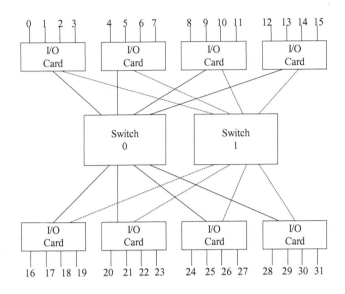

Figure 2-32. Redundant Switching for Fault-Tolerance

Each input is sent to switches, SW-0 and one SW-1. After identical switching by them, the two copies of input signal is available at the output node where one of the copies is selected based on signal quality. The main advantage of redundant switch design is its simplicity. Duplication has been the consistent philosophy in telecommunications. Transport protection schemes such as 1+1 and UPSR (Chapter 9) are essentially a fall out this philosophy- 'Keep it simple and keep the fail over and switch logic simple and autonomous.' A secondary benefit of redundant switch design is that both active and backup switches are identical and hence inventory control is simple as well.

2.8.1.2 Fault-Tolerant Clos (FTC) Multi-Stage Switching

The second alternative to designing fault tolerant switches is to provide extra paths in the switch. In a 3-stage Clos switch, each first stage switch has k-alternate paths to a given last stage switch. In other words, there is *path diversity* inside the switching network. If a path fails, a call in progress that is affected by that fault can be switched over to the remaining k-1 paths. Obviously, a switching network with high path diversity is more resilient to internal faults since a given call can be supported   other paths not affected by the fault. Whatever may be the reason for faults, if they affect one or more existing connections then those connections need to be rerouted and in many cases that must be done in less than 50 msec. In this context, rerouting is important and the reason for discussing routing algorithms for Clos networks earlier in the chapter.

As discussed earlier, building SSNB Clos switching network can be expensive so it is common to find RNB Clos switching networks with k>n. In such switching networks a faulty middle stage switch or a faulty link between adjacent stages can effectively make k<n. Such faults can result in the switching network loosing non-blocking property. Therefore, reference [Kuo91] proposes a scheme to backup switches in each stage and the scheme is called Fault-Tolerant Clos (FTC) switch.

The original FTC proposal using a 9x9 switch is illustrated here:

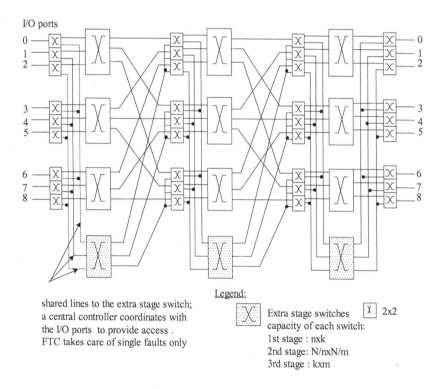

Legend:

shared lines to the extra stage switch;
a central controller coordinates with
the I/O ports to provide access .
FTC takes care of single faults only

Extra stage switches   2x2
capacity of each switch:
1st stage : nxk
2nd stage: N/nxN/m
3rd stage : kxm

Figure 2-33. A (9x9) original 3-stage Fault-Tolerant Clos Switch Construction

It is similar to classical 3-stage Clos construction except that it has an extra switch in each stage as well as a small 2x2 switch at the input of each switch. The I/O card would drive two signals, one signal to the normal switch and another to a bus that is connected to the extra switch. Access to the shared bus is coordinated by some control entity. In the event of fault at the input of a first stage switch, control entity can be informed which can coordinate with the I/O card so that it sends its signal to the extra switch. In addition, the second stage 2x2 switch is configured so that it uses the signal from the extra switch. The assumption here is that the source i.e. the I/O card will not fail but only its transceiver can. If I/O card failures (i.e. equipment failures) are also considered then, it is required that the input itself be duplicated. It is conceivable to provide modifications to the original FTC with source redundancy. Such input duplication can be achieved using passive optical splitters.

2.9. CONCLUSIONS

This chapter introduced basic concepts associated with telecommunications circuit switching. Crossbar and multi-stage Clos-switching techniques used to build circuit switches are explained. Also briefly packet switching was discussed. Next, as part of interconnection networks topologies such as 2-D mesh, 2-D tori, and hypercube were presented. Also discussed are the metrics used to evaluate various interconnection networks. Finally, fault tolerance criterion for interconnection networks is presented. As part of fault-tolerant switch design, redundant and fault-tolerant Clos switching techniques were discussed. Overall, the material presented in this chapter is basic but should help readers seeking simple understanding of switching as applicable to transport networks. Readers can consult [Dal03][Duo02][Pat99] for more in-depth information on switching.

REFERENCES AND FURTHER READING

[Abb]   "Flatland: A Romance of Many Dimensions ", Edwin A Abbot

[Ada87]   "A Survey and Comparison of Fault Tolerant Multistage Interconnection
            Networks", G.Adams, D.Agrawal, and H.Siegel, Computer, June, 1987

[Ann01]   Building SONET/SDH digital-crossconnect systems with STS-1 grooming
            for intelligent optical networks, K. Annamalai, William NG, Integrated
            Communications Design, Nov 2001

[Bel00]   Digital Telephony, John C. Bellamy, Wiley Series in Telecommunications
            and Signal Processing, 2000

[Ben65]   Mathematical Theory of Connecting Networks and Telephone Traffic,
            V.E.Benes, Academic Press, New York, 1965

[Ber03]   Optical Network Control: Architecture, Protocols, and Standards, G.
            Bernstein, et al, Addison-Wesley, 2003

[Bha97]   Engineering Networks for Synchronization, Ccs7, and Isdn, P.K.
            Bhatnagar, Wiley-IEEE Computer Society Pr, 1997

[Bow77]   Bowers, T.L., "Blocking in 3-stage "Folded" Switching Arrays," IEEE Trans. Communications, March 1977

[Bri00]   "A Class of Highly Scalable Optical Crossbar-Connected Interconnection Networks (SOCNs) for Parallel Computing Systems", Brian Webb, Ahmed Louri, IEEE Transactions on Parallel and Distributed Systems, Vol.11, No 5, May 2000.

[Car93]   "A Non-backtracking Matrix Decomposition Algorithm for Routing on Clos Networks", John D. Carpinelli, A. Yavuz oruç, IEEE Trans. on Commun., pp. 1245-1251, Aug. 1993

[Cha79]   Charransol, P. et al, "Development of a Time Division Switching Network Usable in a very Large Range of Capacities," IEEE Trans. Comm., July 1979

[Che97]   "A Tight Layout of the Cube Connected Cycles", G. Chen, F C.M. Lau, IEEE 1997

[Clo53]   Clos, C., "A Study of Non-Blocking Switching Networks," Bell Sys. Tech. Jour., Mar. 1953, pp.406-24.

[Dal03]   Dally, W., Towels, B., Principles and Practices of Interconnection Networks, Morgan Kaufman, 2003

[Din01]   "Nonblocking Routing Properties," Go-Hui-Lin, Switching Networks: Recent Advances, Ding-Zhu Du and Hung Q. Ngo Editors, Kluwer Academic Press, New York, 2001

[Duo02]   Duato, J., Ni. L., Interconnection Networks, Morgan Kaufman, 2002

[Fel88]   "Wide-sense nonblocking networks," P.Feldman, J.Friedman and N.Pippenger, SIAM Journal of Discrete Mathematics May, 1988, 158-173.

[Gok73]   "Banyan Networks for Partitioning Multiprocessor Systems. First Annual Symposium on Computer Architecture", GOKE, L.R., LIPOVSKI, G.J.; 1973, pp.21-28

[Gri83]   Grinsec , Electronic Switching, North Holland, 1983

[Hui90]   Hui, J. , Switching and Traffic Theory for Integrated Broadband Networks, Kluwer Academic Publishers, 1990

[IEEE03] Special Issue on Optical Interconnects, Selected Topics In Quantum Electronics, IEEE, March/April 2003

[Jaj03]   Jajszczyk, A., "Nonblocking, Repackable, and Rearrangeable Clos Networks: Fifty Years of the Theory Evolution," IEEE Comm. Magazine, Oct 2003

[Jaj83]   "Nonblocking Switching Networks Composed of Digital Symmetrical Matrices," A. Jajszczyk, IEEE Trans. Comm., Jan 1983

[Jaj96]   "Broadband Time-Division Circuit Switching," A. Jajszczyk, IEEE Trans. Comm., Feb 1996

[Kar87]   "Input versus Output Queuing on a Space Division Packet Switch", Mark J. Karol, et al., IEEE Trans. Communications, pp. 1347-1356, Dec. 1987

[Kuo91]   "Multistage interconnection Networks: Improved Routing Algorithms and Fault Tolerance", Kuo-Yu-Chen, PhD Thesis, NJIT, 1991

[Max85]   The Manhattan Street Network, M.F. Maxemchuk, Proc. IEEE Globecom, 1985

[MSN91]   Performance Analysis of Deflection Routing in the Manhattan Street

Network, IEEE ICC, 1991.

[Oki03]   "Multi-Stage Switching System Using Optical WDM Grouped Links Based on Dynamic Bandwidth Sharing," E. Oki, et al., IEEE Comm. Magazine, Oct 2003

[Pad90]   "Cube Structures for Multiprocessors", K.Padmanabhan, Communications of the ACM, Jan, 1990

[Pat92]   "Wavelength Division Multiple Access Channel Hypercube Processor Interconnection", Patrick W. Dowd, IEEE Transactions on Computers, Vol.41, No 10, Oct 1992

[Pat99]   Switching Theory: Architectures and Performance in Broadband ATM Networks, Achille Pattavina, John Wiley & Sons, 1999

[Pre81]   "The cube-connected cycles: a versatile network for parallel computation", Preparata, F.P. and Vuilemin, J., Communications of the ACM, May 1981

[Sym]   http://www.symmetricom.com/media/pdf/app_notes/an-stnsdhs.pdf

[Szy90]   An Analysis of "Hot-Potato" Routing in a fiber optic packet switched hypercube, Ted Szymanski, IEEE Infocom, 1990.

[TSS91]   "Time-Slot Sequence Integrity for (Nx64) Kb/s Connection," Electronics and Commun. In Japan, Part-1, vol.71, pp.78-79, 988

[Tur03]   Turner, J.S., Melen, R., "Multirate Clos Networks,", IEEE Comm. Magazine, Oct 2003

[Vid94]   " A New Self-Routing non-blocking switch", Vidya Sagar, Manohar Naidu Ellanti, Venkat J, ITC, 1994, Bangalore, India.

[Web91]   "Multichannel Circuit Switching – Performance Evaluation of Switching networks," A. Weber, et al, IEEE Trans. Comm., Feb 1991, pp.226-232.

[Yu87]   "The Knockout Switch: A Simple, Modular Architecture for High-Performance Packet Switching," Yu-Shuan et al, IEEE JSAC, Oct 1987

[Yua98]   "A Class of Interconnection Networks for Multicasting", Yuanyuan Yang, IEEE Trans. Computers 47(8): 899-906, 1998

[Yuj92]   "Performance Analysis of Non blocking Packet Switch with Input and Output Buffers," Yuji Oie, et al, IEEE Trans. Communications, pp. 1294-1297, Aug 1992

Chapter 3

SONET AND SDH
An Introduction to SONET and SDH transport standards

3.1. INTRODUCTION

The intention of this chapter is twofold. The first goal of this chapter is provide an overview of SONET and SDH for a reader who is unfamiliar either of these technologies. To this end, a brief introduction is provided to the network technology that preceded SONET/SDH and the motivations that lead to the development of the SONET and SDH standards. This material is presented at a level of detail that is intended to help the reader understand material in subsequent sections. The SONET/SDH signal structure, overhead, and payload mappings are described with the key concepts behind SONET/SDH that have made it such a successful technology are discussed along the way.

The second goal of the chapter is to provide a useful reference for those more familiar with SONET and SDH. Table 3-2 and the organization of sections 4 and 5 are central to this goal. Table 3-2 is a quick reference containing a complete list of all the many SONET and SDH overhead bytes, and the section numbers where the reader can find additional detail on the function and format of each byte. Section 4 presents these overhead bytes organized by SONET.SDH layers, while section 5 groups them by function. Section 4 provides details about the overhead functions that are unique to that layer, and refers to 5 for the discussion of bytes that have functions similar to other bytes at other SONET/SDH layers.

The SONET (Synchronous Optical NETwork) standard was developed in North America in ANSI accredited subcommittee T1X1 of Committee T1. In parallel, SDH (Synchronous Digital Hierarchy) was developed by the

International Telephone and Telegraph Consultative Committee (CCITT). To understand and appreciate the value of SONET/SDH to the modern telecommunications network, it is useful to review the history of what preceded it. To summarize, the principal benefits of SONET/SDH over the preceding network are:

1. SONET/SDH provides a standard hierarchy for multiplexing signals in high-speed fiber-optic transmission systems.
   - Prior to SONET/SDH, there were many different proprietary signal formats in used for high-speed fiber-optic systems.
   - The SONET/SDH signal format provides a simpler multiplexing method than the previous hierarchies, especially from the standpoints of being able to form higher rate signals as simple interleaving of the base rate signal, and of being able to directly observe the constituent signals.
2. SONET/SDH provides embedded overhead channels that significantly:
   - reduce the overall costs of configuring (provisioning) and running the network,
   - improve the network reliability, and
   - guarantee the performance of the client signals carried through that network.
3. Due to the cooperation between North American, European, and Asian standards organizations, SONET and SDH are largely compatible.

The common points and differences between SONET and SDH will become clearer in the discussion of this chapter. In many cases, the differences are a matter of terminology rather than technical substance. The convention used here is to state the SONET term with the corresponding SDH term in parenthesis for at least the first appearance of a term. For terms that are used frequently, the SDH term is assumed to be understood and is not repeated in all instances.

The basic organization of this chapter is to first introduce the prior technology, and the motivations that lead to SONET and SDH. This is followed by a general description of SONET/SDH networks and the SONET/SDH frame and signal structure. Next, the signal overhead and its functions are described. As noted above, the overhead channels are first mentioned in the context of the hierarchical level that they serve within the SONET/SDH signal, followed by an expanded description in which the channels are discussed according to their function. Once this understanding of the SONET/SDH network and signal structure are established, the mapping of the various client payload signals into SONET/SDH is described. The chapter provides discussions on more specialized subjects such as synchronization, forward error correction, and sub-rate signals.

3.1.1 The pre-SONET State of the Network – a Brief History

The conversion of the North American network from analog to digital signals began with the introduction of T1 transmission systems in the early 1960s. The conversion in Europe began a few years later. In North America, this network is typically referred to as the asynchronous network, while the Europeans refer to theirs as the Plesiochronous Digital Hierarchy (PDH)[44]. The DS1, which is the base signal for this initial North American digital signal hierarchy, uses a bit rate of 1.544Mbit/s[45]. A DS1 is typically channelized into 24 8-bit voice channels that are referred to as DS0s. The next level of the hierarchy is the seldom-used DS2 signal, which contains four DS1s, and the third level is the common DS3, which contains seven DS2s. The choice of rates for these signals was defined to make the most cost-effective use of the electronic and transmission line technology available at the time the system was introduced. For example, the DS1 signal was designed to use the same cable (shielded twisted pairs) that was used by analog voice signals. Load coils were used on these cables to provide equalization so that analog voice signals could be transmitted over greater distances. The AT&T coil spacing plan was called H88, and placed 88 millihenry coils every 6000 feet. In order for the new T-series carrier system to function, the load coils needed to be removed from the line, and repeaters added to the line. The optimum solution was to replace the load coils with repeaters. The 1.544 Mbit/s rate was about the highest rate that could be transmitted over the desired distance with a practical implementation given the technology of that time. Also, since FDM carrier

---

[44] The reference to Europe on conjunction with PDH and SDH in this chapter is due to their dominant role in the development of many of the standards. The PDH is used virtually everywhere except North America, Japan, and Taiwan. SDH is used everywhere except North America. Japan used DS1 as the base rate fore their asynchronous hierarchy, but developed their own higher rate signals. In part to accommodate their unique hierarchy, the Japanese version of SDH is slightly different than the version of SDH used elsewhere.

[45] A note on a common confusion. A DSn signal refers to the frame format and rate of a signal at level n of the North American hierarchy. A Tn signal is the AT&T transmission facility designation (originating in the days prior to the breakup of AT&T). For example, a DS1 is typically carried on a T1 facility, which is a 4-wire connection that typically uses repeaters. For the European signal hierarchy, En is used to designate the signal frame format and rate.

systems of that day supported 12 channels, the DS1 signal provided a even doubling of their capacity. This emphasis is also evident with the DS3 signal using a prime number of DS2s instead of a more convenient number like eight. (The PDH hierarchy uses multiples of four, with four of their base rate E1 signals being multiplexed to form the E2 signal, and four E2 signals being multiplexed to form the E3 and four E3s multiplexed to form an E4.)

To form a signal of level n of the hierarchy, a group of signals from level n-1 are interleaved in a round-robin, bit-by-bit basis. (The exception here is the DS1 signal in which the 24 DS0s are byte-interleaved within the DS1 frame.) Consider the following example of multiplexing DS1 signals into a DS2. When a DS1 signal is formed by digitizing voice frequency channels, all channels are digitized at exactly the same clock rate with the same digitizing clock, and hence there is no skew between the different DS0 channels in a DS1. When DS1 signals are multipexed into higher rate signals, however, there is no guarantee that each DS1 source used exactly the same clock rate. In other words, the sources of all the DS1 signals may be asynchronous to each other with respect to frequency and framing phase. The actual frequency of a DS1 signal is specified to be within a certain range of a true 1.544 MHz (i.e., ± 77.2 Hz). A multiplexer that combines DS1 signals must be capable of accommodating at least the allowed range of incoming frequencies. What this means in practical terms is that over time, more bits will arrive from one DS1 signal than for another. As a result, it is not possible to simply interleave the bits from each DS1 into the higher-rate signal. The basic technique used in all multiplexers prior to SONET is known as bit-stuffing. A bit-stuffing multiplex uses a bit-rate slightly higher than the combined maximum rate of its incoming tributary signals plus the multiplex overhead. The number of bit positions reserved for each tributary is based on the maximum frequency allowed for that tributary. This allows the multiplexer to accommodate the maximum tributary rate. Some of these payload bit positions, however, are not needed if the tributary's rate is less than its maximum. To accommodate the lower rates, some of the payload bits are designated as "stuff" bits (S-bits). When these S-bit positions are not required for actual tributary data, they are filled with "dummy" data. The receiver must, of course, be notified whether the S-bits contained actual data or dummy data. This notification is accomplished through control bits (C-bits) in the multiplex overhead format. The value of the C-bits are set by the multiplexer to tell the demultiplexer which S-bits contain real data. The bit stuffing technique is illustrated by the following examples.

The DS2 signal format is shown in Figure 3-1. The F-bits provide the masterframe frame alignment pattern while the M-bits indicate the subframe alignment within the masterframe. The information fields consist of a bit-by-bit round-robin interleaving of the four payload DS1s. The bit rate of the DS2 signal is 6.312 Mbit/s ±208 bit/s, which gives minimum, nominal, and maximum bit periods of 158.4232, 158.4284, and 158.4336 ns/bit, respectively. The duration of a DS2 masterframe ranges between (158.4232 ns/bit)(1176 bit) = 186.3056 μs and (158.4336 ns/bit)(1176 bit) = 186.3179 μs. The maximum number of bits that a DS1 tributary can deliver in 186.3179 μs is 287.69 bits, and the minimum number of bits that a DS1 tributary can deliver in 186.3056 μs is 287.64 bits. (These maximum and minimum values take into account the DS1 frequency range.) There are (1176 - 24)/4 = 288 information bit positions in the subframe per DS1 tributary, or 288 - 1 = 287 bit positions if the S-bit position is filled with dummy (stuff) data. Thus, the DS2 signal can take in the DS1 tributary and, on the average, the S-bit position will be occupied with payload data approximately 2/3 of the time. C1, C2, and C3 are set to 1 to indicated that the next S-bit contains a stuff bit (positive stuff), which allows the receiver to perform a majority vote on the three C bits.. (Note that in the DS2 signal, the bits of the second and fourth DS1s are logically inverted in order to give a better balance of data ones and zeros in the resulting DS2 signal.)

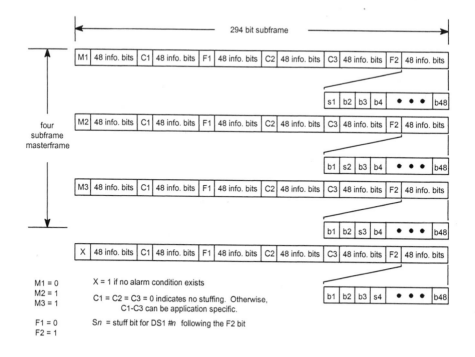

Figure 3-1. – DS2 signal frame format

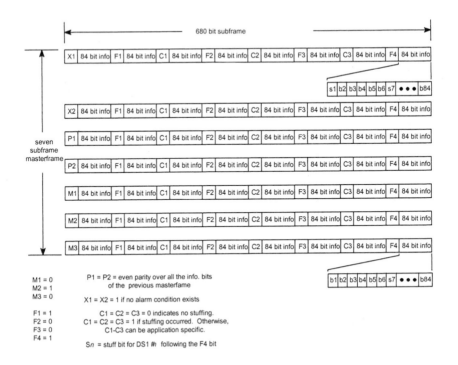

Figure 3-2. DS3 signal frame and M23 multiplexing format

One major drawback to this type of multiplexing approach is that multiplexing must be performed through each hierarchical level when constructing a higher rate signal. For example, as can be seen in Figure 3-2, to multiplex 28 DS1s into a DS3, the DS1s must first be multiplexed into DS2s and the DS2s are then multiplexed into the DS3.[46] This multi-stage multiplexing approach adds complexity to the multiplexer, the demultiplexer, and each point at which someone wants to monitor the constituent lower-rate signals. At the time of the proposal to develop the SONET standard, even though there were no North American standards for rates above DS3, many proprietary systems had already been deployed with

---

[46] AT&T developed a special version of the DS3 signal in part to allow this direct DS1 into DS3 multiplexing. This DS3 is typically referred to as C-bit Parity DS3. It uses the same rate as the original DS3, but re-defines some of the frame overhead bits (i.e., some of the C-bits) and the mapping in order to allow additional OAM&P communication in addition to the direct multiplexing.

different frame formats and rates. Some example rates that existed were roughly 90, 135, 405, 560 Mbit/s, and 1.2 and 1.7 Gbit/s.

Another, more important drawback to the asynchronous/PDH networks was that they provide very little overhead bandwidth for OAM&P (Operations, Administration, Maintenance, & Provisioning) purposes. OAM&P has become increasingly important for two reasons. The first is that OAM&P costs are the dominant costs for a network provider (much more so than equipment costs). The other reason is a result of the divesture of telecommunications networks. In North America, for example, a typical long-distance connection involves at least three carriers: the local exchange carrier connected to the subscriber at each end of the call and the inter-exchange carrier in between. When a problem occurs on a connection, the subscriber wants to know which carrier is responsible so that the problem can be fixed. Depending on the nature of the service and contracts, there can even be a threat of legal action or fines if the problem is not fixed in a timely manner. In this environment, it is extremely important for a carrier to find and fix its problems quickly, and also to be able to prove when the problem is outside its network.

3.1.2 The Origin of SONET/SDH

The SONET standard was developed in the ANSI-accredited Committee T1 that was formed in the wake of the AT&T divestiture. This work began in early 1985. The rates and formats were defined in the T1X1 subcommittee with the OAM&P requirements coming from the T1M1 subcommittee. T1X1 and T1M1 worked together to develop the object-oriented network element models for SONET. Shortly after the SONET work began, the CCITT (Consultative Committee on International Telephone and Telegraph, subsequently renamed the International Telecommunications Union – Telecommunications Sector (ITU-T)) began the development of SDH. The strong desire to maximize compatibility between SONET and SDH led to a series of compromises. Readers interested in the history of this standard development are referred to [BaCh].

3.2. SONET/SDH LAYERING PRINCIPLES

A layered concept was used in the development of the SONET and SDH standards. The use of layers allows sectionalizing the overhead functions so that only those layer functions that are actually required at a given point in the network need to be accessed and processed at that point. Another motivation for layered overhead is that it allows the performance monitoring to be applied at different levels of the connection (trail) through the SONET/SDH network. For example, one set of overhead covers individual spans while a different set covers end-to-end connections, thus eliminating the need to collect and correlate the data from the individual spans to determine the end-to-end performance. A further advantage of layered overhead is that it allows fault information to be localized such that network fault locations can be quickly identified and repaired. The explanation of some of the terminology introduced in this section is left for subsequent, more detailed sections.

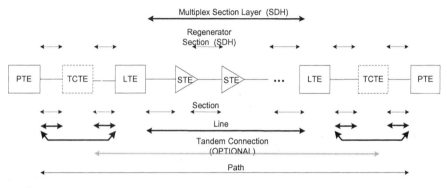

Figure 3-3. SONET layer illustration

The SONET layers are illustrated in Figures 3-3, 3-4, and 3-5. The innermost layer is the Section (Regenerator Section) layer. This layer covers individual facility links, which are typically a fiber connecting two pieces of SONET equipment. (Copper cable and microwave radio links are also specified for some signal rates.[47]) The reason for having a separate Section layer was to allow intelligent repeaters (regenerators) that could be

---

[47] A terminology clarification is appropriate at this point. In SONET, STS-N (Synchronous Transport Signal at level N) refers to the signal rate and format. OC-N is the designation for the Optical Carrier of an STS-N signal (i.e., it is the physical, photonic interface). Similarly, EC-N refers to an electrical interface carrying an STS-N signal, where $N = 1$ or 3. The EC-1 signal has become increasingly popular for intra-office connections between SONET NEs.

maintained and monitored by the network management system through a Section layer communications channel. A NE that only terminates the Section layer is referred to as a Section Terminating Element (STE) and the overhead associated with the Section layer is called the Section Overhead (SOH or RSOH for SDH). All other layers are ignored by an STE, passing transparently through it.

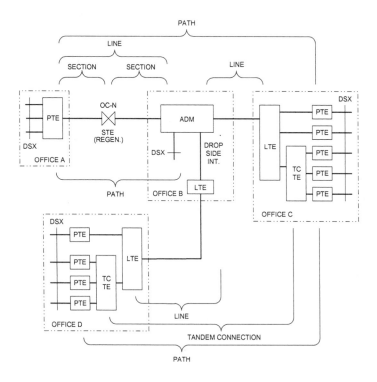

Figure 3-4. Illustration of SONET/SDH layer and network element terminology

The next layer is the Line (Multiplex Section) layer. This layer covers the spans between network elements including intermediate repeaters. A NE that terminates the Line overhead is referred to as a Line Terminating Element (LTE) and the overhead associated with the Line layer is referred to as the Line Overhead (LOH or MSOH for SDH). A LTE must also perform the STE functions. A LTE typically performs the functions of crossconnecting and/or terminating Paths. These functions require the LTE to perform the appropriate network synchronization functions as it deals with signals that may have originated at terminals with different clock sources or that have accumulated differing degrees of jitter and wander in the network. The Line layer is also the layer at which most of the automatic protection switching is performed for network restoration.

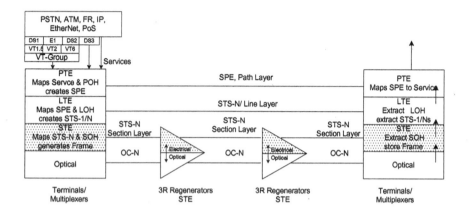

Figure 3-5. SONET layers illustrated in terms of a protocol stack

-   NOTE – The combination of the Section and Line overhead is also referred to as the Transport Overhead (TOH).

The outer-most layer is the Path layer. This layer comprises the end-to-end path that a payload envelope (container) traverses through a SONET/SDH network. There can actually be one or two levels of Path layers in SONET/SDH. In SONET the high order Path consists of N x ≈50Mbit/s (N=1, 3, 12, 48, 192, 768) payload envelopes, and in SDH the high order Path consists of K x ≈150Mbit/s (K=1, 4, 16, 256) payload containers. A SONET high order path is referred to as a STS-1 or STS-Nc, and the SDH high order path is referred to as a VC-4 or VC-4-Nc. A high order Path can either directly contain a tributary signal (e.g., a DS3 in a STS-1 or an E4 in a VC-4, as discussed below), or it can contain a multiplexed group of low order Paths. If the high order Path contains a multiplexed structure of low order Paths, the high order path begins at the point at which this multiplexing occurs and ends at the point at which the demultiplexing occurs. If the Path directly contains a tributary signal, the Path layer begins at the point at which the tributary is mapped into a SONET/SDH signal and ends at the point where the tributary signal is removed from the SONET/SDH signal. A NE that terminates the Path overhead is referred to as a Path Terminating Element (PTE) and the associated overhead is the Path Overhead (POH). A PTE must also perform all the STE and LTE functions.

Low order Paths carry lower rate tributary signals (e.g., DS1 or E1). For SONET, these low order paths are referred to as Virtual Tributaries (VTs) with the specific sizes of VT1.5, VT2, VT3, and VT6. For SDH, the low order paths have the specific sizes of VC-1x, VC-2, and VC-3, where VC is the abbreviation for Virtual Container. Table 3-6 specifies the rates and client signals associated with each VT and VC. An SDH VC-3 corresponds to a SONET STS-1. (Note that the VC-3 can be either a low order or high order Path in SDH, depending on its usage.) SDH uses the term VC-1/2 to collectively refer to those rates and structures that correspond to the SONET VTs.

Having both high and low order Path layers provides some significant advantages. The processing and routing of the lower rate signals mainly occurs in the access or metropolitan area networks, e.g., DS1/E1 signals carrying voice circuits between switches. Long haul networks between metropolitan areas typically provide a bulk transport service and have no need to process the lower rate tributaries. High order Path crossconnects and add/drop multiplexers are substantially simpler and less expensive to buy and manage than equivalent rate equipment with low order Path processing capability. Equipment that must process low order paths must have a substantially larger number of path overhead terminating circuits, larger cross-connect fabrics (due to the smaller granularity), and also require substantially greater provisioning capabilities. As a result, considerable savings can be obtained if the low order Paths are groomed into high order Paths prior to the hand-off to a long-haul network that is mainly compromised of equipment that only processes high order Paths.

Note that SDH, following the formalized network description terminology of G.805, refers to the end-to-end route between where a tributary signal is mapped into SDH to where it is removed from the SDH network as a Trail. In general terms, a Trail is validated information flow. In other words, the end-points of a Trail have the capability of performing some degree of performance monitoring or connection verification in order to validate the correct signal connectivity and its integrity. A Path, in contrast, provides transport services to client layer, or other Path layer signals. Hence, a Path providing transport services to a client layer signal that includes a validation function over the length of the Path is also a Trail. Both the Path and Trail terms are used in SDH, however SONET terminology only refers to Paths.

The Tandem Connection Monitoring (TCM) sublayer was not a part of the original SONET or SDH standards, but was added later when the need for its functions was identified. The need for the TCM sublayer grew primarily from inter-exchange carrier (IEC) requirements. A typical IEC application is to receive a group of high-order tributaries at one side of the IEC network and transport them to another carrier at the other side of the network. The performance of this group of signals must be monitored as it transits multiple tandem line systems (MS sections) within the network. One way to monitor the performance of this group of signals is to monitor the path error detecting bytes for each of the constituent signals in the group and to compare the state of these error-detecting bytes at the beginning and end of this connection. This method is cumbersome, however, and requires a high degree of Network Management System (NMS) communication between the endpoints. TCM inserts overhead to monitor the performance of the signals across the desired portion of the network (i.e., the tandem connection) and also allows a logical grouping of these signal such that a single overhead byte reports the far-end performance for all the signals in that group. The result is a network level performance monitoring capability. The TCM overhead also allows a carrier to identify whether a fault affecting the constituent paths occurred within or outside of that tandem connection. The overhead byte for high order TCM is taken from the Path overhead byte area, and logically sits between the Line and Path layers. A similar capability is available for low order TCM, and its overhead is taken from the VT overhead and logically sits between the STS and VT Path layers. The TCM layers are optional and tend to be used primarily for long-haul applications.

3.3. SONET/SDH FRAME FORMAT

In order to resolve the difficulties associated with multi-stage asynchronous multiplexing, the SONET/SDH frame format was developed to provide a standard method for specifying new, higher-rate multiplexed signals in a consistent manner. Toward this end, each level of the SONET/SDH hierarchy is formed by simply directly interleaving an integer number of lower rate SONET/SDH signals. The interleaving is performed in a byte-by-byte manner rather than a bit-by-bit manner in order to facilitate parallel data path implementations in SONET/SDH processing circuits. Parallel data paths are advantageous since they allow the processing circuitry to run at lower clock speeds than would be required for serial, bit-by-bit processing. The frame format is such that the constituent lower-rate signals can be observed directly at the byte level within the SONET/SDH high-rate signal, with the byte interleaving allowing the visibility to potentially extend down to the DS0 level. This byte level visibility of the constituent payloads greatly simplifies the process of extracting or monitoring the payloads, which in turn has made it possible to economically integrate a substantial amount of switching capabilities into SONET/SDH NEs. To address the need for improved OAM&P capabilities, a number of overhead bytes were added to the SONET frame format to convey various types of OAM&P information. A further refinement in SONET/SDH, as discussed above, was the grouping of these overhead channels at different layers within the frame in order to minimize the amount of overhead processed by intermediate network elements. Another innovation in SONET/SDH is the use of pointers rather than stuff bits to handle clock differences between the constituent multiplexed signals and also between the timing domains of different NEs. Each of these points will become clear in the following sections.

3.3.1   SONET and SDH Frames

The SONET and SDH frame formats are virtually identical, with the differences tied to historical factors. The base SONET signal is called a Synchronous Transport Signal, level 1 (STS-1), and the base SDH signal is called a Synchronous Transport Module, level 1 (STM-1). The North American DS3 signal uses a bit rate of 44.736 Mbit/s, and at the time of the SONET/SDH standards development, this rate was just within the capabilities of CMOS integrated circuit technology. For these reasons, the North Americans preferred a base rate of around 50 Mbit/s. The European E3 and E4 signals used rates of 34.368 and 139.264 Mbit/s, respectively. For that reason, a 50 Mbit/s base signal was a poor fit and they preferred a base rate around 150 Mbit/s. The compromise was that the SDH base rate frame format is essentially the same as three interleaved SONET STS-1 signals with a concatenated payload area. The resulting base signal rates for SONET and SDH are 51.84 and 155.52 Mbit/s, respectively. (See [BaCh] for a historical review of the compromises leading to SONET/SDH.)

The basic SONET and SDH frame format partitioning illustrated in Figure 3-6. The frame is laid out in a two dimensional arrangement in which the transmission order is row by row. In other words, the frame transmission begins with the byte in row 1, column 1, followed by row 1, column 2, through the end of row 1, and then follows the same column transmission order through row 2, and finally ends with the last column of row 9. The numbering conventions for SONET and SDH bytes will be discussed in greater detail below. The time required to transmit a SONET/SDH frame, regardless of the signal rate, is 125 μs. The 125 μs corresponds to an 8 kHz reference clock for the frame, which was chosen since this is the sampling rate used for digitizing voice channels, and as a result is also the frame duration of the DS1 and E1 signals. The distribution of network synchronization information is simplified when everything uses some common reference frequency.

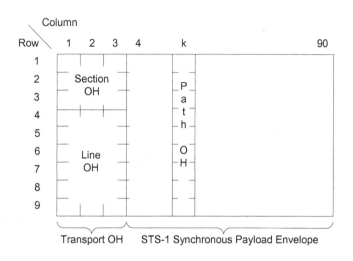

a) SONET STS-1 frame partitioning

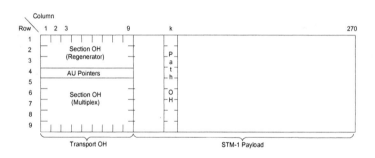

b) SDH STM-1 frame partitioning

Figure 3-6. SONET and SDH base frame formats

For SONET, the active Transport overhead is primarily carried in the TOH columns associated with the first STS-1 regardless of the signal rate. For compatibility, SDH uses columns 1, 4, and 7 of the first STM-1. In other words, for the SONET STS-N, the primary active TOH columns are columns 1, $N+1$, and $2N+1$, and for the SDH STM-M, the primary active TOH columns are columns 1, $3M+1$, and $6M+1$. These locations are illustrated in Figure 3-7 through Figure 3-11. The STS-1 also contains a single STS Path overhead column, the location of which is indicated by the pointer bytes. The STM-1 can contain either 1 or 3 Path overhead columns, depending on the payload structure, and again the location(s) is determined by the pointer bytes.

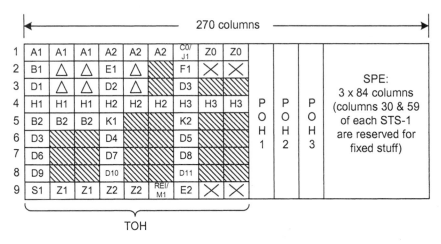

a) *Overhead byte layout in a SONET STS-3 signal*

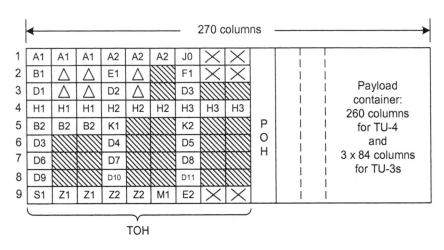

b) *Overhead byte layout in a SDH STM-1 signal*

▨ Unused   △ Media dependent   ✕ National use

Figure 3-7. Illustration of overhead byte locations

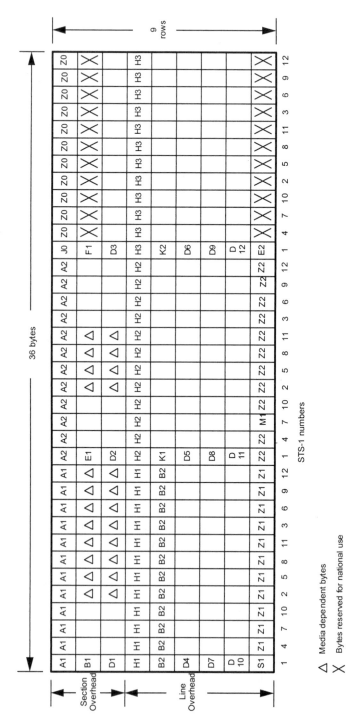

Figure 3-8. STS-12 (STM-4) TOH byte definitions

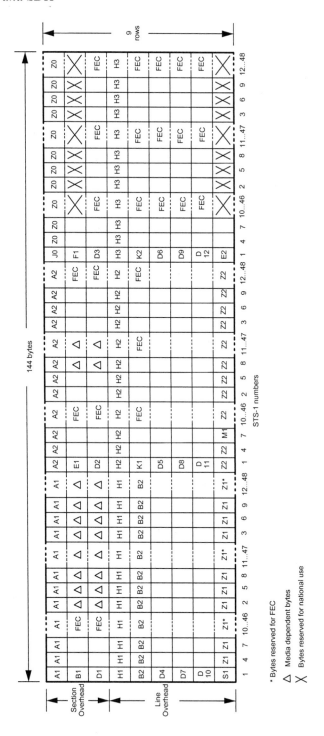

Figure 3-9. STS-48 (STM-16) TOH byte definitions

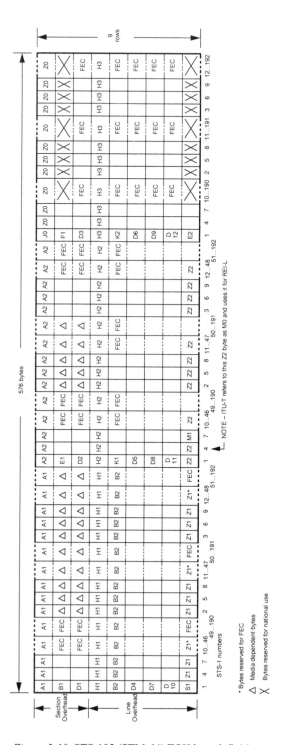

Figure 3-10. STS-192 (STM-64) TOH byte definitions

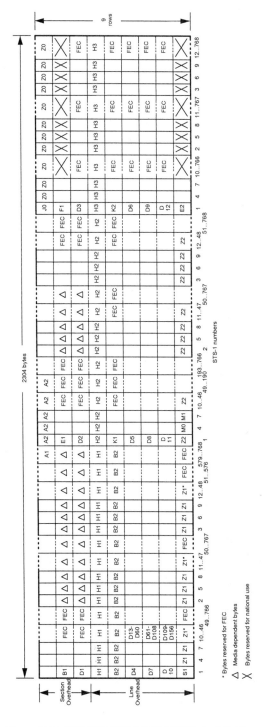

Figure 3-11. STS-768 (STM-256) TOH byte definitions

The SONET and SDH signal rates along with their associated payload rates are shown in Table 3-1. Each higher rate signal is formed by a byte-by-byte interleaving of the base (STS-1 or STM-1) signals so that an STS-N (STM-N) is formed by the byte interleaving of N STS-1s (STM-1s). The multiplexing format for a SONET STS-12 is illustrated in Figure 3-12.a. The two-stage multiplexing shown in Figure 3-12.a is a conceptual illustration, since the same result could obviously be obtained directly in a single stage. The reason for the two-stages and the resulting seemingly awkward interleaving is the desire for compatibility between SONET and SDH, as can be seen from Figure 3-12.b. With SDH, the STM-1s correspond to the STS-3s, and as such would be interleaved with a sequence of 1, 2, 3, 4, 1, 2,…

Table 3-1. SONET and SDH signal rates

OC Level	STS-N	SDH Term	Line Rate	SPE Rate	Effective Payload Rate
OC-1	STS-1	-	51.84	50.112	ATM: 49.536 DS3: 48.384
OC-3	STS-3	STM-1	155.520	150.336	STS-3 3 DS3s or one STS-3c
OC-12	STS-12	STM-4	622.080	601.344	
OC-48	STS-48	STM-16	2488.320	2405.376	
OC-192	STS-192	STM-64	9953.280	9621.504	
OC-768	STS-768	STM-256	39813.12	38486.016	
NOTE – SPE rate includes POH and fixed stuff columns					

As illustrated in Figure 3-13, the bit numbering of the bytes on SONET and SDH follows the tradition for digital telephony. The bits are numbered from 1-8 according to the transmission order, i.e., bit 1 is transmitted first and bit 8 is transmitted last. Bit 1 corresponds to the most significant bit of the byte. As noted above, the transmission order within the SONET/SDH signal is to transmit row by row, with the column 1 byte within a row being the first byte of the row to be transmitted. The bytes are typically labeled in row column format for SONET, as illustrated in Figure 3-14.a. An alternative representation is used for the SDH Section overhead (SOH) bytes. This alternative is illustrated in Figure 3-14.b. The number of a particular SDH SOH byte uses the form S(a, b, c), where the a represents the row number. The meaning of b and c can be understood when we remember the byte-interleaved multiplexing format of the STM-N. The first N columns of an STM-*N* will be first columns of each of the *N* constituent STM-1s, the next *N* columns will be the second columns of the constituent STM-1, etc. The b number, then, designates the number of the constituent STM-1 and the c number indicates the specific column within that STM-1.

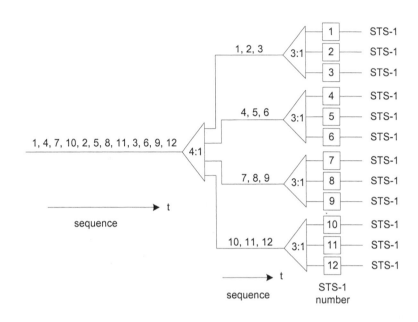

a) SONET STS-12 multiplex interleaving

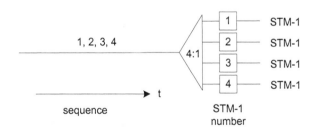

b) SDH STM-4 multiplex interleaving

Figure 3-12. Illustration of the multiplexing format for the 622 Mbit/s SONET and SDH signals

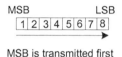

MSB is transmitted first

Figure 3-13. Transmission order

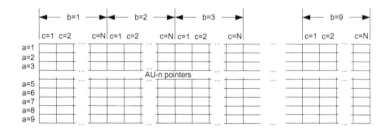

a) *SONET TOH column numbering (STS-1 numbers shown for an STS-12 example)*

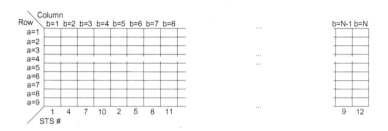

b) *SDH TOH row and column numbering – (S(a,b,c) form)*

Figure 3-14. Numbering of TOH byte locations

3.3.2   Physical Layer Consideration - Scrambling

A non-return-to-zero (NRZ) line code is used for SONET/SDH optical interfaces. NRZ transmits an optical pulse for the duration of the bit interval for a data 1, and transmits no energy for a data 0. The NRZ line code is the simplest to implement and is the most bandwidth-efficient binary line code. The primary drawback to NRZ, however, is that long strings of 1s or 0s in the data cause long periods in which there is no transition in the signal level on the span. During that period, the clock recovery circuits at the receiver can drift out of synchronization with the transmitted data so that the receiver begins sampling data away from the optimum, mid-bit point and ends up mis-sampling the next bit after the long string. In addition, if the string consists of 0s, the automatic gain control at the receiver can begin to perceive the situation as a lower signal level and increase its gain. The result again is non-optimal sampling decision thresholds in the data recovery circuit. If the data stream is used to directly modulate the transmitting laser, an additional problem is that the laser's temperature will change during the long strings, which affects the laser's output signal level. For all these reasons, it is important to have the signal level make frequent transitions and to maintain a roughly balanced number of 0s and 1s that it transmits.

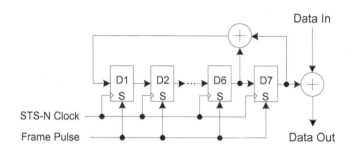

Figure 3-15. Frame synchronous scrambler

For typical streams of data, it is adequate to use a simple frame-synchronous scrambler to randomize the transmitted data enough to obtain a reasonable NRZ transition density and an approximate balance between transmitted 0s and 1s. A frame-synchronous scrambler is a pseudo-random number generator (implemented as a linear feedback shift register) that is set to a known state at a periodic point within the transmission frame. The transmitted bits are the bit-by-bit XOR of the output of the final scrambler stage with the data stream. The 7-bit frame-synchronous scrambler chosen for SONET/SDH uses a scrambler polynomial of x^7+x^6+1, and is shown in Figure 3-15. This scrambler is set to a state of all 1s on the first bit of the payload area (i.e., the first bit after the Nth J0/Z0 byte) of each SONET/SDH frame and is clocked continuously at the STS-N/STM-N line rate. In order to make the framing pattern bytes (see section 4.1) readily visible to the receiver framing circuit, the TOH bytes in the first row are not scrambled. (Note – for STS-768/STM-256, only the A1 and A2 bytes are unscrambled.) At the receiver, the data stream is again XORed with an identical scrambler to recover the original bits. The reason this works can be explained as follows, let st_i and sr_i be the scrambler states i clock cycles into the frame at the transmitter and receiver, respectively, and let di be the data bit that is XORed with st_i at the transmitter. The transmitted bit is $x_i = st_i$ XOR d_i. The output of the receiver descrambler is $r_i = x_i$ XOR sr_i (assuming no transmission channel errors). Since the frame synchronization guarantees that $st_i = sr_i$ then $r_i = st_i$ XOR d_i XOR $sr_i = d_i$.

The frame-synchronous scrambler has proven to be adequate for TDM payload data. Since TDM interleaves the different subscribers on a bit-by-bit or byte-by-byte basis, there is typically no correlation between the data in successive bytes. This situation changes, however, when individual subscribers are given access to the entire SONET/SDH payload area or a long sequence of bytes within the payload area. For example, when the subscriber data is multiplexed on a cell or packet basis into the entire payload area, then a subscriber's data will occupy successive payload bytes up to the length of the cell/packet (ignoring for the moment the presence of the TOH and POH columns). As a consequence, if a subscriber transmitted data that contains the same pseudo-random sequence as the SONET/SDH frame-synchronous scrambler pattern, it is possible that this data will align with the transmitter's frame-synchronous scrambler such that the scrambler output will be a long string of 0s or 1s. (Note that if the scrambler pattern is aligned and XORed with itself it produces 0s at the output of the scrambler.) Hence, if the entire payload area is used as a single payload container, a malicious user sending the frame-synchronous scrambler pattern in his cell/packet payload could potentially cause the receiver clock/data recovery circuit to fail and service to be denied to the other users sharing that cell/packet multiplexed connection until the receiver re-acquires clock and frame alignment. The vulnerability of SONET/SDH to such malicious data attacks was first observed with ATM cell payloads, and becomes much worse for the longer packets used in PoS and GFP (see Chapter 4).

The solution selected to combat malicious user cell/packet data patterns was a self-synchronous scrambler. As illustrated in Figure 3-16, a typical self-synchronous scrambler output is the XOR of the current data with the data that preceded it by n bits. Mathematically, this operation amounts to dividing the data stream by a x^n+1 scrambler polynomial. For SONET/SDH, a $x^{43}+1$ scrambler is applied to the just the payload of cell or packet-oriented payload mappings (with the cell or packet header excluded in some cases to allow faster cell/packet delineation at the receiver). As illustrated in Figure 3-16, the receiver effectively multiplies the payload data by the $x^{43}+1$ scrambler polynomial to reverse the process and recover the original data. The self-synchronous scrambler pattern is much longer than the frame-synchronous scrambler, and since its scrambler polynomial is mutually prime with that of the frame synchronous scramble, the two scramblers combine in a manner that it makes it extremely unlikely that a malicious user's data pattern will cause a problem.

Note that with a frame-synchronous scrambler, a single transmission error will produce a single error after the receiver descrambler. With a self-synchronous scrambler, a transmission channel error will produce two errors after descrambling, since the original erroneous bit is retained in the descrambler register for subsequent XOR with a later bit. This error multiplication is the main drawback to self-synchronous scramblers and is the reason for choosing a frame-synchronous scrambler as the primary SONET/SDH scrambler.

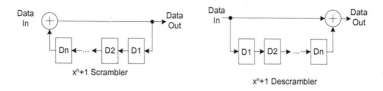

Figure 3-16. Self-synchronous scrambler

3.4. SONET OVERHEAD DEFINITIONS

The overhead bytes used in SONET and SDH are listed in Table 3-2. The functions of these bytes are discussed in more detail throughout this chapter. The term "Line" in Table 3-2 refers to a high-speed interface from a SONET/SDH NE, and "Drop" refers to an intra-office tributary interface.

The format of this section is to discuss the overhead bytes on both a level-by-level basis, with those overhead bytes that perform functions that are unique to a given level are discussed in the description of that level. Section 5 discusses the overhead byte use on a functional basis, with any level-specific functions noted in that discussion.

Table 3-2. SONET/SDH overhead byte definition summary

SONET/SDH Overhead Byte	SONET (SDH) Byte Definition and Function/	Where it is required for Section and Line	Section where it is discussed in this book
	Section (Regenerator Section) Overhead		
A1, A2	Frame alignment bytes. A1 = 11110110, A2 = 00101000	Line, Drop	4.1
B1	Section (Regenerator Section) BIP-8 error detection	Line	5.1.1
J0/C1	This byte was originally designated C1 (for use as an STS-1 identifier), but is currently defined as J0 and is the Section Trace byte	Line (Optional)	5.3
Z0	Section Growth byte in the C1 location of STS-1s 2-N (SDH S (1, 7, 2) to S (1, 7, N))	Line	
D1-D3	Section Data Communications Channel (SDCC)	Line	5.4
E1	Section Orderwire	Line (Optional)	5.6
F1	Section User Byte	Line (Optional)	5.7
	Media dependent bytes		
	Line (Multiplex Section) Overhead		
B2	Line (Multiplex Section) BIP-8 error detection	Line, Drop	5.1.1
H1-H3	H1 and H2 are the ayload (AU/TU) pointer bytes and H3 is the pointer action byte.	Line, Drop	7 and 5.2.1
K1, K2	Automatic Protection Switching (APS) bytes	Line, Drop (Optional)	4.2, 5.2, 5.5, and Chapter 8
D4-D12	Line (MS) Data Communications Channel	Line (Optional)	3.5.4
S1	Synchronization Messaging (4-bits)	Line, Drop	4.2 and 8
E2	Line (MS) Orderwire	Line (Optional)	5.6
M0	STS-1 Line REI – applies only to the first STS-1	Line, Drop	5.1.2
M1	Line REI - used in STS-3 through STS-192	Line, Drop	5.1.2
Z1	Growth		
Z2	Growth		
D13-D15	Extended Line (MS) Data Communications Channel that is used only with STS-768/STM-256	Line (Optional)	3.5.4

Table 3-2 – continued

SONET/SDH Overhead Byte	SONET (SDH) Byte Definition and Function	Where it is required for Section and Line	Section where it is discussed in this book
	Tandem Connection Overhead		
N1	Tandem Connection Maintenance	-	4.3 and.5.8
	High Order Path Overhead		
J1	Path (Trail) trace	-	5.3
B3	Path BIP-8 error detection	-	5.1.1
C2	Path signal label	-	4.2
G1	Path status	-	5.1.2 and 5.2.2
F2	Path user channel	-	5.7
H4	Multiframe indicator	-	4.4 and Chapter 4
Z3/F3	SONET Path growth byte, SDH Path user channel	-	5.7
Z4/K3	SONET Path growth byte, SDH VC-3/VC-4 subnetwork protection	-	5.5 and Chapter 9
	Virtual Tributary (VC-1 and VC-2) Path Overhead		
V1-V3	V1 and V2 are the VT (VC-1/2) payload pointers and V3 is the pointer action byte.	-	6.1, 5.2.1, and 7
V4	Reserved	-	4.5
V5	VT (VC) overhead - multiple functions	-	4.5, 5.1, 5.2.2, 5.2.3,
J2	VT Path trace	-	5.3
Z6/N2	Path growth byte (used in SDH for Low Order Tandem Connection)	-	4.5, 5.3, and 5.8
Z7/K4	Path growth byte – multiple functions in SONET and SDH	-	4.5, 5.2.2, and 5.5

3.4.1 Section (Regenerator Section) Overhead

The Section overhead contains those functions that are required (or desirable) for transmitting a SONET/SDH signal over a link that connects two SONET/SDH NEs. These functions include framing and various OAM functions.

A1 and **A2** framing: These bytes form a pattern for the fast and reliable recovery of the transport frame, with A1 = 11110110 and A2 = 00101000. The A1 and A2 bytes are active in the first and second columns, respectively, of each of the N constituent STS-1s in an STS-N signal, with the exception of the STS-768. SDH uses these bytes in the same manner, with A1 occupying S(1,1,*x*) and A2 occupying S(1,2,*x*), with the exception of STM-256. For the STS-768 (STM-256) case, it was agreed that there was no need to have so many framing bytes. As a result, this signal only uses the A1 byte in columns 705 to 768 (SDH S(1,3,193) to S(1,3,256)) and A2 in columns 769 to 832 (SDH S(1,4,1) to S(1,4,64)). The remaining byte locations that could have been used for A1 and A2 in STS-768 (STM-256) are reserved for future definition. The A1 and A2 bytes are not scrambled with the frame-synchronous scrambler so that they are readily visible for frame recovery.

A factor that was not initially anticipated when the A1 and A2 bit patterns were chosen was the potential affect on the laser for higher rate signals. For an STS-*N* signal, the frame begins with *N* adjacent A1 bytes followed by *N* adjacent A2 bytes. Note that for A1, there are more 1s than 0s, which means that the laser is on for a higher percentage of the time when the A1 byte than for A2, in which there are more 0s than 1s. The A2 byte has fewer 1s than 0s. As a result, if the laser is directly modulated, for large values of *N*, the lack of balance between 0s and 1s causes the transmitting laser to become hotter during the string of A1 bytes and cooler during the string of A2 bytes. The thermal drift affects the laser performance such that the signal level changes, making it difficult for the receiver threshold detector to track. Most high-speed systems have addressed this problem by using a laser that is continuously on and modulate the signal with a shutter after the laser.

J0, B1, E1, F1, and **D1-D3** are discussed below in OAM&P function discussion.

3.4.2   Line (Multiplex Section) Overhead

The Line overhead contains those functions that are necessary or desirable for terminating the transport signal in order to either remove or cross-connect portions of the payload. These functions include providing a multiframe alignment signal for the payload, a signaling channel to protect the payload, and various OAM functions.

K1 and **K2** – Automatic Protection Switching: The K1 and K2 bytes carry the signaling required to protect the signal at the Line level. Protection switching at the Line level involves switching the entire payload from a failed facility to a standby (protection) facility. Protection switching options are discussed in greater detail in Chapter 8. In addition to the APS information, the K2 byte is used to signal the Line level alarms of RDI-L and AIS-L.

S1 Synchronization: In order to minimize the jitter and wander that is generated for the tributary signals, and in order to minimize the associated buffer requirements, SONET/SDH operates largely as a synchronous system. All of the NEs within a timing island are typically tied to the same reference clock. The NEs can either receive their reference timing from an external clock (e.g., a clock distributed from a central timing source in a central office), derive it from an incoming signal, or generate it internally from a free-running clock. The S1 byte indicates the quality of the clock that the transmitting NE is using. The S1 definitions are shown in Table 3-3. See sections 8 for more discussion about network synchronization.

Table 3-3. S1 synchronization byte codes

S1 b5-8	SONET Synchronization Quality Level Description	SONET Quality Level	SONET Synchronization Quality Level Description
0000	Unknown synchronized traceability	2	Unknown synchronization quality
0001	Stratum 1 Traceable	1	Reserved
0010	-	-	G.811
0011	-	-	Reserved
0100	Transit Node Clock Traceable	4	SSU-A
0101, 0110	-	-	Reserved
0111	Stratum 2 Traceable	3	Reserved
1000	-	-	SSU-B
1001	-	-	Reserved
1010	Stratum 3 Traceable	6	Reserved
1011	-	-	Rec. G.813 Option I (SEC)
1100	SONET Minimum Clock Traceable	7	Reserved
1101	Stratum 3E Traceable	5	Reserved
1110	Reserved for Network Synchronization	User assignable	Reserved
1111	Don't use for synchronization	9	Don't use for synchronization

B2, **E2**, **D4-15**, **M0**, and **M1** are discussed further below.

H1-H3 pointer bytes. For SONET, the H1-H3 pointer bytes are considered as part of the Line overhead. SDH associates the H1-H3 bytes with the AU or TU and hence doesn't regard them as part of the Multiplex Section overhead even though they are located in that region. The use and operation of pointer bytes is discussed in detail in section 7.

3.4.3   High Order Tandem Connection Overhead

N1 high order Tandem Connection Monitoring. The TCM function is discussed in more detail below in 3.5.8, including the definition of the N1 bits.

3.4.4   Path Overhead

C2 Signal Label: The Signal Label (payload label) indicates the type of payload that is contained within the STS (VC-3/4) payload. The different C2 values are shown in Table 3-4. If the signal label at the Path terminating NE is different than what it was provisioned to expect, a payload label mismatch (PLM) alarm is declared. Values E3-FC are defined as a Payload Defect Indicator (PDI). As discussed in Chapter 8, if a PTE has the choice of selecting between two alternative received Paths, it can use the PDI in some applications to know health of the payload contained within the one path relative to the other path.

H4 Payload Multiframe: In order to be efficient in its use of bandwidth, the overhead bytes for VTs (VC-1/2) are spread across multiple SONET/SDH frames. By creating a multiframe, the same byte location within the VT/VC can be used for different functions in different STS/STM frames. The basic multiframe signal is a 4-bit counter in the four LSBs of H4, with the count incremented by one for each new H4 transmission. In addition to providing this basic multiframe function, the H4 has byte has recently been defined to carry an extended multiframe and signaling information associated with Virtual Concatenation. Virtual Concatenation is discussed in Chapter 4, and the resulting expanded H4 byte definition is shown in Figure 3-31 of that section.

The **J1**, **B3**, **G1**, **F2**, and **Z3-Z4** bytes are also in the Path over and are described below.

3. SONET and SDH

Table 3-4. C2 Signal Label values

Code (hex)	Payload type	Code (hex)	Payload type
00	Unequipped (note 1)	E1	STS-1 Payload with 1 VT-x Payload Defect
01	Equipped - Nonspecific (note 2)	E2	STS-1 Payload with 2 VT-x Payload Defects
02	Floating VT Mode	E3	STS-1 Payload with 3 VT-x Payload Defects
03	Locked VT Mode (note 3)	E4	STS-1 Payload with 4 VT-x Payload Defects
04	Asynchronous Mapping for DS3	E5	STS-1 Payload with 5 VT-x Payload Defects
05	Mapping under development (note 4)	E6	STS-1 Payload with 6 VT-x Payload Defects
12	Asynchronous Mapping for 139.264 Mbits/s	E7	STS-1 Payload with 7 VT-x Payload Defects
13	Mapping for ATM	E8	STS-1 Payload with 8 VT-x Payload Defects
14	Mapping for DQDB	E9	STS-1 Payload with 9 VT-x Payload Defects
15	Asynchronous Mapping for FDDI	EA	STS-1 Payload with 10 VT-x Payload Defects
16	Mapping for HDLC over SONET (note 5)	EB	STS-1 Payload with 11 VT-x Payload Defects
17	Simplified Data Link (SDL) with self-synchronizing scrambler (note 5)	EC	STS-1 Payload with 12 VT-x Payload Defects
18	HDLC / LAPS	ED	STS-1 Payload with 13 VT-x Payload Defects
19	SDL with use of a set-reset scrambler (note 6)	EE	STS-1 Payload with 14 VT-x Payload Defects
1A	10 Gbit/s Ethernet (IEEE 802.3)	EF	STS-1 Payload with 15 VT-x Payload Defects
1B	Generic Framing Procedure (GFP)	F0	STS-1 Payload with 16 VT-x Payload Defects
1C	Mapping of 10 Gbit/s Fibre Channel Frames (note 6)	F1	STS-1 Payload with 17 VT-x Payload Defects
1D	Asynchronous mapping of ODUk (k=1,2) into VC-4-Xv (X=17,68)	F2	STS-1 Payload with 18 VT-x Payload Defects
20	Asynchronous mapping of ODUk (k=1,2) into VC-4-Xv (X=17,68)	F3	STS-1 Payload with 19 VT-x Payload Defects
CF	Reserved (note 7)	F4	STS-1 Payload with 20 VT-x Payload Defects

Table 3-4 continued

Code (hex)	Payload type	Code (hex)	Payload type
F5	STS-1 Payload with 21 VT-x Payload Defects	FA	STS-1 Payload with 26 VT-x Payload Defects
F6	STS-1 Payload with 22 VT-x Payload Defects	FB	STS-1 Payload with 27 VT-x Payload Defects
F7	STS-1 Payload with 23 VT-x Payload Defects	FC	STS-1 Payload with 28 VT-x Payload Defects, or STS-1, STS-3c, etc. with a non-VT Payload defect (DS3, FDDI, etc.)
F8	STS-1 Payload with 24 VT-x Payload Defects	FE	Test signal, ITU-T O.181 specific mapping (note 8)
F9	STS-1 Payload with 25 VT-x Payload Defects	FF	STS SPE AIS condition (note 9)

NOTES

1 Code 00 indicates STS SPE Unequipped. This code shall be originated if the Line connection is complete but there is no Path Originating Equipment. Any code received, other than code 00, constitutes an equipped condition.

2 Code 01 indicates STS SPE Equipped - Non-specific Payload. This code can be used for all payloads that need no further differentiation, or that achieve differentiation by other means such as messages from an OS.

3 The VT Locked Mode Mapping has been removed from this standard. However, the signal label assigned to this mapping will remain defined in order to ensure backward compatibility between future mappings and equipment that support the VT Locked Mode.

4 Code 05 is only to be used in cases where a mapping code is not defined in the above table. By using this code the development or experimental activities are isolated from the rest of the SONET/SDH network. There is no forward compatibility if a specific signal label is assigned later. If a new assignment is made, the equipment that has used this code must either be reconfigured to use that new specific signal label or be recycled.

5 The value of 16 supersedes the previous IETF assignment of CF.

6 These mappings are under study and the signal labels are provisionally allocated.

7 Previous value assigned for an obsolete mapping of HDLC/PPP framed signal.

8 Any mapping defined in ITU-T Recommendation O.181 which does not correspond to a mapping defined in this Recommendation falls into this category.

9 Code FF indicates STS SPE AIS. It is generated by a Tandem Connection Maintenance source if no valid incoming signal is available and a replacement signal is generated.

3.4.5   Virtual Tributary Path (SDH VC-11/12 and VC-2 Trail) Overhead

There are two sets of overhead bytes associated with each VT. The first set, V1-V4, is outside the payload envelope of the VT. The second set is part of VT SPE.

V1-V3 are pointer bytes and are discussed in section 7. In addition to the pointer information, V1 bits 5-6 also indicate the size of the VT (TU). **V4** is undefined at this time.

V5 is the primary overhead byte. As illustrated in Figure 3-17, it contains several OAM functions. The signal label values, shown in Table 3-5, indicate what type signal is mapped into the VT payload. When bits 5-7 contain 101, the signal label is extended into bit 1 of Z7 (K4). If the signal label at the Path terminating NE is different than what it was provisioned to expect, a payload label mismatch (PLM) alarm is declared.

BIP-2		REI	RFI-V	Signal Label			RDI-V
1	2	3	4	5	6	7	8

Figure 3-17. V5 byte bit definitions

J2 and **Z6 (N2)** are the VT Path (Trail) Trace and Low Order Tandem Connection overhead bytes, respectively, and are discussed in section 5.8.

Z7 (K4) provides an extension to the OAM functions of the V5 byte along with the overhead for virtual concatenation. The bits of Z7 are shown in Figure 3-18. The functions of these bits are discussed in the appropriate sections below.

Table 3-5. VT signal label - Byte V5 (5-7) and Z7/K4 (1)

VT size	Signal label		VT Identification
	V5 bits 5-7	Z7 (K4) bit 1 pattern (hex)	
11 = VT1.5/ TU-11 (for DS1)	000	-	Unequipped
	001	-	Reserved (note 1)
	010	-	Asynchronous Mapping
10 = VT2/ TU-12 (for 2.048 Mbit/s)	011	-	Bit-Synchronous Mapping for DS1 and 2.048 Mbit/s (note 2), unassigned for DS1C and DS2
01 = VT3 (forDS1c)	100	-	Byte-Synchronous Mapping for DS1 and 2.048 Mbit/s, unassigned for DS1c and DS2
00 = VT6/ TU-2 (for DS2)	110	-	Test signal, ITU-T O.181 specific mapping (note 3)
	111	-	VT SPE AIS (note 4)
		00-07	Reserved (Note 5)
		08	Mapping under development (Note 6)
		09	ATM mapping
xx	101	0A	Mapping of HDL:C/PPP framed signal
		0B	Mapping of HDLC/X.85 framed signal
		0C	Virtually concatenated test signal, ITU-T O.181 specific mapping (note 3)
		0D	GFP
		FF	Reserved

NOTES

1 Originally this code meant "Equipped – non-specific" and has been used in cases where a mapping code wasn't defined in the standard. Now the code "101" and extended signal label "02" should be used for new designs. Interworking criteria between old and new equipment is specified in T1.105 and G.707.

2 While the DS1 Bit-Synchronous Mapping has been removed from the standards, the signal label assigned to this mapping was preserved in order to ensure backward compatibility between future mappings and equipment that support the DS1 Bit-Synchronous Mapping.

3 Any virtually or non-virtually concatenated mapping defined in ITU-T Recommendation O.181 that does not correspond to a mapping defined in this Recommendation falls in this category.

4 Code 111 indicates VT-SPE AIS. It is generated by a Tandem Connection Maintenance source if no valid incoming signal is available and a replacement signal is generated.

5. Codes 00 to 07 are reserved to give a unique name to non-extended and extended signal labels.

6 Code 02 is only to be used in cases where a mapping code is not defined in T1.105/G.707. Using this code isolates the development or experimental activities from the rest of the SONET network with the understanding that there is no forward compatibility if a specific signal label is assigned later. If a new assignment is made, the equipment that has used this code must either be reconfigured to use that new specific signal label or be recycled.

Extended signal label	Virtual concat. overhead	UNASSIGNED		(Enhanced) RDI-V			Un-assigned
1	2	3	4	5	6	7	8

a) Bit definitions for Z7 (K4)

Bit number

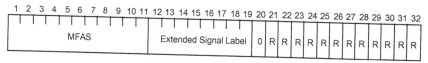

MFAS = Multiframe alignment signal
R = Reserved

b) Bit 1 multiframe

c) Bit 2 Virtual concatenation multiframe

Figure 3-18. Z7 (K4) byte bit definitions

3.5.   OAM&P FUNCTIONS AND CAPABILITIES

As noted, a key motivation behind the SONET/SDH standard is enhanced OAM capability. The OAM capabilities that are incorporated into the standard are discussed in this section.

3.5.1   Error Detection and Indication

3.5.1.1 Forward Error Detection *(B1, B2, B3, V5)*

Error detection is performed with interleaved parity bits at each layer of the signal (i.e., at each level of network connectivity). The error detecting codes are from the family known as *r*-bit bit-interleaved parity (BIP-*r*) codes, where the BIP-8 is also sometimes referred to as byte interleaved parity. The B1, B2, and B3 bytes provide byte-interleaved parity over the Section, Line, and STS Path layers, respectively. With byte-interleaved parity, bit *y* of the B*x* byte provides even parity over all the bit *y* bits of the bytes covered by that B*x* byte. For example, bit 1 of B1 provides even parity over all the bit 1s of the entire signal. The B*x* bytes are calculated over the data in the previous block and are inserted in the current block.

Specifically, the B1 is calculated over all the bytes of the previous frame after the scrambling operation has been performed on that frame. The B2 byte is calculated prior to scrambling over all the bytes in that STS-1 for the previous frame, excluding the SOH bytes. Note that due to SDH frame structure, the STM-1 contains three contiguous B2 bytes that combine to be regarded as a single BIP-24 rather than three BIP-8s like SONET. The B2 BIP-24 is calculated in a manner consistent with three interleaved SONET STS-1s. In general, the B2s of a SONET STS-*N* signal can be considered equivalent to a BIP-8*N*, similar to SDH. The B3 is calculated over all the bytes of the previous STS SPE (SDH VC-4-Xc/VC-4/VC-3) prior to scrambling.

In the case of VTs, a BIP-2 code is located in the V5 byte. The first bit of the BIP-2 provides even parity over all the odd numbered bits in the VT and the second bit provides even parity over all the even numbered bits in the VT, (excluding bytes V1-V4).

In the case of Tandem Connection and in-band forward error correction (FEC), additional overhead information is inserted and removed from the signal in between the end-points of a layer. As a result, the B*x* byte covering that layer must be recalculated and compensated. In the Tandem Connection case, data can only be inserted into the N1 byte when the Section and Line overhead have been terminated, so only the B3 is affected, since it also covers N1. When the B3 is recalculated to include the new N1 values, the recalculation is performed such that any bits in the incoming B3 that indicated errors will continue to indicate errors after the recalculation. In this manner, the B3 at the Path termination point will correctly indicate the errors that have occurred on that path regardless of how many times Tandem Connection data has been inserted or removed from N1 along the path. The same approach is taken with the V5 BIP-2 and N2. In-band FEC uses Section and Line Transport overhead bytes, but logically resides between the Section and Line layers. As a result, the B2 needs to be compensated, but the B1 does not. The B2 compensation follows the same approach as the B3 for Tandem Connection.

NEs typically have provisionable error rate thresholds. When the error rate that is calculated from the incoming BIPs exceeds this threshold, a defect may is declared and some form of protection switch may be initiated.

3.5.1.2 Reverse Error Indication (REI) *(M0, M1, G1, V5)*

It is advantageous for each end of a connection to know the performance of each direction of transmission. For example, the two ends of a Path may lie in different carriers' domains, and each carrier would want to know the performance of the Path in each direction. This capability is known as single-ended performance monitoring. The mechanism used in SONET/SDH to provide single-ended performance monitoring is the Remote Error Indicator (REI). REI is provided at the Line layer in the M0 and M1 bytes, at the Path layer in the G1 byte, and at the VT layer in the V5 byte. The REI field contains a binary value that is the count of the number of bits with incorrect parity in BIP-r fields corresponding to that REI that were received in the last BIP-r. For example, the B3 can indicate 0-8 parity errors; so 4 bits are assigned in the G1 byte (see Figure 3-20) for the REI to communicate the number of parity errors to the far end. The Line REI provides a count of the individual parity bit errors in all the N B2 bytes in an STS-N ($3N$ B2 bytes in an STM-N) signal. Note that for STS-48 (STM-16) and higher rate signals, the REI field is not long enough to indicate all the possible errors, so the largest REI value is sent if more that that many parity bits are seen to be in error.

3.5.2   Defect Indication

A critical factor in reducing the cost of running a network is to be able to quickly identify the location of problems or faults in the network. The general philosophy that has been adopted in the public telephone network is that only the NE that detects the problem should report it as an alarm. Other NEs in the network should report it as a status indication, if at all. The implications of this philosophy are different for the two directions of transmission on a circuit. Figure 3-19 illustrates the information passed in the two directions.

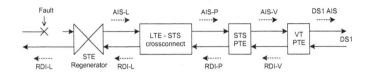

Figure 3-19. Example of AIS and RDI propagation

REI-P				RDI-P			Unassigned
1	2	3	4	5	6	7	8

Figure 3-20. G1 byte bit definitions

3.5.2.1 Alarm Indication Signal (AIS) *(K2, H1-H3, V1-V3)*

Consider first the direction of propagation for the signal encountering the problem. The problem could be the loss or degradation (i.e., high bit error rate) of the signal. If the NE that sees the problem is a regenerator for that signal level, allowing the signal to propagate unaltered would cause the next NE to see the problem and to also report the alarm. The network operator would not be able to tell whether there were one or two problems since there would be two NEs reporting alarms. A similar and more serious problem occurs when there is problem on a multiplexed signal. The NE performing the demultiplexing would propagate a bad signal onto each of the constituent demultiplexed signals. The solution to this problem is to send an Alarm Indication Signal (AIS) that is essentially a distinct keep-alive pattern. When a NE receives an AIS signal at a particular layer, it will propagate that AIS signal at that layer and/or the lower layers if that NE performs demultiplexing. For example, consider a fiber cut that causes a loss of signal ahead of a LTE that only processes high-order paths. This LTE will propagate AIS-P downstream on each of the constituent high-order paths. A subsequent NE that performs VT processing will see the AIS-P in the high-order path and will propagate AIS-V on each of the VTs that it demultiplexes. If one of the constituent payload signals was a DS1, the demapper that outputs the DS1 signal at the edge of the SONET network will generate DS1 AIS. In this manner, each NE subsequent to the one that saw the original signal failure would be aware of the upstream alarm, but would know that it had been reported by an upstream NE and would only note it as a status condition. Line AIS (AIS-L) is indicated by a code in the K2 byte. Path AIS is indicated by sending all 1s in the payload area and in the pointer bytes (H1-H3 for STS (VC-3/4) and V1-V4 for VT (VC-1/2)) associated with that Path.

3.5.2.2 Remote Defect Indication (RDI) *(G1, V5, Z7/K4)*

In the opposite direction, each NE that sees either the original failure or the AIS signal will generate a RDI to report the defect condition to the far end. The RDI thus ties into the single-ended performance monitoring discussed above. When a NE receives an RDI that indicates a defect, it will freeze all updates to its REI calculations. Thus, the RDI also provides a mechanism to insure the consistency of the BER calculated at both ends of the connection.

The original versions of RDI (sometimes referred to as Far-End Receive Failure (FERF)) only indicated that some type of problem had been detected at the far end. Later, RDI was enhanced in order to communicate something about the type of fault that had been detected (Enhanced RDI (ERDI)). The types of faults were broken down into the following three categories, shown in order of highest to lowest priority:

- Server defects, in which there is a problem beneath that layer (e.g., AIS or Loss of Pointer at that layer).
- Connectivity defects, in which a connectivity problem is detected at that layer (e.g., Trace mismatch or a Path indicating that it is unequipped).
- Payload defects, in which there is a problem extracting the payload data from the incoming signal (e.g., payload (signal) label mismatch).

The distinction between categories of remote faults allows more accurate diagnostics to be performed, which in turn lowers the cost and time to respond to the problem. For example, receiving an ERDI indicating a connectivity or payload defect identifies some type of provisioning problem in the network, which may have occurred at the NE receiving the ERDI. Another advantage of ERDI is that the REI count can remain active in the presence of connectivity or payload failures, thus providing continuous performance monitoring during problems that are not associated with the transmission channel.

RDI and ERDI for the STS Path layer are communicated in the G1 byte (see Figure 3-19). At the VT layer, RDI is communicated in the V5 byte with ERDI carried in a combination of V5 and Z7 (K4). See Figure 3-21.

G1 values			Interpretation	Trigger
bit 5	bit 6	bit 7		
0	0	0	No remote defect (Used by pre-ERDI equipment)	No defects
0	0	1	No remote defect (Used by STE)	No defects
0	1	0	Remote payload defect (Used by STE)	PLM-P
0	1	1	No remote defect (Used by pre-ERDI equipment)	No defects
1	0	0	Remote server defect (Used by pre-ERDI equipment)	AIS-P, LOP-P
1	0	1	Remote server defect (Used by STE)	AIS-P, LOP-P
1	1	0	Remote connectivity defect (Used by STE)	TIM-P, UNEQ-P
1	1	1	Remote server defect (Used by pre-ERDI equipment)	AIS-P, LOP-P

a) G1 bit definitions

V5 bit 8 and Z7 (K4) bit 5	Z7 (K4)		Interpretation	Triggers
	bit 6	bit 7		
0	0	0	No remote defect (Used by pre-ERDI equipment)	No defects
0	0	1	No remote defect * (Generated by VT PTE.)	No defects
0	1	0	Remote payload defect * (Generated by VT PTE.)	PLM-V (note 3)
0	1	1	No remote defect (Used by pre-ERDI equipment)	No defects
1	0	0	Remote server defect (Used by pre-ERDI equipment)	AIS-V, LOP-V
1	0	1	Remote server defect* (Generated by VT PTE.)	AIS-V, LOP-V
1	1	0	Remote connectivity defect * (Generated by VT PTE.)	UNEQ-V (note 3)
1	1	1	Remote server defect (Used by pre-ERDI equipment)	AIS-V, LOP-V
* These values are also used in SDH. The interpretation of other values in these ERDI bits differs somewhat from SONET.				

b) V5 bit 8 and Z7 (K4) bit definitions

Figure 3-21. RDI and ERDI bit definitions

3.5.2.3 Remote Failure Indication (RFI) *(V5)*

The V5 byte also includes a RFI bit for use with byte-synchronous mapped DS1 payloads. The RFI is different from RDI in that it requires that the failure be present for a certain period of time before the RFI is declared and absent for a certain period of time before the RFI is cleared. RFI corresponds to the Yellow Alarm that is used in DS1 signals. In the byte-synchronous mapping, which is discussed below, the RFI bit is used to map the DS1 Yellow Alarm across the SONET portion of the connection and restore it at the DS1 egress point.

3.5.3   Connectivity verification – the Trace function *(J0, J1, J2)*

One of the powerful overhead capabilities of SONET/SDH is the ability to verify that signals have the correct connectivity at that layer. Each Trace byte carries a multi-byte pattern that is provisioned by the network provider and is unique relative to the other entities that can exist at that layer. The Section Trace (J0) is used to confirm that the correct signal is being sent over the correct fiber (or wavelength). This is obviously helpful in a WDM system, but it is also useful where a single signal per fiber is used. The primary example here is the case of repairing a backhoe fiber cut. The fiber repair crew can quickly determine which fiber ends to re-connect by examining the J0 on each fiber. At the Path level (J1 for STS and J2 for VT), the Trace message allows the receive end to confirm that its provisioning is consistent with the transmitting end by confirming that the correct trace message is received on the correct Path (Trail). Two options have been used for the J0 byte message format. The original format for SONET was to use a single, repeating 8-bit value. In order to be compatible with SDH, however, the preferred J0 format is to use a repeating 16-byte message as shown in Figure 3-22. This 16-byte format is also used in SDH for the J1 and J2, and in SONET for the J2 bytes. SONET uses a user programmable 64-byte repeating message string for J1.

Byte #	Value (bit 1, 2, … ,8)						
1	1	C_1	C_2	C_3	C_4	C_5	C_6	C_7
2	0	X	X	X	X	X	X	X
3	0	X	X	X	X	X	X	X
:							
16	0	X	X	X	X	X	X	X

Note 1 – $C_1C_2C_3C_4C_5C_6C_7$ is the result of the CRC-7 calculation over the previous frame.

Note 2 – Bytes 2-16 typically carry ASCII values as specified in E.164.

Figure 3-22. 16 byte Trace message formats using

3.5.4 Data Communications Channel *(D1-D3, D4-D12/156)*

One of the most powerful OAM features of SONET/SDH is the provision for packet-oriented data communications channels.

The 192 kbit/s Section data communications channel (SDCC in D1-D3) is the workhorse of SONET/SDH OAM communications. Although the physical connectivity for the SDCC is point-to-point between nodes, the protocol standards provide for the routing of SDCC messages between nodes. This allows a NMS connected to any NE on a SONET/SDH network (known as a Gateway NE (GNE)) to communicate with any SONET/SDH NE on that network. The message traffic between NEs can include remote provisioning commands from the NMS to the NE, alarm and status reporting from NE to NMS, and the download of new software upgrades to an NE. The SDCC protocol stack is specified in [T1055].

The Line data communications channel (LDCC in D4-D12/156) is a point-to-point data link between LTEs (i.e., the NEs do not route LDCC messages). There are no functions standardized for the LDCC at this time. Proprietary implementations have used this channel for software download from one NE to another. For STS-1 through STS-192, the LDCC rate is 576 kbit/s. An additional 144 LDCC bytes were added to STST-768 (STM-256) frame to allow a 9.792 Mbit/s LDCC.

3.5.5 Protection Switching Functions *(K1, K2, K3, K4)*

Bytes K1 and K2 in the Line layer are used to coordinate Line layer protection. The types of the protection topologies supported in the current

SONET standards are 1+1 linear, 1:1 linear, 1:*n* linear, and bi-directional Line-switched rings (BLSR). In general terms, the NE detecting the fault condition (or external command from the NMS) sends a request for switch action on the K1 byte along with an indication of the nature of the fault/command and its priority. The K2 generally communicates the status of the NE receiving the switch request on the incoming K1 bytes. Depending on the protection topology, this status includes the protection switch mode for which that NE is provisioned, whether a bridge and/or switch is active at the NE, the channel being protected, whether Extra Traffic is being sent over the protection Line, and the NE address on the ring. In all protection modes, K2 is used to communicate Line AIS and Line RDI. Note that for 1+1 or path switched protection, if a unidirectional switch topology is chosen, no K-byte communication is required (i.e., the receiver simply chooses the best ring signal and does not inform the transmitting NE). Automatic protection switching is discussed in greater detail in Chapter 8, including the SDH terminology and SDH-specific architectures.

SDH has allocated K3 in the VC-4/VC-3 Path and K4 in the VC-1/VC-2 Path layers for use in subnetwork Trail protection. The use of these bytes is for further study.

3.5.6 Orderwire Function *(E1, E2)*

The orderwire function is a voice channel over a facility connecting pieces of equipment. The orderwire allowed a simple voice port connection that could be used by the craftspersons at each end of the facility who were installing or repairing the equipment to communicate with each other during their work. SONET/SDH provides an orderwire at the Section level (E1) and the Line level (E2). In metropolitan areas, the prevalence of wireless telephone service provides the craftspersons with a more convenient and universal method of communicating with each other, even when the SONET/SDH span was inoperative. The orderwire function is still useful, however, when craftspersons are working on long, repeatered spans in remote areas where there is no wireless service.

3.5.7 User Byte Function *(F1, F2)*

The user byte was originally envisioned to provide a function similar to the orderwire except that it would be a data connection rather than a voice connection. It was also recognized that the User Byte could be used for passing information between NEs that was proprietary to either the network operator or to the equipment vendor. The uses for the User Byte remain proprietary, with no specific use being standardized. The point-to-point nature of the User Byte at the Section layer (F1) and high order Path layer (F2 for SONET, and F2 and F3 for SDH) limit its usefulness.

3.5.8 Tandem Connection Maintenance *(N1, Z6/N2)*

As discussed above in section 2, the optional TCM sublayer is used to monitor the performance of group of high-order or low-order Paths through multiple tandem Line (MS-Section) links in a network by treating the group a single entity. For high-order Paths, N STS-1 (STM-1) signals may be grouped such that N may take on only those values for which a STS-N (STM-N) signal is defined. STS-Nc and STM-Nc Paths can also be combined into a Tandem Connection bundle, subject to the same constraint that the bundle rate must be the same as one of the specified line rates.

For high-order TCM, bits 1-4 of N1 for each constituent high order Path (STS/VC-3/4) contains an Incoming Error Count (IEC). The IEC is the binary count of the number of errored bits in the B3 byte for previous frame of that Path as it enters the TC. The insertion of this information into the N1 byte necessitates the recalculation of the B3 bytes. As discussed above, the B3 bytes are compensated such that the recalculated B3 values will have parity errors in the same bit locations in which errors were detected prior to the N1 insertion. At the termination of the TC link, each B3 is again examined. The number of B3 bits with incorrect parity is compared to the IEC value in order to determine how many additional parity errors occurred in the TC link. If an incoming Path has failed (including the case where the TCTE source receives AIS-P on the incoming Path), the TCTE source inserts an IEC value of 1110 to indicate the incoming signal failure (ISF). This ISF capability allows the TCTE sink to determine whether a Path with AIS-P failed within or outside the TC.

In the case of SONET (T1.105.05), N1 bits 5-8 of the first STS in the TC bundle are used for a 32 kbit/s LAPD data link. This N1 definition is shown in Figure 3-23.a. The data link is used to communicate performance information from the far end including error events and AIS or Loss of Pointer detection for any of the constituent Paths within the bundle. Thus, the collective performance of the entire TC bundle can be determined by examining the LAPD performance messages coming from the far end. SDH refers to this data format as TCM Option 1.

SDH also specifies a TCM Option 2 in which N1 bits 5-8 are used in a bit-oriented manner rather than as a LAPD data link. This option has also recently been added to the SONET standard. Figure 3-23.b illustrates the corresponding N1 byte definition. For Option 2: TC REI is an indication that an error has been detected on the TC link, OEI is an indication that the Path being output from the TC has an error (that occurred either prior to, or within the TC link), TC API is an access point identifier unique to that TC bundle, TC RDI is an RDI for the TC link, and TC ODI is an indicator that a defect was detected at the termination point and AIS is being output from the link. While Option 2 offers some simplification due to its use of a bit-oriented protocol, it adds a management complexity over Option 1 since the TC source has to correlate the N1 byte information for all TC bundle members rather than seeing it directly from the first member's data link message.

SONET and SDH also allow for low-order Tandem Connection using the Z6 (N2) byte for the TC overhead as illustrated in Figure 3-23.c. The BIP-2 here covers the TC link.

N1 of the first constituent Path							
b1	b2	b3	b4	b5	b6	b7	b8
IEC				TC data link			

N1 of the remaining constituent Paths							
b1	b2	b3	b4	b5	b6	b7	b8
IEC				Unassigned			

a) N1 byte definitions for SONET and SDH Option 1

N1 of all constituent Paths							
b1	b2	b3	b4	b5	b6	b7	b8
IEC				TC REI	OEI	TC API, TC RDI, ODI	

b) N1 byte definitions for SDH Option 2

N2 of all constituent Paths							
b1	b2	b3	b4	b5	b6	b7	b8
BIP-2		1	Incoming AIS	TC REI	OEI	TC API, TC RDI, ODI	

c) N2 byte definitions for SONET and SDH

Figure 3-23. Tandem Connection overhead byte definitions

3.6.   PAYLOAD MAPPING

The lower-rate asynchronous signals for SONET are mapped into Virtual Tributary (VT) structures and higher-rate asynchronous signals are mapped into an STS-1 or STS-*N*c. For SDH, the legacy plesiochronous signals are mapped into Virtual Containers (VC).   Table 3-6 shows the asynchronous/plesiochronous signals and the container type into which they map. SONET draws the dividing line between low-order and high-order mappings between VT6 and STS-1, while SDH draws the dividing line between VC-3 and VC-4.   For practical purposes, the main difference between a low-order and a high-order mapping is that a low-order mapping has an additional set of pointers to locate its path overhead within high-order container. Pointers are discussed in more detail in section 7.

Table 3-6. Asynchronous/plesiochronous client signals and their corresponding SONET/SDH containers

North American asynchronous signal	CEPT plesiochronous signal	Client signal rate (Mbit/s)	Virtual Tributary or STS	Virtual Container	Container Payload Rate (Mbit/s)
DS1	-	1.544	VT1.5	VC-11	1.6
-	E1	2.048	VT2	VC-12	2.176
DS1C	-	3.152	VT3	-	3.328
DS2	-	6.312	VT6	VC-2	6.784
-	E3	34.368	-	VC-3	48.384
DS3	-	44.736	STS-1	VC-3	48.384
DS4NA	E4	139.264	STS-3c	VC-4	149.76

A relatively recent addition to the SONET and SDH standards is the ability to virtually concatenate an arbitrary number of VTs, STSs, or VCs such that their individual payload envelopes (containers) are combined into a single larger capacity envelope. Virtual Concatenation and the associated Link Capacity Adjustment Scheme (LCAS) are discussed in Chapter 4.

The hierarchy for mapping and multiplexing tributary payload signals into SONET is shown in Figure 3-24, and the corresponding SDH hierarchy is shown in Figure 3-25. The differences in the hierarchical diagrams, aside from the naming conventions, come primarily from three sources. Beginning on the left side of the figures, SDH uses the term "Container" (C-

n) to designate the payload area of their VC (where the SDH VC is equivalent to a SONET SPE). SONET does not have a special designation for just the payload portion of a SPE. The second difference is the previously discussed difference in the base rates with SONET using a 51.84 Mbit/s STS-1 base rate and SDH using the 155.52 Mbit/s STM-1. (Note that the STM-0 SDH signal is equivalent to the STS-1, except that in the SDH world it was primarily defined for transmission over radio links.) The third difference, which grows out of the second, is that SDH has an optional intermediate construct at the VC-3/STS-1 level called a level-3 tributary unit group (TUG-3). As discussed below, a TUG-3 contains its own Path payload pointers that allow three TUG-3s to be multiplexed into a VC-4.

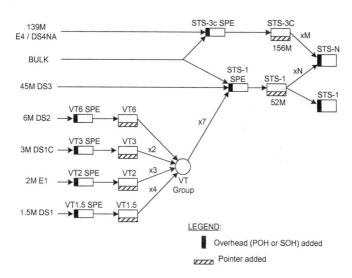

Figure 3-24. SONET mapping and multiplexing hierarchy

The concept employed to map an asynchronous signal into a SONET/SDH SPE/Container is essentially the same as that used for the asynchronous/plesiochronous hierarchy. The SPE payload rate is somewhat higher than that of the tributary signal that it carries. Fixed stuff bits are used to match the SPE and payload signal rates to a value close enough to allow controlled stuffing. Controlled stuff bits are again used to pad the tributary signal up to the rate of the SPE. A majority vote on a set of stuff control bits indicates whether a particular stuff opportunity bit contains a stuff (dummy) bit or a payload data bit.

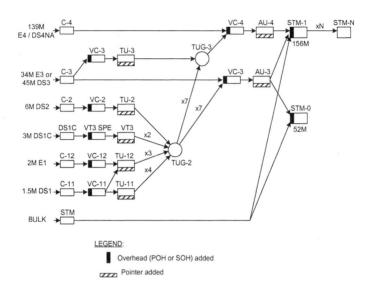

Figure 3-25. SDH mapping and multiplexing hierarchy

3.6.1   Lower-Rate Asynchronous/Plesiochronous Tributaries

The primary rate (DS1/E1) mappings into SONET/SDH are shown in Figure 3-26. Note that SDH VC-11/12 (denoted collectively as VC-1) and VC-2 are the equivalent of SONET VTs. When a general statement is made concerning SONET VTs, the SDH terminology VC-1/2 is sometimes used. Each VT has essentially the same structure. Each contains a set of four, regularly spaced Path overhead bytes, with V5 being the start of the VT. For bandwidth efficiency, the VT frame is spread across four STS frames, with the frame phase indicated by the H4 byte.

VTs are arranged in groups for the purposes of being inserted into an STS-1 (VC-3). Each group has a nominal rate of around 7 Mbit/s, which means that a VT Group can consist of four VT1.5s, three VT2s, two VT3s, or one VT6. Although each VT Group within an STS-1 can contain different types of VTs, all of the VTs within an individual VT Group must be of the same type. Figure 3-27 illustrates the mapping VTs of different sizes into an STS-1. When the VTs are multiplexed into the STS-1 SPE, the multiplexing is done on a byte-wise basis, using a round robin, first with the VT Groups and then with the VTs within the VT Groups. The end result is that each VT occupies a fixed set of columns within the STS-1 SPE, and all the bytes within each individual column are associated with the same VT.

(As will be explained in section 7, a higher order pointer adjustment will cause a split of the VT columns within that frame.) When a VT is specified within an STS-1 (for example, in an NMS system) it is typically specified in terms of the VT Group number in which it resides (1-7) followed by its VT number within the Group.

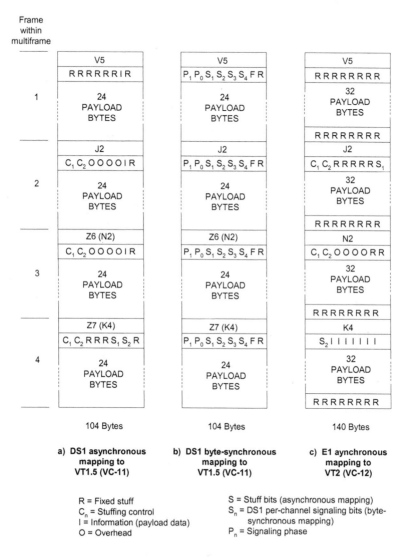

Figure 3-26. Primary-rate tributary mappings

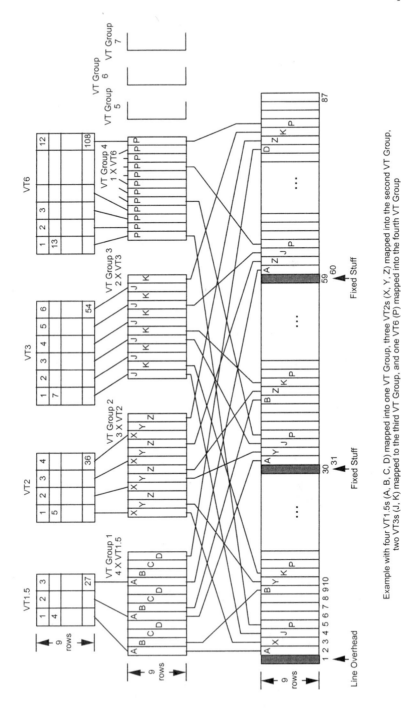

Figure 3-27. Illustration of the mapping of VTs (VC-1/2s) and VT Groups into an STS-1 (VC-3)

There are two methods available for mapping a DS1 signal into a VT1.5. The first and by far the most common, is the asynchronous mapping. This mapping, shown in Figure 3-26.a, is typical of all asynchronous VT mappings in that the VT SPE is a simple channel (i.e., "raw bandwidth"). There is no correlation between the DS1 signal frame and VT frame structure in this mapping. The primary advantage of this mapping is its simplicity. The other DS1 mapping is the byte-synchronous mapping shown in Figure 3-26.b. Here, the DS0s of the DS1 signal are byte-aligned with the SPE bytes of the VT1.5. Each VT1.5 frame contains the DS1 framing bit (in a fixed location) and the 24 DS0 channels of that DS1 frame, also in fixed locations. The channel-associated signaling information and associated signaling bit phase information are carried in the SPE bytes that were used for stuffing and stuff control in the asynchronous mapping.

The E1 signal also has similar asynchronous and byte-synchronous mapping options, with the asynchronous mapping shown in Figure 3-26.c. The E1 signal is already comprised of an integer number of bytes with the signaling information contained in one of those bytes. This makes the byte-synchronous E1 mapping straightforward.

The DS1c and DS2 asynchronous mappings are similar to those for the DS1.

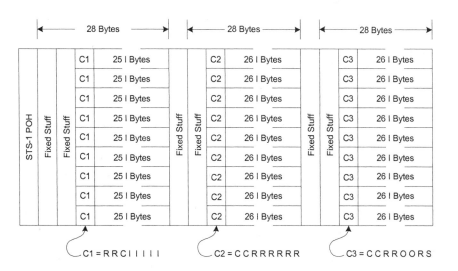

I = Information (Payload Data)
R = Fixed Stuff bit
C = Justification Control bit
S = Justification Opportunity (Stuff) bit
O = Overhead bit

Figure 3-28. DS3 asynchronous mapping into a STS-1 (VC-3)

3.6.2   Higher-Rate Asynchronous/Plesiochronous Tributaries

The mapping of a DS3 signal into a STS-1 (VC-3) is shown in Figure 3-28 and the mapping of the 139.264 Mbit/s E4 (DS4NA) signal into a STS-3c (VC-4) is shown in Figure 3-29. These mappings are similar to the VT (VC-1/2) asynchronous mappings except that they occur at the high order path layer.

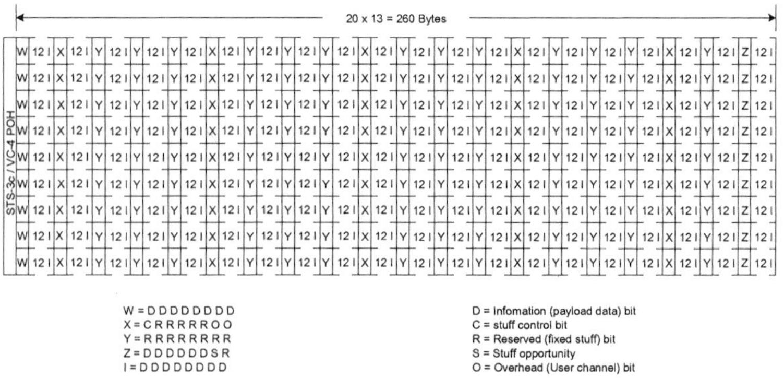

W = D D D D D D D D   D = Information (payload data) bit
X = C R R R R R O O   C = stuff control bit
Y = R R R R R R R R   R = Reserved (fixed stuff) bit
Z = D D D D D D S R   S = Stuff opportunity
I = D D D D D D D D   O = Overhead (User channel) bit

Figure 3-29. 139.264 Mbit/s E4 (DS4NA) signal into a STS-3c (VC-4)

3.6.3   Packet or Cell-Oriented Data Tributaries

Over the past 20 years, the public network has been making a steady transition to being dominated by the transport of voice channels to the transport of data traffic. This trend had already become obvious and was accelerating at the time SONET/SDH deployment began. Since SONET and SDH form the backbone of the public network, there has been a steady drive to increase the efficiency and ease with which it can be used to carry data traffic. The first major step in this direction occurred with Asynchronous Transfer Mode (ATM). ATM was a bold attempt to take all types of traffic (i.e., both constant bit-rate traffic like voice and packet-oriented traffic) and map them into fixed length cells. The fixed cell length simplified the implementation of fast, hardware-based switches with the technology that existed around the early 1990s. The next major step was to create a technique optimized for data packet transport that used an HDLC variant. This technique is commonly known as Packet over SONET/SDH (PoS). The limitations of ATM and PoS lead to the recent development of the

Generic Framing Procedure (GFP). ATM is the subject of extensive books, so no attempt will be made to go into any details here. Those interested readers may want to refer to [SeRe]. Since GFP is new, it will be covered in more detail in Chapter 4. PoS is simple enough that its essential details will be explained in this chapter. The basic idea for each of these mappings is to treat a SONET payload envelope (SDH container) as a stream of bytes into which the packet stream can be mapped.

3.6.3.1 ATM (Asynchronous Transfer Mode) Mapping

ATM was developed at a time prior to the development of virtual concatenation (VCAT), and so it only uses contiguous concatenation. (See Chapter 4 for the discussion of VCAT.) The term asynchronous distinguishes it from synchronous transfer mode in that ATM cells for a given payload are not sent in a regularly time-interleaved stream. An ATM cell consists of five overhead bytes and 48 payload bytes (53 bytes total). The input data is adapted into ATM cells using an appropriate adaptation technique. (The adaptation techniques are referred to as the ATM Adaptation Layer (AAL).) The ATM overhead contains the channel and path identification that is used by ATM switch equipment to route cells between the appropriate source and destination points. The mapping of ATM cells into a STS-1 (VC-3) and a STS-3c (VC-4) are illustrated in Figure 3-30. [Note that for VC-4-Xc or VC-4-Xv, the fixed stuff columns are grouped together immediately following the POH (i.e., the columns that would have been POH for the other VC-4s).]

As discussed in section 3.2, an important practical consideration that separates STM from ATM multiplexing is that a given customer's data can occupy several consecutively transmitted bytes in transmitted stream. As a result, a malicious user could potentially send the SONET self-synchronous scrambler pattern in order to create a situation that could cause the receiver to lose synchronization and declare a loss of signal, thus denying that link to all users during the resynchronization interval. The self-synchronous scrambler described in section 3.2 was added to the mapping to alleviate this potential problem. (See Figure 3-16.)

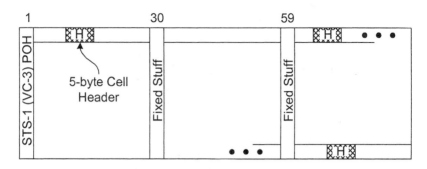

a) ATM cell mapping into a STS-1 (VC-3)

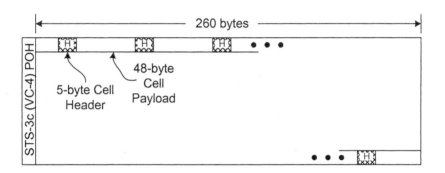

b) ATM cell mapping into a STS-3c (VC-4)

Figure 3-30. ATM cell mapping into STS-1 and STS-3c

While ATM is extremely powerful in it ability to map and carry any type of payload, it suffers from an important truism. Any thing that is flexible enough to be able to handle anything is poorly suited to handle any specific thing. For constant bit rate (CBR) data (e.g., carrying DS1 signals within a stream of ATM cells), the biggest issue is jitter. When the cells are multiplexed with cells from other sources at each ATM switch node, their spacing in time becomes less regular. This situation is especially true when CBR traffic cells compete with the bursts of cells that are typical with packet data. When the cells arrive at their destination, it is relatively difficult to reconstruct the original CBR payload stream and meet its jitter and wander requirements. All in all, chopping up a nice steady stream into cells and reassembling it is a cumbersome proposition. For packet data, a different problem is encountered. Long packets must be broken into multiple cells (48 bytes per cell, depending on the adaptation used). This means that ATM imposes roughly a 10% bandwidth expansion on the transport of packet data. Neither of these drawbacks is insurmountable, and a number of carriers have made extensive use of ATM in the core of their networks. The statistical multiplexing gains that can be achieved with the packet data, and the management efficiencies of treating all payloads as an ATM cell connection provide significant advantages. For the specific case of data, however, the bandwidth expansion from the ATM cell overhead was a prime factor leading to the development of PoS and GFP.

3.6.3.2 HDLC / Packet over SONET/SDH Mapping (including X.85 and X.86)

As just noted above, ATM is inefficient for transporting data packets. It was also desirable to avoid the adaptation process required for ATM. One solution developed in the 1990s by the IETF [Sim] and enhanced in ANSI Subcommittee T1X1 was to adopt HDLC, an OSI Layer 2 protocol. HDLC, however, is character-oriented and relies on special characters to indicate the start/end of packets. For example, the start and end of a data frame (and the idle fill between the frames) consists of the "01111110" character. All special characters have long strings of 1s. Unfortunately, these types of patterns can occur within the customer data within the payload of the HDLC frame. In normal serial transmission of HDLC, a technique called bit stuffing is used in which the transmitter inserts a 0 immediately after each string of five 1s that it encounters in the frame's payload field. The receiver, knowing that rule, can remove the stuffed 0s and restore the original data. The desire with PoS was to maintain a byte alignment between the 8-bit HDLC characters and the bytes of the SONET payload. Instead of bit stuffing, which would destroy the byte alignment, a byte stuffing technique was used. Here, whenever a byte is encountered in the HDLC payload data that has the same value as a control character, an 8-bit special "escape" character is inserted in the byte in front of that offending character. (A pattern in the data that looks like an escape character is also handled by putting an escape character in front of it.) The bit and byte mapping of PoS is shown in Figure 3-31.

- NOTE: The IETF RFCs and ITU-T X.85/X.86 lack a precise, clear definition of some of the bit and byte mappings. Figures 3-31 and 3-32 are intended to remove any ambiguity. The labeling conventions used in Figures 3-31 through 3-33 attempt to follow the conventions of the associated standards.

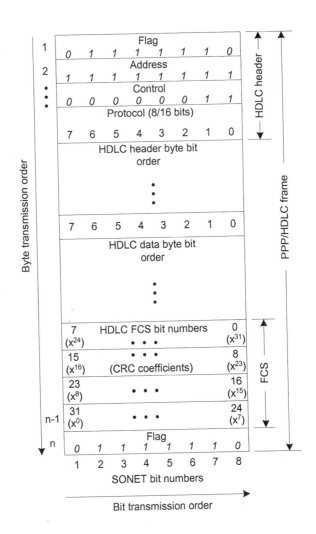

Figure 3-31. Bit and byte mapping of Packet over SONET/SDH HDLC frames into SONET/SDH

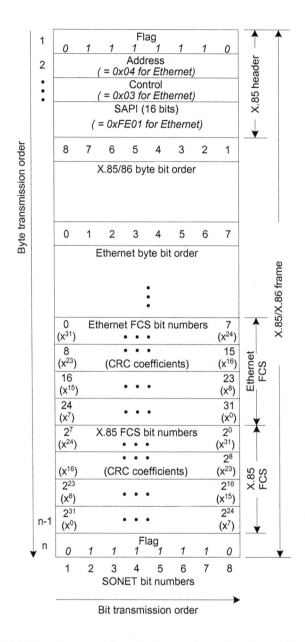

Figure 3-32. Bit and byte mapping of X.85/X.86 frames into SONET/SDH

PoS data can be mapped into the contiguously concatenated structures (similar to ATM) or into VCGs. The VCAT mappings can happen at either low order or high order. A key point overlooked in the original IETF specification is that the same scrambling requirement that we encountered for ATM is even more important for PoS. Since a single user's data occupies even more successive bytes than an ATM cell (due to the whole packet being transmitted rather than one cell at a time), the likelihood of successful attack increases. The vulnerability of PoS to these attacks was verified in a number of laboratories. T1X1.5 took up the issue and determined that the same self-synchronous scrambler used for ATM would also be effective for PoS. T1X1 then adopted the PoS mapping with the self-synchronous scrambler into the T1.105.02 SONET mappings standard, with the ITU-T following the same approach for SDH.

X.85 and X.86 both grew out of work projects in China. X.85 is a technique for carrying IP packets over a SONET network. The IP packets are inserted into an HDLC frame where the HDLC Service Access Point Identifier (SAPI) specifies multiple logical links in order to encapsulate IPv6-based, IPv4-based, PPP and other upper layer protocol packets. The process of inserting the byte-stuffed HDLC frame into a SONET signal is referred to as Link Access Procedure – SDH (LAPS), and is essentially the same as PoS except for some of the bit ordering. X.86 is an extension of X.85 for carrying Ethernet MAC frames (i.e., X.86 specifies the mapping of Ethernet frames into X.85 frames). The Ethernet MAC frame is inserted into the payload of an HDLC frame after removing the Ethernet frame preamble and Start-of-Frame delimiter. LAPS is then used for inserting the HDLC frame into the SONET/SDH signal. The X.85/X.86 bit and byte mapping into SONET/SDH is shown in Figure 3-32. The bit labeling for X.85 and X.86 is rather unique. They use the traditional 1-8 bit numbering of transmission systems, but transmit bit 8 first (i.e., they don't use the numbering as the transmission order). The FCS bits are labeled as $2^0 - 2^{31}$ rather than $0 - 31$, which can create confusion for the reader as to whether 2^0 corresponds to FCS field bit 0 or to the CRC coefficient x^0. (The X.85 convention is that bit 2^0 is bit 0 of the FCS field, which is the x^{31} CRC coefficient.)

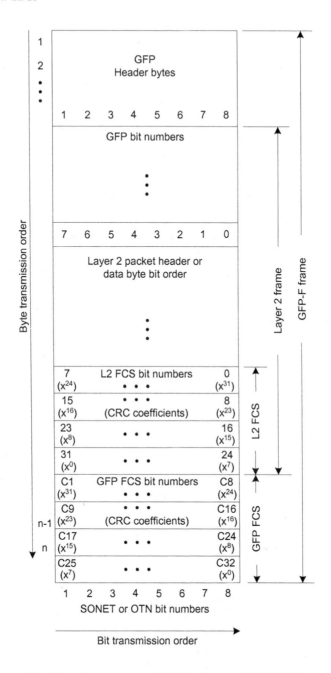

Figure 3-33. Bit and byte mapping of GFP frames into SONET/SDH

3.6.3.3 Generic Framing Procedure (GFP) Mappings

The mapping of GFP frames into SONET/SDH is discussed here, while GFP is presented in detail in Chapter 4. There are two general categories of payload mappings into GFP frames: frame-mapped (GFP-F) and transparent (GFP-T). Both provide a simple data encapsulation and use a framing technique that allows a deterministic transmission bandwidth that is independent of the client data patterns. GFP-F encapsulates a client signal data frame within the GFP frame, and GFP-T encapsulates a number of client data characters into a GFP frame. The GFP frames are then mapped in a byte-wise manner into the SONET payload (SDH container) in the same manner as PoS. The specific bit and byte assignments are shown in Figure 3-33. The same self-synchronous scrambler is used for GFP as for ATM and PoS. As discussed in Chapter 4, however, GFP uses a Core header for frame delimiting. In order to simplify the framing process, this Core header is not scrambled with the self-synchronous scrambler. Instead, the bytes of the Core header are exclusive ORed with a Barker-like code that guarantees good 0/1 transition density during idle periods on the GFP link.

3.6.3.4 10 Gbit/s Ethernet and Fibre Channel WAN Signal Mappings

While it is possible to map Ethernet or Fibre Channel signals at any rate into SONET/SDH via GFP (see Chapter 4), IEEE 802.3 chose to define a wide area network (WAN) signal that can map directly into the payload container of a SONET/SDH signal. (The T11 committee responsible for Fibre Channel chose to use the same physical layer mapping for 10Gbit/s Fibre Channel.) Specifically, while the 10 Gbit/s Ethernet local area network (LAN) interface (10GBASE-R/X) is defined as using a 10 Gbit/s MAC frame bit rate, the WAN interface (10GBASE-W) uses a 9.95328 Gbit/s STS-192c (VC-4-64c) interface. The Ethernet MAC data stream is encoded with a 64B/66B line code and then mapped into the SONET/SDH payload envelope/container. With the 64B/66B encoding and SONET/SDH payload envelope rate of 9.58464 Gbit/s, the MAC data rate is 9.2942 Gbit/s.

Figure 3-34 illustrates the mapping of the Ethernet data into the SONET/SDH payload bytes. This figure uses a 16-bit data unit coming from the 64B/66B block encoder. The LSB to MSB transmission order of the Ethernet data is maintained in order to simplify the Ethernet FCS checking. Note that the 64B/66B line code does not allow a byte-aligned mapping between the original Ethernet MAC frame bytes (i.e., before line coding) and the SONET/SDH payload bytes.

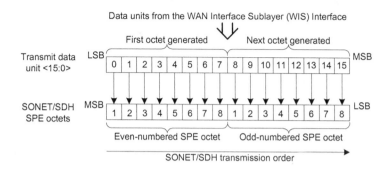

Figure 3-34. 10GBASE-W mapping bit ordering

The actual 10GBASE-W interface uses the SONET/SDH TOH and POH bytes in the manner defined for SONET signals in ANSI T1.416. (T1.416 defines the subset of the TOH and POH that are applicable to access UNIs.) G.707 defines the same payload mapping, but with the full NNI TOH and POH being used. Other than the minor difference in TOH use, the only other difference between the 10GBASE-W interface and the G.707 signal are clock accuracy and jitter specifications. IEEE 802.3 was very reluctant to move beyond the typical Ethernet ±100 ppm clock accuracy out of concern for cost to the interface. In order to guarantee reliable transport of data through a SONET/SDH network, a clock accuracy of ±4.6 ppm is specified with a ±20 ppm clock allowed in SONET as a minimum clock accuracy for ensuring that the TOH can be sent. The 'compromise' reached between IEEE 802.3 and T1X1/ITU-T SG15 was to use a ±20 ppm clock for the 10GBASE-W interface. This accuracy is fully adequate when a 10GBASE-W signal is carried through a G.709 optical transport network (see Chapter 4), it causes problems when interconnecting to a SONET/SDH transport network. The ±20ppm clock inaccuracy can result in excessive pointer adjustments in the transport network. Since excessive pointer adjustments are commonly used as a mechanism to identify synchronization problems in the network, the 10GBASE-W signal could hence generate unwanted alarms. It is not practical for carriers to disable these alarms.[48]

---

[48] Many 802.3 members at the time of the 10GBASE-W development incorrectly concluded that relatively high cost of SONET/SDH interfaces was due to their clock accuracy requirements. In fact, there is so little difference between the ±100, ±20, and ±4.6 ppm oscillators at similar volumes that the gain in interface flexibility by specifying a ±4.6 ppm clock would have far outweighed the small cost savings.

3.6.4   G.709 ODUk Transport Mappings

The ITU-T standard for next-generation optical dense wave-division multiplexed (DWDM) transport systems, G.709, is summarized in chapter 4 along with other new technologies and standards. Deployment of G.709 optical transport networks (OTNs) is expected to follow a slow, gradual process. As a result, G.709 OTNs will initially be deployed in islands. A mapping to carry G.709 OTN format signals within SONET/SDH was developed n order to allow the interconnection of G.709 OTN islands across SONET/SDH networks.

The digital signal format for the G.709 OTN signal carried over each wavelength is called an Optical Data Unit, of level k (ODUk). The extended ODUk frame illustrated in Figure 3-35 is used for mapping into SDH. The details of the ODUk signal frame format are shown in Chapter 4. The mappings into SDH[49] virtual containers are defined for the 2.5 Gbit/s ODU1 and the 10 Gbit/s ODU2. The rates for the ODUk and the associated SDH C-4-Xv are shown in Table 3-7. With the G.709 clock accuracy being ±20ppm and the lack of a convenient contiguously concatenated container, pointers weren't a viable option for adapting the ODUk rate with the SDH container rate. Instead, an asynchronous method was used that is essentially the same as the one presented above for multiplexing into DS2 and DS3 signals. The ODU1 and ODU2 mappings are shown in Figure 3-36. The OUD1 and ODU2 mappings use essentially the same structure except for the total number of columns and that the 884 octet blocks of the ODU2 are structured with 13 subblocks of 68 octets and the ODU1 uses 17 subblocks of 52 octets. For both mappings, five control (C) bits determine whether the S byte contains data, for CCCCC=00000, or a negative justification (value = 0000 0000) for CCCCC=11111. The receiver can perform a majority vote on the C values to guard against bit errors.

---

[49] Although the ODUk mappings have not yet been added to ANSI T1.105.02 for SONET signals, the mapping would be the same as for SDH signals.

Figure 3-35. Extended ODUk frame used for the mapping into SDH

Table 3-7. Rates and capacities associated with mapping an ODUk into an SDH C-4-Xv

OTN Entity	Nominal ODUk bit rate (kbit/s)	C-4-Xv value of X	Nominal bit rate, C-4-Xv (kbit/s)
ODU1	≈2 498 775.126	17	2 545 920
ODU2	≈10 037 273.924	68	10 183 680

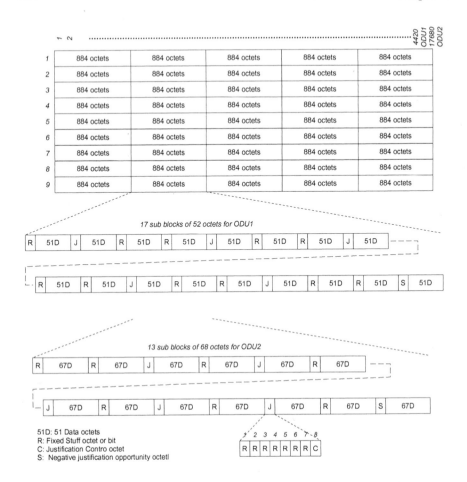

Figure 3-36. Block structure of the ODU1 for mapping into a C-4-17v and ODU2 into a C-4-68v

3.7. POINTERS AND JUSTIFICATION

As the SONET Path signals move from node to node in the network, they can encounter two synchronization issues. Each Digital Crossconnect System (DCS) or Add/Drop Multiplexer (ADM) node uses its own internal frame reference to align the signals that are being combined from multiple sources. For example, in an ADM that has SONET tributary signal interfaces and SONET paths received from the Line interface that are passed through that NE, it must establish the phase relationship between these signals when they are multiplexed for transmitting. The phases of the incoming SONET Lines that carried these Paths will typically be different from one another. A similar situation occurs for VTs that are multiplexed into an STS-1 Path. Also, the NE that terminates the Path must be able to determine the phase of the Path relative to its received Line signal. One mechanism for determining and accommodating these phase differences would be to buffer each signal until they are aligned. SONET, however, accomplishes this phase adjustment and phase tracking by using a pointer mechanism. The pointer at a given level contains the numerical offset in bytes of the start of the SPE (VC) from a fixed reference point. When a Path goes through an intermediate NE, that NE can place the Path directly into the multiplexed signal it creates (i.e., without buffering) by adjusting the pointer value to the Path's phase offset in the new signal. Thus, the pointer mechanism avoids requiring numerous large buffers, which are particularly undesirable with high rate signals.

The other synchronization issue that a Path can encounter at an intermediate NE is that not all NEs use the same timing reference. Hence, when an NE multiplexes Paths that originated in NEs with different timing sources, their phases will slip relative to each other. Again, the pointer mechanism is used handle the phase adjustment. If the clock source for an incoming Path has a lower rate than that for the NE, the NE's input buffers for that Path will move toward underflow. The NE accommodates this situation by incrementing the pointer value when the underflow-warning threshold is crossed. This pointer increment, referred to as a positive pointer justification, is illustrated in Figure 3-37 for the case of an STS-1 Path. The positive justification slides the Path phase back relative to the STS-N frame. In the process, one byte of the SPE is left unfilled (i.e., there is one less byte of the slower rate Path placed into the SPE for that frame, which allows the input buffer to recover from the near underflow condition). This unfilled byte immediately follows the pointer action byte. Similarly, if the clock source for an incoming Path has a higher rate than that for the NE, its input buffers for that Path will move toward overflow. The NE accommodates this situation by decrementing the pointer value when the overflow-warning threshold is crossed. This pointer decrement, referred to as a negative pointer justification, is illustrated in Figure 3-38 for the case of an STS-1 Path. The negative justification advances the phase of the Path relative to the STS-N frame. One additional Path byte is carried in the frame with the negative justification, and this byte is carried in the pointer action byte.

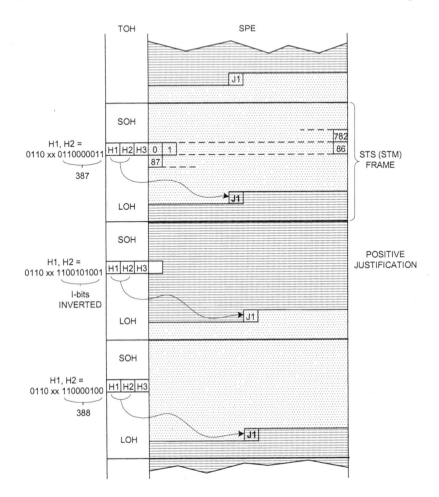

Figure 3-37. High-order positive pointer justification example

The format of the H1/H2 bytes are illustrated in Figure 3-39 and the format of the V1/V2 bytes are illustrated in Figure 3-40. The N-bits are a new data flag that is set when there is a completely new pointer value (e.g., after Section framing is achieved). The I- and D-bits combine to carry the binary value of the offset in number of bytes. As illustrated in Figures 3-37 and 3-38, the I-bits are inverted to indicate a positive pointer adjustment (increment) and the D-bits are inverted to indicate a negative pointer adjustment (decrement). The N-bits are not set to indicate a new pointer value when a simple increment/decrement occurs. In the V1/V2 pointer, the S-bit field indicates the size of the VT (e.g., VT-1.5 (VC-11), VT-2 (VC-12), VT-3, or VT-6 (VC-2)).

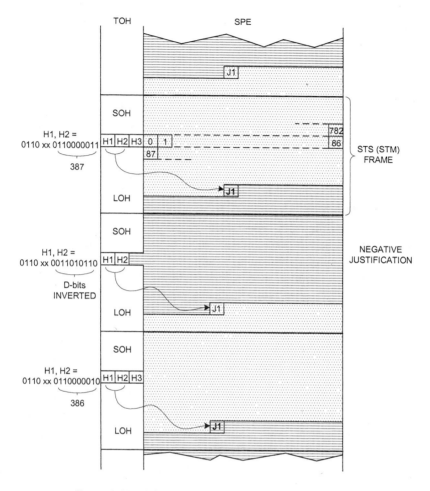

Figure 3-38. High-order negative pointer justification example

N = New Data Flag (= 0110 for normal operation, increment,
    and decrement, and = 1001 for a new pointer value)
S = unspecified
I/D = 10-bit pointer offset value
I = Increment bits (inverted to indicate a positive stuff)
D = decrement bits (inverted to indicate a negative stuff)

Figure 3-39. H1/H2 pointer byte definition

| H1 | 1001 SS11 | 1001 SS11 | H2 | 1111 1111 | 1111 1111 | H3 | H3 | H3 |

Figure 3-40. Pointer byte field for STS-3c (STM-1)

The value contained in the high order pointer, Y, is the offset of the J1 location in terms of Y x N bytes for SONET STS-Nc signals and Y x $3N$ for SDH STM-N signals. Figure 3-40 illustrates the location of the H1, H2, and H3 bytes for a STS-3c/STM-1. H1 is located in the first column, H2 is located in column N+1 ($3N$+1 for SDH), and there is an H3 byte in each of columns 2N+1 through 3N ($6N$+1 through $9N$ for SDH). For STS-Nc or STM-N, the bytes between H1 and H2 contain a concatenation value of 1001 SS11, and the bytes between H2 and the H3 contain all 1s. The concatenation indication prevents these bytes from being incorrectly interpreted as STS-1 pointers. High order pointer adjustments, then, are done in increments of Y x N bytes for SONET STS-Nc signals and Y x $3N$ for SDH STM-N signals.

One of the very powerful concepts in SONET/SDH is the use of two-levels of pointers. The high order pointer contained in the Transport overhead columns allows ADMs and DCSs in the network to operate exclusively on high order Paths (STS / VC-4) without needing to consider the low order pointers. For NEs requiring access to constituent low order signals, the low order pointers allow the processing of the low order Paths. As shown in Figure 3-41, the VT (VC-1/2) pointers operate in a manner directly analogous to the higher order pointers. The V1/V2 byte pointer indicates the offset of the V5 Path overhead byte. Note that within the SONET/SDH frame, the V1 bytes for each VT (VC-1/2) follow immediately after the J1 byte in the first frame of the VT multiframe, the V2 bytes follow the J1 for the second VT multiframe, the V3 bytes for the third, and the V4 bytes for the fourth. For SDH, there is one other type of pointer. SDH networks typically consider the VC-3 (i.e., the STS-1) to be a low order container instead of a higher order container as it is in SONET. As a result, SDH requires a low order pointer for the VC-3. The pointer associated with the VC-3 is located in the fixed-stuff columns of the TUG-3, as illustrated in Figure 3-42.

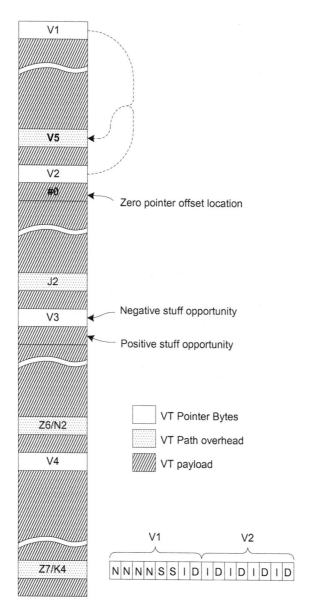

Figure 3-41. VT (VC-1/2) pointer illustration

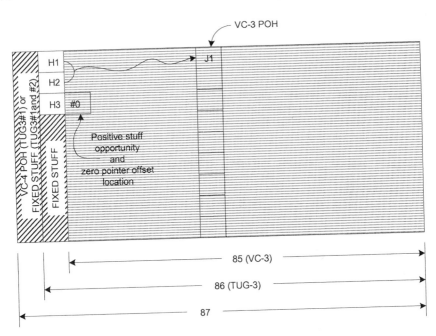

Figure 3-42. Illustration of the pointers for multiplexing a TU-3 into a TUG-3

3.8. SYNCHRONIZATION

As powerful as the pointers are, they have the drawback of adjusting the data alignment by a byte (or multiples of bytes) at a time. These phase movements are difficult to accommodate at the desynchronizers that extract the payload signal from the SONET signal. A desynchronizer must output the payload signal at a clock rate that is reasonably constant, within required parameters for that payload signal. Payload signals (e.g., DS1 and E1) are able to tolerate small perturbations in their instantaneous clock rate, which are essentially phase variations. Jitter is the name given to high-speed clock rate variations, and wander is the name given to relatively slow rate (i.e., below 10 Hz.) variations. The job of the desynchronizer, then, is essentially to track the frequency of the payload signal source, while using an elastic store buffer to absorb any phase perturbations that occur in the transmission path. The desynchronizer typically adjusts its output frequency and phase based on the amount of data in the elastic store buffer (i.e., its fill level).

Small perturbations due to imperfections in regenerator circuits or the effects of bit stuffing removal from asynchronous/plesiochronous multiplexing along the path are relatively easy to accommodate. The large adjustments due to SONET/SDH pointers are much more difficult. In a fully synchronous network, i.e., one in which all NEs are tied to the same reference clock source, clock phase variations will only come from such things as regenerator circuit imperfections or temperature variations affecting the physical length of the fiber. Hence, pointer adjustments will be reasonably rare in a fully synchronous network, giving the desynchronizer ample time to slowly adjust the output clock to accommodate the elastic store fill change between pointer adjustments. SONET and SDH further specify constraints on how often pointer adjustments are allowed, including a minimum time between pointer adjustments. Unfortunately, a fully synchronous network is not always practical. For example, each carrier typically relies on their own internal reference clocks (i.e., their synchronization network), and hence carrier-to-carrier connections typically involve two clock domains. When different NEs use different clock sources, the difference in the resulting clock rates will cause relatively regular pointer adjustments. The larger the difference in the reference clocks, the more frequent the pointer adjustments. For this reason, SONET and SDH specify different qualities of clock sources in terms such as the fraction that they are allowed to be away from the ideal reference clock frequency.

For SONET, Stratum 1 is the highest quality clock and has an accuracy specified at $\pm 1 \times 10^{-11}$. Stratum 2 is next at $\pm 1.6 \times 10^{-8}$, then Stratum 3 at $\pm 4.6 \times 10^{-6}$. SONET Minimum Clock (SMC) has an accuracy of $\pm 20 \times 10^{-6}$. Stratum 1 and 2 require an atomic clock source, while Stratum 3 and SMC can be achieved by conventional oscillators. Due to the relatively low cost of the receivers, it is common today to derive a Stratum 1 clock from the GPS satellite network. SMC is not guaranteed to allow suitable data transmission (and indeed can be an AIS trigger), but is adequate for the communication of the SONET TOH information, including alarms and the SDCC. When a NE loses its highest quality clock source and falls back to a lower quality clock source (e.g., an internal oscillator) this secondary clock source should ideally continue to run at the same frequency as the last known value for the higher quality clock source. This minimizes the pointer adjustments that will occur from the clock source change as well as those from switching back the higher quality clock when it is restored. Holdover is the term used to specify how quickly the back-up clock source can drift from this frequency of the higher-quality reference, and is specified in terms how much deviation can occur in a 24 hour period. The main difference between Stratum 3E and Stratum 3, for example, is the tighter holdover requirements for Stratum 3E. The output phase transients of the clocks are also specified in terms of Maximum Time Interval Error (MTIE).

The S1 byte of the SONET/SDH TOH is used to communicate the quality of the reference clock being used by the transmitting NE. (See Table 3-3.) The receiving NE can then determine whether the received signal is acceptable for use in deriving that NEs own reference clock. When a NE derives its timing from a received SONET signal, it is referred to as line timing, or in some cases loop timing.

An excellent and thorough discussion of jitter, wander, and network synchronization is available in [SeRe]. The authors provide both an intuitive treatment of the different sources for jitter and wander, along with a mathematical analysis for the reader wishinig to delve deeper into the topic.

3.9. SUB-STS-INTERFACES

Although not as widely used as the OC-*N* signals, SONET/SDH interfaces exist at rates lower than the 51.84 Mbit/s OC-1 signal. There are two basic categories of these interfaces, with each having been developed for a different application. The first is the VT1.5 electrical interface and second is an optical VT Group interface.

3.9.1   Electrical VT1.5 interface

The VT1.5 electrical interface was introduced primarily to allow intra-office interconnection between SONET NEs without having to remove the client DS1 from its VT1.5. Maintaining the VT1.5 on this connection means that the VT Path remains intact. Ideally, a network provider wants a VT Path to remain intact from the point where the client signal enters its network to where it exits the network. Whenever the VT Path is terminated within the network, the performance information for that Path must be calculated. The end-to-end performance of the client signal through the network must then be determined as the combined performance of each of the individual VT Paths that client signal used in the network. If end-to-end connectivity can be preserved, then the one VT Path performance information covers the entire Path (Trail) through the network.

DS1 signals are typically connected between optical NEs as DSX-1 electrical signals that are patched through a DSX-1 cross connect panel. In order for an electrical VT1.5 interface to be practical, it would have to be capable of being routed through the same DSX-1 cross connect panels without interfering with the existing DS1 signals. The solution was achieved through the creative use of line codes. As shown in Figure 3-43, the VT1.5 is mapped into a 2.048 Mbit/s frame. This 2.048 Mbit/s portion of the signal is then encoded with the 4B/3T line code, which maps four binary bits into three ternary signal levels. (See [Bel] for details of the 4B/3T line code.) The 4B/3T mapping reduces the symbol rate from $(2.048)(3/4) = 1.536$ Mbit/s, which is the same as the DS1 payload rate. An Alternative Mark

Inversion (AMI) framing bit is then added to this signal every 192 symbols to give a final symbol rate of 1.544 Mbit/s. (Note that this framing bit is added in a fixed location relative to the underlying 2.048 Mbit/s frame so that locating this added framing bit is sufficient to know the 2.048 Mbit/s carrier signal frame alignment.) The framing bit pattern is identical to that used for a DS1, so the resulting signal is compatible with the DS1 in terms of rate, signal levels, and framing pattern. Hence, it can go through a DSX-1 cross connect panel with other DS1 signals, and the receiver can used DS1 framing circuit to recover the signal's frame (which in turn simplifies the 4B/3T synchronization recovery).

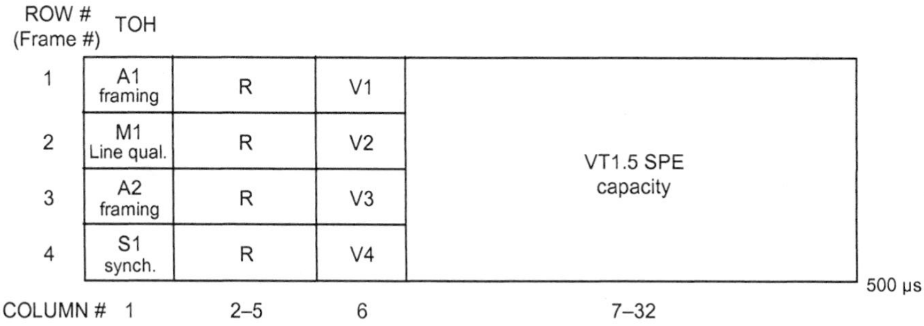

a) VT1.5 interface frame format

Frame	Overhead byte	Byte value
1	A1	1001 1001
2	M1	x10x xxxx
3	A2	0001 1011
4	S1	x10x xxxx

b) TOH framing pattern definition

Figure 3-43. Frame format for the electrical VT1.5 interface signal

As Figure 3-43 indicates, the TOH for the VT1.5 interfaced includes the capability to communicate basic signal quality and network synchronization information. The bits designated with an "x" are defined in T1.105.07 to provide a BIP-2 forward error indication, an REI, and an RDI in the M1 byte and the same 4-bit code in the S1 byte that is used in the S1 byte of an STS-N signal.

Two important points are worth noting here:

- While the electrical VT1.5 signal and a DSX-1 signal both use the same symbol rate and both use a ternary line code, the AMI/B8ZS line code used by DSX-1 signal will have somewhat different spectral characteristics than the 4B/3T encoded VT1.5.

  As a result, although the electrical VT1.5 signal may pass transparently through some DS1/T1 repeaters, there is no guarantee that it will do so.

- While the choice of a 2.048 Mbit/s frame for transporting the VT1.5 was made for bit rate reasons, it also allowed defining the overhead within that frame to be compatible with the 2.048 Mbit/s E1 signal. As a result, if no 4B/3T line code is present, a standard E1 framer can recover the VT1.5 signal within the 2.048 Mbit/s carrier signal.

  This fact allows a VT1.5 to be carried within any network that supports the transport of an E1 signal. One of the more important of these could be the recently developed HDSL2 signal. Encapsulating a VT1.5 within a 2.048 Mbit/s signal carried over HDSL2 would allow SONET VT1.5 connectivity all the way to the customer premise, which would remove all ambiguity over where any errors occurred on the access link.

3.9.2   Optical VT Group

The optical VT Group (OVTG) interface, specified in T1.105.07 was developed to allow an inexpensive optical access link for delivering multiple DS1s to a subscriber. The interface has also been used for connections to wireless base stations. While an OC-1 could have been used for these applications, it introduced a number of factors that raised the cost of the interface, not the least of which was the cost associated with processing the full set of OC-1 overhead. The idea behind OVTG was to have a very simple interface. The basic OVTG interface frame structure is shown in Figure 3-44. A single, 9-byte overhead column was added to the VT Group in order to provide basic TOH functions. The VT's V1-V3 pointers are still used for the pointer functions. The contents of the VT Group can be anything that is allowed in a VT Group (i.e., VT1.5, VT2, VT3, or VT6, but not more than one type in a given VT Group). T1.105.07 initially specified a signal with only one VT Group. Subsequently, the interface was amended to allow up to five VT Groups to be carried in an OVTG signal. When multiple VT Groups are carried, the interleaving of the VT Groups follows the same format as for an STS-1 (see Figure 3-26). The advent of virtual concatenation presents some interesting potential OVTG in transporting data signals such as 10Base Ethernet.

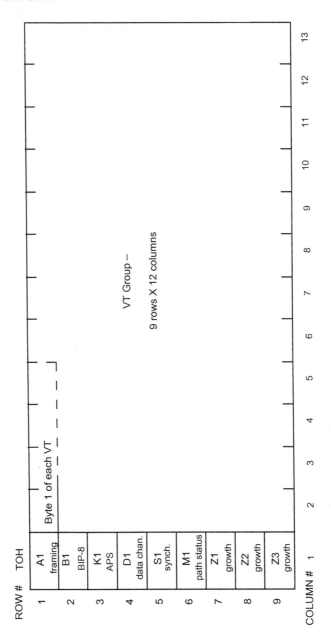

Figure 3-44. Optical VT Group frame format (shown for a single VT Group payload)

The overhead includes forward error detection (B1), path status information (REI and RDI in M1), network synchronization information (S1), a basic protection switching capability (e.g., 1+1 in K1), and a 32 kbit/s DCC in D1.

The SDCC definitions for SONET NEs are based around a full protocol stack, and this type of DCC is one of the options for the OVTG. Since the typical scope of applications for the OVTG is rather simple (e.g., DS1 extension), there was a strong desire to specify an optional, very simple, bit-oriented LAPD message set that could be used instead of a full-featured DCC. This simple message set is included in an informative annex of T1.105.07.

3.10.   IN-BAND FORWARD ERROR CORRECTION

For OC-48 (STM-16) and higher rate SONET/SDH signals a group of TOH bytes can optionally be used to provide forward error correction (FEC).[T1058]   The primary motivation for incorporating FEC into the SONET/SDH signal was to allow upgrades of existing facilities to the next higher rate (e.g., OC-12 to OC-48) on those facilities that were not otherwise capable of error-free performance at the higher rate.   In some cases, FEC was required to allow other signals to be mixed on the same fiber through wave division multiplexing (WDM).

The code chosen for this application is a systematic, shortened version of BCH (8191, 8152) code. This BCH code allows the correction of up to three transmission channel errors (i.e., it is a BCH-3 code). A systematic code is one in which the error correction overhead bits are appended to the original information bits without changing the values of those information bits. A BCH code is a type of linear cyclic code, and the operation of linear cyclic codes can be conceptually described as follows. The block of information bits to be protected is regarded as being a binary polynomial, which is then divided by another polynomial called a generator polynomial.   The remainder of this division is the check bits that are added to the signal for that block. The FEC decoder performs the same division and compares the received check bits to the remainder that it calculated.   The difference between the received and calculated values provides a syndrome that is unique for all error patterns with up to three errors for a BCH-3 code. This uniqueness property allows the decoder to know which error pattern has occurred and to correct the affected bits. (Note that the code construction is such that the polynomial formed by the information bits and the check bits is divisible by the generator polynomial, which means that any remainder calculated by the decoder that is other than zero will be the error syndrome. For more information on the specifics of how BCH and other linear cyclic codes work, see for example [Wic].)

The set of bytes that may be used for FEC are shown in Figures 3-9 through 3-11. In order to minimize the delay, the FEC is applied on a row-by-row basis. Accomplishing row-wise FEC required some creative arrangements, since not all rows had an adequate number of unused TOH bytes available for FEC. As a result, the BCH check bytes for some of the rows are actually located in the subsequent row.

The FEC is a Line layer function, and as such the FEC overhead (check symbol) bytes are covered by the B2 bytes. Hence, as described above, the B2 bytes have to be compensated appropriately for the parity value changes that result from the insertion of the FEC overhead at the transmitter. Another consequence is that the error correction at the receiver will correct errors in the B2 bytes.

A triple error correcting code is adequate for many applications, but more powerful error correcting codes are desirable in dense WDM systems where wavelength cross talk becomes an issue, or in applications with very long spans between repeaters. The desire for strong FEC was one of the motives behind developing the G.709 Optical Transport Network recommendation in the ITU-T.

3.11. CONCLUSIONS

SONET and SDH have proven to be a huge success for a number of reasons, including the following:

- SONET and SDH provide a standard interface and multiplexing format that allows significant cost reductions through economies of scale for equipment and component vendors, through reducing the amount of craft training within each carrier network, and through simplifying the interface requirements between carriers (including international carrier connections).
- The development and release of the SONET/SDH standards were well timed with respect to the development of the optical networking, integrated circuits, and optical component technologies.
- The overhead of SONET/SDH has allowed a substantial increase of the level of performance guarantees that carriers can maintain, while at the same time reducing their operating expenses through automatic monitoring, reporting, fault identification and isolation, and protection switching.
- SONET/SDH has proven to be very versatile in terms of handling legacy telephony signal hierarchies, and now being extendable to efficiently handle various data communications signals.

- The layered approach to the SONET/SDH signal and its overhead allowed significant savings in terms of both equipment and network provisioning complexity (and therefore cost).

SONET and SDH have been so successful as a core / backbone network technology that it promises to retain is dominant position for many years to come. Even the ITU-T G.709 OTN standard for all-optical networks is built around the transport of SONETSDH signals as its payload, with the assumption that payload mapping and most of the multiplexing will take place at the SONET/SDH level.

REFERENCES

[BaCh]	Ralph Ballart, Yau-Chau Ching, *"SONET: Now It's the Standard Optical Network"*, IEEE Communications Magazine, 1989, Vol.27, reprinted in the May 2002 issue.
[Bel]	J. C. Bellamy, *Digital Telephony*, John Wiley, New York, 1982, pp. 181-182.
[Fra]	P. A. Franaszek, "Sequence-State Coding for Digital Transmission," *Bell System Technical Journal*, December 1976, pp. 143-157.
[I707]	ITU-T Recommendation G.707 (1996), *Synchronous Digital Hierarchy Bit Rates.*
[I783]	ITU-T Recommendation G.783, *Characteristics of SDH Equipment Functional Blocks*
[I825]	ITU-T Recommendation G.825, *The Control of Jitter and Wander Within Digital Which are Based on the Synchronous Digital Hierarchy (SDH)*
[I7041]	ITU-T Recommendation G.7041/Y.1303 (2001), *Generic Framing Procedure (GFP).*
[IE83ae]	IEEE 802.3ae (2002): *Information technology—Telecommunications and information exchange between systems—Local and metropolitan area networks—Specific requirements—Part 3: Carrier sense multiple access with collision detection (CSMA/CD) access method and physical layer specifications. Amendment: Media Access Control (MAC) Parameters, Physical Layers, and Management Parameters for 10 Gb/s Operation*
[II239]	ISO/IEC 13239-2002, *Information Technology – Telecommunications and information exchange between systems – High-level data link control (HDLC) procedures*
[SeRe]	M. Sexton and A. Reid, *Broadband Networking: ATM, SDH, and SONET*, Artech House, 1997
[Sim]	W. Simpson, "PPP over SONET/SDH," IETF RFC 2615, June 1999
[Sta]	W. Stallings, *Local and Metropolitan Area Networks*, MacMillan, New York, 1993
[T107]	T1.107-1995, Digital Hierarchy - Formats Specifications.
[T231]	T1.231 Digital Hierarchy – In-Service Digital Transmission Performance Monitoring.
[T269]	T1.269-2000, Information Interchange – Structure and Representation of Trace Message formats for the North American Telecommunications

System.
[T646]   T1.646-1995, Broadband ISDN – Physical layer specifications for user – network interfaces including DS1/ATM.
[T1050]   T1.105-2001 Synchronous Optical Network (SONET) Basic Description including Multiplex Structure, Rates, and Formats
[T1051]   T1.105.01-2000 Synchronous Optical Network (SONET) Automatic Protection Switching
[T1052]   T1.105.02-2001 Synchronous Optical Network (SONET) Payload Mappings
[T1053]   T1.105.03-1994 Synchronous Optical Network (SONET) Jitter at Network interfaces
[T1054]   T1.105.04-2001 Synchronous Optical Network (SONET) Data Communication Channel Protocols and Architectures
[T1055]   T1.105.05-2003 Synchronous Optical Network (SONET) Tandem Connection Maintenance
[T1056]   T1.105.06-1996 Synchronous Optical Network (SONET) Physical Layer Specifications
[T1057]   T1.105.07-2001 Synchronous Optical Network (SONET) Sub STS-1 Interface Rates and Formats Specification
[T1058]   T1.105.08-2001 Synchronous Optical Network (SONET) In-band Forward Error Correcting Code Specification
[T1059]   T1.105.09-2002 Telecommunications – Synchronous Optical Network (SONET) Timing and Synchronization
[Tel253] Telcordia Technologies, *Synchronous Optical Network (SONET) Transport Systems: Common Generic Criteria*, GR-253-CORE, Issue 3, Sept. 2000
[Tel496] Telcordia Technologies, *SONET Add-Drop Multiplexer (SONET ADM) Generic Criteria*, GR-496-CORE, Issue 1, Sept. 1998
[Wic]   S. Wicker, *Error Control Systems for Digital Communications and Storage*, Prentice Hall, Upper Saddle River, New Jersey, 1995

Chapter 4

NEXT GENERATION TRANSPORT TECHNOLOGIES

4.1. INTRODUCTION

Some recently standardized technologies have the potential to play a significant role in next generation transport networks. While an exhaustive discussion of next generation transport network technologies is beyond the scope of this book, an overview of some of these key technologies is given in this chapter. The technologies chosen for inclusion in this chapter are ones that have shown the potential for significant deployment within the next five years.

The amount that carriers have invested in their transport networks is enormous. This investment includes not only the transport equipment, but also the operations systems that support them and the investment in training the craftspeople to operate these networks. As a result, new matter how much potential improvement a new technology promises, it has to be phased in over time. No carrier can afford to replace its transport networks with a completely new technology in a short time frame. A key factor for any new transport technology, then, is its ability to use and interwork with existing network technology. The first technologies discussed in this chapter are those that build on, and hence extend the life of the SONET/SDH networks by allowing them greater flexibility in carrying packet data and broadband payloads. Specifically, these technologies are virtual concatenation, with its Link Capacity Adjustment Scheme (LCAS), the Generic Framing Procedure, and Resilient Packet Rings. Virtual concatenation and LCAS capability has recently also been defined for the older generation asynchronous (i.e., DS1 and DS3) and plesiochronous digital hierarchy signals in order to further extend their useful life. Since this is a new capability, it is included as part of the section on virtual concatenation that otherwise primarily focuses on SONET/SDH signals. The last technology discussed is one that combines SONET/SDH signals onto separate wavelengths for transport through a wave-division multiplexed network.

4.2.   VIRTUAL CONCATENATION (VCAT) AND THE LINK CAPACITY ADJUSTMENT SCHEME (LCAS)

Virtual concatenation and LCAS began as technologies to add flexibility to SONET/SDH. Since then, they have been applied to the new G.709 optical transport network technology, and have also been defined for the legacy asynchronous/PDH signals (i.e., DS1, E1, DS3, and E3). This section begins with a description of the their development and application in the context of SONET/SDH and then describes their application to the other technologies.

4.2.1   Virtual concatenation

As seen in Chapter 3, SONET uses a fixed number of fixed container sizes for carrying client payload data. This fixed set of containers imposed constraints on SONET when it became desirable to carry a variety of data and video signals. Since the infrastructure of most of the worlds' carriers was based on SONET/SDH, it was critical to find a mechanism to extend SONET's capability for these increasingly important payloads. One solution would have been to explicitly concatenate multiple VTs or STSs into a single, contiguous larger channel (i.e., allow an arbitrary value for N in STS-Nc and add a similar explicit concatenation for VT1.5 and VT6). Since the constituent members of explicitly concatenated signals are required to be contiguous, explicit concatenation is typically referred to as contiguous concatenation. This requirement is also the main drawback to explicit concatenation since all intermediate cross-connect points must be able to handle a concatenated group as a single entity. The constituent members' alignment and relative phase must be preserved through the cross-connect. Hence, adding a new concatenated channel size would require the upgrade of all intermediate cross-connects. Also, in order to simplify cross-connecting, the existing contiguously concatenated signals have boundary restrictions with respect to where they are allowed to begin. Arbitrary concatenation sizes would add significant complexity to the cross-connect. The answer, which had been around for some time, was virtual concatenation.

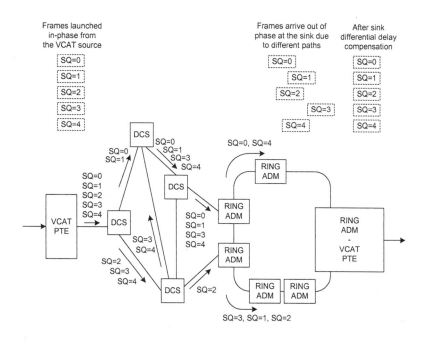

Figure 4-1. Virtual concatenation example illustrating the potential differential delay

Virtual concatenation combines VT1.5s, VT6s STS-1s, or STS-*N*cs, (SDH VC-1/2/3/4 or VC-4-*N*cs) to create larger payload channels. (Only one VT/STS/VC type is allowed within a given group.) The combined group of VT/STS/VC members is referred to as a virtually concatenated group (VCG). The required overhead information for high order virtual concatenation (HOVC) is encoded in the H4 byte, and the overhead information for low order VCAT (LOVC) is encoded in bit 2 of byte Z7 (K4) as a 32-bit string distributed over a 32-frame multiframe. The VCAT source node places the payload data into the constituent members on a byte-by-byte round robin basis in accordance to a sequence assigned by the source. (This type of approach is sometimes referred to as inverse multiplexing.) The source places the associated sequence numbers into the H4 or Z7/K4 bytes of each member. The intermediate NEs that cross-connect the members do not have to be aware of the presence of the VCAT signal. In fact, as illustrated in Figure 4-1, each member can take an entirely different route through the network between the VCAT source and sink. The VCAT sink recovers the payload by examining both the sequence numbers and the H4 or Z7/K4 multiframe phase information. The multiframe phase information is used to re-align the individual VCG members with each other, thus accommodating the differential delay that may have occurred in the network. (See Figure 4-1.) The H4 VCAT encoding is shown in Figure 4-2 with the fields in italics showing the additional overhead that is used for LCAS.

Concatenation overhead octet definition							
Bit 1	Bit 2	Bit 3	Bit 4	Bit 5	Bit 6	Bit 7	Bit 8
Control Packet				MFI1			
MST bit 1	*MST bit 2*	*MST bit 3*	*MST bit 4*	1	0	0	0
MST bit 5	*MST bit 6*	*MST bit 7*	*MST bit 8*	1	0	0	1
0	0	0	*RS-ACK*	1	0	1	0
Reserved (0000)				1	0	1	1
Reserved (0000)				1	1	0	0
Reserved (0000)				1	1	0	1
SQ MSBs: bit 1	bit 2	bit 3	bit 4	1	1	1	0
SQ LSBs: bit 5	bit 6	bit 7	bit 8	1	1	1	1
MFI2 MSBs: bit 1	bit 2	bit 3	bit 4	0	0	0	0
MFI2 LSBs: bit 5	bit 6	bit 7	bit 8	0	0	0	1
CTRL				0	0	1	0
0	*0*	*0*	*GID*	0	0	1	1
Reserved (0000)				0	1	0	0
Reserved (0000)				0	1	0	1
C_1	C_2	C_3	C_4	0	1	1	0
C_5	C_6	C_7	C_8	0	1	1	1

Figure 4-2. H4 overhead used for VCAT and LCAS

Interworking between VCAT and contiguously concatenated channels is under study. For interworking between STS-Nc-Xv and other STS-Nc rates, the interworking is reasonably straightforward. For STS-1-Nv and STS-Nc, however, an STS-1-Nv has $Nx84$ payload columns with the STS-Nc has $(Nx87)$-1 payload columns, which complicates interworking.

4.2.2 LCAS

Since data signals can vary in their bandwidth needs, it was desirable to have an efficient method for changing the size of a VCG. This goal was accomplished with the Link Capacity Adjustment Scheme. The LCAS protocol handles the change in bandwidth in a manner that avoids hits in the payload data. "Hitless" is used in a limited sense here. Obviously, a bandwidth change to a constant bit rate service causes a disruption in the data. For packet data, however, packet-level buffering at the ingress and egress to the VCG can handle the bandwidth change as long as both ends know exactly when the bandwidth change in the transport channel is taking place. The LCAS protocol accomplishes this handshake through control codes (CTRL) passed from the VCG source to sink and message status (MST) bits returned from the sink to the source. The CTRL codes indicate whether a given member is IDLE (i.e., is not currently in use), being added (ADD), contains the highest sequence number of the VCG (i.e., is the end of the sequence (EOS)), or is a normal traffic-bearing member (NORM). Throughout the following discussion of the LCAS protocol, keep in mind that the LCAS protocol works independently in each direction (i.e., each direction proceeds in a unidirectional manner). The VCG size in the two directions of transmission is required to use the same member type, but there is no requirement that the number of members be the same in each direction. The member status (MST) information is sent from the sink back to the source to tell it the OK/FAIL status of each the VCG members sent by the source. The sink places same MST information for all members into all of the members of that VCG's return channel. This allows the source to monitor any of the return members to determine the status of the members it sends for that VCG.

A planned bandwidth change begins with the provisioning of the VCG end nodes (e.g., via the NMS) to add or remove a member from the VCG. The sequence of events for adding a member is shown in Figure 4-3. The VCG source node sends a CTRL=ADD to the sink, and assigns a new sequence number to this new member. If that member is okay from a performance and provisioning standpoint, the sink responds to the source through the MST bits that the member with this sequence number is OK. (Note that the MST bit fields are time shared across a multiframe in order to conserve bandwidth.) The source then changes the CTRL=ADD in the new member to CTRL=EOS (and changes the previous member with CTRL=EOS to CTRL=NORM). In order to avoid having bit errors in the

VCAT/LCAS overhead information lead to spurious bandwidth changes, the overhead is protected with a CRC error detecting code. For LOVC, the 32-bit string is covered by a CRC-3, and for HOVC, the group of 16 H4 bit 1-4 nibbles forms a control packet protected by its CRC-8. When the source changes from CTRL=ADD to CTRL=EOS, it will begin to transmit data in the new member during the STS (VT) frame immediately following the frame that carried the last nibble of the CRC-8 (CRC-3). If the CRC check is good, then the sink will be expecting this new data, and the addition can be made without a hit to any client data packets.

When a single new member is added, its initial sequence number will be the next one available in the VCG's sequence and will become that member's permanent value if the sink agrees that this new member is okay to add. It is possible to attempt to add multiple members simultaneously. In this case, different response delays may cause the source to ultimately assign different sequence numbers to the new members than it used when it sent the initial CTRL=ADD code since the source will assign the permanent sequence numbers in the order in which the MSTs are received. An alternative method to adding multiple members is to allow the NMS to initiate their addition, but have the LCAS source handle their actual addition one at a time.

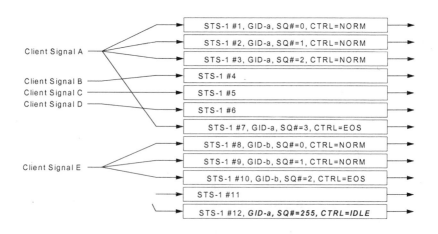

a) Initial condition for adding STS-1 number 12 to VCG a

Figure 4-3. LCAS example for the addition of a new member

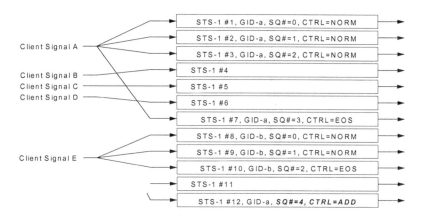

If the new member is okay, the sink responds that member of a is OK

b) Source indicates its desire to add the new member

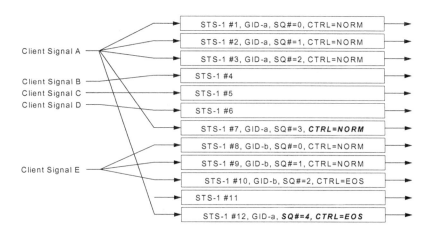

When the sink sees and accepts the change in the VCG bandwidth, it sends a RS-ACK

c) The source completes the addition

Figure 4-3 – continued

The planned removal of a member requires no handshake between the source and sink since the source does not need to know the status of this member at the sink in order to delete it. As illustrated in Figure 4-4, when the source provisioning is changed to remove a member, it simply sends a CTRL=IDLE in that member's control packet. The bandwidth change again takes place in the VT/VC/STS frame immediately following the frame containing the end of the control packet's CRC. The sink can thus anticipate the bandwidth change and implement it without corrupting the payload data. If the member being deleted does not have the highest sequence number, the sequence numbers of all members higher in the sequence will be decremented at the same time that the CTRL=IDLE is sent so that there are no resulting gaps in the member sequence numbers. If the removed member has the highest sequence number, the member with the next highest sequence number will become the one carrying CTRL=EOS.

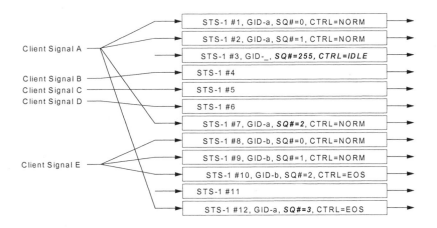

Figure 4-4. Example of removing a member with LCAS

4.2.2.1 Partial restoration with LCAS – a new paradigm

Another powerful feature of LCAS is the method by which it can handle a failure of a subset of the members by falling back to the healthy subset of members during the failure period. This provides a powerful new paradigm for protecting those data services that can tolerate the reduced bandwidth. A service provider can now offer three levels of service resiliency, each with its own price point. Of course, it's always possible to offer service with no protection at a discount. Full protection is a premium service and the subject of Chapter 8. The current telephone network is built around full protection since constant bit rate (CBR) services such as voice can't tolerate anything less than their full rate. Although, there are more efficient techniques, the most common approach to full protection is to reserve the same amount of bandwidth for protection channels as is used by the working channels that they protect. With LCAS partial restoration, there are no reserved protection channels. The partial restoration approach works by routing the different VCG members over diverse routes so that no single failure can affect all members. The amount of route diversity determines fraction of the service bandwidth that can be affected by a single fault. Partial protection is thus as efficient as non-protected service but still allows service to continue in the presence of faults. Several laboratories and universities are investigating VCG route selection algorithms.

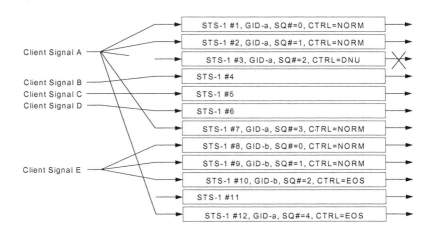

Figure 4-5. Example of an unplanned bandwidth reduction with LCAS

The specifics of the LCAS restoration technique are as follows. If the different members of the VCG are routed through the network on different physical links, a failure of one of those links disrupts the members that traverse it. Since the payload data is byte-interleaved across all the members, the failure initially disrupts the entire VCG. To restore the service, the sink first indicates the failure of those members to the source by sending MST=FAIL for the associated sequence numbers and begins taking data from only the healthy members. The source then sends CTRL=DNU for those failed members and stops placing payload data into them. This situation is illustrated in Figure 4-5. The sink will now be receiving uncorrupted payload data at the reduced bandwidth available from the healthy members. At this point, one of two things can happen. These failed members can be removed from the VCG through a provisioning change. Alternatively, the VCG can continue to function with the reduced bandwidth until the failure is cleared. This second scenario is the one for which the CTRL=DNU field is used. When the failure clears, the sink will see the CTRL=DNU as an indication that the source is still not placing data into these members. When the source receives the MST=OK for the restored members, it will then change back to CTRL=NORM/EOS and begin placing data in them again. The advantage of such a scheme is that it allows using unprotected links through the network (at a reduced cost for the user) with a fallback to lower bandwidth in the event of a failure rather than a complete loss of the connection. This mechanism is discussed further in Chapter 8.

4.2.3   Application of VCAT and LCAS to G.709 OTN

Virtual concatenation is also specified for the Optical channel Payload Units (OPUs) of the G.709 Optical Transport Network (OTN). The OPU1 and OPU2 are approximately 2.5 and 10 Gbit/s, respectively. This is discussed in more detail in section 4.5 of this chapter as part of the description of OTN.

4.2.4   Application of VCAT and LCAS to Asynchronous / Plesiochronous Digital Hierarchy Signals

4.2.4.1   Background

A section devoted to asynchronous / plesiochronous digital hierarchy signals (e.g., DS1 and E1) may seem out of place in a chapter on next generation transport technologies. These networks, however, are still ubiquitously deployed and are still more common than SONET/SDH signals for enterprise access applications. Among the reasons for their ongoing prevalence in the enterprise access networks is that many of the access interfaces are still delivered over copper wires. For compactness, in this section PDH will be used to refer to both the North American DS*n* and ETSI E*n* signals.

At least as important as the availability of PDH signals is the artificial advantage they have due to the U.S. regulatory unbundling of services. U.S. incumbent local exchange carriers (ILECs) are required as part of unbundling to offer DS1 and DS3 access links to other carriers, such as competitive local exchange carriers (CLECs) or interexchange carriers (IECs), for lower tariff rates than the cost of equivalent SONET interfaces. The result of the tariff advantage and the effectively ubiquitous availability of DS1 and DS3 connectivity is that when IECs or service providers lack their own facilities to connect to their enterprise subscribers, they typically lease DS1 or DS3 connections through the ILECs. An example of this network configuration is shown in Figure 4-6.

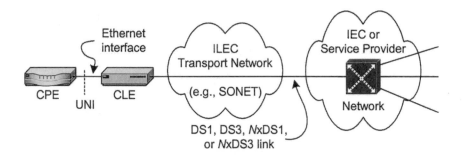

Figure 4-6. Example of PDH access through an ILEC

This PDH situation combined with the growing interest in providing native Ethernet connectivity led inevitably to a desire for a mapping of Ethernet into PDH signals. Although a number of proprietary implementations existed, there were no standards for mapping native Ethernet into DS1 and DS3, or NxDS1 and NxDS3 signals. In order to provide Ethernet connectivity to their enterprise customers over DS1/DS3 connections, the major U.S. IECs asked for GFP mappings into DSn and En signals. (As discussed later in this chapter, GFP provides an encapsulation of native Ethernet frames in order to carry them through a transport network.) The resulting mappings were specified in the new ITU-T Rec. G.8040.

Initially, just the GFP mapping into a single DS3 signal was standardized. More study was desired for choosing the appropriate mappings into DS1, E1, NxDS1, and NxE. Carriers wanted to have the NxDS1 and NxE1 connections and use $N=1$ for mapping into single signals. Subsequently, interest developed for similar NxDS3 and NxE3 signals (e.g., for carrying data from 100Base Ethernet interfaces). Ideally, the NxDS1/E1/DS3/E3 should operate at Layer 1, providing transparent transport of Layer 2 protocol frames, independent of which Layer 2 protocol is being carried. The only existing non-proprietary solution was the Multilink Point-to-Point Protocol (ML-PPP defined in IETF RFC 1990), which performs inverse multiplexing[50] at the packet level.[51] Since ML-PPP

---

[50] Inverse multiplexing refers to taking the payload from a higher rate channel and transporting it by distributing it over multiple lower rate channels. The granularity used for assigning the payload data among the lower rate channels can be at the bit, byte, or packet/cell level.

is a Layer 2 protocol, it requires terminating the Ethernet signal in order to remap the packets into ML-PPP (i.e., change between the two different Layer 2 protocols). Another potential solution would have been to modify the Ethernet Link Aggregation protocol which inverse multiplexes Ethernet frames over multiple Ethernet physical links. The main modification would have been to include the DSn/En physical interfaces. This, again, would also be a Layer 2 specific solution. No byte level inverse multiplexing schemes such as VCAT existed since DS1 and DS3 signals lacked sufficient overhead to support VCAT, and reserving an entire payload channel for the overhead was too much capacity to loose[52]. Table 4-1 shows a comparison of the different candidate technologies that were considered.

4.2.4.2 Technical solution

The solution, specified in the new ITU-T Rec. G.7043, was to adopt byte-level inverse multiplexing, but to borrow one octet from the signal's payload area once per multiframe to carry the per-link overhead information rather than permanently reserving the entire time slot. This approach makes it possible to use VCAT and LCAS in a manner very similar to SONET/SDH. Figure 4-7 shows the resulting multiframe formats for the DS1, E1, DS3, and E3 signals.

---

[51] Of course, ATM solutions existed, including Inverse Multiplexing over ATM (IMA). The carriers requesting the new mapping did not favor an ATM solution for this application due to its overhead inefficiency and it being another full layer to provision.

[52] Another potential solution existed from the Bandwidth ON Demand Interoperability Group (BONDING) consortium. Inverse multiplexing here is performed at the byte level rather than the packet level. An initialization sequence is sent on all the constituent lower-rate channels in order to synchronize the source and sink. While this technique requires no per-packet or per-link overhead, the channel must be disrupted for a long period of re-initialization when the channel size is changed.

Table 4-1. Comparison of technologies for inverse multiplexing into NxPDH signals

OPTION	ADVANTAGE	DISADVANTAGE
Layer 2 frame inverse multiplexing	• Proven technology exists for ML-PPP and Ethernet Link Aggregation • No overhead required for each individual E1/DS1/DS3/E3 link • Easy to add or remove links (trivial control protocol) NOTE – Layer 1 (i.e., GFP) packet interleaving was also considered, with at least one proprietary solution existing. Although it provides the Layer 2 transparency, it otherwise has the same advantages and disadvantages as Layer 2 packet interleaving.	• Layer 2 technology specific – It either enforces a Layer 2 approach or requires re-mapping client data packets. • Requires additional per-packet overhead (e.g., for packet sequence numbering) • Egress queue management more complex due to the need to re-align the packets from the different links in the correct sequence. • When there is a light load, a single link (or subset of links) is used for each packet rather than the entire set. This results in increased latency for lightly loaded cases. • Under any load condition, the egress queue management will tend to introduce additional latency.
Byte inverse multiplexing with overhead barrowing	• Relatively simple, similar to common SerDes tecchniques. • Uses no additional per-link or per-packet overhead.	• Changing the number of links (members) requires taking the connection down for a link re-synchronization.
Byte inverse multiplexing with permanent overhead channel	• Simple (trivial) egress buffer since out-of-order packet arrival is not possible. • Can directly re-use SDH virtual concatenation technology. • No additional per-packet overhead. • Consistency with VCAT and LCAS provides operational consistency and network predictability for the carrier.	• Requires per-link overhead. • Control protocol for adding and removing links is more complex (same complexity as LCAS).

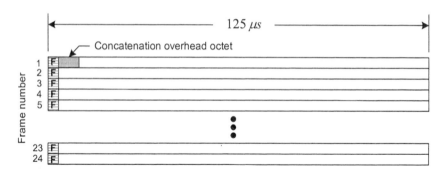

a) DS1 multiframe format for virtual concatenation

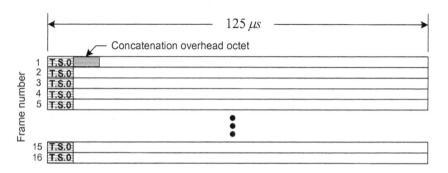

b) E1 multiframe format for virtual concatenation

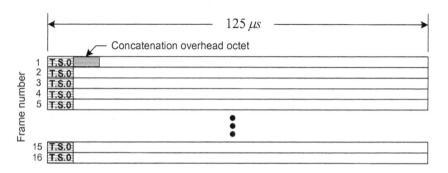

c) DS3 multiframe format for virtual concatenation

Figure 4-7. Multiframe format for the virtual concatenation of asynchronous / PDH signals

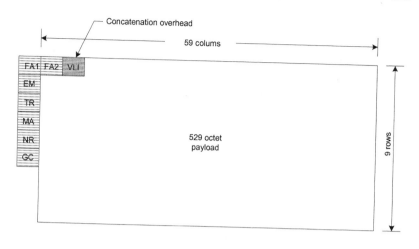

d) E3 multiframe format for virtual concatenation

Figure 4-7 – continued

As shown in Figure 4-8, the VCAT overhead carries the same type of LCAS control packet as is used in the SONET/SDH H4 byte. (See Figure 4-2.) The PDH and SONET/SDH control packet format and bit definitions are identical except for the number of bits used in the sequence number (SQ) and the specific multiplexing of the member status information into the MST bits of each control packet. The transmission order of the control packet in Figure 4-8 is left to right for the bits, and top to bottom for the octets. In all of the fields, the MSB is the first bit to be transmitted. In the case of the SQ, SONET/SDH allows a maximum of 256 members and hence uses a two-nibble (8-bit) SQ field. The maximum number of members is 16 for DS1/E1, and is eight for DS3/E3. Hence, they require a SQ field of 4-bits and 3-bits, respectively. Since the SQ values are justified to the LSBs with the upper, unused SQ field bits set to 0, the SQ field use is still consistent for SONET/SDH (H4), DS1/E1, and DS3/E3 member types.

For DS1/E1, DS3/E3, and SONET/SDH member types, the member status is multiplexed into the MST field based on the multiframe count in the MFI1 and MFI2 fields. When MFI2.MFI1 = 0000.1000, MST field bits 1-4 will carry the member status values of the members with SQ 0-3, respectively. The next MST nibble will contain the member status of members 4-7. Similarly, each subsequent MST nibble reports the member status of the members with the next four higher SQ values until the member with the maximum SQ value is reported. At that point, the member status of members 0-3 is sent again and etc. In the case of DS3/E3 signals, the status of all eight members is sent in each control packet. The DS1/E1 member status requires two control packets to communicate. The result of this modulo frame count is that the LSB of MFI1 and LSB of MFI2 are adequate to identify which members' status is being carried in a given MST nibble.

Concatenation overhead octet definition							
Bit 1	Bit 2	Bit 3	Bit 4	Bit 5	Bit 6	Bit 7	Bit 8
Control Packet				MFI1			
MST (bits 1-4)				1	0	0	0
MST (bits 5-8)				1	0	0	1
0	0	0	RS-ACK	1	0	1	0
Reserved (0000)				1	0	1	1
Reserved (0000)				1	1	0	0
Reserved (0000)				1	1	0	1
Reserved (0000)				1	1	1	0
SQ bits 1-4				1	1	1	1
MFI2 MSBs (bits 1-4)				0	0	0	0
MFI2 LSBs (bits 5-8)				0	0	0	1
CTRL				0	0	1	0
0	0	0	GID	0	0	1	1
Reserved (0000)				0	1	0	0
Reserved (0000)				0	1	0	1
C_1	C_2	C_3	C_4	0	1	1	0
C_5	C_6	C_7	C_8	0	1	1	1

Figure 4-8. LCAS control packet format for PDH signals

4.2.4.3 Mapping GFP frames into virtually concatenated asynchronous / PDH signals

The encapsulation mechanism currently defined for carrying data frames in these virtually concatenated DS1, E1, DS3, and E3 signals is the Generic Framing Procedure (GFP). GFP is presented in the next section of this chapter, and the mapping, which is defined in ITU-T Rec. G.8040, is described here.

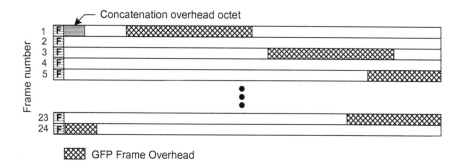

a) *Octet-aligned mapping for GFP into the DS1 signal*

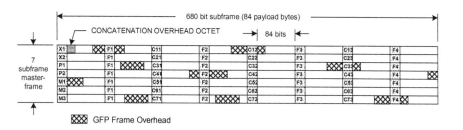

b) *Octet-aligned mapping for GFP into the DS3 signal*

Figure 4-9. Examples illustrating the mapping of GFP frames into PDH signals

When GFP frames are mapped into DS1, DS3, E1, or E3 signals, the GFP frame octets are mapped octet-by-octet into the signal payload container octets. DS1 and E1 signals were originally designed to have octet-structured payload containers, and an octet-oriented payload container structure was added to the E3 signals in ITU-T Rec. G.832. In the case of DS3 signals, the payload structure was originally bit-structured.. When ATM cell mappings were defined for DS3 payloads, they exploited the fact that there are 84 payload bits between DS3 framing bits and defined a nibble-wise mapping with the ATM bytes divided into 4-bit nibbles and 21 nibbles mapped into the 84 payload bits. The GFP mapping took advantage of there being an even number of nibbles per subframe and chose an octet-oriented structure for the payload mapping in order to be consistent with the octet-oriented concatenation overhead channel[53]. These mappings are illustrated in Figure 4-9 for the DS1 and DS3 signals.

4.3.   GENERIC FRAMING PROCEDURE (GFP)

GFP was developed to allow efficient transport of packet data through SONET/SDH networks, making use of the new virtual concatenation and LCAS technologies for creating flexible-sized transport channels. A competing approach to broadband data transport that was explored by some start-up carriers was to use a pure packet network (e.g., an Ethernet WAN). The conversion from a SONET/SDH backbone network to an Ethernet backbone would not only have been cost prohibitive, data networks such as Ethernet have not typically provided the integrated OAM&P capabilities that have made SONET/SDH so valuable to carriers. A technology such as GFP that allowed efficient packet transport within the existing SONET/SDH backbone is thus very attractive. GFP is covered in depth in this chapter since it is not covered in any existing books.

4.3.1 GFP background

GFP was originally born in T1X1 to address some limitations of ATM and PoS (Packet over SONET/SDH). Subsequently, it was decided that the in order to expedite international agreement, it was better to let the IUT-T publish the final standard. GFP is now specified in ITU-T Rec. G.7041.

As described in Chapter 3, ATM suffers from bandwidth inefficiency due to the amount of cell overhead relative to payload capacity. Another perceived drawback to ATM is that ATM implementations typically include a great detail of complexity that is not required for relatively simple packet data interconnections (e.g., the ATM signaling protocols). The drawbacks to PoS are its non-deterministic bandwidth requirements and its requirement that the client data packets be mapped into the IETF Point-to-Point Protocol (PPP) and HDLC for Layer 2. The first PoS drawback is a consequence of the fact that escape characters must be added for each client data byte that looks like an HDLC control character (see Chapter 3 section 3.6.3.2). In the worst case (which could potentially be created by a malicious user), the byte-stuffed HDLC packet payload field could be twice as long as the

---

[53] The nibble-oriented mapping was originally chosen for the GFP mapping into a DS3 signal. Since virtual concatenation is less complex to implement with an octet-oriented structure than with the nibble-oriented structure, the mapping switched to octet-orientation after virtual concatenation was defined for DS3s. .

original payload. The requirement to map into PPP is a distinct drawback if a carrier wants to offer a simple Ethernet extension. Here, the Ethernet MAC frame would need to be terminated so that the client data could be re-mapped into PPP with HDLC and a new Ethernet MAC frame generated on the other end of the PoS link. With over 90% of the data transported through the public WAN originating and/or terminating on an Ethernet at the customer's premises, Ethernet extension is a natural desirable service for the carriers to provide.

GFP allows the simple, efficient transport of a variety of data signals through SONET/SDH, asynchronous/PDH, and G.709 OTN networks. The protocol and its capabilities are the subject of this section.

4.3.2 GFP frame structure

The basic structure of a GFP frame is illustrated in Figure 4-10. The four-byte Core header consists of a 16-bit Payload Length Indicator (PLI) with a CRC-16 to protect the PLI. The PLI is the binary count of the number of bytes in that GFP frame's payload field. The Core header is the key to eliminating the bandwidth expansion problem of PoS. Rather than relying on a special character for frame delimiting, the GFP receiver begins the framing process by looking for a 16-bit field that is followed by a correct CRC-16 for that field. When such a 32-bit pattern is found, it is very likely to be the Core header at the beginning of a GFP frame. (The probability of such a pattern randomly occurring in the client data is 2^{-32}.) The GFP receiver then uses the PLI information to determine where the end of that GFP frame is. The next Core header will occur immediately after the end of that frame, so if another valid 32-bit pattern is found in that location, the receiver can be sure that it has acquired framing for the GFP stream. The CRC-16 is adequate to provide a simple, single error correction capability for the PLI, which increases the robustness of the GFP framing. (In contrast, an error in an HDLC start/end flag has no error correction, which makes HDLC framing more vulnerable to transmission errors.)

The GFP payload area consists of payload header information, a payload field that contains the client data, and can optionally contain a CRC-32 over the payload field. The payload headers are broken down into two types. The payload Type header is used in all GFP frames except Idle frames, and consists of 2 bytes of overhead information protected by a CRC-16. If additional payload overhead information is needed (e.g., a channel identifier to identify different GFP connections when GFP frames from different sources are frame multiplexed together), then an Extension may also be used. Specifically, the Type header contains a Payload Type Indicator (PTI) to indicate the type of GFP frame, an indication of whether the optional payload frame check sequence is included in the payload field (PFI), an indication of what type of Extension header, if any, is present (EXI), and a User Payload type indicator (UPI) to indicate the type of payload in the payload field. The only PTI types identified so far are client data frames and client management frames (CMFs). For the client data frames, the UPI indicates the protocol type of the payload that is mapped into the GFP frame, and whether that mapping is frame-based or transparent. The only Extension header defined at this time is the Linear Extension header, which consists of an 8-bit channel identification field, an 8-bit spare/reserved field, and a CRC-16 over the header. Other Extension headers are under consideration that would potentially expand the identification field so that flows within channels could be identified.

Any idle time between GFP frames carrying client data is filled with GFP Idle frames. A GFP Idle frame consists of only a Core header with a PLI=0.

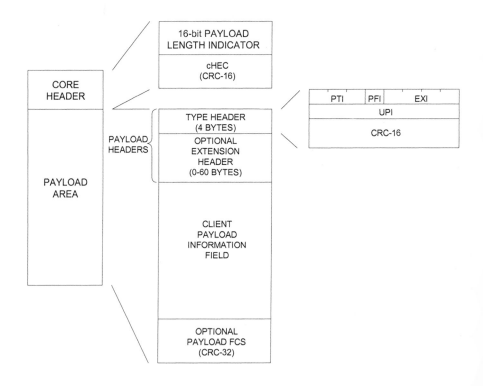

Figure 4-10. GFP frame structure

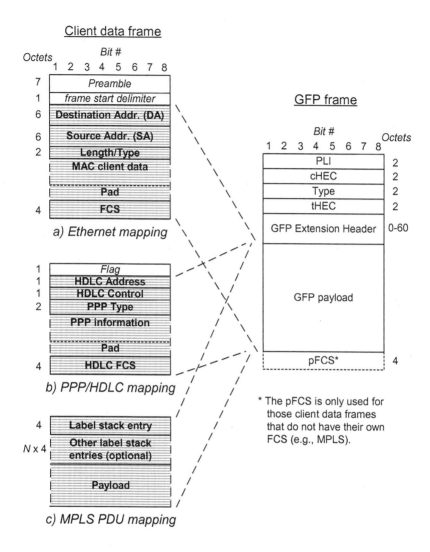

Figure 4-11. Example client data mappings into GFP-F

4.3.3 Frame mapped GFP (GFP-F)

The initial GFP work focused on GFP-F. The basic concept is to take a client frame (typically a Layer 2 frame), remove its unnecessary overhead (e.g., frame delimiting characters or preamble), and then simply place the remainder of the client data frame into the GFP Payload Field. Hence, there is a one to one mapping between client data frames and GFP-F frames.

Figure 4-11 shows some example mappings with the shaded portions of the client data frame indicating the fields that are placed into the GFP payload field. The specific bit alignments between the client data, the GFP-F frame, and the SONET/SDH payload are illustrated in Chapter 3.

The example of the Ethernet mapping is shown in Figure 4-11(a). Note that for the Ethernet mapping, it is assumed that the optional payload FCS is not used.[54] When the client data frame is known to have its own frame FCS, the GFP payload FCS is somewhat redundant since a GFP-F demapper will typically have the capability to examine the client data FCS. The GFP payload FCS would typically only be used if the client data frame either had no FCS of its own or a weaker FCS (e.g., a CRC-16).

The simplicity of the GFP-F mapping is a key to GFP's flexibility. While most of the initial focus on GFP-F was for carrying Ethernet payloads, there is growing interest in using GFP-F for other payloads. As a result, there have been several new mappings proposed. The mappings into GFP-F as of the initial publishing date of this book are listed in Table 4-2.

A Point-to-Point Protocol (PPP) mapping was part of the original G.7041 standard. (See IETF RFC 1661 for the definition of PPP.) As seen in Figure 4-11(b), this mapping includes the HDLC-like framing overhead specified in IETF RFC 1662 ("PPP in HDLC-like Framing"), to be consistent with the PPP over SONET/SDH (PoS) mapping (see Chapter 3). In PoS, the HDLC flags are also retained so that the HDLC framing can be used for frame delineation. The HDLC Address and Control bytes are defined in RFC 1662 to carry fixed values (1111 1111 and 0000 0011, respectively), so for the GFP mapping the HDLC header contains no useful information. It is possible to use a PPP Link Configuration Protocol (LCP) optional procedure to remove the HDLC header and replace the PPP type field with a 1-2 byte PPP protocol field. There is also a proposal under active consideration to have an optimized PPP mapping into GFP-F that always removes the HDLC header and FCS, and uses the GFP pFCS for error detection. At the same time, there has been discussion about whether there are enough applications to justify a generic HDLC into GFP-F mapping. At the time of this printing, these topics are on the G.7041 "Living List" for further study.

[54] In fact, ITU-T Recommendation G.8012 (2004), Ethernet UNI and Ethernet NNI specifies that the GFP pFCS is not used when Ethernet is carried in GFP-F.

Table 4-2. List of client payloads mapped into GFP-F (accepted and proposed)

GFP-F client payload type	UPI value <7:0> (PTI = 000)
Ethernet	0000 0001
PPP	0000 0010
Multiple Access Protocol over SDH (MAPOS)	0000 1000
IEEE 802.17 Resilient Packet Ring	0000 1010
Fibre Channel FC-BBW	0000 1011
MPLS direct mapping	0000 1101
MPLS (multicast)	0000 1110
IS-IS	0000 1111
IPv4	0001 0000
IPv6	0001 0001
PPP direct mapping	(note)
HDLC	(note)
Reserved for proprietary use.	1111 0000 through 1111 1110

Note: A proposal under active consideration is to have a PPP mapping that omits the HDLC header and replace the HDLC FCS with a GFP pFCS in order to simplify processing and improve efficiency. As part of this discussion, there has been a proposal to add a generic mapping for any HDLC-framed client.

GFP-F was originally defined as a Layer 1 (1½) transport encapsulation mechanism for carrying Layer 2 frames. It was appreciated at the time of its development that the existence of the PTI, UPI, and pFCS fields along with the channel identification field in the linear Extension Header gave GFP a rudimentary set of Layer 2 type capabilities. There had been some reluctance to exploit (or enhance) these features since there was general agreement that GFP should not be ballooned in complexity to become a full switching layer similar to ATM.[55] On the other hand, it made sense to exploit these capabilities to improve network bandwidth efficiency.

---

[55] While some have seen particular advantages to being able to send GFP frames along a pre-provisioned route based on the ID fields in an Extension header, no one wanted to create the equivalent of an ATM switched virtual circuit (SVC) network with its own signaling protocol, etc. One of the proposals on the current G.7041 Living List is a new linear

Recently, some of these GFP Layer 2–like capabilities have been exploited for Multiprotocol Label-Switching (MPLS) frame transport. (See IETF RFC 3031, RFC 3032, and RFC3270.) MPLS adds a header to a client data packet that contains a label field to identify what is called a forwarding equivalency class (FEC). MPLS label-switched routers (LSRs) use these labels to route packets through the MPLS network. All packets sharing the same label value are then treated in the same manner by a LSR. The MPLS labels are not global, but are assigned with local significance between two adjacent LSRs. IP or IS-IS frames are typically used to communicate the information required to establish the MPLS label assignments among MPLS network nodes and the hop-by-hop routes that will be taken for packets with each MPLS label (i.e., provide the MPLS control plane signaling)[56]. MPLS, which is nominally somewhere between a Layer 3 and Layer 2 protocol, typically also uses a Layer 2 such as PPP. The Layer 2 protocol provides a protocol identifier field that identifies its payload as being either MPLS or IP (or other protocol) frames and an error check over the frame. At the request of multiple carriers, a direct mapping of MPLS into GFP with no Layer 2 protocol overhead bytes was added. (See Figure 4-11(c).) These carriers look to MPLS as a key technology for their core networks and are concerned about bandwidth efficiency and about reducing the number of layers/protocols that need to be provisioned and administered in their networks. The GFP PTI/UPI codes identify the MPLS payload, and the GFP pFCS provides the frame error check. To complement the direct MPLS mapping, direct mappings into GFP-F were also added for IP (IPv4 and IPv6) and IS-IS in order to carry the MPLS control plane information. These mappings are essentially the same as the MPLS mapping (i.e., they use the GFP pFCS for their error detection). With the addition of these mappings, Layer 2 can become a null layer when MPLS is carried over GFP. It should be noted that although the direct IP mappings into GFP-F were primarily added to support MPLS control plane traffic, they could also be used as a generic mapping for IP networks. Some carriers are considering this possibility.

Extension Header with an expanded address (flow ID) field and other fields similar to an MPLS header for this pre-provisioned routing. The intention was to allow mapping MPLS header information into the GFP frame to allow the transport provider to route frames without having to go up to the MPLS layer. It would also allow simultaneous routing of non-MPLS packets on the same link. The future of this proposal is uncertain.

[56] The Label Distribution Protocol (LDP) and Resource Reservation Protocol for Traffic Engineering (RSVP_TE) are the two protocols outlined in IETF RFCs 3031 and 3270, respectively, for communicating the label assignments among MPLS LSRs. Other, non-IP-based protocols (e.g., IS-IS) have also been used by some vendors and networks.

4.3.4 Transparent GFP (GFP-T)

Several important data protocols use the 8B/10B line code at the physical layer. The protocols and their associated UPI codes are shown in Table 4-3. Fibre Channel, Fibre Connection (FICON) and Enterprise System Connection (ESCON) are commonly used in modern storage area networks (SANs), and Digital Video Broadcast – Asynchronous Serial Interface (DVB-ASI) is popular for video distribution. The 8B/10B line code maps the 2^8 possible 8-bit byte values into one or two of the 2^{10} possible 10-bit values. One goal of this mapping is to maintain a running balance between the number of 1s and 0s that are transmitted. Since only a limited number of 10B characters can have five 1s and five 0s, a given 8-bit data value is mapped into two different 10B values, each with a complementary number of 1s relative to the other. The difference in this running 0/1 balance at the end of each transmitted 8B/10B character is referred to as the running disparity, and it is used to choose which of the two potential 10B values is used for next 8B data value to be transmitted. Each of the above protocols exploits the fact that there are leftover 10B characters that can be used as control codes (e.g., to indicate the start and end of a data frame or to trigger some type of link initialization). For some protocols, a control code flags the beginning of a sequence of characters that communicate control information. Such sequences are called primitive sequences.

If the frames from these protocols were carried with GFP-F, the 8B/10B control code information would be lost. Another consideration for the SAN protocols is that they are typically very sensitive to transmission delay (latency) between the two SAN nodes. With GFP-F mappings, the one drawback to using the Core header with its PLI is that an entire client data frame must be buffered prior to transmitting the GFP-F frame in order to determine the required PLI value for that frame. In most applications, these issues are not a concern. For these 8B/10B-encoded clients, however, the control code transparency and latency issues can become important. A tribute to the versatility ("genericness") of the GFP protocol is that a variation of the protocol was developed that resolved these transparency and latency issues while providing a significant transmission bandwidth savings over what would be required to send a stream of the native 8B/10B characters. Due to its transparency to 8B/10B codes, this GFP mapping came to be called transparent GFP (GFP-T).

Table 4-3. List of client payloads mapped into GFP-T

GFP-T client payload type	UPI value <7:0> (PTI = 000)
Fibre Channel	0000 0011
FICON	0000 0100
ESCON	0000 0101
Gbit Ethernet	0000 0110
DVB-ASI	0000 1001
Asynchronous Fibre Channel (see note)	0000 1100
Note: Asynchronous Fibre Channel is asynchronous in the sense that the transport container is allowed to have less capacity than the Fibre Channel interface.	

The desires for transmission bandwidth efficiency and the transparent communication of the control codes were solved by transcoding the character stream into a new block code. The first step in the GFP-T mapping process is to decode the 8B/10B characters into their original 8-bit data values or the associated control characters. These data and control characters are then re-encoded into a 64B/65B code. Each 64B/65B code carries the information of eight of the 8B/10B characters. For data characters, the original 8-bit data values are mapped directly into the payload bytes of the 64B/65B codes. The 8B/10B control codes are re-encoded and mapped into their own bytes. The details of the 64B/65B code construction are illustrated in Figure 4-12.

The first bit of the 65B block indicates whether there were any 10B control codes among the 8 characters encoded into that 64B/65B code. If control codes are present, they are placed into the first byte(s) of the 64B/65B code payload. Since there are 12 or fewer control codes used for any 8B/10B client, a 4-bit code is adequate to represent them. A 3-bit address indicates where the control code occurred in the client character stream relative to the other characters encoded into that 64B/65B block. The MSB of a control code byte indicates whether this is the last control code in that 64B/65B block or whether the next byte also contains a control code. Figure 4-13 gives an example of the mapping of data and control characters (designated as D and K characters, respectively) into a 64B/65B code.

Input Client Characters	Flag Bit	64 bit (8 octet) payload field							
		Octet 0	Octet 1	Octet 2	Octet 3	Octet 4	Octet 5	Octet 6	Octet 7
All data	0	D1	D2	D3	D4	D5	D6	D7	D8
7 data, 1 control	1	0 aaa C1	D1	D2	D3	D4	D5	D6	D7
6 data, 2 control	1	1 aaa C1	0 bbb C2	D1	D2	D3	D4	D5	D6
5 data, 3 control	1	1 aaa C1	1 bbb C2	0 ccc C3	D1	D2	D3	D4	D5
4 data, 4 control	1	1 aaa C1	1 bbb C2	1 ccc C3	0 ddd C4	D1	D2	D3	D4
3 data, 5 control	1	1 aaa C1	1 bbb C2	1 ccc C3	1 ddd C4	0 eee C5	D1	D2	D3
2 data, 6 control	1	1 aaa C1	1 bbb C2	1 ccc C3	1 ddd C4	1 eee C5	0 fff C6	D1	D2
1 data, 7 control	1	1 aaa C1	1 bbb C2	1 ccc C3	1 ddd C4	1 eee C5	1 fff C6	0 ggg C7	D1
8 control	1	1 aaa C1	1 bbb C2	1 ccc C3	1 ddd C4	1 eee C5	1 fff C6	1 ggg C7	0 hhh C8

Legend:

- Leading bit in a control octet (LCC) = 1 if there are more control octets and = 0 if this payload octet contains the last control octet in that block

- aaa = 3-bit representation of the 1st control code's original position (1st Control Code Locator)

- bbb = 3-bit representation of the 2nd control code's original position (2nd Control Code Locator)

...

- hhh = 3-bit representation of the 8th control code's original position (8th Control Code Locator)

- Ci = 4-bit representation of the ith control code (Control Code Indicator)
- Di = 8-bit representation of the ith data value in order of transmission

Figure 4-12. 64B/65B block code structure

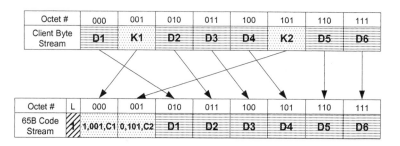

Figure 4-13. Example of mapping client data and control characters into a 64B/65B code

The 64B/65B codes have two drawbacks, however. The first is that the 65-bit code length would prevent a stream of 64B/65B characters from being byte-aligned with the SONET payload bytes. Byte alignment, which is a feature of all the other SONET/SDH payload mappings[57], has the advantages of allowing direct payload data byte observability within the SONET/SDH stream and of allowing the use of parallel data path implementations. Parallel data paths within devices and systems allow lower device/system clock rates. The lower clock rates in turn allow the use of more power-efficient technology such as CMOS. The second drawback of the 64B/65B code is its relative lack of error detection capability. (Error considerations are treated in greater detail below.) Both of these drawbacks were addressed by the combining of eight 64B/65B codes into a superblock with an appended error check. The superblock structure is illustrated in Figure 4-14. The payload data bytes of the eight constituent 64B/65B codes are placed into the superblock in transmission order, with the eight leading (flag) bits of these codes grouped together in a trailing byte. Two more trailing bytes are then added as a CRC-16 error check code over the information in that superblock.

---

[57] The one exception is the 10 Gbit/s Ethernet mapping discussed in Chapter 3. Since it uses a 64B/66B block code, it has no byte alignment with the SONET/SDH payload bytes.

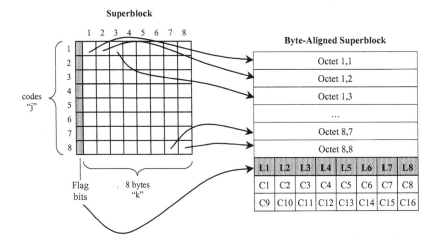

where: Octet j, k is the kth octet of the jth 64B/65B code in the superblock

Lj is the leading (Flag) bit j^{th} 64B/65B code in the superblock

Ci is the ith error control bit

Figure 4-14. GFP-T superblock structure

The final step in the GFP-T mapping is to insert an integer number of superblocks into a GFP frame. The process of going from the 64B/65B code to a GFP frame is illustrated in Figure 4-15. The GFP-T frame is identical to a GFP-F frame except for the specific PTI and UPI values that are used in the Type header. Since GFP-T has CRC-16 error checks per superblock, the payload FCS is typically not used. The number of superblocks that are grouped into a GFP-T frame is a function of the difference between the client data rate, the channel rate of the SONET/SDH channel, and the amount of "spare" bandwidth that is desired for client management frames. The minimum number of superblocks for various clients is shown in Table 4-4. The details for calculations regarding the number of superblocks and the resulting spare bandwidth are shown in an appendix of G.7041.

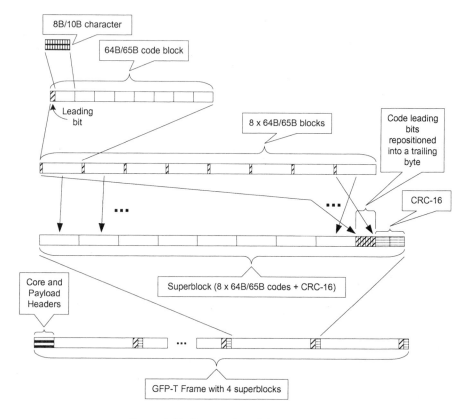

Figure 4-15. Construction of the GFP-T frame

Latency reduction was another consideration for GFP-T. Since GFP-T encapsulates a number of client data characters rather than a whole client data frame there is no implied correlation between the boundaries of the GFP-T frames and the client data frames. As a result, it was possible to pre-determine the GFP frame PLI value, thus eliminating the need to buffer an entire GFP frame prior to transmitting it. In order to make this work, however, a rate adaptation problem needed to be solved. The problem is that the SONET/SDH channel necessarily has a slightly higher bandwidth than is needed to carry the client data in the GFP-T stream. The client signal and the SONET/SDH VCAT channel rates required to carry the full client signal rate are shown in Table 4-4. The higher bandwidth of the transmission channel means that the ingress buffer at the GFP-T mapper will periodically underflow as the mapper extracts bytes at a somewhat higher rate than the client data stream supplies them. Unless additional buffering (with its associated latency) was added at the mapper, the underflow situation would be a problem if the GFP frame were already being transmitted. In order to avoid this additional latency, a special 64B/65B control character was defined as a 65B_PAD that is inserted into a 64B/65B block code whenever the mapper ingress buffer approaches an underflow threshold. As illustrated in Figure 4-16, the 65B_PAD character is inserted into the 64B/65B code in exactly the same manner as if an 8B/10B control code had been received in the client data stream. The GFP-T demapper recognizes this character as dummy padding and discards it.

Table 4-4. Virtually concatenated channel sizes for various transparent GFP clients

Client Signal	Nominal Native (Unencoded) Client Signal Bandwidth	Minimum Virtually-Concatenated Transport Channel Size	Nominal Transport Channel Band-width	Minimum Number of Super-blocks per GFP Frame	Worst/Best-case Residual Overhead Bandwidth[1]	Best case client management payload bandwidth[2]
ESCON	160 Mbit/s	STS-1-4v / VC-3-4v	193.536 Mbit/s	1	5.11 Mbit/s / 24.8 Mbit/s	6.76 Mbit/s
Fibre Channel	850 Mbit/s	STS-3c-6v / VC-4-6v	898.56 Mbit/s	13	412 Kbit/s / 85.82 Mbit/s	2.415 Mbit/s
Gbit Ethernet	1.0 Gbit/s	STS-3c-7v / VC-4-7v	1.04832 Gbit/s	95	281 Kbit/s / 1.138 Mbit/s	376.5 kbit/s

NOTES
1. The worst-case residual bandwidth occurs when the minimum number of superblocks is used per GFP frame. The best case occurs for the value of N that allows exactly one Client Management frame per GFP data frame. A 160-bit Client Management frame was assumed for the best case (with a CRC-32). For both cases, it was assumed that no Extension headers were used.
2. The best-case client management payload bandwidth assumes 8 "payload" bytes per Client Management frame and the best-case residual overhead bandwidth conditions.
3. Fibre Channel also supports rates of 425, 1700, and 3400 Mbit/s (unencoded). The SONET/SDH channel size scales linearly for these rates, and the minimum number of superblocks is 13 for all cases.

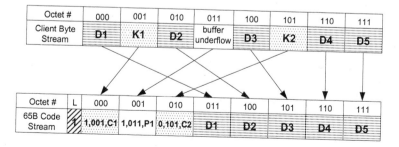

Figure 4-16. Example of 65B_PAD insertion

It is typically not cost effective, however, for a customer to subscribe to a full-rate channel through the WAN (see Table 4-4). There are often idle periods between data packets such that the actual average client data rate is a fraction of the full rate. In order to allow a sub-rate connection while still preserving transparency to 8B/10B control codes, Committee T11, working with ITU-T SG15, has included an asynchronous, sub-rate GFP-T option as part of their new FC-BB-3 standard. The idea, illustrated in Figure 4-17, is to have a circuit between the full-rate client signal interface and the GFP-T mapper. This circuit removes idle characters from the full-rate stream and outputs a sub-rate 8B/10B encoded stream to the GFP-T mapper. Similarly, a circuit inserts idle characters into the sub-rate stream output from the GFP-T demapper in order to create the full-rate egress stream. The sub-rate would correspond to the desired WAN channel rate (e.g., DS3 or STS-3). Due to the need to buffer client packets for the process of removing the idle characters, there is an increased latency with sub-rate GFP-T over that of full-rate GFP-T. The latency of sub-rate GFP-T is still typically less than that required for a GFP-F mapping.

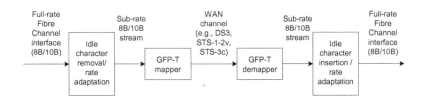

Figure 4-17. Sub-rate GFP-T example for Fibre Channel

4.3.5   Performance Considerations and Client Management Frames

At the time of writing for this book, the OAM aspects of GFP are being defined. The GFP demapper can check such parameters as errors detected in any of the GFP headers, a mismatch from the expected values in the payload Type (or Extension) header, errors in the payload field (if the optional GFP payload FCS is used), and loss of GFP frame synchronization. These faults are reported to the network management system by the NE containing the GFP demapper. Some carriers have indicated a desire to perform single-ended performance monitoring similar to SONET/SDH paths/trails. (I.e., have the far end report its performance parameters so that each end knows both the parameters that it calculates directly and the parameters calculated by the far end.) If this capability is adopted for GFP, the demapper reports will be sent in GFP Client Management Frames (CMFs). CMFs use a different PTI value and their own set of UPI values, but are otherwise identical to the GFP client data frames. The one application that is currently standardized for CMFs is the client signal fail indication that is discussed in the next section.

4.3.6 Error handling considerations

Transmission errors can occur on the client data stream either outside or within the SONET/SDH network. GFP-F and GFP-T handle these errors and faults differently due to the manner in which each performs its encapsulation process.

Client frame delimiting is part of the GFP-F mapping process, and a typical part of frame delimiting is checking the client data frame's FCS. If this check is bad, the GFP mapper discards that client data frame rather than encapsulating it. Similarly, if the GFP-F demapper receives a GFP frame with a bad check sequence in the any of the GFP headers, the GFP payload FCS (if it is being used), or the client data frame itself, that client data frame is discarded.

GFP-T focuses on the character level rather than the frame level. Since a limited number of 10-bit values are used for 8B/10B codes, and since the running disparity is updated with each character, transmission errors can typically be detected in the 8B/10B stream. When the GFP-T mapper detects either an illegal 8B/10B character or a running disparity error, it encodes the offending character as a special 10B_ERR control character in the 64B/65B code. When the GFP-T demapper decodes the 64B/65B code it encodes the 10B_ERR code into a particular illegal 8B/10B character so that the client receiver will be aware of the error.

Most of the redundant information that makes error detection possible is removed when the 8B/10B characters are transcoded into the GFP-T 64B/65B codes. The 64B/65B codes are particular vulnerable to errors in the leading (flag) bit or the last control code indicator bits if control codes are present. Errors in these particular bits can cause the GFP-T demapper to misinterpret multiple data/control characters, which could in turn create a burst error condition that is beyond the guaranteed detection ability of the client FCS (typically a CRC-32). The CRC-16 code that was added to each superblock provides a strong triple error detection or (optional) single error correction capability with double error detection. When the GFP-T demapper sees a bad CRC-16 for a superblock, it treats all 64 client characters in that superblock as if they had been received as 10B_ERR codes. (Single errors can be simply corrected if that option is used.)

NOTE – The self-synchronous scrambler that is used on the GFP payload field inherently multiplies the transmission errors by creating a duplicate error 43 bits after the original error. This error multiplication would weaken the performance of each of the existing standard CRC-16 codes. As a result, a new CRC-16 was developed for the GFP-T superblock that preserved its error detection and correction capability in the presence of the descrambler, and was optimized for this specific block length. The interested reader can see [3] or [40] for more details.

In the case of a client signal failure prior to the GFP mapper, both the GFP-F and GFP-T mappers send a periodic client signal fail (CSF) frame to the GFP demapper. This signal failure could be the complete loss of the incoming signal or the loss of character synchronization for those clients using a character-oriented line code. The CSF message is sent in a Client Management Frame every 100-1000ms. The action taken by a GFP demapper is client specific. In the case of GFP-T, the output 8B/10B stream will contain the same characters used to indicate a bad received 8B/10B character at the mapper. (Note that some client signals have other options that could be used in the situation.)

4.4.   RESILIENT PACKET RING (RPR)

There has been growing interest in RPR as a technology to extend the capability of SONET/SDH access and metro networks for packet data access. The RPR protocol is currently under development in IEEE 802.17 with the initial version of the standard recently published. The essential concepts of RPR will be described in this section. A set of definitions for some of the abbreviations used in RPR is provided in Table 4-5.

There were two basic motivations behind RPR. From the data communications side, there was a desire for a broadband MAN/WAN protocol that could utilize rings in order to provide higher reliability by having two directions available for routing traffic, and to bring some of the protection switching and OAM facilities of the telecommunications world to a data communications architecture. From the telecommunications side, there was a desire to extend the concept of a TDM add/drop multiplexer to the packet world. Specifically the telephone operating companies wanted to extend their SONET infrastructure with data access NEs that could be provisioned as part of their SONET network. Ideally, the packet add/drop could be accomplished through upgrades to their existing SONET ring equipment.

Table 4-5. Some RPR terminology abbreviations

CIR	Committed Information Rate	MTU	Maximum Transmission Unit
EIR	Excess Information Rate	OAM	Operations, Administration, and Maintenance
FRTT	Fairness Round Trip Time	PDU	Protocol Data Unit
FS	Forced Switdh	PRS	packetPHY Reconciliation Sublayer
GRS	GFP Reconciliation Sublayer	PTQ	Primary Transit Queue
HEC	Header Error Check	RRTT	Ring Round Trip Time
LAPS	Link Access Procedure - SDH	RS	Reconciliation Sublayer
LLC	Logical Link Control	SD	Signal Degrade
LAPS	Link Access Procedure - SDH	SF	Signal Fail
LLC	Logical Link Control	SDU	Service Data Unit
LOC	Loss of Connectivity failure	SRS	SONET/SDH Reconciliation Sublayer
LRTT	Loop Round Trip Time	STQ	Secondary Transit Queue
MAC	Medium Access Control	WAN	Wide Area Network
MAN	Metropolitan Area Network	WTR	Wait-to-Restore
MS	Manual Switch		

The resulting protocol is essentially a 2-fiber ring-topology network consisting of packet add/drop NEs (RPR stations). The data is transmitted in the opposite on each of the two fibers, which allows the ring to automatically self-heal from any fiber-related failure. Such a ring, known as a counter-rotating ring, is illustrated in Figure 4-18. Packets can be sent between nodes in either direction around the ring based on availability of an insertion opportunity. A fairness protocol is used to ensure that upstream stations do not starve downstream stations of insertion opportunities. SONET/SDH is typically used for the fiber links themselves, although it is possible to use optical Ethernet-like links instead of SONET/SDH links.[58]

---

[58] RPR uses just the PHY layer of the optical Ethernet interfaces to transport RPR MAC packets rather than the complete Ethernet interface (i.e., MAC, management, etc.). To distinguish it from the full Ethernet interface, the 802.17 group created the new term "packetPHY" to designate these interfaces.

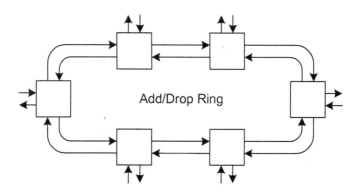

Figure 4-18. Illustration of a counter-rotating add/drop ring

4.4.1   Media access

The technique that is used for adding and dropping packets from the ring (i.e., the media access control (MAC) protocol) is known as buffer insertion or register insertion. The basic concept is illustrated in Figure 4-19(a). As a packet enters a node from the ring, it is placed bit-by-bit into a transit buffer (queue). The buffer fills until the destination address for that packet is available, at which point the node decides whether to extract the packet (if the station is that packet's destination) or to forward the packet. The packet is forwarded by shifting it back out onto the ring. If a node must buffer the entire frame before shifting it out, it is referred to as store-and-forward transit operation. If the node is able to begin shifting out the frame after the frame's header has been buffered, it is referred to as cut-through transit operation. RPR allows both types. When an RPR node has a data packet to insert onto the ring, it must wait until has enough space in its transit buffer to hold at least one packet. It can then stop the flow of data out of its transit buffer at the end of transmitting a frame and begin shifting its own packets onto the ring. (The RPR approach is an enhancement to the simple buffer-insertion rings in which a node is required to wait for an empty transit buffer before it is allowed to insert its data.) A packet that arrives on the ring while the station is sending its own packet is buffered in the transit buffer and sent as soon as the node's own packet is sent. Buffer insertion rings are inherently quite efficient and reasonably fair in that a node with lots of data to send can continue to use the ring when no other nodes have packets to send. When other nodes are also inserting packets, a sending node will see these packets appear in its transit buffer and will have to wait before it can transmit another packet. Care must be taken, however, to insure that an upstream node doesn't consume too much bandwidth if there are downstream nodes with data to send. This bandwidth-hogging problem is addressed by the RPR distributed fairness algorithm.

The RPR protocol enhances the basic insertion buffer idea with three levels of packet transmission priority and an associated enhanced fairness protocol. A node that uses a single transit buffer, as shown in Figure 4-19(a), can only insert packets onto the ring when its transit buffer is empty (or empty enough to store a packet arriving from the ring). In order to support different packet priority levels, an RPR station can also be implemented with two transit buffers, the Primary Transit Queue (PTQ) and Secondary Transit Queue (STQ). (See Figure 4-19(b).) The ways in which the different priority packets use these buffers is explained below in the discussion of packet priorities.

The destination node for a frame removes that frame from the ring. Once this frame is removed, its bandwidth is available for that station or downstream stations to use. This approach is known as spatial re-use, since the same effective bandwidth can be used by different frames on different portions of the ring. In the event that no node recognizes the destination address as being its own, a Time To Live (TTL) approach is used to remove the packet from the ring so that it doesn't circulate forever[59]. The TTL field is set to its initial (Base) value by the packet's source node. The TTL field is then decremented by each node the packet passes through. When a node sees the TTL field reach zero, it discards the packet from the ring. The source node also sets the TTL Base field in the packet to be equal to the initial TTL value. Each node can then determine the number of hops traveled by that packet by comparing the TTL and TTL Base field values.

In addition to the normal unicast transmission of frames, which allows for the spatial re-use, RPR also allows multicast and broadcast frames. These frames propagate either all the way around the ring in one direction, or halfway around the ring in both directions.

---

[59] At the time this book was published, IEEE 802.17 was addressing the details of sending Ethernet frames over the RPR between addresses that were either associated with nodes that are off the ring. When the destination address is off the ring, the RPR source node must flood the frame onto the ring. IEEE 802.17b will specify that TTL be used to insure the removal of the frame from the ring after it is forwarded by the station that can bridge to that destination address.

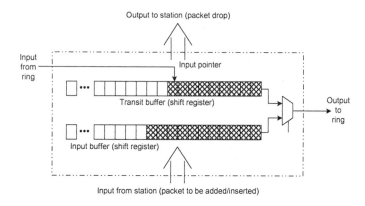

a) Basic single queue

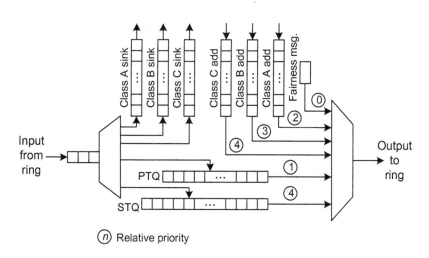

b) Dual priority queue with three input priority queues

Figure 4-19. Illustration of buffer insertion

4.4.1.1 Traffic classes

The different RPR traffic classes are summarized in Table 4-6. The ring bandwidth for Class A0 is reserved. If, for some reason, an RPR node (station) has does not have enough data to fill its entire Class A0 bandwidth, that unused bandwidth cannot be reclaimed for other traffic classes and would thus be wasted. Class A1 and B-CIR traffic bandwidth is pre-allocated; however if that bandwidth is not used, it can be reclaimed by other traffic. Class A1, B-CIR, and B-EIR can also make use of ring bandwidth that is not pre-allocated, subject to the fairness algorithm.

Table 4-6. RRP traffic classes

Class	Sub-class	Description
A		Low latency and low jitter for real-time service, non-Fairness Eligible (non-FE)
	A0	• Reserved, pre-allocated bandwidth
	A1	• Pre-allocated, reclaimable bandwidth
B		Predictable latency and jitter for near real-time service
	B-CIR	• Committee Information Rate • –Non FE, Pre-allocated, reclaimable bandwidth • Bounded jitter
	B-EIR	• Excessive Information Rate • Opportunistic, FE bandwidth use • Unbounded jitter
C		• Best effort service • Unbounded jitter • Opportunistic, FE, reclaimable bandwidth

In a dual-transit queue node, Class A frames are queued in the PTQ and Class B and C frames are queued in the STQ. (See Figure 4-19(b).) Frames in the PTQ have the highest priority for being inserted onto the ring. The next highest priority is Class A frames that a node needs to add to the ring. The result is that Class A frames are added to the ring with priority over all other classes of add or transit frames, and once added to the ring experience a minimum of delay. Class B frames can be added to the ring when there are no frames queued in the PTQ and can be added with priority over Class B and C transit frames as long as the STQ is reasonably empty. This allows Class B-CIR frames to be added with low latency and jitter. Class C frames can be added when there are no other frames in the PTQ, STQ, or other add queues. Since Class B and C frames share the STQ, they experience similar delay around the ring.

An important feature of RPR is that preemption of lower priority frames is not allowed. The MAC protocol guarantees this through the relative priority of add and transit packets, however it requires an adequately large STQ to prevent it from overflowing during the transmission of higher priority frames. This is done to eliminate the possibility of packet loss on the ring; once a packet is injected on the ring, it is guaranteed to be delivered to its destination.

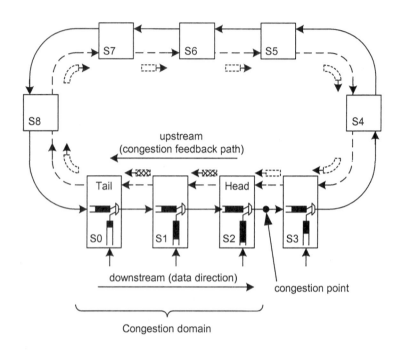

non-full rate (fair rate request) fairness message
full-rate fairness message

Figure 4-20. Illustration of an RPR ring with a congestion condition

4.4.1.2 RPR fairness algorithm

RPR does not discard packets in order to resolve congestion, regardless of the packet's priority. Instead RPR relies on ingress shaping and a distributed fairness protocol as part of the MAC function. An RPR node performs shaping and smoothing of the incoming traffic prior to inserting it into the insertion buffers in order to prevent over-committing the ring's bandwidth. The shaping and smoothing parameters are agreed to apriori with the customer. When the amount of data that the nodes have to transmit on the ring exceeds the bandwidth of the ring, the RPR fairness algorithm works to distribute the available bandwidth fairly among the ring nodes. Since Class A0 bandwidth is reserved, it is always transmitted and unaffected by the fairness algorithm. The bandwidth eligible for the fairness algorithm (i.e., the FE bandwidth) is that bandwidth not reserved for Class A0, and not used by the pre-allocated Class A1 and B-CIR data.

Since a node can only insert FE data onto the ring when the transit queue is (sufficiently) empty, an upstream node can potentially starve downstream nodes by continually inserting packets. The basic concept of the distributed fairness algorithm is that when a downstream node detects a congestion condition that is preventing it from sending data, it sends a notification (fairness request) to the upstream nodes to request bandwidth. The upstream nodes are then required to reduce their injected traffic, in order to provide the requested bandwidth. This approach allows much more dynamic bandwidth sharing than approaches that re-assign ring bandwidth on either a permanent or master/slave request/grant basis.

A congestion condition and the associated fairness signaling are illustrated in Figure 4-20. If a node uses a single transit queue, it detects congestion on an outgoing link when it either detects an excessive transmission rate relative to the transmission link capacity, or when it takes a long time for it to be able to add its packets to the ring. If the ring uses a dual queue, it detects congestion when its secondary transit queue becomes too full, thus causing long delays in forwarding transit frames. The node detecting the congestion first calculates its estimate of what the "fair" rate should be, and then sends this value upstream (i.e., on the fiber in the opposite direction of the of the congested link) in a fairness message. This node is referred to as the "head" of the congestion domain, and is represented by node S2 in Figure 4-20. The fair rate estimation methods are discussed below. (Fairness messages are sent by each node at preset time intervals. If a node is not detecting congestion, it sends the default value that allows upstream nodes to send data at the full link rate, i.e., a full-rate fairness message.) When the first node upstream from the congested node receives the fairness message (node S1 in Figure 4-20), it decreases its add traffic in order to provide the requested fair bandwidth rate to the congested node. If the transmit traffic from the upstream nodes is great enough that this node still can't provide the requested fair rate, this node propagates the fairness request message upstream to the next node. This process continues until a node is reached that can either provide the fair rate requested by the head end congested node, or until a node is reached that is head end of another congestion domain. This node is called the tail of the congestion domain (node S0 in Figure 4-20). The congestion domain, then, is the congested node and the set of upstream nodes that acting on its fairness request messages. Of course, if a node's response to a fairness message causes it to experience congestion (i.e., excessive time to add its own packets) it will detect the congestion and send its own fairness message upstream, thus becoming the tail of the one congestion domain and the head of a new one.

There are two methods for estimating the fair rate: the Aggressive method, which attempts a rapid response to the congestion, and the Conservative approach, which attempts to respond more carefully.

The Aggressive method estimates the fair rate as the rate at which it has recently been able to add data to the ring. Due to the congestion, this rate is less than the ultimate desired rate for the station, but it serves as a starting point for an iterative approach. If the bandwidth made available by the upstream nodes alleviates the congestion, the previously congested node will begin transmitting the default value in its fairness messages. If the resulting increase in traffic from the upstream nodes renews the congestion condition, the temporary relief will mean that the congested node's recent add rate is higher than it was when it sent its prior congestion message. These iterations continue until the fair rate estimate of the congested node is closer to the real fair rate and it is able to catch up on adding its traffic to the ring. Clearly this approach responds quickly at the expense of some oscillation in the traffic adding patterns among nodes.

The Conservative method can either estimate the fair rate as (a) dividing the total available FE bandwidth by the number of upstream stations that are currently sending frames through that node, or (b) using that node's own add rate. In order to provide stability, a node will continue to send this fair rate estimation until the upstream nodes have the opportunity to respond to its original fairness message. This response time is referred to as the Fairness Round Trip Time (FRTT), and is checked with a timer. The FRTT is calculated as follows. A Loop Round Trip Time (LRTT) is calculated when the ring it initialized (or its topology changes) as the time required for a Class A frame to travel from a congested node to the node at the other end of the congestion domain and back (i.e., it passes through the congestion domain). The Fairness Differential Delay (FDD) is the LRTT minus the time required for a Class C frame to make the same round trip. The congested node computes the FDD by sending the Class C FDD frame when it detects the congestion. (The LRTT, and FDD are special types of control frames.) By using the LRTT timer, the Conservative method can observe the effects of its fairness requests before making further fair rate decisions. This results in fewer adjustments than the Aggressive method. The Conservative method is a better fit for nodes with a single transit queue.

The details of the fair rate computations and traffic shaping are too complex to be presented here, but are specified in the 802.17 standard.

Each node has a local weight for the traffic shaper associated with the amount of FE traffic that it is allowed to add to the ring during congestion conditions. The weight can be equal among the nodes or unequal with an unequal rate allowing some nodes to send a higher percentage of their add traffic than other nodes during a congestion condition. The purpose of the local weighting is to prevent a node from unfairly throttling downstream nodes by using a disproportionate amount of the rings capacity. The traffic shaper uses the fairness frames that it receives in order to determine its insertion rate for fairness eligible traffic. The basic fairness frame is referred to as a single-choke fairness frame and carries the fair rate request of the downstream node that is being affected the most by the transmissions from the node receiving the fairness frame. Another option is called a multi-choke fairness frame and is broadcast periodically by all stations. The stations can then use this information as part of the fairness processing.

4.4.2 RPR physical layer

The types of physical layers supported by RPR for the transport of RPR MAC frames are summarized in Table 4-7. In the case of SONET/SDH interfaces, the RPR MAC frames are encapsulated prior to insertion in the SONET payload (SDH container) area. The reconciliation sublayers for the packetPHY[60] interfaces are specified in 802.17.

Table 4-7. Summary of RPR physical layer interfaces

Physical Interface	Options
SONET/SDH	(For any SONET/SDH rate) • GFP encapsulation (G.7041) • PoS-like encapsulation (RFC 2615 &1662) • LAPS encapsulation (X.85)
packetPHY	• 1 Gbit/s over optical Ethernet PHY • 10 Gbit/s over optical Ethernet PHY

---

[60] Rather than defining its own physical layer interfaces, RPR refers to other standards. As noted above, 802.17 uses the term "packetPHY" to designate those RPR MAC transport interfaces that use just the PHY layer of the corresponding optical Ethernet interface.

Table 4-8. Summary of RPR frame fields

Field	Definition	Function discussed in section:
baseControl	Control byte containing ri, fe, ft, we, and p fields	-
control Data unit	Control information carried in a RPR control frame	-
control Type	Type of control frame	4.4.3.2, 4.4.1.2
control Version	Version of control protocol being used	4.4.3.2
da	Client RPR station MAC destination address	-
daExtended	Destination address extension (used only with extended frames)	4.4.3.1
ef	Indicator of extended frame	4.4.3.1
extended Control	Control byte containing ef, fl, ps, and so fields	-
fairness Header	Header information specific to a RPR fairness frame	4.4.3.3
fair Rate	The fair rate advertised by the originating node to upstream nodes	4.4.1.2
fcs	Frame check sequence (CRC-32)	-
fe	Indicates a frame is fairness eligible	4.4.1.2
fi	Flooding indicator	4.4.5.2
ft	Frame type indicator	Table 4-9
hec	Header error check (CRC-16 over preceding header bytes)	-
p	(odd) parity over the base Control fields	-
protocol Type	Indicates what protocol the service Data unit is	4.4.3.1
ps	passed source (used with wrap protection)	Chapter 8
ri	Ringlet indicator (i.e., the ring direction a frame was originally transmitted)	-
sa	Local RPR station MAC source address	-
saCompact	Source address field in a fairness frame (given a different name due to having a different location in the header than other frames)	-
saExtended	Source address extension (used only with extended frames)	4.4.3.1
sc	Service class indicator	Table 4-9
service Data unit	Client data packet carried in an RPR data frame	4.4.3.1
so	Indicates whether a frame requires strict ordering	-
ttl	Time to Live counter	4.4.1, 4.4.5.2
ttlBase	Base value for the ttl count	4.4.1
we	Indicates whether a frame can be wrapped during a wrap condition. (This bit is ignored by steering stations.)	Chapter 8

4.4.3   RPR packet formats

There are four categories of RPR frames, each with their own format:

- data frame
- control frame
- fairness frame
- idle frame

Table 4-8 provides a summary of the frame fields used in the different frames along with pointers to the sections within this chapter where the functions of those fields are discussed.

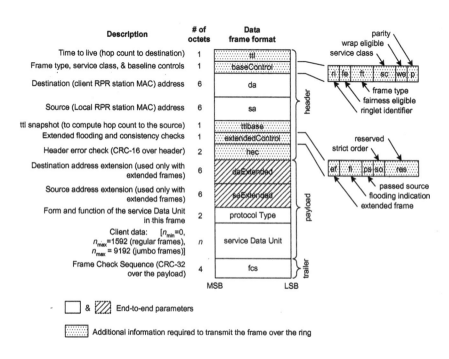

Figure 4-21. RPR Data frame format

4.4.3.1 RPR data frame

The data frame is illustrated in Figure 4-21, with the fields summarized within the figure. Extended data frames, which are indicated by the ef field, include the extended source and destination address fields, while the basic data frames do not. The extended data frames are used when the MAC source or destination node is not directly attached to the RPR ring. As explained in the bridging and routing section, the extended SA/DA fields are used to carry the MAC address of the source/destination node that is off the RPR ring. The protocol type field is essentially the same as the equivalent Ethernet frame field with values less than 1536 indicating the frame length, and values greater than or equal to 1536 using the IEEE Type Field Register values to indicate the MAC client protocol. Table 4-9 shows the definitions for some of the base Control octet fields.

The maximum service Data unit size for a regular frame is 1500 octets. A maximum service Data unit size of 9100 octets can be negotiated in order to support jumbo frames. For both maximum sizes, up to 92 octets can be reserved in addition to the client data.

Table 4-9. Definitions for the ft and sc baseControl fields

ft		sc	
values	MEANING	values	MEANING
00	Idle frame	00	Class C
01	Control frame	01	Class B
10	Fairness frame	10	Class A, subclass A1
11	Data frame	11	Class A, subclass A0

4.4.3.2 Control frames

The RPR control frame format is shown in Figure 4-22. The functions of the control frames include station attribute discovery, topology and protection, link round trip time measurement, fairness differential delay determination, and various OAM functions. A version field indicates the version number corresponding to the control type field. The control data unit can range between 24 and 1616 bytes.

4.4.3.3 Fairness frames

Fairness frames, which are illustrated in Figure 4-22, are sent to upstream stations to communicate the fair rate that a station can accept from its upstream nodes. A station will indicate full-rate in the fairness message if it is not experiencing congestion. When congestion occurs, it sends its estimate of the fair rate that it needs to see from the upstream nodes in order to resolve the congestion condition. The fairness frame can either be sent to a station's immediate upstream neighbor, or broadcast to the entire ring. A fairness frame is defined to be 16 bytes in order to minimize its jitter and bandwidth impacts. The fairness header is a 16-bit field with the three

MSBs indicating whether single choke (000) or multi-choke (001) fairness is being used. The remaining MSB values and remaining bits are reserved. Note that a fairness frame is only eligible to be wrapped by a wrapping station and not by a steering station during protection conditions.

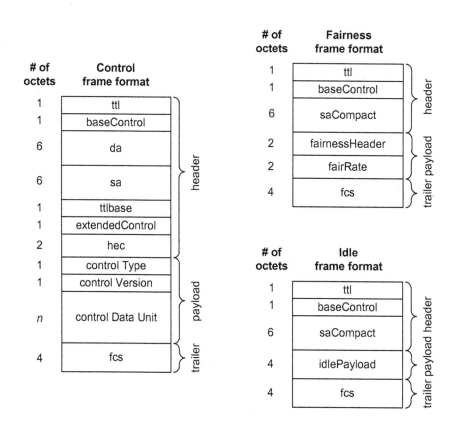

Figure 4-22. Other RPR frame formats

4.4.3.4 Idle frame

Idle frames, as shown in Figure 4-22, are 16-byte frames sent to neighboring nodes. Although they can be used on rings where all nodes are synchronized to the same clock source, their primary purpose is to provide transmission rate synchronization between nodes that operate asynchronously from each other (i.e., rely on locally-generated transmit clocks rather than a global ring clock)[61]. This synchronization prevents a node using a faster local transmit clock from causing a PTQ overflow in a downstream node with a slower local transmit clock. It is accomplished by having each node provide Idle frames that can be discarded by a slower downstream neighbor node. If Idle frames are supported, they are generated at a nominal rate of 0.05% line rate. If a node's primary transit queue fills beyond a threshold value, the node will begin sending the Idle frames at 0.025% of the line rate until the queue fill is sufficiently reduced. This function is referred to as an idle shaper. Note that Idle frames are transmitted hop-by-hop rather than to a specific destination address.

4.4.4 RPR protection switching

RPR supports two protection modes referred to as "wrapping" and "steering" modes. The discussion of RPR protection is left to Chapter 8, which is devoted to subject of survivability.

4.4.5 Topology, bridging, and routing

One of the goals of RPR was the type of 'plug-and-play' operation that has made Ethernet so popular. This capability applies primarily to the ring itself, but also to the ring's connections to other networks. The interconnection can either be made at Layer 2 with the RPR node acting as a bridge to other IEEE 802 networks, or at Layer 3 (e.g., using IP routers).

4.4.5.1 Automatic ring topology discovery

RPR can seamlessly accommodate changes to the topology of the ring such as the addition removal of a node, changes in the capabilities of the nodes, changes in the status of links, and the existence of a protection state. In order to accomplish this, each node maintains a database that is update through messages that are either sent periodically or when a node detects a change in the ring. The database includes the topology of the ring (the addresses of the ring nodes, their relative position around the ring, the propagation delay between each station, and the propagation delay around the ring); the attributes of each node; and the protection status (whether a

---

[61] All ring nodes nominally use the same transmit clock rate, but these clocks will differ slightly if they are locally generated. For SONET nodes, the clocks difference will typically be in the ±4.6 ppm range, but RPR allows for a clock difference of ±100 ppm in order to accommodate the Ethernet PHY clock tolerance.

signal degrade or signal failure exists on each link). This database allows the dynamic insertion and removal of nodes in the ring.

4.4.5.2 Bridging and Routing

Any RPR node can serve as a bridge to other IEEE 802 networks, including Ethernet or other RPR networks. The RPR network appears as a single, broadcast-enabled LAN segment to the spanning tree protocol, which allows spanning trees to work across the RPR network. Unknown unicast frames are flooded (broadcast) to all LAN segments in order to ensure that the frame can be delivered. The flooding can either be unidirectional or bi-directional, with the flooded frame removed from the ring either by the source or when its TTL count expires.

RPR behaves like any other Layer 2 technology as far as allowing Layer 3 routers to work on top of it, using RPR to interconnect the routers. Since RPR allows different quality of service classes, it would be possible for the Layer 3 protocols to make use of them to provide different quality of service guarantees.

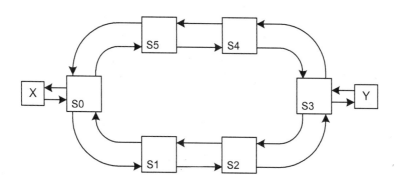

Figure 4-23. RPR bridging / routing example

When a packet's destination is off the ring (i.e., not an RPR station), the initial version of the RPR standard only allows for reaching that destination by broadcasting (flooding) the RPR frame onto the ring. Since flooding does not allow spatial reuse, it is desirable for the RPR source node to be able learn which RPR node is connected to that destination address so that it can send those frames directly to that RPR node. (This is referred to as spatial awareness.) IEEE 802.17b, which was under development at the time this book went to press, is addressing this situation. The basic learning mechanism works as follows. Consider the example network of Figure 4-23.

When RPR node S1 needs to send a frame to destination Y, which is not on the ring, it floods the RPR frame to all nodes using an extended frame with the address of Y in the extended DA field. RPR node S3 forwards the packet to Y. When Y sends a packet back to S1, S3 will send the packet in an extended frame with its own address in the SA field and Y as the extended SA. When S1 receives this RPR frame, it will update its forwarding database. The next time S1 needs to send a packet to Y, it will send an extended RPR frame with the DA being S3 and the extended DA being Y.

4.4.6   RPR Conclusions

The carriers expressing the most early interest in RPR see it primarily as a technology for the access or metropolitan portion of their network. Most don't see RPR as scaling well to fit the needs of their mesh-connected core networks. One of the key advantages to RPR in the access/metro applications is that the RPR MAC protocol allows them to provide a relatively simple fair mechanism for variable bandwidth demands from their customers. The statistical multiplexing gains increase the efficiency of the fiber facility utilization and it also allows customers the possibility of higher burst bandwidth rates during times when the ring traffic is light. Another key attraction of RPR is that RPR technology enables carriers to aggregate large numbers of relatively low-volume best-effort customers in a metro area without incurring the substantial provisioning and network management costs that would result if each individual customer received a provisioned virtual circuit of some type. The carriers can then rely on traffic engineering to ensure some minimal level of service guarantee and rely on the RPR fairness protocol to ensure enforcement during high-load conditions.

4.5.   OPTICAL TRANSPORT NETWORKS (OTN)

During the "telecom bubble" era, there were high hopes and speculation that all-optical networks would quickly become prevalent. There were dreams of relatively simple backbone networks where client signals were optically (wave division) multiplexed and switched without the optical network elements having to do any electrical (and hence client signal dependent) processing of the client signals. In many ways, it appeared to be the ultimate integrated network. Realistically, however, it will be several years before there are significant deployments of the ITU-T standard OTNs discussed in this section. The primary early deployments are in Japan and among some of the European carriers, with relatively little interest among

North American carriers.[62] Three factors contribute to this slower growth. First, as discussed frequently in this chapter, carriers have huge capital investments in their existing networks and lack money to replace or over-build them with a new type of OTN and its associated new network management systems. Second, a number of SONET/SDH-based proprietary wave division multiplexed (WDM) solutions have already been developed that, while not ideal, are adequately serving the needs of many carriers. In fact, the ITU-T Rec. G.709 standard discussed in this section is very similar to SONET/SDH in many ways. Third, the combination of the existing WDM equipment and the large amount of fiber currently deployed in the backbone networks mean that carriers have little need in the short-to-medium-term for the large bandwidth increases provided by OTNs. In fact, the most compelling reason to deploy the new OTNs is for point-to-point links where the enhanced forward error correction capability allows longer spans or higher data rates. As a result, this section will only provide an overview of OTN without going into the level of detail provided for SONET/SDH in Chapter 3. Note that the physical layer discussion in section 4.5.2 is equally applicable to SONET/SDH-based WDM systems as to the OTN-based systems. The terminology for the OTN standards is summarized in Table 4-10. The OAM&P channel terminology is defined in section 4.5.5.

4.5.1 Background

As optical component technology improved, it became clear that the best way to increase the traffic sent over a fiber was to send multiple signals, each on its own wavelength rather than increasing the rate of a single signal (e.g., sending 16 OC-48 signals, each on their own wavelength rather than a single OC-768). Such multiplexing is referred to as wave-division multiplexing (WDM). WDM also held the promise of being able to send each signal in its native format rather than mapping it into the payload of another signal such as SONET/SDH. It is difficult, however, for a network operator to provide OAM&P for each signal if it uses its native signal format since this would require multiple, client signal dependent management systems. This problem is especially true for analog signals (e.g., TV channels), which have a very different set of channel requirements than digital signals. More will be said on this topic in the discussion of the OTN signal architecture.

---

[62] It is, however, becoming common for carriers to request OTN support or upgradeability in new WDM NEs.

Table 4-10. OTN terminology

3R	Reamplification, Reshaping and Retiming	ONNI	Optical Network Node Interface
CM	Connection Monitoring	OOS	OTM Overhead Signal
IaDI	Intra-Domain Interface	OPS	Optical Physical Section
IrDI	Inter-Domain Interface	OPU	Optical Channel Payload Unit
JOH	Justification Overhead	OPUk	Optical Channel Payload Unit-k
MFAS	MultiFrame Alignment Signal	OPUk-Xv	X virtually concatenated OPUk's
MSI	Multiplex Structure Identifier	OSC	Optical Supervisory Channel
naOH	non-associated overhead	OTH	Optical Transport Hierarchy
OADM	Optical Add-Drop Multiplexer	OTM	Optical Transport Module
OCC	Optical Channel Carrier	OTS	Optical Transmission Section
OCh	Optical channel (with full functionality)	OTU	Optical Channel Transport Unit
OCh	Optical channel with full functionality	OTUk	completely standardized Optical Channel Transport Unit-k
OCI	Open Connection Indication	OTUkV	functionally standardized Optical Channel Transport Unit-k
ODU	Optical Channel Data Unit	OXC	Optical cross-connect equipment
ODUk	Optical Channel Data Unit-k	PSI	Payload Structure Identifier
ODTUjk	Optical channel Data Tributary Unit j into k	PT	Payload Type
ODUk-Xv	X virtually concatenated ODUk's	TC	Tandem Connection
OH	Overhead	TCM	Tandem Connection Monitoring
OMS	Optical Multiplex Section	TxTI	Transmitted Trace Identifier
ODTUG	Optical channel Data Tributary Unit Group	vcPT	virtual concatenated Payload Type
OMU	Optical Multiplex Unit		

One reason for developing a new signal format for WDM signals (instead of just using the existing SONET/SDH signals) was the possibility to add new overhead channels that would give the added functionality required to efficiently perform OAM on the WDM network. Another reason for developing a new standard was to provide a means for more powerful forward error correcting (FEC) capability. As discussed in Chapter 3, a relatively modest FEC capability was added to SONET/SDH. As signals are combined with WDM, however, the signal to noise ratio is decreased. Since the carriers were hoping to increase the transmission distances and the bit rates per wavelength, the SONET/SDH FEC is not always adequate. In order to better support the multiplexing of a substantial number of signals onto a single fiber, the ITU-T developed a set of new standards covering the wavelengths and signal formats. These signal format and hierarchy standards cover digital signals and include the OAM&P overhead as part of the signal format. In the context of this chapter, OTN refers to networks using the ITU-T Rec. G.709 standard for WDM signals.

4.5.2 Physical layer

A full discussion of lasers, receivers, and the characterization of fiber optic channels is beyond the scope of this chapter. The interested reader can find more detail in books such as [Ka00]. This section will only introduce some of the basic physical layer concepts so that the reader can appreciate some of the decisions that were made in defining the OTN and its signal formats.

4.5.2.1 Introduction to WDM

Glass fibers are not fully transparent to light, and in fact typically have three wavelength windows where the light attenuation is lowest. The first is in the 820-900 nm wavelength region, which is easily generated by inexpensive silicon lasers, but does not allow single-mode transmission[63]. The next window is 1280-1350 nm, which has substantially lower loss than the 820-900 nm region and supports single mode transmission. The third window is the 1528-1561 nm region, and has the lowest attenuation, but also requires lasers that are more expensive than those for the other two regions.[64]

---

[63] Light propagates through a fiber by the process of total internal refraction in which the light going through the core of the fiber is refracted back into the core when it hits the cladding that surrounds the core. (The cladding has a lower index of refraction than the core.) In multi-mode transmission, the light "bounces" through the fiber as it encounters the cladding in such a manner that the portion of a light pulse that encounter the fewest bounces has a shorter path than the one that has the most bounces. The result is a time spreading of the pulse at the receiver that limits possible spacing between pulses (i.e., the possible data rate for a digital signal). In single mode transmission, only the light that goes directly through the core is able to propagate, thus minimizing any pulse spreading.

[64] This section will follow a common practice of referring to the regions by their center wavelength, i.e., 1310 and 1555 nm.

An ideal laser would output light with a single wavelength, which would appear a single line on a spectral graph. In practice, however, physical realities mean that a laser outputs light with some spread around its central wavelength. Since this spread looks like a fatter line on a spectral graph, a laser's wavelength spread is often referred to as its line-width. Clearly, the line-width of the lasers determines how many lasers' signals can be combined in a wavelength window with WDM. If the wavelength of two lasers overlaps, they will interfere with each other at their respective receivers. In addition to the laser's line-width, there is also some spreading that occurs due to the modulation of the laser output by the signal it carries. The line width can also be spread by some of the laser modulation techniques and by interaction of the fiber's refractive index and the signal amplitude.

Although the lasers for the 1555 nm region are more expensive than those for the 1310 nm region, the good news is that it is also possible to manufacture these lasers with relatively narrow line widths that are compatible with the erbium-doped fiber amplifier (EDFA) optical amplifiers discussed below. Multiple-quantum well (MQW) lasers with distributed feedback (DFB) can achieve line widths of a few hundreds of kHz (a few millionths of a nm).

In Recommendation G.694.1 and G.694.2, the ITU-T has defined "grids" of wavelengths that can be used for WDM. These grids specify the wavelengths that lasers can use, but does not specify which of these may or should be used within a WDM system.[65] When the wavelengths in a WDM system are close together, it is referred to as dense WDM (DWDM). When the wavelengths are far apart it is referred to as coarse WDM (CWDM).[66] As wavelengths closer together on the grid are used, crosstalk can become a problem. The primary way to minimize or eliminate cross talk is to have lasers with sufficiently narrow line width and low enough drift from their center wavelength that their modulated signal does not overlap with adjacent channels. Another phenomenon that creates crosstalk is called four-wave mixing in which the signals from three nearby channels (or two in some cases) interact with each other such that a signal is created on another wavelength that may align with another channel. These crosstalk and mixing problems are familiar to people who are experienced in frequency

---

[65] The wavelengths in the grid are called the C-band and are on evenly spaced wavelengths of 0.39 nm (50 GHz) starting at 1528.77 nm. The grid was originally assembled largely as a collection of the wavelengths supported by the various laser vendors. In addition to the C-band, DWDM systems can use the L-band (1561-1620 nm). The S-band (1280-1350 nm) is typically used for single wavelength rather than DWDM applications.

[66] A common, extreme example of CWDM is to use just two wavelengths, one at 1310 and the other at 1555 nm.

division multiplexed (FDM) systems, since wavelength modulation is essentially a form of frequency modulation. Bit rate, fiber type, and fiber length are also factors in determining how many channels are possible in a DWDM system.

4.5.2.2 Optical signal regeneration

There are three aspects to the regeneration of optical signals that are named by three "Rs":

- Re-amplification of the optical signal
- Re-shaping of the optical pulses
- Re-timing/re-synchronization

An all-optical amplifier is a 1R amplifier. A 2R regenerator both amplifies and re-shapes the optical pulses. A 3R regenerator also performs clock recovery on the incoming signal and re-times the outgoing signal in order to remove jitter on the pulses. Both the 2R and 3R regenerators convert the optical signal to an electrical signal (OE conversion) and create a new optical pulse (EO conversion) after amplification (and re-timing for 3R regenerators).

Clearly, a 3R regenerator needs to be aware of the client signal (e.g., SONET/SDH), at a minimum having the ability to perform clock recovery at that client signal rate, but possibly requiring the ability to frame on the signal (e.g., to function as a SONET STE / SDH RS terminating NE).

An alternative that has become possible is to amplify the signal in the optical domain (i.e., a 1R amplifier), thus removing any requirements on the regenerator concerning the rate of the client signal. This is accomplished through optical amplifiers, of which the EDFA is the most common. In an EDFA, the signal passes through a section of fiber that has been doped with erbium. A strong signal from a pump laser is coupled into this fiber segment. The 980 (or 1480) nm wavelength energy of the pump laser excites the erbium atoms, and the presence of 1555 nm signals causes the erbium atoms to transfer their energy to the 1555 nm signal through stimulated emission. Gains of over 20 dB are possible. Other types of amplifiers exist, such as the Raman amplifier that can amplify a much wider range of wavelengths (1300-1600+ nm). A full discussion optical amplifier technology is beyond the scope of this book.

4.5.2.3 Optical switching

One of the key motivations for developing all optical networks is that if signals are kept in the optical domain, the network equipment can be agnostic to the client payload signals and eliminate the circuitry required for conversions between electrical and optical domains. For example, switching SONET signals requires STS-1 electronic cross connect fabrics and the OEO functions to convert the signal between the optical and electrical domains. Switching in the optical domain has the promise of lower equipment and provisioning costs at the expense of large granularity in the switched signals. The OTN can still use cross-connects that switch in the electrical domain of the client signal whenever it is desirable to groom the signals that are placed on the wavelengths. Such switches are referred to as hybrid switches.

A number of different technologies exist for switching in the optical domain, including solid-state devices (e.g., directional couplers that are combined to form multi-stage switch fabrics), free-space techniques (e.g., waveguide grating routers), and micro-electrical-mechanical switches (MEMS). MEMS technology allows the construction of an array of mirrors in silicon where each mirror's reflection angle can be controlled by an electrical signal. The optical input signals to the MEMS array can be steered to the appropriate output ports. MEMS switch times are in the order of microseconds. For fast switching, $LiNbO_3$ solid-state switches can achieve switch times in the order of nanoseconds. See [Ka00] for additional information on switching component technologies.

Equipment with extensive cross-connect capability is known as an optical cross-connect (OXC). Simpler equipment that is capable of adding or dropping wavelengths is known as an optical ADM (OADM). As illustrated in Figure 4-24, an OADM filters an incoming wavelength(s), removing it from the incoming signal and steering it to a drop port. At the transmitter of the OADM, the signal from the add port is then optically merged back into the outgoing signal. OADMs can either add/drop fixed wavelengths or dynamically select which wavelengths to add/drop.

It should be noted that a typical OXC or OADM implementation would have an EDFA at the input to the NE that acts as a pre-amplifier prior to the cross-connect fabric. The pre-amplifier boosts the signal amplitude to compensate for the attenuation over the fiber, and sets it to an appropriate signal level for the switch fabric. Another EDFA is present at the output of the OXC or OADM to amplify the signal for transmission.

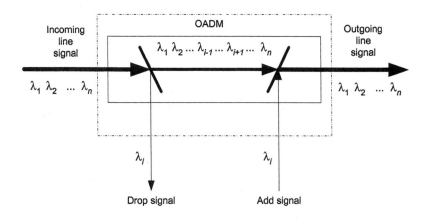

Figure 4-24. Optical Add/Drop Multiplexer illustration

4.5.3 WDM multiplexing approach and architecture

A number of different approaches had to be examined at the outset of the WDM standardization work, with numerous tradeoffs to be considered. Ideally, any type of native signal could be carried on any of the wavelengths with extensive OAM capabilities for each signal. This ideal is difficult to achieve in practice, however. Broadly speaking, the approaches fell into two categories: The first is to send the client signal essentially in its native format (with the exception of its normal wavelength) and add OAM capability in some type of separate channel. The second approach is to treat the client signal as a digital payload signal and encapsulate it into a frame structure that includes channel-associated OAM overhead channels.

The approach of assigning each client signal to its own carrier wavelength and carrying it in its native format creates the question of how to create the overhead channel(s). One option is to have the OAM information carried on a separate wavelength. Having the client signal and its associated OAM channel on separate wavelengths has some serious disadvantages, however. First, the OAM channel won't necessarily experience the same impairments as the client signal channel. ·Second, it is possible for provisioning errors to properly connect the OAM signal but not the client signal channel. Another option that received serious consideration was using sub-carrier modulation to create the OAM channel. In this approach, the carrier frequency (wavelength) is modulated with a low frequency signal that carries the OAM channel and could be removed through low-pass filtering at the termination point. There was some concern that this approach would be too complex, including its impact on jitter performance.

The approach of carrying the client signals as the payload of a digital frame was referred to as a "digital wrapper" approach.[67]   The digital wrapper, which contained the various OAM overhead channels, is conceptually similar to SONET/SDH.

In the end, a hybrid approach was chosen. The digital wrapper approach was chosen for the basic encapsulation and channel-associated OAM overhead for the client signals. Once this digital signal is transmitted over a wavelength, additional overhead wavelengths are assigned to carry other optical network overhead.

The basic signal architecture is illustrated in Figure 4-25. The client signal is inserted into the frame payload area, which, together with some overhead channels, becomes the Optical Payload Unit (OPU). An OPU is conceptually similar to a SONET/SDH Path. OAM overhead is then added to the OPU to create the Optical Data Unit (ODU), which is functionally analogous to the SONET Line (SDH Multiplex Section). Transport overhead (e.g., frame alignment overhead) is then added to create an Optical Transport Unit (OTU), which is the fully formatted digital signal and functionally analogous to the SONET Section (SDH Regenerator Section). The OTU is then transmitted on a wavelength. The client signal through OTU layer signal frame relationships are also illustrated in Figure 4-27. This OTU and the associated overhead data stream that is transmitted on a separate wavelength comprise the Optical Channel (OCh). A wave division multiplexed group of optical channels with an overhead channel (on its own wavelength) that is carried between access points constitutes the Optical Multiplexed Section (OMS). The Optical Transport Section (of order n) consists of an OMS (of order n) and an overhead channel (on its own wavelength). The OTS defines the optical parameters associated with the physical interface. The OCh, OMS, and OTS overhead channels provide the means to assess the transmission channel quality, including defect detection, for that layer. The OCh and OTS overhead also provides a means for connectivity verification. The OCh, OMS, and OTS layers are described in ITU-T Rec. G.872, and will not be discussed further here.

---

[67] For this reason, the G.709 OTN frame is sometimes referred to as a "digi-wrapper."

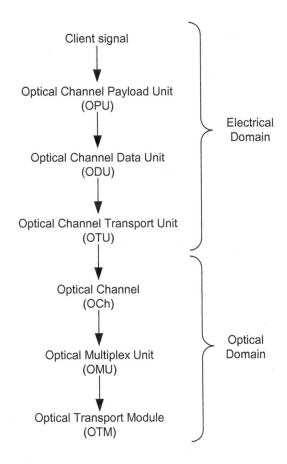

Figure 4-25. Information flow illustration for an OTN signal

Figure 4-26 shows an example OTN with the different layers and their relative scope. The IrDI is the inter-domain interface and is specified as having 3R regenerator processing at both sides of the interface. The IrDI is the interface that is used between different carriers, and can also be useful as the interface between equipment from different vendors within the same carrier's domain. Since the IrDI is the interface for interworking, it was the focus of the initial standard development. The IaDI is the intra-domain interface that is used within a carrier's domain. Since the IaDI is typically between equipment of the same vendor, it can potentially have proprietary features added such as a more powerful FEC.

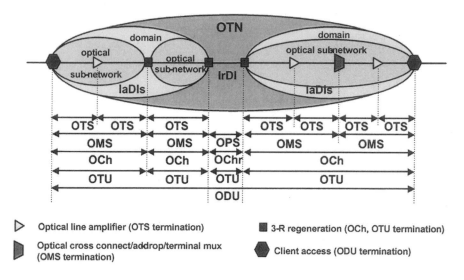

Figure 4-26. Illustration of OTN network layers[68]

Once the choice was made to use a digital wrapper approach, the next choice was what client signals should be allowed. Clearly the digital wrapper approach restricts the clients to being digital signals. Although it would have been ideal to allow both analog and digital clients in the same OTN, the main problem is that analog and digital signals have very different channel requirements. A channel that may be very adequate for a digital signal can have an unacceptably low signal-to-noise ratio or too much distortion for an analog signal. This makes it very difficult, especially administratively, to deploy mixed analog/digital networks in a DWDM environment. The next decision was what types of digital signals to include. Originally, there was a strong desire to carry optical data interfaces such as Gbit/s and 10 Gbit/s Ethernet in addition to SONET/SDH signals. In what appeared to be uncharacteristic shortsightedness, the decision was made to limit the constant bit rate (CBR) clients to the SONET/SDH signals[69]. The assumption was made that other signals could be mapped into SONET/SDH first, with these being mapped into the OTN. This decision not to directly support native Ethernet clients, while potentially simplifying the frame structures, is already proving to be a significant handicap to wide-scale

---

[68] Used by permission from Maarten Vissers

[69] As explained in section 4.5.4.2, it is possible to map other CBR signals into the OPUk payload, but OPUks are sized for SONET/SDH clients and any other client must do its own rate adaptation to fit into the OPUk.

deployment of G.709 OTNs[70]. In addition to CBR signals, mappings are defined for placing ATM or GFP frames directly into the OPU payload area (i.e., with no SONET/SDH frames). Payload mappings are discussed further in section 4.5.4.2.

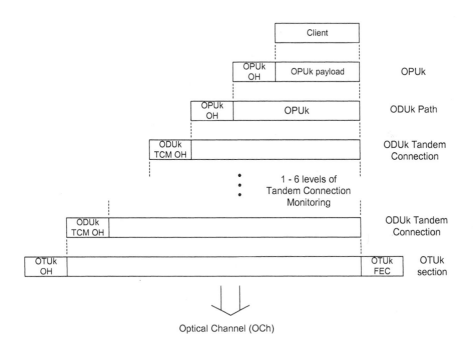

Figure 4-27. Information containment relationships for the electrical signal portions

---

[70] The primary reason for not supporting the full 10 Gbit/s payload was the 12.5 Gbit/s bandwidth constraint imposed by undersea cable systems. Supporting the full 10 Gbit/s payload rate would not leave an acceptable amount of overhead bandwidth for FEC. IEEE 802.3, with some initial reluctance, attempted to salvage the situation by defining the 10 Gbit/s Ethernet signal to have a WAN PHY rate of 9.58464 Gbit/s so that it could map directly into a SONET STS-192c payload envelope rather than the 10.3125 Gbit/s PHY rate used for the LAN. This mapping is described in Chapter 3. There have also been recent attempts to define a mapping for 10 Gbit/s Ethernet LAN-rate signals into OTN using GFP-F encapsulation to provide the frame delineation and eliminate the need for inter-packet Idle characters. No consensus has been reached on this approach, however.

Table 4-11. OTN signal and payload rates

k	OTUk signal rate	OPU payload area rate
1	255/238 × 2 488 320 kbit/s	2 488 320 kbit/s
2	255/237 × 9 953 280 kbit/s	238/237 × 9 953 280 kbit/s
3	255/236 × 39 813 120 kbit/s	238/236 × 39 813 120 kbit/s
Note: All rates are ±20 ppm.		

4.5.4   Signal formats and frame structure

This section describes the signal format for the digital portion of the OTN signal. The containment relationships of the client, OPU, ODU, and OTU layers and their overhead are shown in Figure 4-27. Figure 4-27 also illustrates the existence of multiple levels of Tandem Connection Monitoring (TCM), which will be described below. It can also be seen that the FEC is added at the OTU level, which is the last step before the optical transmission of the signal.

There are three currently defined OPU/ODU/OTU rates. When referring to an OPU, ODU, or OTU of a particular rate, it is referred to as an OPUk, ODUk, or ODUk with k = 1, 2, or 3. The respective signal and payload rates are shown in Table 4-11.

The OPU, ODU, and OTU frame structure is shown in Figure 4-28, including the overhead for each level. The frame is structured as four rows by 3824 columns, regardless of the signal rate. The OPU payload area consists of columns 17-3824 for all four rows. The overhead information for the OPU is contained in the D and E areas of Figure 4-28. The OPU overhead is similar in function to the SONET/SDH Path overhead, covering the OPU from the point at which the client signal is mapped into the OPU until it is extracted at the OPU termination point. As shown in Figure 4-28, the OPU overhead contains indicators for the payload type (PT) and multiframe structure (MSI), and frequency justification information (for adapting the client signal into the payload area). Unlike SONET/SDH Paths, however, it relies on the next lower level (ODU) for end-to-end error detection. When virtual concatenation is used, its overhead is located in the E area of Figure 4-28. Otherwise this area is reserved.

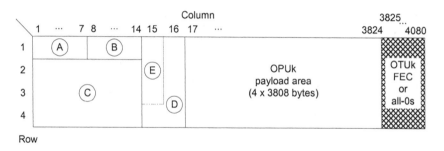

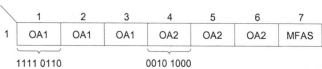

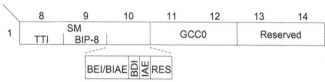

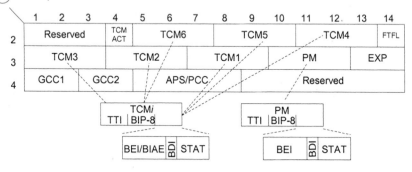

Figure 4-28. G.709 OTN signal frame and overhead structure

D OPU specific overhead area

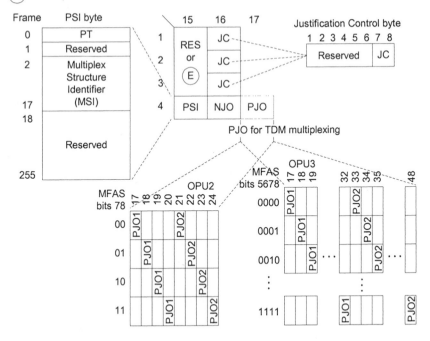

E Virtual concatenation overhead

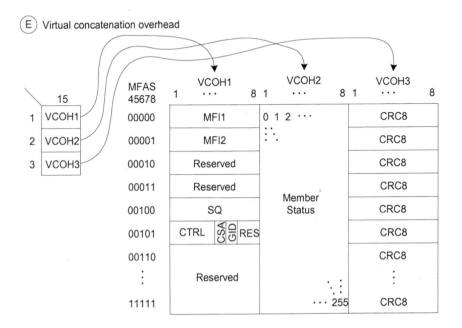

Figure 4-28 - continued

The ODU consists of the OPU and the ODU overhead, which is functionally similar to the SONET Line (SDH Multiplex Section) overhead. The ODU overhead is area C in Figure 4-28. It contains the overhead for path performance monitoring (PM), fault type and fault location (FTFL), two generic communications channels (GCC), an automatic protection switching and protection communications channel (APS/PCC), six levels of tandem connection monitoring (TCM), and a set of bytes reserved for experimental purposes. The PM and TCM overhead consists of trail trace identifier (TTI, similar to SONET Path trace for connectivity fault detection), a BIP-8 for error detection, status information (to indicate whether this is a normal signal or a maintenance signal), and backward error indication (BEI). Similar to the SONET/SDH REI, the BEI is sent by the ODU sink to the ODU source as a (binary) count of the number of errors detected by previous BIP-8. The TCM overhead also contains a backward incoming alignment error indicator (BIAE) that is sent in the upstream direction to indication the detection of a frame alignment error. The backward defect indicator (BDI) is used by the sink to inform the source that it is seeing a signal failure (similar to SONET/SDH RDI).

The OTU consists of the ODU, the OTU overhead and the FEC, if used. The OTU overhead is shown as the A and B areas in Figure 4-28. The A field contains the frame alignment pattern and the multiframe alignment signal (MFAS). The MFAS field is a binary counter that shows the phase of the current frame within the 256-frame multiframe. Those fields in Figure 4-28 that are defined as spreading across the multiframe (e.g., the PSI and virtual concatenation overhead) use the MFAS to determine the meaning of the byte during that frame. The B area of Figure 4-28 provides a GCC and service management (SM) information for the OTU. The SM fields include the TTI, BIP-8, BEI, and BIAE that were discussed for the ODU. In addition, the SM overhead for OTU includes an incoming alignment error (IAE) indicator. The IAE indicates that a frame alignment error was detected on the incoming signal, with the BIAE informing the source that an IAE was seen. The IAE and BIAE are used to disable the error counting in their respective directions during frame alignment loss conditions.

Note that the final step before transmitting the OTU on the optical channel is to scramble it in order to assure adequate transition density for reliable receiver clock recovery. The scrambling is performed on all OTU frame bits, including the FEC bytes, but excluding the framing bytes. A frame-synchronized scrambler is used with polynomial $x^{16}+x^{12}+x^3+x+1$ that is reset to all 1s on the MSB of the MFAS byte.

Table 4-12. Payload mapping code points for OTN signals

Hex code (Note 1)	Interpretation
01	Experimental mapping (Note)
02	Asynchronous CBR mapping
03	Bit synchronous CBR mapping,
04	ATM mapping
05	GFP mapping
06	Virtual Concatenated signal
10	Bit stream with octet timing mapping,
11	Bit stream without octet timing mapping,
20	ODU multiplex structure
55	Only present in ODUk maintenance signals
66	Only present in ODUk maintenance signals
80-8F	Reserved codes for proprietary use
FD	NULL test signal mapping
FE	PRBS test signal mapping
FF	Only present in ODUk maintenance signals

NOTE – Experimental mappings can be proprietary to vendors or network providers. If one of these mappings/activities is standardized by the ITU-T and assigned a code point, that new code point is used instead of the 01 code point.

4.5.5 Payload mapping

G.709 supports either constant bit rate (CBR) client signals and cell/packet based signals. The payload type (PT) overhead definitions for the defined mappings are shown in Table 4-12.

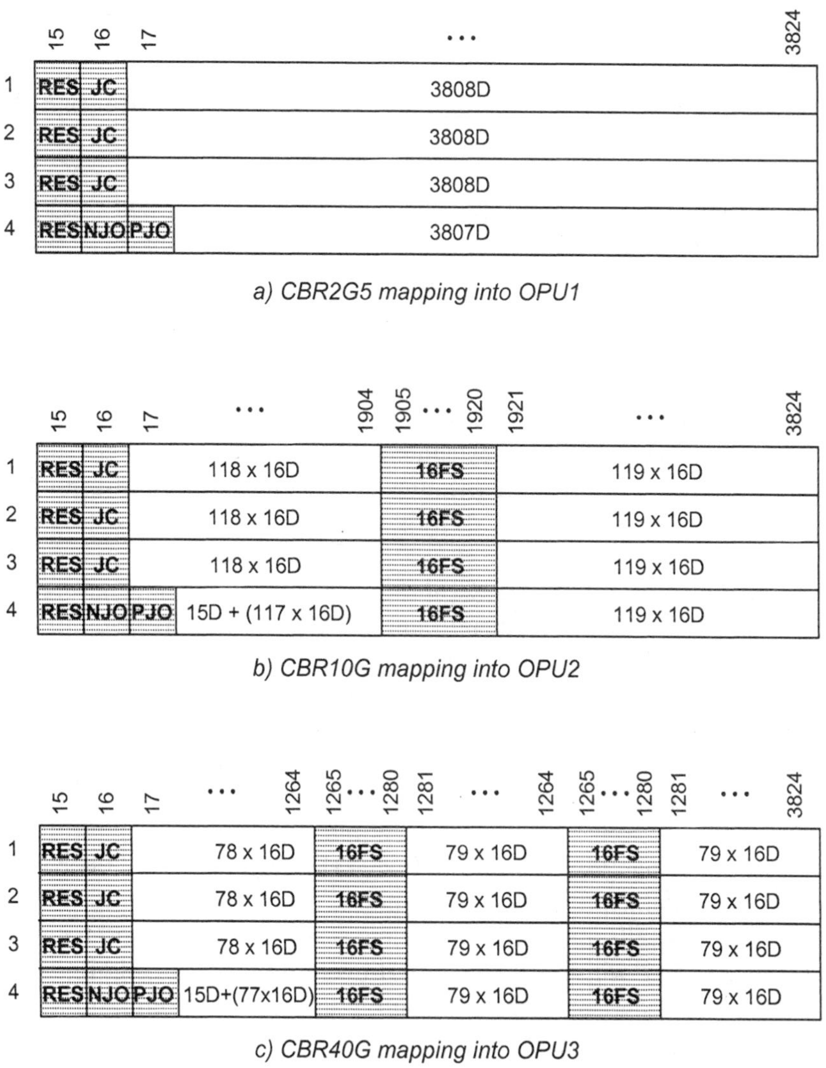

a) CBR2G5 mapping into OPU1

b) CBR10G mapping into OPU2

c) CBR40G mapping into OPU3

Figure 4-29. Mappings of CBR (SONET/SDH) signals into OTN

4.5.5.1 CBR mappings

As discussed in section 4.5.3, the initial set of CBR client signal mappings defined for G.709 OTN were SDH STM-16, STM-64, and STM-256 (SONET STS-48, STS-192, and STS-768), which are referred to as CBR2G5, CBR10G, and CBR40G, respectively. The CBR2G5, CBR10G, and CBR40G are in turn respectively mapped into the OPU1, OPU2, and OPU3. The OPUk payload area structures associated with these mappings are shown in Figure 4-29 where D indicates a payload data byte and FS is a fixed stuff byte. There are two methods for mapping the CBR signals into the OPU:

- **Asynchronous mapping:** With asynchronous mapping, the OPU clock is generated locally. The adaptation between the OPUk payload rate and the client signal rate is performed through the use of the justification control (JC) bytes and their associated Negative Justification Opportunity (NJO) and Positive Justification Opportunity (PJO) bytes.
- **Bit synchronous mapping:** With the bit synchronous mapping, the OPU clock is derived from the client signal clock (e.g., CBR10G signal). Because the OPU is frequency and phase locked to the client signal, there is no need for frequency justification. The JC bytes contain fixed values, the NJO contains a justification byte, and the PJO contains a data byte.

In addition to the CBR2G5, CBR10G, and CBR40G, G.709 also allows for a non-specific client bit stream into the OPU. In this mapping, a client signal (or set of client signals) is encapsulated into a CBR stream at the rate of (i.e., synchronous to) the OPU payload area. Any rate adaptation must be performed within the CBR bit stream as part of the process that creates it.

4.5.5.2 GFP and ATM mapping

Direct mappings for GFP frames and ATM cells into the OPU payload area are also defined. In these mappings, a continuous stream of GFP frames or ATM cells are mapped in an octet-aligned manner into the whole OPU payload area with no SONET/SDH overhead (and no OPU fixed stuff columns). The mapping is illustrated in Figure 4-30. The delineation of the GFP frame or ATM cell boundaries is performed using the header information of these protocols. GFP Idle frames or ATM Idle cells are send when there is no data to send.

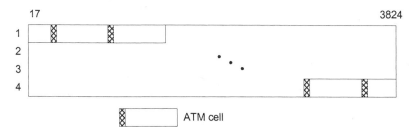

a) ATM cell mapping into OPUk payload area

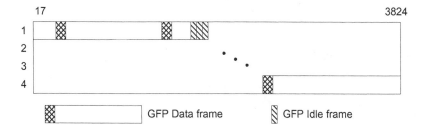

b) GFP frame mapping into OPUk payload area

Figure 4-30. Mapping for GFP frames and ATM cells into the OPU

4.5.6 TDM multiplexing

Originally, there was no intention to provide TDM within the OTN frame. The thinking was that the OTN signal should be kept as simple as possible, and that any TDM multiplexing could be performed within the SONET/SDH client signal prior to the OTN. In the end, however, TDM multiplexing was added. Among the driving applications was that of the "carrier's carrier." Carrying each of the original carrier signals on its own wavelength was regarded as too inefficient when these are OC-48s. Multiplexing the original signals into a higher rate SONET/SDH signal, however, requires terminating the incoming SONET Section and Line (SDH RS and MS) overhead, including the Section Data Communications Channel (SDCC). Some proprietary schemes accomplished Section and Line overhead preservation by shifting these bytes to some otherwise unused transport overhead byte locations, but mapping the SDCC is not trivial since the SDCC source clocks will be different than the multiplexed signal SDCC clock. As a result, SDCC packet store and forward buffering may be required. The solution was TDM multiplexing within the OTN.

The OTN TDM multiplexing hierarchy is shown in Figure 4-31. An asynchronous multiplexing technique is used (rather than the pointer-based technique used in SONET/SDH) with the frequency justification performed using the JC, NJO, and PJO bytes as discussed in 4.5.6. The multiplexing is performed at the ODU level. Four ODU1s can be multiplexed into an OPU2. An OPU3 can contain a multiplexing of four ODU2s, 16 ODU1s, or a mixture of ODU1s and ODU2s. Beginning with column 17, the OPUk is partitioned into tributary slots on a per-column, round-robin basis. (Column 17 is used for tributary slot (TS) 1, Column 2 is used for TS 2, etc. with Column 3824 used for TS 4 in an OPU2 or TS16 in an OPU3.) When an ODUj is multiplexed into an OPUk, it is first structured as an Optical channel Data Tributary Unit (ODTUjk). The ODTU is a justified ODU that includes (i.e., is 'extended with') the framing bytes. An ODTU12 is mapped into one of the four OPU2 TSs, and an ODTU13 is mapped into one of the 16 TSs of the OPU3. An ODTU23 is mapped into any four arbitrary TSs of the OPU3 (i.e., the TSs do not need to be contiguous or aligned to a fixed boundary).

The multiplex structure identifier (MSI) overhead is carried in frames 2-17 of the PSI byte (see Figure 4-28). The first two bits (MSBs) of each MSI byte indicate whether the OPU is an OPU1 (00), OPU2 (01), or OPU3 (10). The six LSBs of each MSI byte contain the number of the tributary port associated with each OPU TS. The first MSI byte (i.e., the byte carried in frame 2 of the PSI multiframe) contains the tributary port number that is mapped into TS 1, the second MSI byte contains the tributary port number associated with TS 2, etc. While it is possible to use a fixed mapping between tributary ports and time slots, the MIS can thus be used to increase the flexibility of the assignments.

Since there is only room for one set of justification overhead (JOH) bytes in each frame, (area D in Figure 4-28), it is necessary to time-share these bytes among the different ODTUs. As shown in Figure 4-28, the first OPU frame of the multiframe carriers the JOH for TS 1, the second frame carries the JOH for TS 2, etc. It should also be observed that two PJO bytes (PJO1 and PJO2) are required with TDM multiplexing in order to accommodate the various clock rate tolerances. In the case of ODU1 to ODU3 multiplexing, an additional fixed stuff column (Column 119) is also required.

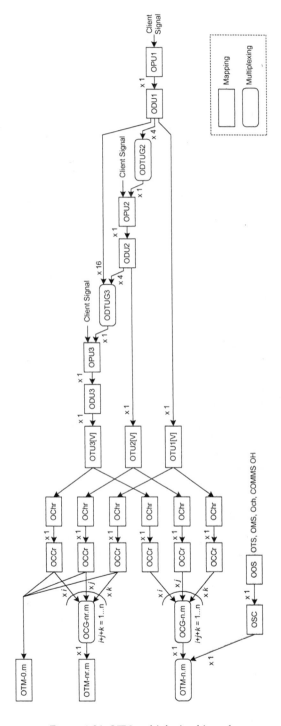

Figure 4-31. OTN multiplexing hierarchy

Table 4-13. Payload type values for virtually concatenated payloads (vcPT)

Hex code)	Interpretation
01	Experimental mapping (NOTE 1)
02	asynchronous CBR mapping
03	bit synchronous CBR mapping
04	ATM mapping,
05	GFP mapping
10	bit stream with octet timing mapping
11	bit stream without octet timing mapping
55	Present only in ODUk maintenance signals
66	Present only in ODUk maintenance signals
80 - 8F	reserved codes for proprietary use (NOTE 2)
FD	NULL test signal mapping
FE	PRBS test signal mapping
FF	Present only in ODUk maintenance signals

NOTE 1 – Experimental mappings can be proprietary to vendors or network providers. If one of these mappings/activities is standardized by the ITU-T and assigned a code point, that new code point is used instead of the 01 code point.

NOTE 2 – Proprietary mappings are similar to experimental mappings. If the mapping subsequently becomes standardized, the new code point is used. ITU-T has rules associates with this specified in G.806.

4.5.7 Virtual concatenation

Virtual concatenation is specified in G.709 for OPUk channels. In concept, OPUk virtual concatenation works the same as described above for SONET/SDH and PDH signals. The group of OPUks is launched with the same frame and multiframe phase. Each OPUk is allowed to take a different, independent path through the network with the receiver using the multiframe information (including the LCAS overhead MFI) to perform the compensation for the differential delay between the members. A virtually concatenated channel is referred to as an OPUk-Xv, where X is the number of OPUks that are concatenated. Each OPUk is placed into its own ODUk with the X ODUks being referred to as an ODUk-Xv. The virtually concatenated OPUk has no fixed stuff columns, giving a payload capacity of $X * 238/(239-k) * 4^{(k-1)} * 2\,488\,320$ kbit/s $\pm$ 20 ppm.

The virtual concatenation and LCAS overhead location and structure is shown in Figure 4-28. The individual fields in the overhead are the same as those discussed previously for SONET/SDH and PDH. The actual structure differs in that there is a member status (MST) byte and a CRC-8 included for each of the other bytes. Allowing for up to 255 members seems somewhat optimistic about the future progress of optical components since the OPU1 is about 2.5 Gbit/s.

The payload mappings into an OPUk-Xv are identified in the PSI byte with the values shown in Table 4-13. The first byte of the PSI byte frame (PT) specifies that this OPU is part of a virtually concatenated group. The second byte (frame 1, shown as Reserved in Figure 4-28) is used as the virtual concatenation payload type indicator (vcPT), with the values as defined in Table 4-13. The mapping techniques into virtually concatenated channel are essentially the same as discussed above for non-concatenated channels.

4.5.8 OAM&P

The key to saving network operational costs is having an effective OAM&P capability built into the signal format. The lack of this capability has been one of the reasons that Ethernet has not taken hold as a carrier technology.[71] The OTN OAM&P overhead was built on the experience gained from the SONET/SDH overhead.

4.5.8.1 Types of overhead channels

The different OAM&P overhead channels and their functions are summarized in Table 4-14. Most of these overhead functions (e.g., BIP-8 and TTI) have been discussed in the context of SONET in Chapter 3. The OTN BDI, BEI, GCC, and OA are functionally equivalent to the SONET/SDH RDI, REI, DCC and A1/A2 overhead channels, respectively. The functions that are unique to G.709 OTN are the following:

---

[71] Carrier-type OAM&P capability is being added to Ethernet through activities in IEEE 802 and ITU-T SG13. Any Ethernet OAM&P, however, must travel in-band as an Ethernet frame in the same channel as the client data frames. This means that Ethernet OAM&P frames consume client signal bandwidth and require all NEs that make use of this OAM&P information to be capable of removing and inserting the OAM&P frames from the client data stream.

Table 4-14. OAM&P channel definitions

OAM&P channel	Used in	Function
APS / PCC	ODU	Automatic Protection Switching / Protection Communications Channel – Similar to the SONET/SDH K1 and K2 bytes, but with potential for additional capability. The APS/PCC byte is time-shared across the multiframe to create channels for the control of sub-network connection protection at the ODUk Path and each TCM level.
BDI	OTU, ODU PM, each TCM	Backward Defect Indication – Sent from the overhead sink to the source to indicate that a defect has been detected in the forward direction. (Similar to SONET/SDH RDI.)
BEI	OTU, ODU PM, each TCM	Backward Error Indication – A binary count of the number of BIP-8 bits indicating errors, sent from the overhead sink to the source. (Similar to SONET/SDH REI.)
BIAE	OTU, each TCM	Backward Incoming Alignment Error – Indication sent from the overhead sink to the source that it received an IAE.
BIP-8	OTU, ODU PM, each TCM	8-bit Bit Interleaved Parity- Used in the OTU, ODU PM, and each level of TCM overhead.
FTFL	ODU	Fault Type and Fault Location – A 256 byte message with the first 128 bytes applying to the forward direction and the last 128 to the backward direction.
GCC	OTU, ODU	Generic Communications Channel – Similar to the SONET/SDH DCC. One available in the OTU overhead and 2 in the ODU overhead. GCC1 and GCC2 in the ODU are clear channels whose format is not specified in G.709.
IAE	OTU	Incoming Alignment Error – Indication sent downstream to inform the receiving NEs that a framing alignment error was detected on the incoming signal.
MFAS	OTU	Multiframe Alignment Signal – Binary counter used to establish the 256-frame multiframe that is used for the time-shared overhead channels that spread their content over the course of a multiframe.
OA	OTU	Optical Alignment – Frame alignment signal for the OTU. OA1 = 1111 0110 and. OA2 = 0010 1000
TCM ACT	ODU	Indication that TCM is being used on the ODUk.
TTI	OTU, ODU PM, each TCM	Trail Trace Identifier – Similar to the Trace identifiers used in SONET/SDH.

Table 4-15. APS/PCC mutliframe definition

MFAS bit 678	Level to which the APS/PCC applies
000	ODUk Path
001	ODUk TCM1
010	ODUk TCM2
011	ODUk TCM3
100	ODUk TCM4
101	ODUk TCM5
110	ODUk TCM6
111	ODUk SNC/I APS

- **FTFL** – OTN networks can potentially be much more complex than SONET/SDH networks due to the mixture of TDM and WDM technologies. For this reason, it is very advantageous to have better fault type and fault location indication capability. As noted in Table 4-14, the FTFL information is spread across the 256-byte multiframe with the first 128 bytes pertaining to the forward direction and the last 128 bytes to the reverse direction. The first byte of the 128-byte frame is the fault indication field, the next 9 bytes are an operator identifier field (country and carrier codes), and the remaining bytes 118 bytes are an operator-specific field. No fault (00_H), signal fail (01_H), and signal degrade (02_H) are the only currently defined fault types.

- **IAE** and **BIAE** – The IAE gives a specific indication that a frame alignment error was detected on an incoming signal. This indication allows the receiving NE to distinguish between a loss of signal and a loss of frame alignment when AIS is received. In SONET/SDH, only AIS is available for both types of failures. BAIE is the reverse IAE indication. Reporting BAIE in each of the TCM channels gives greatly enhanced fault location capability compared to an end-to-end indication like the BDI or SONET/SDH REI.

- **APS/PCC** –SONET/SDH use the K1 and K2 bytes for a Line (MS-Section) protection channel and reserves K3 and K4 for Trail protection. OTN, however shares a common protection channel to allow subnetwork connection protection at the level of the ODU and each TCM level. As shown in Table 4-15, the protection channel is time multiplexed across the signal multiframe. See Chapter 8 for more discussion of subnetwork connection protection.

4.5.8.2 Maintenance signals

The maintenance signal set for OTN is somewhat richer than the simple AIS of SONET/SDH and PDH, which is a reflection of the added wrinkles of the TDM/WDM mixture. These maintenance signals are summarized as follows:[72]

- **OCI** – Open Connection Indication. OCI is provided at the OCh and ODU levels. The OCI indicates to the OCh or ODU termination point that, due to management command, no upstream signal is connected to their corresponding source. This allows the termination point to distinguish the intentional absence of a signal from an absence due to a fault condition. The ODUk-OCI is carried in the STAT field of the PM byte and each active TCM byte in order to allow monitoring each of these points.
- **AIS** – The AIS signal is sent at the OTU and ODU levels in response to upstream failures. ODUk-AIS is an all-ones pattern in the OPUk (payload and overhead) and ODUk overhead except for the FTFL byte. OTUk-AIS fills the entire OTUk frame with the "generic AIS" pattern, which is defined as the pseudo random sequence generated from the $1+x^9+x^{11}$ polynomial (PN-11, as defined in ITU-T Rec. O.150). The PN-11 gives better signal transition characteristics than an all-ones signal.
- **LCK** – Lock condition. The ODUk-LCK is a downstream indicator that the upstream signal is "locked" and no signal is passing through. It is carried in the STAT field of the PM byte and each active TCM byte in order to allow monitoring each of these points.

4.5.8.3 Tandem Connection Monitoring (TCM)

As discussed in Chapter 3, TCM allows the insertion and removal of performance monitoring overhead at intermediate points in the network that correspond to some administrative boundary. It is done such that their insertion and removal don't destroy the performance monitoring overhead that traverses that region. The OTN TCM is illustrated in Figure 4-32. The PM byte is used for end-to-end path performance monitoring. Here TCM1

---

[72] It should be noted that other maintenance signals have been defined for use in the optical domain. PMI (Payload Missing Indication) applies to the OTS and OMS layers and indicates the absence of an optical signal. FDI-O and FDI-P provide a Forward Defect Indication for the payload (server layer) and overhead layers, respectively at the OMS and OCh levels. At this time, neither of these overhead signals has been implemented in commercial equipment.

is used by the user to monitor the connection QoS. TCM2 is used by the primary network operator (Operator A) to monitor the connection from ingress to egress of its network, including the link through Operator B. TCM3 is used by each operator to monitor the connection through its own subnetwork and for the connection between the operator domains. Operator B uses TCM4 to monitor the connection through the protected facility.

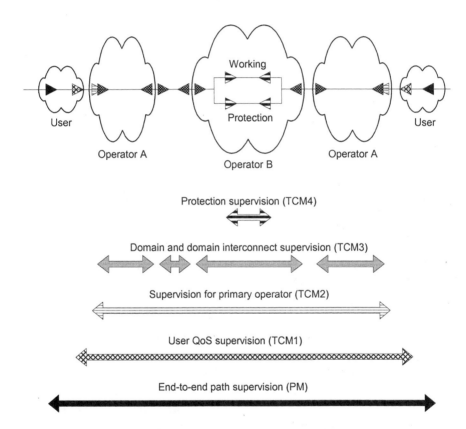

Figure 4-32. Illustration of TCM domains

Recall that the SONET/SDH TCM BIP is part of the field covered by the Line BIP-8 (B2) and hence B2 had to be compensated whenever the TCM BIP was modified (i.e., on insertion or removal). This was due to TCM being added after the SONET/SDH signal was defined. In the OTN, however, each TCM only covers the OPUk payload area, and hence no compensation is required.

4.5.9   Synchronization and mapping frequency justification

4.5.9.1 Synchronization

One of the key decisions for the OTN was that it would not be required to transport network synchronization as part of the OTN signal. Since the client signals such as SONET/SDH can transport this synchronization, there was no compelling reason to add the extra complexity and stringent clock requirements to the OTN signals. The only constraint was that the OTN justification scheme for mapping SONET/SDH clients had to guarantee that these clients could be carried without causing them to violate the ITU-T Rec. G.825 jitter and wander specifications.[73]

Table 4-16. Comparison of PDH, SONET/SDH, and OTN frequency justification

Hierarchy	Technique	Adjustment increment
PDH	Positive justification (stuff)	Single bit
SONET / SDH	Positive/negative/zero (pnz) justification (via pointers)	Single byte for SONET VTs and STS-1 (SDH VC-1/2/3). N bytes for SONET STS-Nc, 3 bytes for SDH VC-4, and 3N bytes for SDH VC-4-Nc.
OTN	Positive/negative/zero justification	Single byte

4.5.9.2 Justification for mapping and multiplexing

Frequency justification in OTN is required for some of the CBR mapping techniques and for TDM multiplexing. As indicated in Table 4-16, the justification technique is a hybrid of the techniques used for asynchronous/PDH networks and SONET/SDH. Similar to the PDH networks, the justification is based on an asynchronous technique with justification control fields rather than the pointer-based approach of SONET/SDH. Like SONET/SDH, however, it provides for both positive and negative byte-wise adjustments rather than the bit-oriented positive adjustments of PDH.

---

[73] The jitter and wander requirements for OTN network interfaces are specified in ITU-T G.8251.

The justification overhead (JOH) in the OTN is the Justification Control (JC), Negative Justification Opportunity (NJO), and Positive Justification Opportunity (PJO) bytes. As illustrated in Figure 4-28, these bytes are part of the OPUk overhead. The NJO provides a location for inserting an additional data byte if the client signal is delivering data at a faster rate than the OPUk payload area can accommodate. The PJO provides a stuff opportunity if the client signal is delivering data a lower rate than the OPUk payload area can accommodate. The NJO is thus analogous to the SONET/SDH H3 byte and the PJO to the SONET/SDH positive stuff opportunity byte. The demapper ignores the contents of the NJO or PJO bytes whenever they carry a justification byte. Bits 7 and 8 of JC are used to indicate the contents of the NJO and PJO, somewhat analogous to the SONET/SDH H1 and H2 or the PDH C-bits. The mapper assigns the same value to each of the three JC bytes in an OPUk frame so that the demapper can perform a two-of-three majority vote for error correction.

Table 4-17 shows the definitions of JC, NJO, and PJO for CBR mappings. Here, PJO is a single byte in the OPUk payload area. As noted in section 4.5.4.2, justification is only required for asynchronous mapping since the OUPk clock is generated independently of the client signal clock. Since the bit-synchronous mapping uses an OPUk clock derived from the client signal, it can used fixed assignments for NJO and PJO.

Table 4-17. Justification Control and Opportunity definitions for CBR mappings

JC [78]	Generation by asynchronous mapper		Generation by bit-synchronous mapper		Interpretation by a demapper	
	NJO	PJO	NJO	PJO	NJO	PJO
00	justification byte	data byte	justification byte	data byte	justification byte	data byte
01	data byte	data byte			data byte	data byte
10	not generated		not generated		justification byte	data byte
11	justification byte	justification byte			justification byte	justification byte
NOTE – Since the mapper never generates the JC [78] = 10, the interpretation by the demapper is based on the assumption that an error has corrupted these bits.						

In the case where TDM multiplexing is used in the OPUk, the JOH structure is modified from the non-multiplexed case. Each ODU tributary that is being multiplexed into the OPU requires its own justification. Since there is only a single set of JC bytes and NJO byte in each OPUk frame, they must be shared among the tributary ODUs. This sharing is done based on the frame number within the multiframe. Figure 4-28 illustrates this sharing of the OPUk JOH overhead. In order to provide the appropriate frequency range accommodation, two PJO bytes, PJO1 and PJO2, were defined. The PJO bytes, of course, need to appear in the column associated with the other data bytes for that tributary, which results in the structure shown in Figure 4-28. The JOH use and interpretation with TDM multiplexing are given in Table 4-18.

Table 4-18. Justification Control and Opportunity definitions for TDM mappings

JC [78]	NJO	PJO1	PJO2	Interpretation by the demapper
00	justification byte	data byte	data byte	no justification (0)
01	data byte	data byte	data byte	negative justification (-1)
10	justification byte	justification byte	justification byte	double positive justification (+2)
11	justification byte	justification byte	data byte	positive justification (+1)

It is important to note that while the SONET JOH bytes are located within the transport overhead, the OTN JOH bytes are located within the OPUk overhead, which is analogous to the SONET Path overhead. This JOH location choice has an important implication: Retiming an OTN signal requires demultiplexing back to the client signal.

4.5.10 Forward Error Correction (FEC)

One of the primary benefits to the G.709 OTN is that it provides for a strong FEC code. As shown in Figure 4-28, a four-row by 256-column area at the end of the OTUk frame is reserved for FEC. The FEC code specified in G.709 is a Reed-Solomon RS(255,239). The RS(255,239) symbol size is 8-bits, and the code is implemented as a byte-interleave of 16 separate RS(255,239) codes. The advantage to interleaving is that it allows the resulting interleaved codes to detect/correct burst errors of a length up to the interleaving depth (i.e., 16 bytes here). The Hamming distance of the RS(255,239) code is 17, which allows each code to correct up to 8 symbols (for the error correcting mode) or detect up to 16 symbol errors (in error detection mode). When used for error correction, this leads to a coding gain of 6.2 dB for systems with an operating BER of 10^{-15}. This coding gain can be used to allow higher rates over existing facilities, longer span lengths, higher numbers of DWDM channels, or relaxed parameters for the system optical components.

If FEC is not implemented, the FEC field contains all zeros.

In the case of an IaDI, the signal remains within a single domain where there will often be longer spans than would exist for the typical IrDI interface. As a result, there has been some desire for an even stronger FEC option for these IaDI applications. Since such applications will typically have the same equipment vendor's equipment on each end of the link, there is no compelling need to standardize this FEC. Vendors, then, are free to develop their own FEC to gain a competitive advantage. Different implementations place the FEC bytes in different locations. Some use the frame structure of the OTU frame in Figure 4-28 except that the FEC field extends for additional columns. Others intersperse the FEC bytes throughout the frame. Various types of FEC codes have been proposed and/or used, and several of these are listed and defined in ITU-T Rec. G.975.1. As with all FEC codes, there are tradeoffs between the error correction/detection capability of the code and the decoder complexity plus the extra transmission bandwidth required to carry it.

4.5.11 OTN Conclusions

The ITU-T OTN hierarchy has many merits and advantages for DWDM systems. The G.709 OTN frame provides some very significant OAM&P capabilities, especially in the nesting of multiple TCM overhead channels. The rigid tie between the payload rates and the SONET/SDH signals is a definite drawback, especially seen in its inability to transparently carry a full-rate 10 Gbit/s Ethernet LAN signal. At the same time, the increased complexity of supporting TDM multiplexing and virtual concatenation take away from its potential cost/complexity advantage relative to SONET/SDH DWDM systems. The initial focus of G.709 OTN systems was in core long haul networks. Since the current network capacity bottlenecks are in the metro and especially the access networks, there is relatively little need for extensive new capacity increases in the long haul networks, however. The decreasing cost of some of the components is starting to open applications for G.709 in metro networks (e.g., where the FEC capability is required to achieve higher rates over existing facilities). As a result, there is ongoing interest in G.709, but it is not clear when there may be large-scale G.709 OTN deployments.

4.6.   GENERAL CONCLUSIONS

The transport network underwent a major revolution beginning in the early 1990s with the introduction of SONET/SDH. The previous generation had very little in way of embedded OAM&P capability. The extensive SONET/SDH OAM&P capabilities allowed carriers to offset the cost of building a new backbone network with the substantial ongoing OAM&P cost reduction. OAM&P costs are in fact the primary expense for carriers. Also, when SONET/SDH was introduced, carriers were just in the early stages of deploying their fiber optic networks, so there was relatively little equipment to replace and it made immanent sense to use a standard interface rather than vendor proprietary interfaces. Now that the SONET/SDH backbone has be widely deployed throughout the world along with the managements systems to handle the OAM&P, it is foundation of the carrier networks. Any new transport technology seeking large-scale deployment has to prove itself to have substantial ongoing cost benefits over SONET/SDH. Carriers are not in a position to make huge capital investments in new networks without clear cost justification in a reasonable timeframe. In the meantime, SONET/SDH has continued to prove its value by virtue of its ability to evolve to handle new applications and client data types. This prolonging of the useful life of the SONET/SDH networks and their associated OAM&P systems is extremely beneficial to the carriers. In summary, the carriers would much prefer an evolutionary approach to providing new services and capabilities rather than a revolutionary approach that requires huge new capital investments.

The first three new technologies presented in this chapter fall into the evolutionary category. VCAT and LCAS provide the needed flexible channel sizing capability to support the various data and video services. GFP provides the efficient and flexible encapsulation mechanism to map data into the SONET/SDH networks. RPR provides the ability to add Layer 2 MAC functionality to SONET/SDH access and metro networks.[74] These three technologies have received a great deal of worldwide interest, especially VCAT, LCAS and GFP, and should be very important in the coming years. The last technology, OTN is largely a revolutionary technology in that it would replace SONET/SDH in DWDM systems. It is also evolutionary in the sense that it is optimized to carry SONET/SDH clients. Ironically, this close tie to SONET/SDH client signals may is also a major drawback due to its limitations with Ethernet client signals.

---

[74] Of course, GFP is also applicable to OTN and RPR can use Ethernet links instead of SONET/SDH links. Their primary initial benefit, however, is the value they add to the existing SONET/SDH infrastructure.

The final question to address is what the future will bring for transport networks and transport network technology. Two factors will be very important. The first is the evolution of the access network. If end-user applications drive a demand for ever-increasing access bandwidth, the access networks will be upgraded to provide it. Increased customer traffic coming into the network will in turn necessitate increased bandwidth in the core transport networks.

The other factor for the next generation is the potential integration of multiplexing and switching technology for all service types. ATM was the first attempt to use a packet/cell-based technology for multiplexing and switching. In the next generation of networks, both Ethernet and MPLS have their advocates. By virtue of being encapsulation and multiplexing technologies, MPLS and Ethernet can be considered as transport technologies. Ethernet shows great promise as an access technology, especially for business users. Guaranteeing a specified quality of service (QoS) for every service type is one of the most difficult challenges in cell/packet-based networks. As part of the QoS guarantee, OAM&P capabilities are again critical. Ethernet has lacked the inherent OAM&P capability to make it a viable contender for large-scale public networks. The IEEE and ITU-T are currently adding OAM&P capability to Ethernet, however, there will always be some drawbacks to having to insert the OAM&P overhead into the same channel (and hence bandwidth) as the client payload signal. MPLS is receiving more serious consideration as a core network multiplexing and switching technology, especially with Voice over IP (VoIP) gaining ground. The largest two North American long distance carriers and at least one of the major European operators have announced their intention to move toward an MPLS-based core network. Most people expect to see SONET/SDH (enabled by VCAT, LCAS, and GFP) to remain the underlying Layer 1 transport technology for MPLS. While hardcore Ethernet fans favor Ethernet as its own Layer 1 technology, in all likelihood SONET/SDH will also be the most common Layer 1 technology for Ethernet WANS.

REFERENCES

General next generation transport references

[GoWi02]   S. Gorshe and T. Wilson, "Transparent Generic Framing Procedure: A Protocol for Efficient Transport of Block-Coded Data Through the SONET/SDH Network," *IEEE Communications Magazine*, May 2002, pp. 88-95

[ITU7042]   ITU-T Recommendation G.7042/Y.1305 (2004), *Link Capacity Adjustment Scheme for Virtual Concatenated Signals*

[ITU7041]   TU-T Recommendation G.7041/Y.1303 (2004), *Generic Framing Procedure. (GFP)*

SONET/SDH related references

[ITU707]   ITU-T Recommendation G.707 (1996), *Synchronous Digital Hierarchy Bit Rates.*

[SeRe97]   M. Sexton and A. Reid, *Broadband Networking: ATM, SDH, and SONET*, Artech House, 1997

[T10500]   ANSI/ATIS T1.105-2001 Synchronous Optical Network (SONET) Basic Description including Multiplex Structure, Rates, and Formats

[T110502]   ANSI/ATIS T1.105.02-2001 Synchronous Optical Network (SONET) Payload Mappings

Data related references (transport and client)

[ETS8839]   ETSI (CENELEC): EN 50083-9 (1998), *Cable distribution systems for television, sound signals and interactive multimedia signals; Part 9: Interfaces for CATV/SMATV Headends and Similar Professional Equipment for DVB/MPEG-2 transport streams (DVB Blue Book A010), Annex B, Asynchronous Serial Interface.*

[IEEE3]   IEEE Standard 802.3-2002, *Information Technology – Telecommunications and Information Exchange Between Systems – LAN/MAN – Specific Requirements – Part 3: Carrier Sense Multiple Access with Collision Detection (CSMA/CD) Access Method and Physical Layer Specifications.*

[IEEE17]   IEEE 802.17-2004, *Information technology – Telecommunications and information exchange between systems – Local and metropolitan area networks, Specific requirements – Part 17: Resilient Packet Ring Access Method & Physical Layer Specifications* (RPR)

[IET1195]   IETF RFC 1195, December 1990 "Use of OSI IS-IS for Routing in TCP/IP and Dual Environments

[IET1661]   IETF RFC 1661, July 1994 "Point-to-Point Protocol (PPP),"

[IET1662]   IETF RFC 1662, July 1994 "PPP in HDLC-like Framing,"

[IET1990]   IETF RFC 1990, August 1996, The PPP Multilink Protocol (MP).

[IET3031]   IETF RFC 3031, January 2001 "Multiprotocol label switching architecture"

[IET3032]   IETF RFC 3032, January 2001 "MPLS label stack encoding"

[IET3270]   IETF RFC 3270, May 2002 "Multi-Protocol Label Switching (MPLS) support of Differentiated Services"

[INC342]   ANSI INCITS 342-2001, *Information Technology – Fibre Channel Backbone (FC-BB).*

[INCBB]   ANSI INCITS xxx-2004[75], *Information Technology – Fibre Channel Backbone (FC-BB-3)*

[INC372]   ANSI INCITS 372-2003, *Information Technology – Fibre Channel Backbone (FC-BB-2).*

[ITU8012]   ITU-T Recommendation G.8012 (2004), *Ethernet UNI and Ethernet NNI*

[OSI10589]   OSI/IEC TR 10589 "Intermediate System to Intermediate System Intra-Domain Routeing Exchange Protocol for use in Conjunction with the Protocol for Providing the Connectionless-mode Network Service (ISO

---

[75] This standard was in the process of formal approval at the time of publishing, so the number was not available.

	8473)," ISO DP 10589, February 1990
[St93]	W. Stallings, *Local and Metropolitan Area Networks*, MacMillan, New York, 1993
[X3230]	ANSI X3.230-1994, *Fibre Channel – Physical and Signaling Interface (FC-PH)*.
[X3296]	ANSI X3.296-1997, *Information Technology – Single-Byte Command Code Sets CONnection (SBCON) Architecture*.

OTN standards and references

[Ka00]	S. Kartalopoulos, *Introduction to DWDM Technology*, IEEE Press, Piscataway, HJ, 2000
[ITU692]	ITU-T Recommendation G.692 (1998), Optical interfaces for multichannel systems with optical amplifiers
[ITU709]	ITU-T Recommendation G.709 (2003), *Interfaces for the Optical Transport Network (OTN)*
[ITU798]	ITU-T Recommendation G.798 (2004), *Characteristics of optical network hierarchy equipment functional blocks*
[ITU8251]	ITU-T Recommendation G.8251 (2001), T*he Control of Jitter and Wander within the Optical Transport Network*
[ITU872]	ITU-T Recommendation G.872 (2001), *Architecture of optical transport networks*
[ITU6941]	ITU-T Recommendation G.694.1 (2002) *Spectral grids for WDM applications: DWDM frequency grid*
[ITU6942]	ITU-T Recommendation G.694.2 (2003) *Spectral grids for WDM applications: CWDM frequency grid*
[ITU9751]	ITU-T Recommendation G.975.1 (2004), Forward error correction for high bit rate DWDM submarine systems

Other standards and references related to next generation technologies

[Be82]	J. C. Bellamy, *Digital Telephony*, John Wiley, New York, 1982, pp. 181-182.
[BOND]	Interoperability Requirements for Nx56/64 kbit/s Calls, version 1.0, from the Bandwidth ON Demand Interoperability Group (BONDING) Consortium, 1992
[Fr76]	P. A. Franaszek, "Sequence-State Coding for Digital Transmission," *Bell System Technical Journal*, December 1976, pp. 143-157.
[Go02]	S. Gorshe, "CRC-16 Polynomials Optimized for Applications Using Self-Synchronous Scramblers," *Proc. of IEEE ICC2002*, pp. 2791-2795
[ITU704]	ITU-T Recommendation G.704 (1998), *Synchronous frame structures for 1544, 6312, 2048, 8448, and 44736 kbit/s hierarchical levels*
[ITU832]	ITU-T Recommendation G.832 (1998), *Transport of SDH elements on PDH networks – Frame and multiplexing structures*
[ITU7043]	ITU-T Recommendation G.7043 (2004), *Virtual Concatenation of Plesiochronous Digital Hierarchy (PDH) signals*
[ITU8040]	ITU-T Recommendation G.8040 (2004), *GFP frame mapping into Plesiochronous Digital Hierarchy (PDH)*
[T1107]	ANSI/ATIS T1.107-1995, Digital Hierarchy - Formats Specifications.
[Wi95]	S. Wicker, *Error Control Systems for Digital Communications and Storage*, Prentice Hall, Upper Saddle River, New Jersey, 1995

Chapter 5

MANAGEMENT ARCHITECTURES
Architectural Components

5.1. INTRODUCTION

Architecture and framework for Telecommunications Management Network (TMN) defined by ITU has been in existence for over ten years. The initial architectural components have evolved based on both deployment of products and business process framework developed by TMF. Another influencing factor for management framework in general is the introduction of transport technologies such as SDH and OTN, These technologies have introduced control plane activities that migrate traditional management functions into the control plane thereby introducing the need for coordination between the two planes. The physical, functional, and information architectural components of TMN are summarized briefly to recapture the information in the literature. This chapter also introduces recent work on the how the architectural components of TMN are influenced with the introduction of SDH and OTN.

Because TMN architecture had elements that gave the appearance of being tightly associated with OSI[76] (and CMIS/P[77] together known as CMISE), several articles in the literature and in practice this was held as the strict definition within Telecommunications industry. This chapter brings out the general architectural building blocks that do not necessitate the tight coupling in an attempt to dispel these incorrect perceptions of TMN. In

---

[76] Open Systems Interconnection, a legacy and old standard for open systems management
[77] Protocol details of CMIS/P are discussed in Ch7

addition the chapter addresses inaccurate criticism targeted at the architecture regarding (a) logical layered architecture (LLA) (b) lack of integration of Internet, CORBA, and Java technologies, and (c) lack of distributed computing framework in the architecture.

5.2.   WHAT AND WHY OF TMN

Management of Telecommunications network has always been a key area for service providers. Even though there has been a constant evolution of transport and switching technologies, introduction of new value added services and the onset of next generation network blurring the boundaries between traditional data and telecommunications networks; the need for managing and maintaining the promised quality of service to the customer does not change. The goal of TMN, (which was first introduced in 1992) is to offer an architecture and set of principles to meet these goals. The fundamental principles are still applicable irrespective of the changes in the telecommunications network environment even though the details may have undergone many changes.

5.2.1   Why TMN?

An end-to-end communications service is offered today with multiple technologies linking the different networks from multiple service providers[78]. Unlike ten or fifteen years ago, in the current environment the communication path may traverse combinations of wireless, and land based facilities using different technologies. In addition to traditional service providers such as an Local and Inter-exchange carriers, several new carriers have been offering services. As an example, cable MSOs also started offering telecommunications services using cable/Hybrid Fiber Coax network.

When TMN was initially formulated, the predominant services were telephony, narrow-band ISDN, broadband-ISDN (ATM), Frame Relay, X.25, and private line services such as T1/T3 (E1/E3). Several new services, such as DSL, Cable Modem, EtherNet, wireless etc. are now being offered with the increased need for bandwidth, mobility etc. Private line services are being offered using SONET/SDH and DWDM to connect remote locations for geographically dispersed businesses.

---

[78]See Ch1 for an explanation of the term MSO and others

Such broad and diverse service offerings results in network consisting of many different types of equipment - customer premises equipment, DLCs, Channel Banks (CSU/DSU), local class-5 switches, intra and inter office transmission and switching equipment, DSLAMs, ATM/FR data switches, and so on. For a wireless service network, the entities include base station equipment and systems, storage modules for data bases such as Home Location Registers. For HFC networks the entities include Head End broadcast systems, 2-way coaxial return systems (CMTS, Pay per view systems, etc), Fiber Nodes and Coax/Fiber Hubs where residential feeder plant is concentrated. Many of these equipment types and network applications have been introduced in Ch1. Besides, equipment supporting traditional telecom services, equipment supporting new and emerging services is also being deployed. Such new services include storage area networking (SAN) services, value added non-transport services, such as Data Hosting, Application Hosting, VPNs, etc[79].

The equipments in the telecommunications network as well as the management systems today are software intensive to support the level of intelligence required for the services. The management systems located in network operations centers are usually reliable computing platforms with sophisticated software. These systems are referred to generically as Operations Support Systems (OSS or simply OS).

Figure 5-1 illustrates the various actors - equipment vendors, operators, and system vendors (ISVs)[80] involved in specification, implementation, and use of one or more pieces of telecom software.

---

[79] One can notice that telecommunications networks are very dynamic with changing equipment and services. Given this, one may ask - is TMN specification general and abstract enough that it can continue to be relevant? We don't claim to provide answer to this but let the reader be aware that such critical questions are being asked of TMN.

[80] In some cases, carriers may have in-house software development effort that addresses their needs and hence may not use Independent Software Vendors (ISVs).

Chapter 5

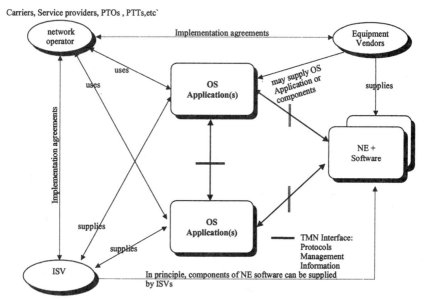

Figure 5-1. Multiple Actors of Telecommunications Environment

Given multitude of equipment and systems running different types of software, from OSS perspective (i.e. from management perspective), naturally leads to new issues that were not visible prior to the era with a single major supplier of all these components. Two key issues drove the need and introduction of TMN standardization.

The first issue relates to how OSS interfaces to equipments from multiple suppliers in order to monitor and control the network. As an example, consider the provisioning of a private line service between two locations of a business in different service provider networks. The private line service requires establishment of subnetwork connections, assignment of link connections, establishment of access from the customer premises equipments etc. Assuming a degenerate case of a single service provider with different equipment, OSS would interface with the various network elements to establish the path (private line service.). The OSS exchanges management information using possibly differing protocols and message structure to issue the commands for example to establish cross connections. A standard approach reduces the interoperability and training required to mange these different vendor equipments.

The second issue relates to communication interfaces between OSs, both within a single service provider network and across different administrative boundaries of multiple service providers. Because of the complexity and the breadth of functions to be supported in network management applications, multiple OSs work collectively to provide a complete solution. Further complications arise with OS-OS interaction, between two different providers, required when services are transported across multiple networks, owned by different providers.

Even though the two issues address different aspects of management, the need for interoperability first at the semantics level and syntax at the adjunct level is common to both issues. To overcome these issues the architecture and interface definitions were introduced in TMN standardization efforts. When products are built to the standardized interfaces, the service providers have the option to select solutions from marketplace based on price performance competitive landscape instead of monopoly. The downside of standardization often is that it inhibits innovation. In addition, standardization can become tangential or irrelevant if technological changes outpace standardization. This is the most frequent comment on TMN with regards to new technologies that emerged in IT, Internet, Java, and Distributed Processing (e.g. CORBA). Not withstanding such criticisms and despite being an old standard, TMN continues to be relevant to all types transport networks (TDM, Optical, ATM, FR, IP, MPLS, Ethernet, etc) as many management systems are built and deployed to the guiding principles of TMN even though the exact details may not be as defined in the initial set of standards.

5.2.2 What is TMN?

Figure 5-2, illustrates an abstract representation of telecommunication network consisting of management, transport, and control planes. The 'transport plane' represents that part of telecommunications network that supports customer services, such as voice or transport services (T1/E1, E3/T3, SONET/SDH, WDM, Frame Relay, ATM, etc.). This plane is also called the service or data plane that carries the user traffic.

The control plane in support of both transport and switching addresses functions for connection setup and teardown, bandwidth management, topology discovery, etc. This plane is referred to as the Control or Signalling Plane and discussed in detail in Ch 9. The control plane entities are also subject to management similar to the network elements.

The 'management plane' manages the 'transport plane' and provides management services. A 'management network' or 'management plane' refers to that part of network, and associated systems that are involved in monitoring and controlling the entities used to provide services.

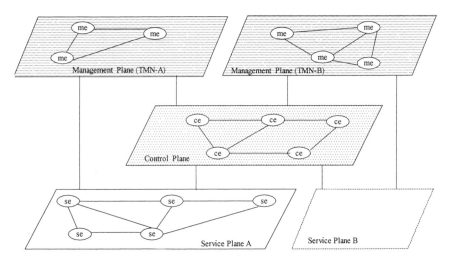

Figure 5-2. Transport, Management and Control Planes of a Telecommunications Network

CE: Control Plane Entity
ME: Management Plane Entity
TE: Transport Plane Entity

The management plane not only manages the entities used to provide services but it also provides facilities and services to manage entities of the management plane itself. The important thing to be kept in mind, and should be apparent by now, is that management plane is nothing but a network of entities, and hence the title – telecommunications management network (TMN). Entities that are part of the management network are discussed later in the context of the physical architecture.

5.3. ARCHITECTURES OF TMN

Recommendation M.3100 describes TMN using three architectures. These represent the different aspects or dimensions of a TMN. They are called functional (logical), information and physical architectures. The first dimension provides a view for organizing the management entities and loosely coupled to the other two architectures. The latter two are tightly coupled and addresses interoperability of the systems. The subsections below describe in detail these architectures.

5.3.1 Functional Architecture

The functional architecture introduces two different concepts. The first concept defines a number of function blocks that may be combined in one physical system. The second concept relates to the logical layering and can be related to the process groupings defined as part of Telecommunications Operations Map from TMF and discussed below.

5.3.1.1 Functional Blocks

Figure 5-4 shows the functional blocks of TMN along with the reference points between the functional blocks. The term "functional block" is used to emphasize the separation between logical and physical entities, a prevailing theme in the architecture definitions. The realization of the functional block requirements in one or more systems (for example a network element that includes mediation function versus two systems) is an implementation decision. Table 5-1 is a list containing the definitions of the functional blocks.

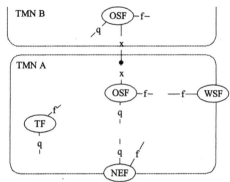

TF   Transformation Function   OSF Operations Systems Function
NEF Network Element Function   WSF Workstation Function

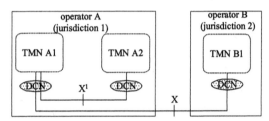

Note 1: Within one jurisdictional boundary, it may be possible to
use proprietary interaction and protocols across an x referene point

Figure 5-3. Functional Blocks and Reference Points

Table 5-1. Functional Blocks Definitions

Function Block	Definition
NEF	Implements Network Element function (i.e., the NE Agent)
OSF	Implements Operations System Function
WSF	Implements workstation (GUI or command-line interface) function
TF/MD	Implements protocol translation or mediation device function

There can be multiple instances of OSF, WSF, NEF, and TF within one instance of a TMN. A physical TMN is realized by packaging multiple instances of OSFs into one application or system, or multiple instances of WSFs into one application or system. One can also combine one or more instances of WSF plus one or more instances of OSFs into one single application or system.

Each *functional block* may contain one or more support *functional components* - user interface support, database support, message communication support etc. These functional components and how a function block is composed of these components are described in later sections.

The functional blocks support management applications known in the industry by the acronym FCAPS introduced as part of Recommendation X.700 on OSI Management. Table 5-2 provides a description of these categories. Historically prior to the introduction of FCAPS, (this is true even today) the terms OAM&P has been used specifically in ANSI to capture the management functions. Table 5-2 provides mapping between OSI FCAPS and ANSI/T1 OAM&P, though there is no strict and formal definition of such mapping[81].

Table 5-2. Definition of FCAPS and Mapping to ANSI Terminology

OSI Management Application	Description	ANSI/T1 OAM&P
Fault	Alarm reporting, log control, alarm processing, fault correlation, testing	Maintenance
Configuration	Entity installation, configuration, provisioning	Provisioning, Maintenance
Performance Monitoring	Facility/Resource Performance Collection, Monitoring, Quality of Service management	Administration, Operations
Security	Security for the management activities such as user account management, operator personnel privileges to configure/provision entities in the management and service plane	Administration
Accounting	Operator User accounts, customer accounts, billing, customer service usage meters etc.	Not clearly mapped to ANSI OAM&P. However, billing and customer account management is treated as part of Operations

According to M.3010, OSF is the smallest deployable unit of TMN management functionality. A given OSF may provide some or all functions

---

[81] ANSI T1.210-1993 does have some details on mapping between OSI FCAPS and ANSI OAM&P.

of FCAPS. For instance, there can be an OSF that supports only Fault Management Application function. As mentioned before, a set of OSFs may then be deployed to provide overall management functions of FCAPS.

As will be explained later, an operator may deploy layered management systems consisting of EMS[82], NMS etc. If a particular vendor supplies multiple OSFs then the vendor may package them in to a single executable or supply them as different executables.

Each type of service network will have domain specific management aspects that are best handled by management systems specially built for that domain. For instance, management systems specific to voice switching, DWDM transmission, SONET/SDH transmission and switching may exist. Recommendation M.3200 introduced the concept of managed areas and some examples are listed below.

- Switched telephone network
- Mobile network
- Switched data network
- IMT-2000 – 3rd generation wireless mobile telecommunication
- Transport Network – DWDM, 3/1 DCS, SONET/SDH etc

Domain specific management systems may make use of logical layering principle where by resources managed are separated into levels of abstraction. In such a scheme, Element Management Systems (EMS) are concerned with managing the element level information (circuit pack) and network management systems with network level information (end to end path provisioning). An OSF-CircuitSwitchPathManager that provides services to TMN users for circuit path management may use the services of EMS or NMS to perform path provisioning and maintenance. Domain specific systems may implement all or some areas of FCAPS though OSI system management functions tend to stay at the lower layer components such as OSF-EMS or OSF-NMS. Management services for each managed area are further expanded using function sets and functions within each set.

---

[82] EMS uand NMS are physical realizations of the logical functions included in OSF at different layers of abstraction.

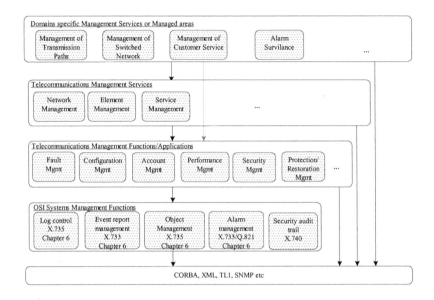

Figure 5-4. Mapping between managed areas and functions

Not withstanding domain specific aspects, the various operations performed by domain specific management systems can still be classified as belonging to one or the other areas of FCAPS. Figure 5-4 shows relationship between *managed areas* and general FCAPS functions (applications).

5.3.1.2 Distributed versus Centralized Functional Block implementation

As an architectural concept, Recommednation M.3100 does not impose implementation options for the functional blocks. For example, Operations Systems Function (OSF) can be implemented in a centralized manner running on one machine or on a set of machines. It is also possible that the OSF function is distributed meaning different entities in the management plane do different functional components of OSF. For instance, one OSF implementation may include only PM and another realization includes Alarm Correlation and so on. Another choice is to use horizontal and vertical partitioning. Horizontal partitioning involves splitting network into groups or subnetworks and assigning one or more subnetworks to be managed by one system. Vertical portioning is to deploy systems for managing element level information different from managing network level.

5.3.1.3 Decomposition of Function Blocks

M.3400 specification defines and lists, for each FCAPS application, a broad set of functions using *function set groups* and *function sets,* some shown in Figure 5-5[83]. The multiple layers of grouping were introduced to categorize the numerous atomic functions and there is no implication on implementation requirements.

---

[83] Some function sets in PM and Alarm Mgmt areas arbitrarily just to give some examples. Reader should not assume the listing is exhaustive. For all areas of FCAPS reader can consult M.3400 to see full list of function set groups and function sets.

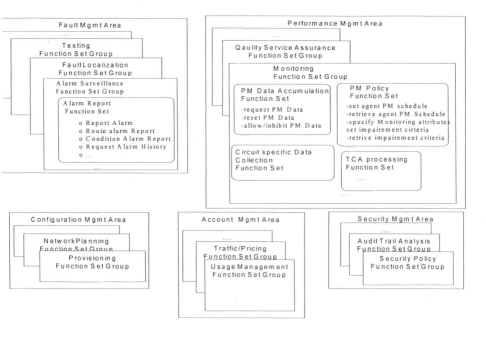

Figure 5-5. Management Function set groups, function sets and functions

The managed entities defined as part of the information architecture include capabilities that support the functions listed in M.3400. For example, a circuit pack entity includes behaviour for reporting a communication alarm when one of the ports supported fails to detect an incoming signal. An NE or OS implements many of these functions.

5.3.1.4 Functional Components

A functional block comprises of one or more *functional components*. The components are:

- Management Application Function (MAF)
- Message Communication Function (MCF)
- Security Function (SF)
- Directory Access or Client Function (DAF)
- Directory System or Server Functions (DSF)

Figure 5-6 illustrates an example on how different functional blocks can be composed from the above components. Because these functional components are closer to implementation considerations, these are now documented in Recommendation M.3013.

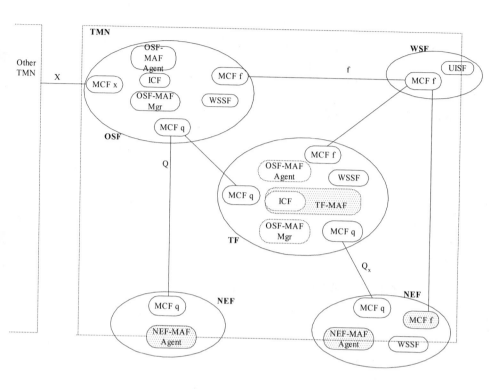

Figure 5-6. Functional Components

MAF is an application process that participates in system management. Both NEF and OSF should have MAF and all TMN messages are originated and terminated in MAF. In NEF, the MAF contains agent process that interacts with FCAPS applications. The DSF can be implemented in a stand-alone system or as part of OSF, TF, or NEF. Examples for DSF include name servers, Name to address translation service, etc.

5.3.2 Logical Layering

Logical layering within the functional architecture addresses the levels of abstractions in the management of telecommunications network and services offered by that network. Whether viewed from business process perspective discussed later or from the separation of concerns, the logical layers are useful in portioning the complex problem into manageable groupings.

The logical layers defined in Recommendation M.3100 are as follows:

- Network Element Layer (NEL)
- Element Management Layer (EML)
- Network management Layer (NML)
- Service Management Layer (SML)
- Business Management Layer (BML), and in some cases
- Customer Management Layer (CML)

The capabilities of each of these layers are easy to surmise based on the names. NEL includes functions that are required to manage the elemental level information. A network element has the primary function of providing transport or switching functions that support the services offered by the service provider. However it is also necessary to manage the network elements so that services can be provisioned and failure of HW or SW components can be monitored and corrected resulting in no or minimal service interruption. Thus the network element includes capabilities that allow monitoring and control by management systems.

An Element Management layer includes functions of a management system to monitor and control network elements or a group of them (an SDH ring for example). A system with element management function will be required in practice to understand the internals of the implementation even though there are generic standards available for management information models. The initial goal in defining TMN was to support the supplier of the network element and the management system from a different supplier. However, this has not proven to be the case for reasons mentioned above.

A network management layer considers is concerned with the level of abstraction suitable for example in a path set up scenario. To set up a path between two ends within one service provider network, subnetwork connections are to be made including the degenerate form that establishes cross connection in a fabric within a network element. NML can abstract the internal details of the network element and can address at the subnetwork level.

The choice of vertical partitioning depends on software vendor, as well as on network operator preferences. Some vendors may choose not to partition. Some may do partial partitioning, such as a single system that combines EML[84] and NML (no SML and BML). In some cases, OSS vendors may implement all layers in monolithic systems and connect to NEF using Q (old Q3) interface[85]. Alternatively, network element vendors may provide northbound OSS support from their EMS or NMS products. It is also pointed out that the layering principles led some to deploy EMS, NMS, and SMS as physically separate systems and thereby causing alarm propagation issues Since the focus of the book is engineering aspects of transport networks, service management, business management and customer management are not discussed.

These layers are as discussed above an approach to divide and conquer the complex management problem. It is not to be considered as a prescriptive for developing or deploying management systems but a reference to be used in placing a management system in terms of the functionalities supported relative to these layers.

5.3.3 Reference Points

The functional architecture specifies three primary reference points, namely q, f, x and two non-standard reference points – m and g. When a reference point introduced in the functional architecture is realized on a physical interface then it becomes an *interface*. Lower case letter is used to refer to a *reference point* and upper case letter refers to the corresponding interface. ($q \rightarrow Q, f \rightarrow F, x \rightarrow X$).

---

[84] EML and EMS are used interchangeably both here and in the industry Similarly, assume interchangeability of NML with NMS; SML with SMS; and BML with BMS.

[85] Most OSS vendors provide interface adapters that provide SNMP, Q3, and CORBA, Http support. Customization is limited to modifying the interface adapters so that a given equipment information is suitably converted for internal use by the OSS. Generally OSS solutions combine EMS + NMS. Equipment vendors typically provide EMS. Network operators' OSS can connect to EMSs or to NEs using interface adapters.

The Q interface is the interface between NE and the management system or between management systems within the same administrators domain (thus security requirements are less stringent than across X between service providers). The f reference is not very strictly defined though some functions are listed in ITU documents. This is further discussed in a later section. An X interface is used between two OSSs of the same operator or different operators. X interface provides all features of Q plus security.

Reference points are useful in developing standards and avoid implementation decisions. However in reality they do not offer solutions for the interoperability issues which was a primary motive for introducing TMN depends on standardizing the information across interfaces.

5.3.4 Physical Architecture

Based on the logical function blocks and the reference points it is possible to design a management network combining the functions into systems and introducing interfaces between the systems. One possible instantiation of TMN Physical architecture is shown in Figure 5-7 taken from Recommendation M.3100.

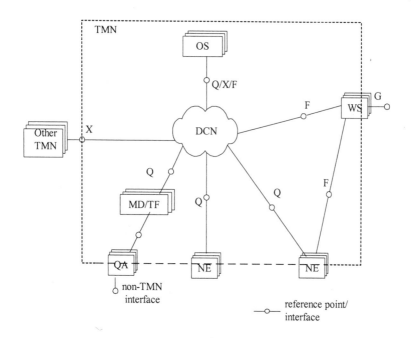

QA: Interface Adaptor
WS:WorkStation/Craft Terminal
MD:Mediation Device (M.3010/92)/TF: Transformation Function
Q reference point/interface (Q3 or Qx from the M.3010/1992)

OS: Operations Systems
NE: Network Element
DCN:Data Communication Network

Figure 5-7. Example Physical Architecture

TMN physical architecture consists of as a minimum Network Elements (NE), Operation Systems (OS), also referred to as management systems at different levels of abstraction, Work Stations and Data Communication Network (DCN). The systems interact through TMN interfaces such as F, Q, and X using the DCN.

Physical architecture is an instance or a view of a deployed or *would be* deployed management infrastructure. Usually, based on carrier business and operational needs, functional requirements are drafted for management infrastructure. For a given set of functional requirements, multiple physical deployment architectures might be possible. All such possible physical deployment architectures can be described as having generic layout as shown in Figure 5-7. It is not necessary that the entities shown must exist as individual systems of the management network. It is also not necessary that some of these systems are necessary for any carrier management infrastructure. It is more a generalization or superset of possible most granular physical implementations. It is possible to find similar implementation in an existing management infrastructure of a carrier, implemented before TMN publications[86].

The sections below describe briefly the functions supported by the systems shown in the physical architecture.

5.3.4.1 Operations Systems (OS)

Operation Support Systems (OSSs) or simply operation systems (OSs) are an integral part of any carrier network. The operation systems are management systems and depending on the resources managed they may perform some of the functions allocated to Element Management and Network management systems. These systems perform configuration and monitoring of resources used in providing the services in the transport and control planes. OSs can also be involved in management of entities of the management plane also. Typical applications implemented in OSs include billing, customer care, alarm and fault correlation, service monitoring etc. Monitoring includes detecting failures and collecting performance metrics of facilities and systems in the Telecommunications network. Figure 5-8 shows activities of Operation Systems that work at the service and network management level. The figure is based on Telecom Operations Map produced by TMF, which uses a process model to map activities of these two layers. In TMN, they would map to SML and BML, which were explained earlier when we discussed TMN layers in 5.3.2.

---

[86] Such as PDH and Voice switch network management infrastructure which predates TMN

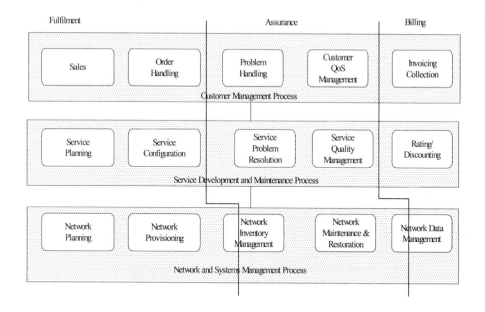

Figure 5-8. TMF Telecom Operations Map

The OS physical block shown in TMN's *physical architecture* is a generic term for management systems. Historically this term has been used to represent large systems that performed different management functions. When multi-vendor network elements have become the normal mode of deployment in Telecommunications network, terminology such as EMS for element management system to manage a collection of network elements and those that perform specific functions such as customer relation management system (CRM) have been introduced. Systems supporting CRM functions may have indirect role, while EMS and NMS are directly responsible for the operational management of telecommunications service network. Some deployments may have legacy or specialized operations systems for provisioning, fault management, or inventory management. An example for inventory management system is the Trunk Inventory Record Keeping System (TIRKS) product from Telcordia used by many North American telecommunications service providers.

The functions performed by management systems have been included as stated in the functional architecture in Recommendation M.3400. Few vendors both because of complexity and the need to maintain several legacy systems entrenched for several years in the networks largely dominate the product space for Operations systems.

5.3.4.2 Transformation/Mediation Device (MD/TF)

Transformation/Mediation Devices is used to encompass multiple capabilities some or all of which may be present in the system. These capabilities include protocol conversions between different management protocols (discussed in Chapter 7) or information processing elements such as fault correlator or subnetwork alarm managers. Depending on the need, MD may or may not be present in a deployed TMN. It is interesting to see that ITU has kept place for different protocols to exist within a TMN and hence the need for TF in TMN architecture.

It is not required to include the transformation functions performed in a separate physical device. As an example, Element Manager (EM) may include the protocol conversions if different protocols are used to interface to the NEs and higher-level management systems (for example network management system). This is typically the case when EMS is managing NEs that support only TL1 while OSSs need to interact with the same set of NEs using SNMP or CORBA. In such cases the EM would provide mediation by implementing TF/MD function.

5.3.4.3 Data Communication Network (DCN)

DCN supports communication between different entities in the management plane. A strict definition of DCN, as per TMN, is that it connects NEs, TFs[87], and OSs at the Q interface. Communication traffic is a result of interaction between various entities as mentioned before. Besides, management and G.8080 Control Plane (Ch9) applications, other applications may be users of DCN.

The various protocol layers based on OSI reference model and the functions and features to be supported across the different interfaces of the TMN are specified in Recommendations Q.811 and 812. The protocol stacks include three approaches for management protocols. These are CMIP, SNMP and CORBA based models over IIOP. These protocols are discussed in Ch7. DCN requirements to support layers 1 to 3 of the OSI Reference Model are included in Recommendation Q.811.

The network elements forming the DCN provide L3 functions such as routing, Gateway NEs are sometimes considered to be part of DCN. Gateway NE (GNE) may act as a proxy for other NEs that are within the same subnetwork (a ring or some such defined boundary). GNEs include functions such as address translation to route messages to a specific network element supporting the telecommunications service.

DCN may be implemented using X.25 links (a separate network), private communication lines (dial-up, ISDN, etc), or embedded operations channels (EOC) or whatever mechanisms are available in the service network, which is being managed by TMN. If the reader can recall from Chapter 3, SONET/SDH transport networks support EOC (using Section and Line DCC Overhead bytes) for management traffic. For OTN networks, GCC overhead bytes are available to construct EOC. In HFC networks, bi-directional management channel can be realized using some bandwidth in the 5-40Mhz range (upstream) and any of the 6Mhz bands in 50-750Mhz spectrum (downstream). In general, how DCN is constructed is not part of original TMN specification. More detailed requirements for DCN are specified in the context of control plane requirements for SDH and OTN in Recommendation G.7712.

An example DCN deployment is shown in Figure 5-9.

---

[87] Earlier MDs and QAs

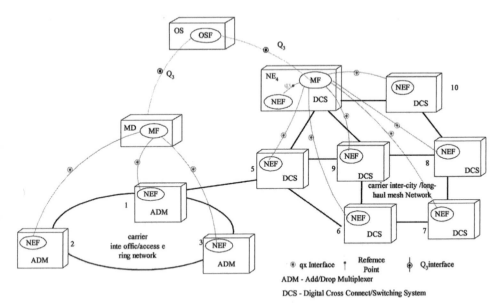

qx Interface   ↑ Refernce Point   φ Q₃ interface

ADM - Add/Drop Multiplexer

DCS - Digital Cross Connect/Switching System

In this example, NEs belonging to the ring network are managed through an explicit and external mediation device. All NEs on the ring network are represented by the MD, which implements MF. The OS accesses NE1, NE2 and NE3 by sending requests to MD only. Internally, the MD forwards the requests to respective NEs.

The figure includes an inter-city mesh network where NE4 (the GNE) implements both NEF and MF. When management traffic is received at the NE4, it forwards that traffic to the appropriate NE. Reader can notice that even though NE4-MF and NE5-NEF are in different equipments these two blocks interface is characterized as Qx and not Q. It is the same with NE6, NE7 and so on. This is to illustrate the fact that NE4, the mediation device, can hide the details of NE-to-NE communication. However, OS may still have Q interface view of NE5, NE6, NE7, NE8, NE9, and NE10 (the remote NEs or RNEs). In general, GNE to non-GNE communication can be proprietary with the OS still having a standard interface view of the non-GNEs (i.e. the RNEs).

The above example should not be interpreted to mean that a ring network always needs an external MD. Similarly, a mesh network does not require that one of the NEs should integrate MD besides NEF. In general, integration of one or more components shown in the physical architecture is possible but such integrated components still perform their assigned roles and interact with other components at the *reference points*. For instance, NEF and MF when integrated in a physical entity such as an NE interact across Q_x reference point; NE4-MF and NE4-NEF interact across Q_x reference point. On the other hand, if function blocks, such as NEF, MF, OSF, and WSF, in different physical entities (NEs, OSs) interact with each other then such interaction takes places across *TMN standard interfaces*.

5.3.4.4 Work Station (WS)/Craft Terminal

Work Station Function (WSF) provides maintenance and operations personnel friendly and intuitive presentation of management information pertaining to the telecommunications service network. It also provides commands to manipulate the NEs and their management information.

WSF can be realized using command line interface (CLI) or Graphical Interface User Interface (GUI). Many transport and data NEs (routers and L2 switches) support CLI. In order to ease the selection of choices and reduce the complexity of knowing the syntax for CLI, there is a strong preference to using GUI. It is therefore common with new equipments to offer both GUI and CLI interfaces. The syntax, semantics, as well as features, such as command completion etc, are specific to the area of management. In transport networks, the most widely used syntax for CLI is the Transaction

Language one (TL1). Routers typically implement their own syntax for CLI such as the one used with Cisco and Gated routers. Even though historically CLI and TL1 are treated as separate, in this book TL1 is considered as a class of CLI without loss of generality.

Recommendation M.3100 even though includes the need for WSF, detailed requirements are not include in any of the TMN standards. As such it is possible to define and implement different capabilities in WSF supporting different dimensions such as specific to an application or domain. For example, a WSF for DWDM networks, a WSF for voice circuit switching and so on. One can also imagine WSF associated with each physical block. WSF may also be associated with an NEF, EML-OSF, NML-OSF, etc. WSF associated with NEF is commonly called a craft interface terminal (CIT). The CIT allows operations and maintenance personnel to access and manage just one NE. The CIT may either use syntaxes and messages widely accepted in the industry or proprietary specifications. The key considerations from service provider perspective relates to training time or allocating experts for different vendor specific CTI interfaces. To minimize the impacts, a common look and feel as well as industry accepted syntax is necessary

An approach used by operators and vendors to minimize the different workstation functions is to package and integrate multiple WSFs into a single executable or application. For instance, an equipment vendor supplying multiple equipment types (such as DWDM Muxes, SONET/SDH ADMs, Digital Cross-Connects) can integrate WSF for each of the associated NEF type into one monolithic WSF. This is to maintain some commonality so that the training time required by the operations personnel is minimized. Such integration or packaging of one or more WSFs, the functional blocks in the physical architecture, into one physical entity, as far as TMN is concerned, is the choice of the carrier or the system vendors.

WSF communicates with its associated entities across an *F* interface. Where a single WSF has multiple associations, it communicates with each of the associated entities using a separate f reference point. WSF displays the received information to the user at the *G* interface. Additional processing may be performed before presenting the data to the user. As discussed later, the details of G interface has not been standardized.

There are two parts to the information exchanged at the f interface point. The first part deals with protocols such as TCP/IP, HTTP or SNMP etc. The second part deals with what is exchanged and what type of information is exchanged.

Even though WSF is defined as a separate block in the architecture, the not many details are included. The major reason for this lack of specification is a vendor who supplies OSF or NEF also supplies its associated WSF. As noted earlier it is the supplier and service providers choice on how one or more functions are included in the management system or how multiple NEs are managed not specifying all the details is an appropriate decision to make in the standards.

Figure 5-6 shows TF, OSF and NEF function blocks having a workstation support function (WSSF). A support function is nothing but some special set of functions to support WSF. The functions can include downloadable code such as Java applets or some presentation images such as network graphs or equipment images. It could also include any persistent information that WSF wants to keep on the NEF, TF, or OSF so that next time the same WSF connects to them it can retrieve that information. This will be similar to the cookie concept popular with Web browsers. Such persistent information can be shared across multiple WSFs if that is useful for a given implementation. Similar to WSF, these functions are not standardized and left as an implementation choice.

5.3.4.5 Network Elements (NE)

Chapter 1 introduced many types of equipment used in telecom networks. From TMN perspective, all these entities are network elements (NE) that are monitored and managed by OS. Thus NEs reside both in the transport plan and in the management plane. In the transport plane, each NE has its assigned role or function, such as switching, routing, etc. TMN is not concerned with such function but only the management aspects. For example provisioning a switch is a management plane activity whereas when a call is received the switching function itself is taking place in the transport plane. The resources managed are represented using information models discussed in Chapter 6. In the management plane, NE is treated as a managed entity and managed by one or more system playing the managing roles. In some cases, an NE may be managing other NEs (such NE is typically called subnetwork controller). Such an NE can be considered to have both the managing role as well as managed role.

5.3.4.6 Q Adapter (QA)

QA provides mediation services between non-TMN complaint entities that many not be able to implement a full TMN stack or has an alternate management stack. For instance, a TMN entity with OSI protocol stack and a non-TMN entity with a proprietary stack.

Even though the concept of QA was initially included in Recommendation G.7718 (draft) has introduced the transformation function to include the functions performed by QA and MD.

5.3.5   Information Architecture

Managing the telecommunications network requires monitoring and controlling the resources used in providing the telecommunications service. The management information definition principles are defined as part of the information architecture. A common understanding of the information is a basic interoperability requirement between the managing and managed systems. There are three basic components associated with the information architecture. These are the interaction model between the managing and managed systems, information model that specifies the semantics and syntax of the management data and unique identification of managed entities.

The repository of management information is referred to as the Management Information Base. There are different uses of the term MIB depending on the paradigm used for the interface protocol. The usage in the TMN recommendations is to show that a MIB represents the schema of the management information. The approach used for defining the schema is based object oriented concepts.

Entities in the management plane of TMN can interact with each other using different paradigms. The interaction model uses the manager/agent role taken by the system. This definition using roles emphasizes the point that the roles are associated with per interaction between the two systems. In some cases such as NE –EMS interfaces NE always takes the role of the agent whereas an EMS assumes the manager role. On the other hand when an OS-OS interaction model with an X interface is considered it is possible for two systems to reverse roles for different interactions.

The information models have been defined applying a number of concepts such as inheritance, state transitions, and attributes. The initial information models were specified using the object oriented models and the definitions of the managed resources were specified following the principles defined as part of a specific protocol called Common Management information Protocol. Even though the protocol itself did not reach the expected level of wide deployment, several of the models developed in that context have been adapted for use with other paradigms. One of the concerns noted when adapting the models to different paradigms is the lack of well-documented requirements and representation of management information without necessarily requiring the use of a specific protocol.

5.3.5.1 Modeling Methodology

In response to these concerns, the initial version of Recommendation M.3020, Methodology for TMN specification was modified to use industry-accepted approach. The name used in the current version of Recommendation M.3020 for the methodology is called Unified TMN Requirement, Analysis and Design (UTRAD). Recently published Recommendations follow this approach where requirements are documented in terms of use case descriptions, an analysis that provide the managed entities and the information associated with them and protocol specific design. An example is found with Alarm Report Control function discussed in Chapter 6.

The UTRAD methodology does not mandate a specific technique for documenting requirements and analysis. However the recommended and used approach is the application of Unified Modeling Language (UML) that has been widely used in SW and system designs. Several books are available on the topic and reader is encouraged to get the syntax and semantics of the language from the books and the standard available from Object Management Group (OMG). In the in the context of Telecommunications Management information definition this chapter or the book is not discussing the capabilities offered by UML in general but how it is applied to this specific application. Examples of using this technique are illustrated in the information models discussed in Chapter 7. Note that UML is defined to use a number of object-oriented concepts that were already part of the models available in ITU Recommendations prior to the introduction of Unified TMN Requirements Analysis and Design (UTRAD) methodology.

5.3.5.2 Requirements Phase

The requirements in UTRAD are specified in terms of use cases. The system being defined consists of use cases and one or more actors interact with these use cases. The actors represent the entities that interact with the use cases bounded by the system. Examples of use cases include telecommunication service providers, management systems, and operator. A use case is represents the functional requirements the actor expects and is defined using an ellipse with a name included inside it. The details of a use case are defined with the information in textual and or tabular form. There are minor differences in the use case definitions amongst different standards and also with defacto standards from groups such as TMF. However these differences are not significant and are not necessary to discuss here. The information included is as follows: a summary field used to explain the purpose of the use case, actors, pre-conditions that specify the necessary steps that must be in place before the use case can begin, description of the expected behaviour along with the trigger for the start of the use case, exception condition when there are failures in the execution of the requests, and post conditions that define for example the state of the entities at the end of completing the various operations associated with the use case.

An example of the use cases and use description is shown below from Recommendation Q.834.3 for testing. Figure 5-10 identifies the use cases and the actors.

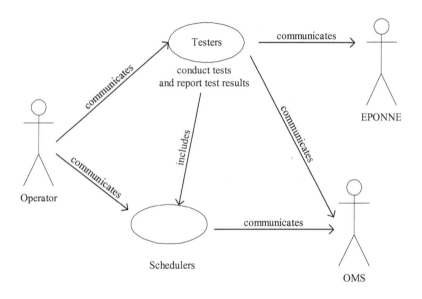

Figure 5-10. Use cases for testing

Multiple actors are noted in the figure. These are different management systems or an operator. The use case may require the inclusion of other use cases. In the above example, scheduler is a use case commonly used in many other use cases. For example other use cases requiring similar features include performance monitoring, traffic management etc. This example represents the use cases for the EMS to NMS interfaces and thus shows the network element as an actor. The system considered for these use cases is the EMS.

Providing a textual description along with pre, post and exception condition is necessary to understand the expected behaviour. One pre condition for testing is the NE resources are installed and provisioned before testing can be initiated. The description includes the different types of possible tests, for example physical level tests for quality of data transmission, on demand self test, loop back test etc. The exceptions raised such as unknown resource, invalid time out are identified. The post condition defines the resulting state and in this case the service provider has additional information to resolve a customer trouble.

5.3.5.3 Analysis Phase

The second phase of defining the information model is the analysis. Several UML techniques are used in this step. Note that UML techniques are based on object oriented concepts. Three techniques are used in most analysis. The first step in the analysis is to define the object classes along with their properties. The properties include attributes; operations permitted on them (read, write) and other object oriented operations (perform test with input and output parameters). The second set of diagrams used for objects that have state information is a state transition diagram or state model. They define the events that result in the object transitioning between the different states as a result of events. The third common diagrams seen in the analysis phase is the interaction diagram defining the object-to-object information exchanges in time sequence. Even though UML itself does not include the notion of events or notification, this is a key element in network management. Events are reported as a result of failure, threshold-crossing events etc. that are critical for the service provider to maintain the network operations successfully. To support this in a standard approach for all technologies, Recommendation M.3020 defines keywords and methodology for modeling events in the object. It introduces an object called Notify dispatch and a send keyword (stereo type in UML terminology). Upon the occurrence of various events, the objects send the notifications to the Notify Dispatch object and by registering with this object the events are retrieved by the subscriber of the events.

5.3.5.4 Design Phase

The result of the first two phases is an information model that is not specific to an interface management protocol. The classes, interaction diagrams with the messages exchanged and the state models can be used with different protocols such as Simple Network Management Protocol (SNMP), Common Management Information Protocol (CMIP), Common object Reference Broker Architecture (CORBA) or XML. The design phase is a mapping of the results of analysis on the protocol specific definitions. Consider the use case discussed above on testing. If an object class for a circuit pack is defined with loop back test capability, the mapping in SNMP is a definition of a table with the ability to set a loopback test state that will trigger the testing. In the case of CORBA the circuitpack object class definition can map directly into an object definition. In XML, a message definition or schema will specify the information transferred when the test is requested. Chapter 7 discusses the various network management protocols and as a result the different mappings necessary to meet the protocol definition will become clear to the reader.

5.3.5.5 Naming

The UML methodology has been adapted to the needs of network management. The class definition of UML represents a managed entity also referred to as managed object (MO). Each MO models some physical or logical resource. For instance, MOs represent circuit packs, SONET/SDH ports, cross connects, subnetworks, and so on. In order to uniquely identify a managed object, logical containment is used. Figure 5-11 shows an example of the management information tree that defines a unique name to each managed entity.

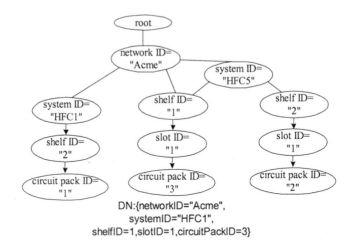

Figure 5-11. Naming Tree Example

In the above example the distinguished name (DN) is the unique name and the tree structure is an approach to define such a name. The circuit pack 2 is named relative to the slot, shelf and system it is contained in thus forming the unique name by traversing down the tree. The containment is not restricted to physical aggregation as shown in this example. The naming and addressing schemes are also dependent on the management information protocol used (SNMP, Http/XML, CMIP, etc). Further details with examples are discussed in Ch7.

5.4. RELATIONSHIP BETWEEN TMN ARCHITECTURES

The preceding sections described the three different architectures forming the overall TMN architecture. Figure 5-12 shows the relationship between these architectures.

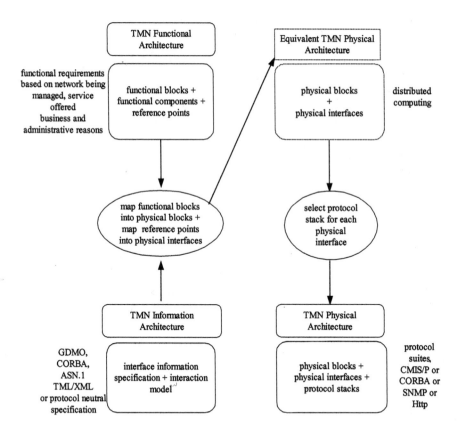

Figure 5-12. Relationship between TMN Architectures

The functional architecture and the information models combined with interface protocols are applied in the physical implementation. The physical architecture and decisions on the exposed and proprietary interfaces are left to the service providers. Thus the specific functions supported by the interfaces and protocol choices, distributed versus centralized management are determined as part of deployment. A mapping is therefore necessary for each deployment scenario from the logical and information architectural components to the physical realization.

5.5.   RELATIONSHIP BETWEEN CONTROL AND MANAGEMENT PLANES

As noted earlier the management architectures and functional specifications were developed prior to the introduction of automatic switched optical networks (ASON). Functions such as discovery, path set up etc are now specified as part of control plane. Because the resources are common to the control and management planes, additional coordination in allocation, monitoring and control of resources is required with the introduction of SDH and OTN that are enablers for ASON.

Recommendation G.7718 (draft at the time of writing) introduces additional functional block to the existing TMN blocks. Figure 5-13, taken from G.7718 draft, introduces the control plane function along with previously defined management blocks in Recommendation M.3010. The control plane functions represented as CF define capabilities required to monitor and control the network element resources used by the control plane. Even though control plane functionality is part of the network element function, it has been separated out in the Recommendation in order to show the need for coordination between the activities of the two planes.

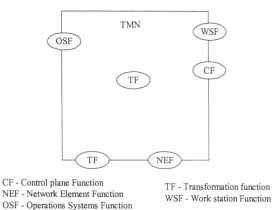

CF - Control plane Function
NEF - Network Element Function
OSF - Operations Systems Function

TF - Transformation function
WSF - Work station Function

Figure 5-13. TMN Functional Blocks in the presence of control plane

CF   Control plane function
NEF Network element function
OSF Operations systems function

TF   Transformation function
WSF   Work station function

Table 5-3 from G.7718 shows the mapping between the functions performed by the control plane and logical layered architecture discussed earlier. A third column is added to denote the managed functional area where this component would have been placed in the absence of control plane.

Table 5-3. Mapping between Control Plane and TMN functional components

ASON Component	Logical Layer	Management Function
Call Controller	S-OSF – N-OSF	Trail set up and tear down
C onnection Controller	N-OSF	Link connection set up and tear down
Discover Agent	E-OSF – N-OSF	Configuration audit (MIB information)
Link Resource Manager	N-OSF	Link set up and tear down
Protocol Controller	Note 1	Note 1
Routing Controller	N-OSF	Route setup and management (pre-provisioning prior to ASON)
Termination And Adaptation Performer	E-OSF	Configuring the ports to support specific rates which results in performing adaptation between client and server layers

5.6.   TMN REALITY CHECK: WHAT HAS GONE WRONG WITH TMN?

Telecommunications networks are spread geographically, and as mentioned before, consist of a variety of equipment and systems. Not withstanding such geographic spread and diversity, various entities of the network work together, under the supervision of a single or multiple management systems, to provide consistent and reliable telecommunications services to customers. Collectively, the network elements, and the management systems that manage them represent a distributed computing environment of a large magnitude. One of the criticisms of TMN has been that such distributed computing nature was not taken into account. However, there is sufficient understanding and interoperability between the entities involved to make the network function as it does today across the globe without using the distributed computing concepts. More discussions of this topic can be found in Ch 7.

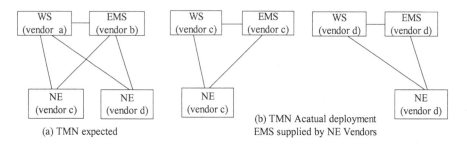

Figure 5-14. Expected versus Actual Deployment Scenarios

Figure 5- 14 shows the original goals with which TMN was introduced and the way it has been deployed today.

The original expectation with the standardization is illustrated in Figure 5-14 (a). Products and solutions from different vendors would be used in a TMN environment with standard interfaces and interoperate seamlessly. In particular, it was anticipated that NEs, supplied by different vendors but each one hosting a TMN complaint NEF, would be managed by OSS solutions from other vendors (including NE vendors). In general, the focus of TMN was interoperability across OS-NE interface and thus considerable TMN standardization effort was spent on this. However because of multiple factors this multi-vendor interoperability for NE-OS interface was never materialized in any large-scale deployments.

NE vendors implemented their own Element Managers to take care of vendor specific configurations. This is primarily because the capabilities of NEs have started to grow. Every NE vendor packaged one or more proprietary capabilities and features to give added value over their competition. In addition some management activities on an NE may require participation of other NEs. As an example, circuit provisioning in a DWDM NE may require proprietary commands to be sent to NEs at the two ends of the DWDM circuit in a proprietary DWDM transport systems. Standardization of such capabilities takes longer than required by the current trend for reduced time to market considerations.

Hardware and software technology advances together with new network technologies, architectures, and applications, had impact on management systems and their evolution. In many cases new equipment was deployed that needed workable management solutions without necessarily taking the TMN track waiting for standaridization. On the software front, 90's saw advances in Computer Science, IT, Internet, and in general, software related technologies. Object Oriented (OO) technologies such as C++, Java, and CORBA became much more easily available enabling one to abstract implementation details. Such abstraction enabled software re-use and modularization. Advances also occurred in GUI with the introduction of MS-Windows, and Java. In addition, TCP/IP protocol family became the default transport protocol of choice for non-telecomm centric equipment, such as routers. Newer telecommunications equipment included support for TCP/IP transport as the default transport protocol. Web-based management tools and services using Http and XML began to influence new management plane designs. Similarly, on the hardware front, following Moore's law[88], larger and faster memories, as well as faster CPUs, at lower cost points became available thus making NEs more intelligent. Memory and CPU power[89] offered by NEs are beginning to equal desktop/servers. As a result, many functions can be integrated into NE. For instance, one can package EMF and NMF on an NE. While this is an extreme case, it is feasible, with ring-based or otherwise topology/geography bounded subnetworks, for one or two designated NEs to fulfill EML and NML roles besides their NEL roles.

Efficiency, flexibility, and productivity attributes became more valuable than software size, performance, and conformance. The combined cost of memory and CPU power can be less than the cost of manpower required to build compact and efficient systems. Software development productivity can be improved by using more memory, faster CPUs, and more importantly

---

[88] Doubling of number of transistors on a chip (CPU or memory) every 18-24 months, articulated first by Gordon Moore.

[89] 400 Mhz CPU in NEs wouldn't be uncommon while Ghz CPUs are being considered as main processor in newer NE designs. Compare this with 2 Mhz 8 bit processors in eighties and perhaps 16-40 Mhz processors in early nineties.

general-purpose tools and technologies compared to using TMN specific ones. Such productivity gains can result in reduced development cost for vendors as well as reduced deployment time and cost for operators. At the same time, operations personnel productivity can be higher if they have to use similar systems, tools, and technologies in IT, Internet, and in network management.

Training costs reduction was considered one of the positive influencers for standard interfaces. However this has not been achieved with proprietary features added by suppliers with the advances in transport technology. Over a period of time the motto became – '*keep management operations transparent and atomic*' with operators needing to know only the function accomplished and not the details. Requiring that operations personnel be trained with ever changing domain specific technologies is not cost effective for carriers. For example within a single domain, such as DWDM transmission or voice switching, many details and variants exist that require training. Service providers may introduce deviations from standards to reduce operational expense (Opex). Also, they might have been forced into deploying new equipment and associated non-TMN based OSs to introduce new services to support increased revenue.

The above factors began to have impact resulting in varying levels of TMN adoption by vendors, and deployment by carriers resembling that shown Figure 5-14 (b).

There is a trend in the industry towards changing the service network architecture that would require less number of OSS. As explained in Ch1, one option would be a move towards converged network architectures (along with using Control Plane). Besides service network simplification, consolidation of billing systems across multiple service networks (PSTN, Wireless, Cable etc) may be required. The new generation operation system and software (NGOSS) as part of TMF effort has defined infrastructure and business components that offer service providers with plug and play features.

5.7.   CONCLUSIONS

TMN Architecture is a standard that has enumerated various entities, their roles, and interactions between them in the management plane. The standard defines *functional architecture* consisting of functional blocks, *physical architecture* consisting of physical blocks, and finally *information architecture* consisting of management information. TMN architecture should be considered as guidelines and principles useful for implementing and deploying effective telecommunications management systems. Vendors using the guidelines and principles can offer interoperable interfaces to the network elements and management systems. Carriers and service providers can use the concepts in the standard to build management plane for managing their networks. Following the guidelines in TMN architecture will enable one to build management infrastructure that will be consistent, reliable and scalable, and one that is capable of managing diversity of service networks. Scalability will allow managing a small local network to managing a large national/international network; with few network elements and systems to hundreds of network elements and systems. Diversity will allow operators to manage different types of service networks such as circuit-switching networks, packet switching networks, storage networks, application service networks, and hybrid next generation networks that provide combination of services - optical transport and switching, TDM transport and switching, data transport and switching, etc. Overall, information in this chapter should help in forming architectural view of management plane and prepare the reader for the information presented in following two chapters as well as control plane discussion in Chapter 9. .

REFERENCES

[ANSI93]   ANSI T1.210-1993, Operations, Administration, Maintenance, and
            Provisioning (OAM&P) – Principles of Functions, Architectures, and
            Protocols for Telecommunications Management Network (TMN)
            Interfaces
[ANSI94]   ANSI T1.215-1994, Operations, Administration, Maintenance, and
            Provisioning (OAM&P) – Fault Management messages for interfaces
            between operations systems and network elements
[ANSI97]   ANSI T1.204-1997Operations, Administration, Maintenance, and
            Provisioning (OAM&P) – Lower Layer Protocols   for
            Telecommunications Management Network (TMN) Interfaces, Q3 and X
            Interfaces
[BCR]   GR-831,Operations Application Messages - Language for Operations
            Application Messages, BellCore
[ITU00a]   M.3400, TMN management functions, ITU-T, 2000
[ITU00b]   M.3020, TMN Interface Specification Methodology, ITU-T, 2000
[ITU00c]   M.3010, Principles for a Telecommunications Management Network, ITU-T,
            2000
[ITU00d]   M.3013, Considerations for a Telecommunications Management Network,
            ITU-T, 2000
[ITU01]   G.874, Management aspects of Optical Transport Network Element, ITU-T,
            2001
[ITU03]   Y.1703, Architecture and specification of data communications network, ITU-
            T, 2003
[ITU97]   M.3200, TMN management services and telecommunications managed areas,

               ITU-T, 1997
[Jos00]   Inter-Domain Management Layer (IDML) for the Management of Hybrid
               Circuit/Packet Networks (HCPN), C. Anthony Cooper, V.Josyula, T1M1.5
               contribution 2000-219.
[Jun98]   "Follow-up to Previous Proposal for a more rigorous definition of TMN
               Architecture", Antonio Rodriguez-Moral, Junao Xue, Contribution to
               T1M1.5, 1998

Chapter 6

MANAGEMENT INFORMATION MODELS

6.1. INTRODUCTION

Chapter 5 introduced three components of management architecture. A key component is the information exchanged between managing and managed systems to monitor and control the elements forming the network. This chapter addresses this component with emphasis for SONET/SDH management. Even though the focus here is management of SONET/SDH network elements in most cases the requirements and models are applicable to also network elements supporting different technologies in transport and switching networks (data, Optical networks etc.). The management information is defined using schemas similar to what has been done for several years in SW development of data base systems. The schema definitions are also referred to as management information models. The schema, in data base models simplistically stated, define using entities and relationships between them. The semantics of the information is specified in addition to constraints on the values, allowed operations. There have been several new trends for information modeling recently to support object oriented databases, software design and deployment. Thus the development of information models is not unique to network management and has proved to be a powerful technique to define the information without being influenced by the development environment such as programming languages, operating system, processor and memory types etc.

The early development of information models discussed in other books and articles in the literature were influenced by what is available in standards. The discussions in this chapter compared to the models from standards differ in two key areas. The models used as examples are to show how the technology aspects of SDH discussed in the previous chapters managed using the information models. These models derive the requirements from SDH architectural definitions in Recommendation G.803, equipment functions in Recommendation G.783 and SDH management in G.784. The second aspect pertains to the description of the models. These were based on a specific technique most suitable for use with specific management protocol. In this chapter, the models are described such that they can be adapted to different protocols. The focus is on requirements, properties and interactions of the managed entities instead of the format used in describing them.

6.1.1 History of Information Modeling

Even though an exhaustive history is not intended here, this is to briefly note the various techniques used in the information models developed in support of network management. The earliest approach was similar to the efforts with data base definitions. The simplest form is a tabular representation of the various properties associated with an entity. The columns of the table correspond to the attributes and the rows when populated define the instances. This simple approach offers a static definition and does not include the dynamic properties. These are also called as "views" in some applications. Different views may be obtained from the table by selecting specific columns. In order to uniquely identify the instances in the table, keys made of one or more column values are used. This simple approach lacks the ability to explain the relationships between entities and how changes to one impact the properties of a related entity. In addition information such as cardinality of the related entities are not specified. An example of this approach is seen in the Memory Administration definitions developed by Telcordia in GR 199. Different views are defined and the data in these views are operated with a small set of commands using a language (also a protocol in this case) called Transaction Language 1 discussed in more detail in Chapter 7.

The entity-relationship (E-R), a diagrammatic approach solved the problem of explaining the different type of relationships between the entities and based on the semantics of the relationship type outcome of a specific operation can be understood. One example is dependency relation. Suppose entity B is dependent on A for its existence, then it is understood that deleting A will imply either deleting or B or A is not deletable until B is deleted. In addition cardinality of the relationship is also expressed in these diagrams. Many of the models developed for network management though used the object oriented definitions, displayed the relationships using E-R like diagrams. The advantage with this approach is simplicity and enhanced understanding from the visual representation. The disadvantage however is the inability to express the behaviour aspects. The early standards including those defined for SDH management discussed in this chapter include E-R diagrams to describe the relationship between entities. These diagrams illustrate both the dependency and key association relationships amongst the entities.

The object oriented methodology was introduced in network management with the development of Recommendation X.720. While some of the modeling tools and aspects of the methodology are easier to apply with some protocols than others, a key element of the methodology necessary for all protocols is the behaviour definition. The behaviour definitions include for example state models of the entity, interactions across multiple attributes of the entity. However these models did not explicitly include interactions between objects, time ordered sequence of information exchanged between the entities and parallel activities that may be relevant in some cases. To overcome this defect the recent efforts in modeling address not only defining the characteristics of the managed entities but also interactions between them as part of the analysis phase. This chapter is designed to include examples of the information models including where appropriate time ordered interactions.

The models described in this chapter are taken from Recommendations initially written to use a specific protocol. The approach taken here in describing the models does not require the reader to understand the specification language but only concentrate on the requirements and how the different entities are defined to meet them. The methodology of a three-phase approach called Requirements, Analysis and Design has now been accepted within ITU and used in many of the new Recommendations as well as in specifications from International Groups such as TMF. The protocol specifics are reflected in the design phase.

Even though not specifically used within SDH management standards, XML is gaining a lot of popularity because of the expected ease of implementation. XML schemas have been defined for some applications and many of the concerns expressed above are also valid for these cases. The schema definitions are specified using XML constructs (discussed in Chapter 7) and while these are simpler than a programmatic language used in SDH and other standards, they lack documentation of requirements and behavioral aspects. Behaviour may be translated into annotations as specified in Recommendation M.3030. With GDMO, the behaviour description has been a subject of controversy. Even though it is merely a text description, the logic should be implemented and executed at run-time.

6.1.2 Current Approach to Information Modeling

The three phase methodology was introduced to address the concerns from user and implementation levels. Most of the early OO models defined them in a specification language and did not document the requirements from what a user (network operator) has to do to monitor and control the network. As a result it was not possible for those not involved in the specification development to determine either the rationale behind the definition of the objects or how combining them in an implementation meets the needs of a management application. The users had to learn the notation in order to understand how the models address the requirements and thus resulted in adopting these definitions. These definitions often did not include the interactions between the objects and thus developers had to analyze these models to identify them. The new methodology with the phases for clearly documented user requirements and analysis of the management information including interactions and state definitions addressed these concerns. This approach is similar to what is used in the industry when developing complex system or software.

One of the techniques for defining the requirement and analysis is to use Unified Modeling Language (UML). The advantage of this technique is that it offers visual modeling of many aspects: object interactions, time ordering for exchanging messages, state transitions, conditional execution of activities etc. Many of the new recommendations and some of the functions described in this chapter apply this methodology for the definition of the information models. Examples include alarm report control, access network management of Broadband passive optical network and Emergency Telecommunications Management Services. The features available with this technique are described in Chapter 7. Many of these features have been applied to the information models discussed in this chapter. Another effort built upon UML called as Model driven Architecture (MDA) is also now prevalent in some applications. A brief introduction of this effort is also included in Chapter 7.

Because this three-phased approach was introduced only in late 1990s many of the models in SDH standards models in G.774 series recommendations do not follow this approach. It should be noted that in describing the information models for SDH/SONET management in this chapter, the existing recommendations are adapted to provide the requirements and analysis from an operational perspective instead of basing it on the GDMO notation used in the standards. In other words, this chapter takes the approach of extracting the requirements and analysis of the model without being concerned with the representation required for a specific management protocol.

6.1.3 Dimensions of managing a SDH networks

Managing any network, specifically transport networks require consideration of managed entities from at least two different perspectives – architecture and technology. The generic and SDH specific transport architectures described in Recommendations G.805 and G.803 form the basis for these perspectives. These recommendations define multiple levels of transport hierarchy using client server paradigm. The management models define entities at the client and server layers and what is managed when client layer information is adapted to the server layer.

Three other recommendations influence significantly SDH management information model. These are equipment functional blocks in G. 783 and SDH management in G.784[90]. The equipment functions represent capabilities required for the transfer of payload on the data plane. However some of these functions requires management plane interactions such as configuring the trail trace identifier or reporting a threshold crossing event to a management system. Recommendation G.784 discusses the following areas: various management architectures such as multiple network elements located in one site; communication interfaces; and the management functions in the different areas such as controlling protection switching parameters. The third Recommendation from which the SDH management functions are derived is Recommendation M.3400 that details the atomic functions for managing the resources that are within an NE and those that span multiple NEs in a subnetwork.

There are two basic dimensions (alternately called as views) addressed by the management models for the transmission technology - node level or network element level and network level. Modeling the network level inherently shows the recursive nature of the network as explained in the architecture. These two dimensions have formed the basis for how management systems are being developed and deployed in many service providers networks. Even though the original standards were developed to allow for different vendor network elements to be managed by third party management system this has not been the reality in the market. Even though vendor specific element management systems and standard interfaces to network management systems is becoming the norm, it is necessary to realize that the information models for element view are still valuable. In

---

[90] For Optical transport technology Recommendation G.874 defines the management requirements.

many management applications the network level management system must be made aware of the element level information for determining corrective actions. An example is reporting alarm when an equipment failure in an NE results in generating a loss of signal versus a cut in the facility connecting the network elements. Based on the alarm reports, a network management system can perform further analysis to determine the root cause of the problems, take actions to correct the problem by dispatching personnel to the location for replacing the failed entity.

The examples below address both these dimensions.

6.1.4 Generic and Technology Specific Models

The information models irrespective of the protocol for exchanging management information have been defined using object oriented design. Recommendation M.3100 forms the basis for most of the managed object classes (MOCs) or managed entities defined in SDH management. The object classes in M.3100 are considered to be generic and this term is used in two different contexts. In the first context the generic classes are at the top levels of the hierarchy such that the properties associated with them are applicable to multiple technologies. These classes are further refined to meet the technology specific requirements. As will be seen with the examples for SDH, the network element models for terminating an SDH signal includes additional characteristics beyond the generic definitions. The second usage of the term is to allow management of the network without being concerned with specific technologies used in the network. This is often used to manage for example end-to-end connections where the various subnetworks that support them may use different technologies. A network management system configuring a service for a customer between two end points may not be concerned about the technologies used but assuring that the quality of service and service level agreements are satisfied.

In addition generic information models in support of functional requirements are also available for use with different technologies. A simple example is the alarm reporting function where object classes such as alarm record are applicable irrespective of the technology. These are not resource object classes like termination points for signals but are required as support infrastructure for nodal and network level management. Other books in the literature have explained these models. The goal of this chapter is to show SDH specific resource models.

6.2. TRANSPORT ARCHITECTURE

This section introduces how the transport architecture defined in G.803 forms the basis for defining both NE and network view managed entities representing the resources. Some key concepts that influenced the identification of the entities such as layering and partitioning are noted in the subsections.

Four topological entities have been identified to describe the network in abstract terms. They form the basic entities used in describing a network. These are layer networks that define characteristics of a layer in the digital hierarchy, subnetworks that are connected together to form the network, links connecting the subnetworks and access groups representing the termination of a trail which can be simply considered to be at the end points of a layer network.

Figure 6-1 from G.805, the generic architecture from which SDH specific extensions in G.803 were defined illustrates a key concept in defining the digital hierarchy. Without going into the details already covered in the previous chapters relationship between the layers of the network is defined in terms of client and server. The information models reflect this relationship between specific entities. The two components to describe the elements of the client server relationships are transport entities, and transport-processing functions.

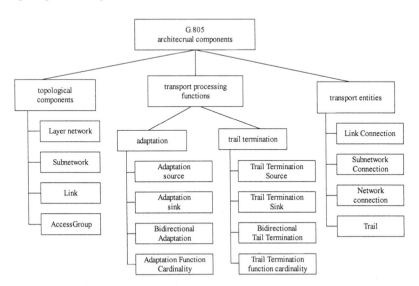

Figure 6-1. G.805 Transport Network Architectural Components

Transport entities support the transparent transfer of information across a layered network and the two basic entities are connections and trails. From client server perspective in simple terms a connection is a client served by a trail. A trail represents the transfer of a client layer signal across a server layer network. An example is a DS3 signal transported across an STS-1 layer network. The client signal is DS3 and the trail is the STS-1 connection setup in the STS-1 path layer network. The STS-1 path layer is the server layer that provides transport services to DS3 client layer network. As discussed in chapter 3, it is possible to monitor a trail using for example trail trace identifier. Even though in general this feature is not present in connection, recent extensions discussed in Chapter 3 introduces the ability to perform tandem connection monitoring (TCM).

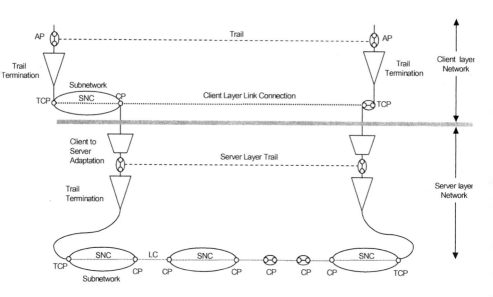

Figure 6-2. G.805 Transport Network Functional Model

The two categories of transport processing functions are adaptation and trail termination functions. The adaptation function is used to map/re-map client layer information for suitable transfer using the server layer trail. For instance, DS3 signal being transported across STS-1 path layer network is first mapped into STS-1 SPE network and such mapping is part of adaptation. Trail termination function uses the adapted information and adds (source) or removes (sink) information required for monitoring. Continuing with the earlier example, once DS3 is mapped into STS-1 SPE, the trail termination source function can calculate the value for B3 path overhead. This value is used bit-error-rate (BER) monitoring. At a destination NE, trail termination sink function extracts B3 and calculates trail bit error rates across the SONET path layer network. Both adaptation and trail termination functions are part of server layer.

In determining the managed entities, it is not required to have a one to one relation with those in G.803/805. If the entity or function in transport architecture is not managed then a managed entity is not required. In some cases, it may be appropriate to combine G.803 entities from what is managed in the network. An example is the adaptation function described above. This function is not visible to management and thus not explicitly modeled. In other words there are no management operations performed on this function and it is included in the definition of trail termination class.

The next two subsections discusses two other concepts that is reflected in the management plane. These are known as layering and partitioning.

6.2.1 Layering concept

Layering is associated with the vertical direction and is based on the client server paradigm discussed above. The earlier chapters introduced various transport technologies and service providers will continue to deploy new technologies without replacing existing ones in the network. In other words the amount of investment in deployed networks does not permit swapping with new technologies as replacements. The layering concept facilitates introducing new services and technologies over existing transport networks. For example, when SONET/SDH signals are used for transport this is a layer 2 technology that can use fiber optics or copper based physical layer 1[91]. Mapping has also been defined for transporting TDM/PDH signals in a SONET/SDH frame. Services offered by Frame Relay, ATM, IP/MPLS may be transported over different layer 2 technologies. New developments

---

[91] Sometimes the terminology used is layer 0 for optics and layer 1 for SDH.

with Generic Framing Procedure (GFP) combined with virtual concatenation discussed in the previous chapters allows flexible mapping between client and server layers.

The relevance of layering concepts, which allows different server layers supporting various clients is represented as follows in the context of network management. There are managed entities defined for each layer both in terms of the network element view and network view. In addition, there are configuration functions that may be necessary in setting up for example virtual concatenation.

6.2.2 Partitioning Concept

This concept is used to capture the fact that any network need not be a monolithic entity but may be composed of multiple number of networks often referred to as sub-networks. This division of networks in terms of constituent sub-networks is considered as partitioning the network in the horizontal direction. There are many aspects to partitioning a network. The sub-networks comprising a network may have different characteristics: operated by different or same service providers, support different transport technologies and have different security policies.

From management perspective, the concept has several implications. In managing an end to end service or connection a number of entities are identified that are controlled and monitored at different levels. In addition, when exchanging information across different administrative boundaries of partitions access privileges and authentication requirements are to be considered. These impact how the management interfaces are defined. In contrast to the layering concept which is driven by the technology perspective, the decomposition into sub-networks is influenced by factors such as deployment architecture, economics and business drivers of the service or network provider. Some of these are reflected in management view and others are outside the scope of standards.

6.2.3 Concept of Planes

While layering and partitioning concepts are described as part of transport architecture, another concept of relevance has been introduced in the architecture developed for Automatic Switched Optical Networks (ASON) in G.872. This is described in terms of "planes" and is not completely new to telecommunications. In switching networks the need for separating user traffic from signaling and management traffic resulted in introducing three planes: transport (data), control and management. With the introduction of Automatic Switched Optical Networks (ASON) in transport architecture, the same concept was introduced in transport networks. ASON facilitates dynamic allocation of resources in real time using signaling mechanisms in contrast to existing approach where configuration is done only may management systems. The control plane is used for signaling a path set up, discovery of connectivity information, routing etc. More extensive discussion of Transport Networks control plane including G.8080, ASON architecture is included in Chapter 9.

Though the planes themselves are not necessary to understand the models for management, there are interactions that must be considered as one moves to the automatic switched optical network. Resource allocation for example need to be coordinated between the control and management planes. Changes to state of the resources allocated by control plane should be communicated to the management plane. Thus the introduction of control plane while supporting a more dynamic method for call and connection control also necessitates additional functions to support proper allocation and control of resources. Chapter 9 discusses the necessary coordination required between the management and control plane for these functions.

6.3. GENERIC INFORMATION MODELS

The previous section introduced key transport technology concepts that impact the management functions to maintain and support the quality of service expected by customers. With this back ground, the next step is to define the entities and information exchanges required between the managed and managing systems.

The telecommunications network management architecture described in the previous chapter emphasized the importance of information modeling in support of one of the three dimensions (communication, management functions and management information models). Before introducing relevant managed entities for SONET/SDH technology, common terminology is expressed in this section.

Information models are built using object oriented modeling concepts. In developing the models in standards, the term "generic models" was introduced. There are two contexts in which this term is applied. In the first context, managed entities include properties that are applicable to different transport technologies. In the second context, managed entities defined are applicable across multiple management functions and technologies.

In the first context, the managed entity represents what is managed but specializations of these entities are defined to include technology specific characteristics. As an example, the transport architecture defines a function for originating and terminating a trail (an end to end path to describe it simply). A trail termination as a generic entity includes technology independent properties such as alarm status; rate and format of the signal etc. The generic definitions are expanded to include technology specific information and new (SDH) trail termination managed entities are developed. SDH trail termination managed entities are modeled to address different layers at which the signal may be generated or terminated such as regenerator section, multiplex section etc.

In the second context, generic entities are used to support multiple management functions. Another term often used for this is "common" entities. These are entities that support multiple technologies and management functions. An example is a log entity that is generic for storing information related to events such as alarms, security violation, performance threshold crossing alerts etc. independent of either SDH or PDH entities are emitting alarms or the event is resulting from fault monitoring or security management function.

Another terminology seen in some of the standards is called "fragment". This is discussed in the next section.

6.3.1 Concept of Fragments

A fragment is used to group together entities that perform similar functions. It is not a term that has implementation significance. For example there is no requirement or even appropriate to expect that an NE should implement all entities grouped in one fragment. Nor is it to be expected that implementing one fragment is all that is required to support a type of managed entity. For example, in Recommendation M. 3100 cross connection fragment brings together all entities that are appropriate for network elements such as digital cross connect performing the cross connection function. However a fragment such as termination point defined in M.3100 is not complete because technology specific termination point classes are included in other Recommendations. Even if it includes various types of termination points it is not necessary that all network elements that terminate a trail will support all termination point managed entities. For example suppose the termination point fragment includes all SDH terminations, a regenerator is expected to support only those termination points relevant for the transport functions it performs and not others such as different types of path terminations.

Considering fragment as grouping of similar entities, there are different approaches used in various standards. As noted above, Recommendation M.3100 includes in each fragment all entities that are similar in terms of properties or those that address a certain feature of the network element. Another example of grouping is using a multi part document series. SDH uses this approach where all performance monitoring entities are included in one recommendation, those needed for multiplex section protection in another document etc.

In summary, grouping or fragment is to be considered as a documentation technique to facilitate readability and ease of use of the numerous models by implementers. There are no other semantic constraints associated with these groupings or fragments.

6.3.2 Example Fragments

To illustrate further the use of fragments, consider Recommendation M.3100. An example as mentioned above of a fragment that is appropriate for implementation by network elements such as digital or optical cross connection is cross connection. Note that in describing the managed entities, the phrase "entity" and "object" are used interchangeably.

The following entities are defined as part of this fragment.
a)   Fabric: represent the matrix in a network element connecting the two end points,
b)   cross connection: represents the relationships between the to and from end points connected by the matrix,
c)   group termination point: represents a defined number of termination points to be connected together such as those required for concatenated signals,
d)   multi-point cross connection: represents a point to multi-point connection
e)   named cross connection: represents those connections that are red-lined or should not be removed without special considerations and
f)   termination point pool: represents a collection of termination points with similar properties such that any one may be used in setting up a cross connection.

In this example, all the entities of this fragment are appropriate for inclusion in a network element used in the transport network for cross connecting input and output signals.

Another example of a fragment is the switching and transmission fragment defined in Recommendation M.3100. The objects included are circuit end point sub group appropriate in modeling trunk groups that interconnect exchanges (central offices), connection and trail representing the transport network views. Even though this fragment collects the network level view entities, it is not likely for the switching NE to include all transport entities such as trail and vice versa. Even if an element management system where network view is relevant, combined management of both trunks and transport connections in one system is not expected. The semantics of this grouping is therefore very different from that of the cross connection fragment.

6.3.3 Example of Grouping in SDH

The grouping in SDH standards is in terms of document structure. In such a grouping again it should not be assumed that there is a need to implement all entities in one document.

Consider the example of G.774.1 that contains definitions for performance monitoring. The entities defined represent the values of performance parameters during a collection interval and historical storage of the collected information for a given number of intervals. The entities defined support parameters collected for the different layers of SDH hierarchy such as regenerator section, multiplex section and path termination. Corresponding to the accumulation of current collection interval entities for previous intervals are included. Even though all entities required for PM of SDH are included in this document, as noted earlier an NE that only has the regeneration feature is expected to support those entities that define the PM parameters for either multiplex section or path termination.

Another example is Recommendation G.774.3 for multiplex section protection. The entities defined here are multiplex section protection group, multiplex section protection unit, protected trail termination and unprotected trail termination in addition to generic entities for protection group and unit. A network element that is for example part of a ring architecture with protection paths is expected to include all of these entities. This document is similar to the example cross connection fragment where a network element supporting a feature is expected to include implementation of all the entities.

The above examples illustrate once again the rationale for grouping is not meant to define conformance of implementations to a specific grouping but from readability and scooping perspective to aid in the implementation.

6.4. SDH INFORMATION MODELS

The models for SDH documented in G.774 for network element level information were developed assuming a specific protocol for the interface and a technique used with that protocol. This technique is referred to Guidelines for the Definition of Managed Objects (GDMO) and uses an object oriented approach. These models were developed to support a number of management functional requirements documented in Recommendation

M.3400, SDH management functions in G.784, and M.2120 for fault detection and localization (includes performance monitoring). While one approach to explain examples of the models is to describe them using the notational technique in the standard, this chapter takes an alternate path for the following reason. Using the notation requires a reader to be familiar with it in order to follow the details. Another more significant reason is that it is necessary to understand the requirements and analyze how to satisfy the requirements in defining the managed entities, their relationships, interactions to capture the essence of the model than be encumbered by the notation. The sections below are therefore written using requirements and analysis instead of the description in the standards.

A number of recommendations are available for SDH at the element level in ITU. These documents are planned to be updated by SG 15 during next year.

Table 6-1 provides a summary of the NE level management models for SDH.

Table 6-1. NE level Management Models

Recommendation	Management areas
G.774	Configuration, fault reporting and other infrastructure elements such as network element required for SDH NE management
G.774.01	Performance Monitoring
G.774.02	Configuration of various SDH adaptation function in support of different payload structure
G.774.03	Management of multiplex-protection function
G.774.04	Management of sub-network protection in support of Automatic protection switching within NEs for low and high order path layers.
G.774.05	Configuration of low and higher order paths for connection supervision
G.774.06	Performance Monitoring of uni directional paths and sections
G.774.07	Configuration of lower order path trace and interface label
G.774.08	Management of Radio relay network elements using SDH
G.774.09	Management of fixed linear protection groups
G.774.10	Management of shared protection switching

In addition, TeleManagement Forum (TMF), an organization with focus on implementation has developed models that are suitable for network level management supported by the EMS- NMS interface. Refer to Chapter 1 for an overview of various standard groups. The requirements are to support multiple technologies. These are documented in TMF 814 A "EML-NML interface for management of SONET/SDH/WDM/ATM transport networks". In ITU, the network level model is defined to be generic and is applicable for many different technologies.

6.4.1 Network Element View

The concept of views is often mixed with the logical layering of TMN architecture. Though there is a strong relation, they are not the same concept using different terms. There is a subtle difference that should be recognized when using the terms. This distinction becomes important when in practice the logical layers have been equated to management systems at that layer. The concept of views has been used to differentiate the scope of the information elements. Consider the example of termination of a signal represented using the entity called termination point. The termination of a signal within an NE belongs to NE view. However a network management system which is often considered to be the physical realization of NML is not just interested in the loss of a end to end connection but the next level of detail as to which termination point of a specific NE failed and the reason for the failure (fiber cut, failure of the circuit pack with that termination etc.). Thus the higher-level management system is concerned with for some applications (fault management) information at multiple logical levels. There are other applications such as configuration management where the details of the HW and SW architecture are not necessary to be visible at the higher level. An alternative way to look at the views is in terms of data stewardship or ownership. Network element view as the name implies is stewarded by the repository in the NE.

With the above distinction, the examples illustrated in the following subsections define the management information from NE view. The models in the literature for the NE view have been derived either by refining generic entities applicable for a number of transmission technologies or defined to meet the specific SDH requirements. Even though initial efforts in SDH models were developed from the generic models, there are models where requirements were initially identified as part of SDH effort. Later it was noted that the requirements are generic and apply to more than SDH technology. This resulted in augmenting the generic models. Alarm report control feature is one such example, which stemmed from the port mode definitions for SDH in G.783. The model for equipment protection switching was based on the concepts introduced in the protection model for termination points in SDH.

6.4.1.1 Examples

The management models have been defined to meet the user needs, user being the network operator or service provider. The management requirements for the different areas of management are defined in G.784. These requirements are in most cases similar to the functions defined in Recommendation M.3400. Though at the generic level they are the same, details such as performance mentoring parameters, possible causes for an alarm are specific to SDH. G.784. In defining the equipment function blocks in G.783, the management points and the input and output are identified. These determine the management capabilities required for SDH equipments. For example, the function block terminating a signal should support monitoring for trail trace identifier, detection of misconnected traffic, monitoring and reporting alarm indication signal etc. These functional requirements in the context of the configuration functions would be translated to an information element for trail trace identifier settable by a management system or a control function that turns on or off detection of misconnection etc. If a modeling technique such as Unified Modeling Language is used, the requirements are described using use case diagrams with the actor being the management system in this case. The managed entities and their properties are identified in the next analysis phase. When analyzing the requirements, in addition to defining the properties of the managed entity, the interactions between the entities are also described.

These interactions include both the information exchanges between the entities as well as sequence in time for the exchanges.

Examples in the following section are grouped into two categories. In the first category, models enhance generic models by including SDH specific requirements. One of the examples is to illustrate the extensions to the resource aspects and the second example is extensions from the functional aspects. The second category of models describes how requirements developed in the context of SDH were made generic and useful for other technologies.

6.4.1.2 SDH Specific Derivations of Generic Models

Two examples are discussed in this section defining the requirements and analysis for managing entities relevant to SDH. The first example addresses requirements that permit configuration and fault monitoring functions of SDH resources corresponding to the server and client layers hierarchies. The second example discusses the performance management functional area. As discussed in earlier chapters, one of the key reasons for the success of SDH is the protection architecture and protocols defined as part of the standards.

6.4.1.2.1 Termination Points

A termination point as defined in generic models represents the source and sink of signals for the various layers of the hierarchy. Two types of terminations have been defined in the transport architecture as described above. Many of the requirements are common across the terminations of trail and connections for all the layers. The requirements included in the model for terminations are associated with functions that support configuration and fault monitoring to determine the health the network.

6.4.1.2.1.1 Requirements

The following list includes the common requirements applicable when origination or terminating connections or trails in the network element. These requirements are illustrative and are not to be treated as all-inclusive.

a. Entities may support one of the following function to the SDH transport: source or sink or bi directional (including both source and sink functions) for the signal.

b. It should be possible to provision and monitor the termination or origination of the signal corresponding to the following layers of SDH hierarchies: physical termination at the optical and electrical level, regenerator section, multiplex section, administrative unit, various lower and higher order path layers such as VC11, VC12, VC2, VC3, VC 4, different types of tributary units etc.

c.   It should be possible to retrieve the rate and format of the managed entity representing the characteristic information for the signal originated or terminated by the managed entity.

d.   The state indicating the health of the entity originating or terminating the signal must be supported. This state is not controlled by the management system but provides the information useful in fault localization and alarm correlation.

e.   The management system should be able to control administratively the operation of the termination. For example the management system should be able to explicitly disallow signal-processing function in order to conduct maintenance activities.

f.   An alarm should be reported indicating the cause of failures such as loss of signal, loss of frame etc.

g.   To support the misconnection detection feature defined in SDH, it should be possible to provision a trail trace identifier value. A misconnected fault cause should be reported when a mismatch is detected between provisioned and received trace identifiers.

h.   It must be possible to control reporting of alarms during initial set up and maintenance activities[92].

i.   Events are to be reported when there are changes to the state as well as when the entity is created or deleted.

j.   All events and alarm reports should include a time stamp with a resolution of 1 second using the local clock of the NE containing the managed entity,

k.   The severity associated with the alarms should be configurable depending on the cause of the failure.

l.   The defects reported should be consistent with the type of termination. For example the physical layer defects are transmit failure and loss of signal. In the case of path layer, the appropriate failure conditions are loss of pointer, trace identifier mismatch, far end receive failure, signal label mismatch, alarm indication signal and loss of multiframe.

m.   It should be possible for a management system to retrieve the state and status information

n.   The name of the entity must be unique and unambiguous in some context

---

[92] Even though this requirement was noted as part of the SDH management function, for historic reasons the detailed requirements and analysis was completed after the introduction of the termination point model.

6.4.1.2.1.2 Analysis

Analysis of the above requirements indicates that there are several properties that are generic and applicable to different transport technology. This can be seen for example with requirement (f) on alarm reporting or defining the state of the entity that reflects its ability to process a signal. Table 6-2 illustrates a generic class called trail termination point where properties common to all technologies is included along with the operations to be supported are identified. Included are also fault events along with the possible causes for the failure. Using the generic entities, specialized classes for SDH are developed for the different layers of SDH. The events generated as per the above requirements are also identified in the notification section of the tables. For the various properties the table includes the allowed operations. In order to trace the entities and the properties, the requirements met by each property are also identified. In addition to the tables, an example of the interaction is illustrated for retrieving the properties and reporting events, Note that the table is a simple representation of the properties and does not capture all the details of some requirements. In a complete analysis with classes and their properties many of the details should be described. For example the requirement k where the severity of the alarms should be configurable is not apparent by the property where a relationship to another class is noted. By setting the severity values in the alarm severity profile object indirectly this requirement is met. This approach was taken because an object that holds all the settings for the severity associated with the various alarms and threshold results in reuse instead of including it in every resource object. Another approach would be to include the severity assignment property in each class and allow both read and write operation on that attribute.

Table 6-2. Common Properties of Termination Point for all SDH layers

Property	Operations	Requirement Trace
Attributes		
Characteristic information (specific values suitable for a SDH layer specific trail and connection terminations)	read	b,c
Signal direction (source, sink or bi-directional)	Read and write	a
Operational state (enabled or disabled)	read	d
Administrative state	read and write	e
Alarm Severity assignment (pointer to another entity containing the assignments)	read and write	k
Name of the termination point	read	n
Events and parameters		
Communication alarm	Failure cause, time stamp, severity ,,,	f,l
State changes	Old and new state values, how the change was triggered, time stamp …	i
Creation and deletion	Name of the entity created or deleted, time stamp and values for attributes if creation event	i

Table 6-3. Additional Properties of SDH layers specific Termination Points

Property	Operations	Requirement Trace
Attributes		
Characteristic Information	read	b (in addition layer specific terminations are defined.
Trail Trace identifier	read and write	g

For SDH, a number of trail and connection terminations are defined in G.774 corresponding to the various layers. These are section physical interface terminations (optical and electrical), regenerator section, multiplex section, various rate path terminations (VC3, VC 4 etc.). Requirement "l" which defines failure conditions appropriate for each layer is reflected in the analysis where for the communication alarm only relevant failure causes are permitted.

Analysis of the requirements not only includes not only the identification of the managed entity and their properties but also how they interact with other entities in the system. These interactions are illustrated using a sequence diagram containing a time ordered message flow between the entities.

A simple example of the interaction diagram is shown in Figure 6-3. A number of requirements listed above relate to reporting events such as alarms and state changes to a management system. In order to support this function, a generic approach has been defined in standards. Even though there are some variations in the details, the fundamental aspects of the model that enables event reporting are the same. An entity called "Event Dispatcher" is used to report events to a management system. A factory type entity is required to generate different instances of managed entities. In the sequence diagram below the management system creates an event dispatcher entity using a factory object for that class of entities. The event dispatcher is configured to include criteria for determining which event should be reported to the management system. The time ordered sequence of message flows describes the creation and deletion as well as administratively suspending and resuming the reporting capability. Once the event dispatcher is created in an SDH network element, a communication alarm from a termination point will be sent to the dispatcher. Based on the criteria set, for example critical or major severities are to be reported, an alarm report is dispatched to the management system. While this example is simple, these interaction diagrams useful for implementations when there are many interactions and different managed entities are required to meet the overall requirement.

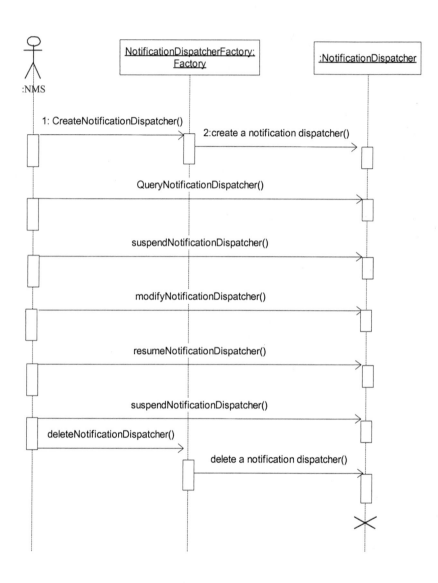

Figure 6-3. Sequence Diagram for administration of event reports

6.4.1.2.2 Performance Monitoring

When considering performance monitoring, there are two aspects to be considered: mechanisms to collect and report performance data and definition of technology specific parameters. Recommendation Q.822.1 defines a set of requirements for collecting performance parameters for different intervals and retrieving this information. This generic set of requirements is applicable irrespective of the technologies. The network operators to characterize the quality of services transported across the network use the values of the technology specific performance parameters collected. The term used in the industry for the agreement between the service provider and customer is "Service Level Agreement" or SLA. From analyzing the PM data, loss of service is avoided by rerouting the traffic until corrective actions are taken to repair the degraded entity.

6.4.1.2.2.1 Requirements

The requirements are classified in terms of technology specific expectations and support environment required to collect and report the data. Examples of both aspects are discussed here.

PM can be either intrusive or non-intrusive as described in Recommendation M.2310. When a test packet is interleaved with normal traffic. This causes additional traffic and it is necessary to avoid congestion with the introduction of test packets resulting in the loss of end user packets. Collecting performance statistics on the real time traffic through the network for specific intervals of time performs the non-intrusive monitoring. The support environment and the SDH specific performance goals discussed in this section are for non-intrusive monitoring.

Requirements for the support environment to collect PM data are extracted from the model in Recommendation Q.822. A subset of the requirements is as follows:

a. Performance monitoring data is to be collected over 15 minutes and 24 Hour intervals.

b. It should be possible to suspend and resume the data collection process.

c. Historical PM data should be stored for a prescribed period of time (typically 8 Hrs) and made available upon request to the management system

d. In order to reduce storage requirements for the data, it is necessary to suppress intervals with all zero values for the parameters and identify the number of intervals with this characteristic.

e. The management system should be able to define threshold values for the parameters and appropriate alerts are to be reported if the threshold is crossed.

f. If the data collected in an interval is deemed to be invalid a flag must be set to indicate the data is suspect.

g. If the threshold value for monitored parameters is crossed a quality of service alarm should be emitted.

In addition the following SDH specific requirements have been identified in G.783 and G.874.

a) The performance parameters collected depends on the layer of the hierarchy. Table 6-4 below defines the relevant parameters and the layers they apply.

b) Threshold values to be for reporting performance crossing alert notification are shown in the table below. An example set of parameters and the values valid for data collected over 15 min interval are included below in Table 6-5.

c) It should be possible to inhibit accumulation of multiplex section layer parameters except for UAS, FC, PSC and PSD during periods of unavailability of the monitored multiplex section. There are several other rules defined for inhibiting collection of these parameters not listed in this section.

d) In order to support transparent monitoring capability, intermediate path PM is required to monitor the incoming transmission signal without terminating it.

Table 6-4. Performance Parameters

Parameter	Section Layer	Multiplex /Line	Path Layer	
Code Violations (CV) CV are BIP errors that are detected in the incoming signal.	CV-S (B1, BIP-8)	CV-L (B2, BIP-8)	CP-P (B3, BIP-8)	CV-P (V5 BIP-2)
Errored Seconds (ES) At each layer, an ES is a second with (a) one or more coding violations at that layer (b) one or more incoming defects (e.g., SEF, LOS, AIS, LOP) at that layer.	ES-S	ES-L	ES-P	ES-P
Unavailable Seconds At the Line, Path, and VT layers, an unavailable second is calculated by counting the number of seconds that the interface is unavailable. At each layer, the SONET/SDH interface is said to be unavailable at the onset of 10 contiguous SESs.		UA-L	UA-P	UA-P
Severely Errored Seconds At each layer, an Severely Errored Second (SES) is a second with x or more CVs at that layer, or a second during which at least one or more incoming defects at that layer has occurred. Sample values of x are listed for SES-S	SES-S	SES-L	SES-P	SES-P

Table within "Severely Errored Seconds" row (SES-S):

Rate	x	Min. BER
OC-1	9	1.5×10^{-7}
OC-3	16	1×10^{-7}
OC-12	63	1×10^{-7}
OC-48	249	1×10^{-7}

Table 6-4 Cont. Performance Parameters (Continued)

Parameter	Section Layer	Multiplex /Line	Path Layer	
Severely Errored Framed Seconds Count of the seconds during which SEF defect was present.	SEFS-S			
Near and Far end Line failure counts Failure event begins when AIS is declared and ends when it is cleared.		FC-L, FC-LFE	FC-P, FC-PFE	FC-P, FC-PFE
PSC and PSD Protection Switching count and duration		PSC PSD		
Pointer Justification Several parameters are included in this row. These specify the positive and negative pointer justification count, count of 1 sec intervals containing one or more pointer justification counts			PPJC-Pdet NPJC-Pdet PPJC-Pgen, NPJC-Pgen, PJCDiff-P PJCS-Pdet, PJCS-PGen	PPJC-Pdet NPJC-Pdet PPJC-Pgen, NPJC-Pgen, PJCDiff-P PJCS-Pdet, PJCS-PGen

Table 6-5. Threshold Values

Parameter	Section	Multiplex/Line	Path
CV	16,383	16,383	16,383
ES	900	900	900
SES	900	900	900
SEFS	900	900	900
PSC		63	
PPJC			1,048,575

6.4.1.2.2.2 Analysis

To reflect the two types of requirements, the analysis describes information modeled for the support environment and the infrastructure design of object classes. The SDH specific requirements are then analyzed using this infrastructure.

Recommendation Q.822.1 describes the model for the support environment in addition to the protocol specific information model design. To support the requirement for collecting PM parameters for different accumulation period the approach chosen was to define an entity that has all the relevant characteristics defined in the requirement. Analyzing the requirements together is sometimes necessary as in the case of PM. These parameters are collected on a monitored entity such as a termination point discussed above for a specific layer of the hierarchy. While it is possible to include as part of the analysis of SDH specific termination points, the PM parameters, this reduces the flexibility for operations. For example, parameters collected may vary depending on the pro active monitoring performed to resolve a specific problem. The strict binding of the PM parameters to be collected to the transport function supported by the termination point results in unnecessary deletion and creation of the management representation of the termination point. To allow for flexible treatment of PM collection and parameter changes, resulted in defining a managed entity called "Current Data" that included properties relevant to the support environment (Table 6-6). Using this generic representation, further refinements are defined that include for example SDH specific parameters.

Table 6-6. Common Properties of Current Data for all SDH layers

Property	Operations	Requirement Trace
Attributes		
Administrative state	read and write	b
Granularity Period	read and write	a
Operational state (enabled or disabled)	read	
History retention period	read and write	c
Zero suppression required	read and write	d
Suspect flag	read	f
Threshold data pointer	read and write	e
Events and parameters		
Quality of service alarm	Monitored parameters, threshold values	g

For collecting SDH specific parameters relevant to a layer as noted in the requirements, various current data entities are defined in Recommendation G.774.1. Table 6-7 is an example of layer specific current data.

The requirement for storing the current data for a period of time automatically implies the historical data is to be collected periodically. In support of this requirement, Q.822.1 defines an entity called "history data file" which is created with the parameter values collected over an interval requested by the managed system. The interactions between the entities to

create and retrieve the history file are shown in Figure 6-4 taken from
Recommendation Q.822.1.

Table 6-7. Regenerator Section Current Data

Property	Operations	Requirement Trace
Attributes		
Errored block	read	a
Errored Second	read	a
Severly errored second	read	a
Out of frame	read	a
Out of frame seconds	read	a
Threshold values	read and write	b
Administrative state (from current data)	read and write	c

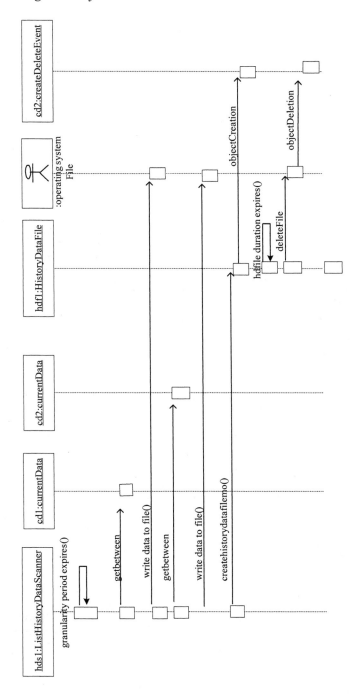

Figure 6-4. Interaction diagram for PM data collection

The interaction diagram in Figure 6-4 shows that a history scanner entity collects the data from the current data managed entities at the end of the granularity period. Once the data is collected a history file is created based using the operating system of the managed element. The creation of the file is notified using a managed entity called history data file that represents the file. The managed system can then retrieve the file using mechanisms such as file transfer protocol efficiently. The various entities involved in creating and deleting the files and the messages exchanged are shown in the figure.

6.4.1.3 Generic Models Based on SDH Requirements

The examples shown above illustrate the direction of deriving and analyzing SDH specific management requirements using the generic models. In some cases, new requirements were identified as part SDH management. Further review of these requirements indicated that the features are not specific to SDH but are applicable to multiple transport technologies. Three examples are discussed in this section. They are structured similar to the previous sections except a new section is included for specific application of the feature in SDH.

6.4.1.3.1 Alarm Report Control

SDH equipment functional block characteristics defined in Recommendation G.783 introduces the concept of port mode where the monitoring function of the termination point can be set to allow or inhibit declaring alarms in some scenarios. These include when connections and trails are provisioned or maintenance activities are performed. Recommendation M.3100 Amendment 3 lists the requirements, analysis and a model using a specific protocol paradigm.

6.4.1.3.1.1 Requirements

A sample list of requirements is as follows:

a. Alarm free set up of sections, lines and paths, modification of payload structures and maintenance activities should be supported.
b. When alarm reporting is turned off, performance monitoring threshold reporting should be disabled.
c. Even though the reporting is turned off, the data collection of PM parameters should be continued.
d. The alarm information shall be available for retrieval even though the reporting is disabled.

e. When an alarm that was sent prior to setting the report control clears, this should be reported even if the control is in effect.

f. Alarm indications and corresponding clears that occur after the control is set should be inhibited.

g. Transitions between alarm reporting turned on or off states should be supported.

h. It should be possible to define a persistent interval at the end of which the reporting state is resumed. This allows automatically turning on the reporting if the operations personnel omits turning the reporting on after the provisioning or maintenance activity.

6.4.1.3.1.2 Analysis

A key element of analyzing the requirements and developing the model relates to the values associated with a state referred to as the alarm report control of the managed resource. The values of the ARC state and the events that transitions between them is defined in Recommendation M.3100 Amendment 3. Instead of defining all the values and events a subset of key values and corresponding event transitions are included here. The description is intended to show that in developing information models, depending on the function a state driven approach is meaningful.

A managed resource when alarm report control feature is included has the following values for ARC state:

a) ALM indicates that alarm reporting is allowed for the resource

b) NALM indicates that alarm reporting is turned off

c) NALM-TI indicates that when a preset time interval expires then transition to ALM where alarms are to be reported

d) Other state values defined include NALM-CD, which is used to specify a persistent interval so that when the failure cause is resolved, the resource is placed in a count down state. The resource when the timer expires and the resource is problem free then alarm reporting begins. While in the count down state, the problem recurs in the managed resource, it transitions to to NALM-NR to indicate that the resource is not ready to report alarms.

Table 6-8. Alarm Report Control State Transition Table

Event\State	ALM	NALM	NALM-TI
Managed resource becomes qualified problem-free	Clear alarm(s) as normal, Remain in ALM	Remain in NALM	Remain in NALM-TI
Qualified problem raised	Raise alarm(s) as normal, Remain in ALM	Remain in NALM	Remain in NALM-TI
Manager request to transition to ALM	Reject Request, Remain in ALM	Report existing alarms raised during ARC, Transition to ALM	Report existing alarms raised during ARC, Transition to ALM
Manager request to transition to NALM	Transition to NALM	Reject Request NALM	Transition to NALM
Manager request to transition to NALM-TI, interval not provided in request	if NALM-TI supported set timed interval to default timed interval and Transition to NALM-TI; otherwise Reject Request and Remain in ALM	if NALM-TI supported set timed interval to default timed interval and Transition to NALM-TI; otherwise Reject Request and Remain in NALM	Reject Request, Remain in NALM-TI
Manager request to transition to NALM-TI, interval provided in request	if NALM-TI supported set timed interval and Transition to NALM-TI; otherwise Reject Request and Remain in ALM	if NALM-TI supported set timed interval and Transition to NALM-TI; otherwise Reject Request and Remain in NALM	Reject Request, Remain in NALM-TI
Timer expires	-	-	Report existing alarms raised during ARC, Transition to ALM
Manager request to modify timed interval default	If current value specifies no adjustment, Reject Request; otherwise Change default, 1st potential use is next transition to NALM-TI	If current value specifies no adjustment, Reject Request; otherwise Change default, 1st potential use is next transition to NALM-TI	If current value specifies no adjustment, Reject Request; otherwise Change default, 1st potential use is next transition to NALM-TI
Manager request to change ARC interval	Reject Request, Remain in ALM	Reject Request, Remain in NALM	Change Timed Interval, re-enter NALM-TI
Manager request to modify ARC probable cause list	Modify list, Remain in ALM	Modify list, send alarms for existing alarms no longer inhibited, Remain in NALM	Modify list, send alarms for existing alarms no longer inhibited, Remain in NALM-TI

For any managed resource that supports alarm report control, the two basic state values are ALM and NALM. Instead of the basic NALM value others that may be used to support a refined form of control are those that prevents fluctuations (hysteresis) or allows the user to specify a predefined time in which the resource is to be left in not alarmed state. The latter is applicable in cases where a provisioning or maintenance activity is taking place and the operations personnel can estimate the time required to complete that activity and set the resource to be not alarmed during that period.

A state transition table 6-8 for the subset of values ALM, NALM and NALM-TI is provided in the table below. This table is an extract from Recommendation M.3100 Amendment 3.

An alarm report includes in addition to the severity of the alarm and other parameters such as a mandatory field that specifies the probable cause or the potential reason for the failure of the resource. The alarm report control may be customized to turn off alarm reports for specific probable cause. For example when a trail is being set up, the cause for raising an alarm is a loss of signal. However if there is an equipment failure during the period the set up is performed then this problem should be reported. Thus while a loss of signal is an expected failure scenario during the set up operation an equipment failure is an unexpected failure. In order to support this scenario it is essential to inhibit only the relevant alarm while reporting others. Depending on the activity in progress the alarm report suppression of the probable cause will vary. The last row of this table introduces the event where a management request is issued to modify the list of probable causes corresponding to which the alarm reporting is inhibited. The effect of changing the list is immediate as noted in the table. Alarms that had not been reported should be issued when the respective probable causes are deleted.

A functional model describing the flow of information within the managed system for detecting the failure, assigning the probable cause and severity and the relationship to alarm reporting control function is described in the recommendation.

The object classes defined include an alarm report control entity that contains the criteria based on which reporting is inhibited and profiles containing the default values of the timers described in the state transition table.

In developing the initial analysis and the corresponding design, the approach used was to include this feature as part of the managed entity. Recently another enhancement was approved in Recommendation M.3100 Amendment 7. In this approach, a separate managed entity was defined called alarm report control manager. This entity can be configured to include the resources where alarms are to be inhibited. This addition was also triggered by SDH model effort. With the number of termination points in SDH corresponding to the different layers of the hierarchy it was appropriate to introduce this new feature so that it is easy to add this capability in existing implementations without requiring updates to original design.

6.4.1.3.1.3 Application in SDH management

Recommendation G.783 discusses the port mode for SDH termination points to prevent the alarms from being reported during the provisioning and maintenance activities. The concept of port mode was introduced for trail terminations. This property has the values MON, NMON and AUTO. As one can see from the state table above these values is related to the subset of the alarm reporting state values. MON value for the port mode has the same behaviour as the ALM for the alarm report control state. This value indicates that the termination point is being monitored and alarm reporting is turned on. When the port mode has the value of NMON it corresponds to setting the alarm report control mode to NALM so that alarm reporting is suppressed. AUTO has the same function as the NMON except when the loss of signal condition is cleared (the trail is set up for example) the termination point automatically transfers to the monitoring mode. There are a couple of differences that should be noted. Even though initial model was based on this three simple values, the concern raised by some service providers was that when the port mode is in NMON state the PM data is not collected. In effect all monitoring functions ceased to exist. It was a requirement to be able to retrieve alarm and PM information even if the reporting is inhibited. In addition, as described above there was a need for specifying persistence interval so that before the reporting is turned on again a wait period was necessary to remove oscillations of the loss of signal.

A simple transition diagram for the MON and NMON state values of port mode taken from Recommendation G.783 are shown below in Figure 6-5.

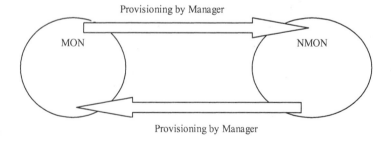

Figure 6-5. Port Mode in SDH

Even though this has not been captured in the existing G.774 Information Model series recommendations, the work plan includes updating them to include this feature. This has already been introduced in G.875 for OTN.

6.4.1.3.2 Equipment Protection Switching

Protection switching is a key feature of SDH and thus it is not surprising the requirements and information models were developed first in that context. The APS protocol and the different architectures were defined for protecting the traffic and thus aimed at trails and connections. In order to support the high availability requirements for telecommunications equipments redundancy is provided at multiple levels in equipments. This is referred to as equipment protection switching. The generic model in M.3100 Amendment 2 was developed using the concepts first developed as part of the protection model in Recommendation G.773.03.

6.4.1.3.2.1 Requirements

A subset of requirements for supporting protection switching for resources is included below. Example of resources include different types of circuit packs such as control cards where management functions are implemented, line cards connecting traffic from and to the network as well as power entry modules, fan control units etc. The group of entities of the same type that include the resources is referred to as protection group or protection system.

a) Different protection scheme should be supported. The most commonly used schemes are 1+1,and M: N. 1:1 where the resources are in active standby mode is a special case of M:N.

b) The protection scheme may be either revertive or non-revertive. The management operation should be able to set the mode to be one of these two values.

c) When supporting a revertive protection scheme, it should be possible to wait for an interval of time before reverting to the previously active resource. This interval is settable by the management system and is used to avoid toggling back and forth between the resources participating in protection.

d) For some scenarios such as performing maintenance activity it should be possible to disable automatic protection switching or force a protection switch.

e) When the protection-switching event occurs it should be reported to the managing system.

f)   The protection group should have a unique name for referencing purpose.

g)   The protection units forming the protection group should be capable of supporting different applications. These are duplication of resources where a functional element is not explicitly defined as the protecting entity, and where a functional entity is identified. Example of the first application is where two controllers are used in active stand by mode and all the applications supported by the failed controller switches over to the stand by controller. In the second application a transport entity such as termination point is protected. In this application both resource level protection model described here as well as the transport layer protection (multiplex section protection discussed later are required. Another application that should be supported is 1:N where one protecting unit is used to protect n units each of the n containing a specific number of functional entities.

6.4.1.3.2.2 Analysis

The concept of protection group was initially introduced as part of SDH protection model to bring together the protected and protecting resource to provide a single identity for explaining the relationship between them. The group includes zero or more instances where one or more standby entities provide protection for one or more active entities.

Given the definition of protection group, a question that may arise is the semantics of having zero protection units. The rationale is to allow the creation of the group and add to the group the relevant resources subsequently. It also caters to the scenarios where some systems may be used either in protection mode with one resource protecting another or as two independent active resources that support load balancing. This approach offers the flexibility to allow existence of the group with zero instances and add based on the application needs appropriate resource representations.

The properties of the protection group to meet the requirements listed above are shown in Table 6.9. It is assumed that the two basic states are appropriate to include even though they are not listed above. The requirement (g) defines applications depending on which the values of some of the properties of the protection unit varies.

Table 6-9. Properties of Protection Group

Property	Operations	Requirement Trace
Attributes		
Protection Type	read-write	a
Revertive	read and write	b
Operational state (enabled or disabled)	read	
Administrative state	read and write	
Wait to restore time	read and write	c
Name of the protection group	read	f
Events and parameters		
Protection Switching occurred	Name of the protecting unit, reason for protection,	d
State changes	Old and new state values, how the change was triggered, time stamp ...	
Creation and deletion	Name of the entity created or deleted, time stamp and values for attributes if creation event	
Actions		
Invoke Protection	Lockout/force switch/manual switch	e
Release Protection	Release lockout/force switch/manual switch	e

In contrast to the scenario shown in Figure 6-5, the application where a functional entity is protected is described in Figure 6-6 (extracted from Recommendation M.3100 Amendment 3). The generic reference to the termination point is replaced with a SDH specific termination point. Specifically this figure shows the relevant properties in the application where the resource such as a circuit back upon failure is switched to another standby entity. In this case all the functions performed by one circuit pack will be resumed by the stand by entity. The reliable and unreliable resource pointers reference no functional entity. This is why the figure shows them to have a null value. The protection unit relates the resource and the entity being protected. Because in this case the two circuit packs back up each other there is no third entity involved as part of the protection scheme.

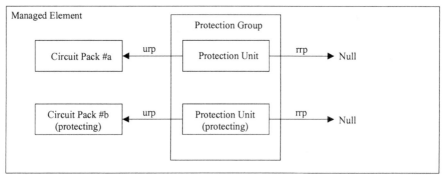

urp Unreliable Resource Pointer
rrp Reliable Resource Pointer

T0412030-99

NOTE 1 – Containment relationship for the Circuit Pack is not shown.
NOTE 2 – If the unit of redundancy is not a single circuit pack but a set of circuit packs, the resource pointers will point to all the circuit packs in the set.

Figure 6-6. Protection Entities in the simple active /standby case

The logical entity being protected in this is a multiplex section trail termination point. In this example the circuit packs are unreliable resources. The protection units (protecting or otherwise) reference them and relate them to the multiplex section bi directional trail termination points that are reliable. The reason for these termination points to be reliable is because the functions are protected by the circuit pack #b. This circuit pack is one protecting both a and c. The figure shows the case where c fails. The switch over to circuit pack b protects the trail termination function supported by the multiplex section (ms) trail termination point entities. Thus the protection units corresponding to a and c reference these terminations as reliable resources to indicate there is no loss of functionality of these terminations when the circuit packs containing them fails because c protects them. In this case when b is repaired or replaced the revertive switching will be required so that another failure is protected. This scenario resolves single failure cases. Before the circuit pack c is replaced with a working one, a fails then this results in double fault and is not protected. This is acceptable practice in the industry.

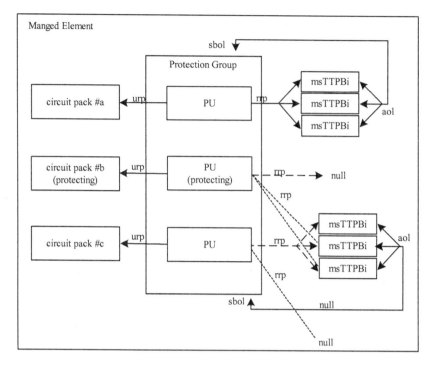

Figure 6-7. N:1 Protection

Legend:
    aol – affected object list
    sbol – supported by object list
    r/urp – reliable/unreliable resource pointer
    PU – protection unit

6.4.1.3.2.3 Application in SDH management

This model is generic and applies to all technologies and different types of equipments that support a high availability platform. The first application where two circuit packs whether they are controller cards, switching modules in a blade applications or a network element the model supports the ability to have a seamless transfer to a protecting entity without loss of service. However this model also implies that both hardware and software support are required to achieve the switchover function.

The second application is a specific need to achieve the level of protection expected in an SDH network. With high bandwidth transport supporting several traffic flows from multiple users, the need for protection becomes critical. Irrespective of the scheme chosen, the logical function for generating and terminating an SDH signal is protected.

Every SDH and optical networking network element designed and deployed today support the resource level protection. The common approach used is to have 1:1 protection for control functions and 1:n protection for traffic carrying circuit packs.

6.4.1.3.3 Enhanced Cross Connection Features

The initial cross connection model in Recommendation M.3100 was enhanced to support two new features referred to as bridge and roll function and supporting interconnected rings at a network element. The bridge and roll functions defines the extensions so that switch over of the connection between two network elements is performed without loss of traffic. The ridge and role process is explained in Recommendation M.3100 Amendment 4. A second extension described in Recommendation M.3100 Amendment 5 is described in this section. Both features are initiated by requirements in SDH and this section discusses the second enhancements for dual homed ring interconnections, which is prevalent in SDH network offering drop and continue functions. With this feature the network element interconnects two rings and supports the signal to be dropped in one node and continue so that it can be sent to other nodes.

6.4.1.3.3.1 Requirements

The requirements are explained using the interconnection diagram taken from Recommendation M.3100 Amendment 5.

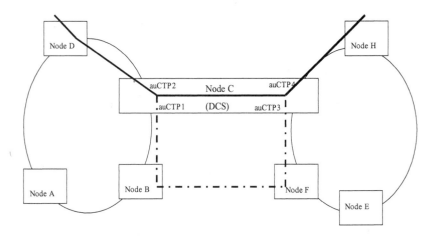

Figure 6-8. Ring Interconnection with dual homed NE

In figure 6.8 the node C is interconnecting two rings thus both the rings. of the network share it. There are two independent protection-switching connections performed by node C depending on the direction of the traffic. The ring architecture offers survivability in a network because of the two different paths the signal can take within the ring. Node C as an interconnecting node between the two rings should perform different uni directional cross connections depending on the direction of the traffic flow. There are two paths shown in the figure. The primary path from Node D to H is through C. The secondary path is the dashed line connecting nodes C, B and F. If the connection from node D to C fails the alternate path chosen will select the signal from Node F. This is because the signal passing through both directions of the ring will travel from node B to Node F and transmitted to Node H using a uni directional cross connection in C between termination points 3 and 4. When primary path is used a bi-directional cross connection is set up between termination points 2 and 4 in node C. If on the other hand a failure is encountered in the direction from H to D, the traffic will be sent to D from B using the F to B connection. This example shows that the requirement to change an existing bi-directional cross connection to uni directional cross connection is required.

Other requirements include

a) the ability to drop and continue traffic,
b) joining uni-directional traffic flows into one bi-directional (join)
c) splitting a bi-directional flow into two uni directional flows
d) convert a uni-directional point to point connection to a multicast connection with a single leg
e) reverse of the above requirement.
f) changes noted above should be performed without impacting the traffic.

6.4.1.3.3.2 Analysis

The initial model using fabric to represent the cross connection was found to inadequate to meet the above requirements. Additional properties were required to support the above requirements.

Setting a cross connection in effect creates a new managed entity with properties such as from and to terminations, state of the cross connection. Using the fabric or switch matrix and specifying the ports involved in the connection, the cross connection management entity is created.

To support the above requirements, additional operations are necessary on the fabric. These are to split the cross connection, join cross connection; convert a point to point to a multicast and vice versa. The connections modified by these operations may be protected or unprotected. In addition join or split of a connection should be between the same termination points.

The parameters for the operation requests identify the cross connection entities that are to be modified.

6.4.1.3.3.3 Application in SDH management

The requirement for dual homed network element interconnecting two rings is even though not specific to SDH, the enhanced cross connection model was introduced generalizing from Recommendation G.774.04. The interconnection of rings using add drop multiplexer where no interaction exist between the rings and each ring may support varying levels of protection. As described in the next section the model to support subnetwork connection protection fabric was defined from fabric with new actions to migrate between protected and unprotected connections, uni-cast and multicast connections etc. The focus of the fabric extensions in SDH some of which are applicable to generic model is on protected connections. However the initial version of the model for the cross connection fabric or the extensions provided by G.774.04 did not support the requirement for unidirectional connections for the example of interconnecting ring shown above. In either of these models it is not possible to migrate from single-homed interconnection (s simple bi-directional connection) to dual-homed, drop-and-continue, same-side interconnection (which uses unidirectional connections) without disrupting traffic.

The development of the generic model shown here followed a different path than the alarm report control or equipment protection switching models. In the first case, the model was developed because of the need for turning off alarm monitoring during a provisioning process. The model to be used in future revision of G.774 and in G.875, OTN would be the generic model discussed here. In the case of equipment protection switching, concepts such as protection group, reliable and unreliable resource references were adopted from SDH protection model. In the enhanced cross connection model, the direction taken was neither the initial generic model nor the protection focused SDH model were adequate to meet all other requirements for interconnected rings. Thus the combined features of the two models were augmented using additional actions discussed earlier.

6.4.1.4 Models Specific to SDH Technology

Information models that have the technology specific features built in are discussed in this section. Many of these features can also be found with OTN. The first example discusses the requirements associated with configuring the adaptation functions to support different payload mappings.

The second example relates too multiplex section protection feature. The former is related to the discussions on the transport architecture with clients adapted to server trail and the latter addresses different protection architectures defined in the standard.

6.4.1.4.1 Configuration of Payload Structure

In discussing the transport architecture the concept of adaptation was introduced. Using different adaptation functions many different clients are transportable by a server layer. The adaptation functions introduce the term administrative unit that defines how the payload in the higher layer of the hierarchy, say a path layer is adapted to the client layer (a path layer adaptation to multiplex section layer).

Recommendation G.783 defines a number of processes associated with the adaptation function. These include scrambling/descrambling, adapting to the different bit rate between the client and server, framing and determining the pointer values, multiplexing and de-multiplexing, recovering time, identifying payload etc.

SDH payload configuration functions are used to configure the various SDH adaptation functions. When equipments support modifying the payload structure, the model defined in Recommendation G.774-02 is used. If only one type of client is supported by the server trail termination, then the model defined in G.774 are applicable.

6.4.1.4.1.1 Requirements

The following list of requirements is to be supported in the model when the network element is capable of supporting the mode where the payload structure is modifiable.

1) When a specific payload type is configured the structure of the terminations below that level are created automatically according to a predefined default configuration.

2) Reconfiguring the multiplex structure should be supported from not multiplexed payload structure to multiplexed configuration. For example it should be possible to change a payload structure from a tributary unit group 3 to a structure that contains 7 tributary unit group 2 (lower rate signals multiplexed to a higher rate).

3) When the payload structure is reconfigured the related client layer terminations are also defined.

4)   Configuration or reconfiguration defines the complete subtrees between the Trail Termination Point of the server layer and the Connection Termination Points of its client.

5)   It should not be possible to change the configuration of the terminations that are already cross connected without disconnecting them first.

6.4.1.4.1.2 Analysis

The section on termination function model defined SDH specific termination and their properties. Depending on the equipment reconfiguration of the payload may or may not be possible. To support this distinction, two different managed entities are required; un-modifiable where the payload structure is predefined and modifiable where the reconfiguration is permitted.

Recommendation G.707 describes the different adaptations possible in a STM-1 and discussed in detail in Chapter 3. An STM-1 may be configured with one VC-4 or 3 VC-3s. If the termination supports the reconfiguration capability then it would be possible to change the payload structure. This change implies that the adaptation function is modified accordingly.

To meet the above requirements the model introduces another set of termination points that parallels those defined in G.774 for the different server layers. The modifiable termination point managed entity includes in addition to the characteristics defined in the trail termination an action with parameters that defines the type of adaptation function.

The mechanism to configure the payload structure by modifying the adaptation function is applicable different types of trail termination points. The modifiable trail termination includes the list of clients supportable by the trail termination. The configured payload structure should be one of the values available in this list. For VC4 trail termination, there are two possible values – tributary unit group 3 and tributary unit group 2. The operation indicates whether the trail termination is submultiplexed or not. If it is not submultiplexed, and the value for the client type has no value then all contained connection termination points are deleted. Because there is no multiplexing involved, a VC 4 connection termination managed entity is created. In the case where multiplexing option is chosen, there are two possible mappings possible using tributary unit group 2 or 3. In the former case seven tributary unit group 2 entities are created and in the latter case two are three tributary unit group 3 entities are created. If the payload structure configuration request succeeds, then depending on the multiplexing appropriate connection termination managed entities are created. An error is reported if the requested configuration is not performed successfully. The reason reported include the following: the requested structure is not supported by the server trail termination entity or that it is not possible to permit the modification because the entity being reconfigured is cross connected to another termination.

The analysis to support this payload configuration function is relatively simpler compared to the other functions. The multiplexing structure is well defined and thus configuration of payload structure requires selecting one of the allowed structures.

6.4.1.4.2 Multiplex Section Protection

Recommendation G.841 defines the types and characteristics of SDH protection. These have been described in detail in chapters 3 and 8. Recommendation G.774.3 defines a model that supports the management of automatic protection switching in a network element for the multiplex section layer. The management is based on the protection function defined in Recommendation G.803.

6.4.1.4.2.1 Requirements

The requirements include both generic protection support and those specific to multiplex section protection. The protection protocol used to perform the select to ion and switching functions is not discussed in this section – only how to manage configuring and performing protection switching by the management system. A subset of the management requirements to configure and initiate switching from the management system is as follows.

a) The protected and protecting resources should be managed together as a group with configuration characteristics for the architecture (m:n or 1+1), type of switching (revertive or non revertive) and time to wait before performing a reversion to the replaced resource.

b) Configure a group of protection units that support the protection switching function together and report to the management system when the operability of the resources forming the protection group changes.

c) The ability to perform the operations for manual switch, forced switch, and lock out a protection or working channel.

d) Configure the priority of switching in a m:n protection architecture so that when there are multiple failures, protection will be available according to the priority scheme.

e)   For linear multiplex protection architecture, configure all protection resources to be in either uni or bi-directional protection switch mode

f)   Report the protection switch status of each of the resource with the status value that indicate whether the switch request is active or pending for automatic switch request

g)   When extra traffic is permitted (reserved protection channels are used for this purpose when not protecting a failed channel) management request to suspend and resume this traffic must be supported. Other options for managing the extra traffic include preventing or allowing the preemption of the extra traffic by higher priority requests and reestablishing it when higher priority request is removed.

h)   Report the following notifications related to protection switching: when a protected resource, having been switched to a protecting resource, is preempted by a higher priority request from the same or another protected resource (applicable in 1:n systems); when a protected resource (unit) is switched onto a protecting resource; and when any switch is released;

i)   Detect a mismatch in the protection architecture between the near and far end configuration of the protection group (1+1 vs. 1:n).

6.4.1.4.2.2 Analysis

Similar to equipment protection switching discussed above the multiplex section also uses the definition of protection group. The properties are the same as those described earlier. The definition of the protection group meets many of the above requirements and is not repeated here. To support the multiplex section protection specific requirements new properties have been added. These are shown in table 6-10.

In these table for some properties the requirements trace is cell is empty. For example all entities have to generate creation or deletion report when they are created and deleted. These are considered to be generic and thus no function specific requirements are identified.

Table 6-10. Properties of Multiplex Section Protection Group

Property	Operations	Requirement Trace
Attributes		
Protection switching mode (uni or bi)	read-write	e
Revertive	read and write	b
Operational state (enabled or disabled)	read	
Administrative state	read and write	
Wait to restore time	read and write	a
Protection mismatch	read	I
Events and parameters		
Protection Switching occurred	Name of the protecting unit, reason for protection etc.	d
State changes	Protection status, Old and new state values, how the change was triggered, time stamp ...	
Creation and deletion	Name of the entity created or deleted, time stamp and values for attributes if creation event	
Actions		
Invoke Protection	Lockout/force switch/manual switch	c
Release Protection	Release lockout/force switch/manual switch	c

Table 6-11. Properties of Multiplex Section Protection Unit

Property	Operations	Requirement Trace
Attributes		
Channel number	read	
Operational state (enabled or disabled)	read	
Extra traffic control using Administrative state	read and write	g
References to the protecting termination points in the	read and write	
Priority order for protection	read	i
Events and parameters		
Protection Switching occurred	Name of the protecting unit, reason for protection etc.	d
State changes	Protection status, Old and new state values, how the change was triggered, time stamp ...	
Creation and deletion	Name of the entry created or deleted, time stamp and values for attributes if creation event	

A protection group includes the protection units that set up the relationship between the protected and protecting connection termination points. The protection unit also identifies the channel used to transmit the protection protocol. For the case of protecting unit this channel is not defined or used.

Table 6-11 shows the characteristics defined for multiplex section protection unit.

6.4.2 Network View

Even though the initial attempts in the industry was to provide a standard management interfaces to network elements so that different vendor supplied NEs and EMSs may be used interchangeably, in reality this did not materialize. The EMS were vendor proprietary and managed only the NEs supplied by the same vendor. The NE specific details were not disclosed or abstracted to the generic level and thus prevents interoperability with a third party supplied EMS.

The trend in the industry during the latter of nineties is to assume that the NE vendor supplies the EMS and it is more appropriate to consider standardization at the network level such as the interface between an EMS and NMS[93]. This section addresses example of requirements for managing the network level entities.

In developing the network level views, two different directions were taken between ITU Recommendations and efforts such as Tele Management Forum. The direction taken by ITU Recommendation adhered to the architectural purity of the G.803 and the models were not optimized to meet implementation issues for levels of abstractions and thus increased number of managed entities. The goal for the direction taken in implementation for a is to develop the models that meet the use cases and combine some of the layers to reduce the number of managed entities thus saving on storage and processing.

This section includes models resulting from both approaches. The generic models for the two use cases take the approach that is consistent with maintaining the architectural separation between the layers and their strict relationships. In discussing the SDH specific network level models, the examples were based on requirements in Telcordia GR 2955 Hybrid SONET/ATM Element Management and TMF Multi technology network management documents.

---

[93] Even though it is not necessary to equate the logical layering defined in the TMN architecture to the systems, it is common industry practice to use the EMS to NMS interface to represent the network level view.

6.4.2.1 Generic Models

Prior to the introduction of the three-phase methodology and use of UML in describing the managed entities a different approach was used in network level models. This was based on open distributed architecture and introduced the equivalent of requirements phase in Enterprise viewpoint. The analysis phase was provided by the combination of information and computational viewpoints. The former is equivalent identifying the managed entities and their properties and the latter addressed the operations performed on them. The various features were considered as communities of interest and these viewpoints were developed. However this approach was not embraced in developing standards or documents in various for a possibly because of the more powerful techniques supported by SW tools became available.

The examples used in this section are based on the viewpoints available in the G.85x series Recommendations.

6.4.2.1.1 Sub-network Connection Management

The concept of subnetworks was introduced earlier in describing the partitioning concept from G.805. A subnetwork in formed in a layer of the network by a collection of termination points. Connections between the termination points forming the subnetwork establishes a subnetwork connection that facilitates the routing of the characteristic information, namely the signal at that layer. Figure 6-9. shown below is taken from Recommendation G.852.2 describing the various transport resources.

· By flexibly connecting the termination points forming the subnetwork different routes can be established through the subnetwork. A subnetwork may be composed of one or more NEs that are grouped together based on reasons such as support for ring architecture and administrative boundaries between different service providers. A cross-connection discussed in the network element model is a most simple form of a subnetwork connection(SNC) between terminations points. A set of link connections shown in the figure connects the subnetworks and the set is referred to as a link. The link with link connections included represents the capacity of the connections between subnetworks.

Management of subnetwork connections involves setting up and releasing subnetwork connections to support the flexible routing of signals. The type of subnetwork connection management addressed here is for simple point-to-point connection. Standards and documents from different groups such as ATM Forum also address requirements for point to multipoint subnetwork connection management.

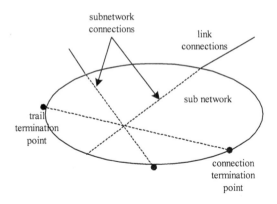

Figure 6-9. Components of a Subnetwork

6.4.2.1.1.1 Requirements

The functional requirements in the management of a subnetwork connection include responding to requests from a network management system to establish and release them.

1) Establishment of Subnetwork Connections

Subnetwork connections are established in a network layer to support both end-to-end customer connectivity services (e.g., DS 3 trail or VC 4 connection) and server layer trails between nodes of the network to support client layer link connections (e.g., where an STS-1 trail is serving multiple DS1 link connections). There are several variations possible in establishing this connection in order to offer flexibility to the service provide and a subset of these are as follows:

a. The selection of the route the connection must use may be specified in the request as static set of termination points. The selection of the route may also be left to the decision made by the element management system.

b. The selection may take into account routing diversity, use of reserved resources or other criteria.

c. The route selection may be in terms of individual network elements and in this case the Element Management System selects and assigns the individual ports, and activates these assigned ports.

2) It should be possible to create a subnetwork connection activated upon creation or at some later time

3) The management system must be notified of successful creation or otherwise, change of state of the subnetwork connection.

4) Either deactivating without deleting or deleting a subnetwork connection should be supported.

5) Change to the route of a subnetwork connection by the management system for reasons such as detecting lower cost route availability or failure should be allowed.

6) In creating a SNC the following parameters are to be supported.

   a. Identification of the two ports which must be part of the SNC
   b. A unique identifier for the duration of the SNC
   c. A user identifier for the requested subnetwork connection that can be used as the unique identifier for a contracted service
   d. Characteristics of the requested transport service (e.g. bandwidth, directionality, route selection criteria, availability, etc.) based on the contract or specific technology. The requested SNC is not created if one or more of the characteristics are not supportable by the subnetwork.

6.4.2.1.1.2 Analysis

A SNC is a type of transport connection and is related to other transport entities as well as other subnetwork connections. These relationships are described in the Figure 6-10 taken from Recommendation M.3100 Amendment 1.

The requirement to create a SNC and activate at a later time leads to defining a state model. Table 6.12 defines the state transition model for SNC. The states are Active, Pending and Partial. In the active state, SNC is created and is capable of passing traffic. When a SNC is in the pending state, it has been created but all the necessary cross-connections to activate it to carry traffic have not been established. The partial state is similar to the pending state except some cross connections have been established in the network elements but all the necessary cross-connections have not been established.

Based on the request from a network management system to create, activate and delete SNC, table 6.12 defines the state transition events and the resulting states.

Table 6-12. State Model for Subnetwork Connection

Event	States				
	Initial	Active	Pending	Partial	Null
Create and activate	=> Active	Active	=> Active	=> Active	-
Create and activate partially	=> Partial	-	=> Partial	=> Partial	-
Create	=> Pending	-	-	-	-
Activate	-	-	=> Active	=> Active	-
Activate some	-	-	=> Partial	=> Partial	-
Delete and deactivate	-	=> Null	=> Null	=> Null	Null
Delete	-	=> Null	=> Null	=> Null	Null
Deactivate some	-	=> Partial	=> Partial	-	-
Deactivate	-	=> Pending	=> Pending	Pending	-

The SNC establishment requirements above identify the set of parameters included in the request. The properties of the SNC according to these requested parameters are:

a) the termination points at the ends of the SNC usually referred to as the A and Z ends;
b) Characteristic information of the signal carried by the SNC
c) Directionality (unidirectional or bi-directional)
d) Composite SNCs if the given SNC is part of another SNC in the same layer of the hierarchy (applicable to partitioned subnetworks)
e) Component SNCs if the SNC is made of other SNCs (applicable in partitioned subnetworks
f) Routing profile or the selected routes
g) User friendly label for reference
h) Operational, administrative and connection states.

Note that the composite and component SNCs are reverse relations in that in the first case the given SNC is part of other connections and in the latter case one or more SNCs are used to form this SNC.

These relationships and how an SNC is related to other topological entities are shown in Figure 6-10. The end points of the subnetwork connection though specified in the above list as single terminations, in practice they can be collection of termination points. This is why the figure refers to the end points of the SNC as a list. Because a subnetwork connection can be part of another SNC or includes within it other SNCs, the figure shows both attributes as part of the same line where one SNC is related to another SNC.

Though not explained in detail, all the relationships between the network level managed entities are illustrated in the entity relationship model shown in the figure.

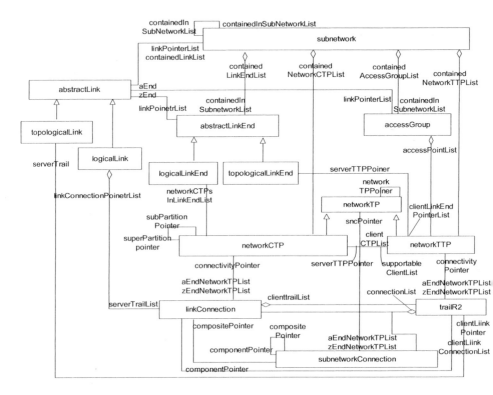

Figure 6-10. Relationship between network level resources (managed entities)

6.4.2.1.2   Route Discovery

Recommendation G.85x.16 specifies the requirements and analysis for determining or identifying the components of a route in support of establishing various connections (trail, SNC etc.). When there are multiple routes available additional criteria such as the protection scheme is used in selecting the candidate.

6.4.2.1.2.1 Requirements
In establishing the trails, SNCs or tandem connections, the following parameters should be supported:
- a)   A route as an ordered series of link connections or connection termination points belonging to a specific layer
- b)   The routes should support various protection schemes
- c)   Returned route set containing route components should support the required service contract offered to the user of the service which is being provisioned.
- d)   The request for finding the resources should include as a minimum the ends of the route.
- e)   The route may have to meet conditions that are either technology dependent or independent. Examples of the latter are the maximum number of nodes, diversity amongst one or more routes. Examples of the former include maximum average spare capacity, bandwidth properties of the nodes, specific protection scheme (1+1, 1:1, m:n etc.).
- f)   The route provided based on the request condition may include the constituent route components.
- g)   The route provided should only be composed of components that are free and thus available for pre-provisioning.
- h)   A notification shall be sent containing the route ends and the route set to inform of the route discovery.

6.4.2.1.2.2 Analysis

A key element of the information defined as part of discovering the route is not only the end points of the route but the components that make up the route. Figure 6-11 below from G.852.16 depicts the relationships between the route components and the route. Both cases of unprotected route and protected route are shown in this figure.

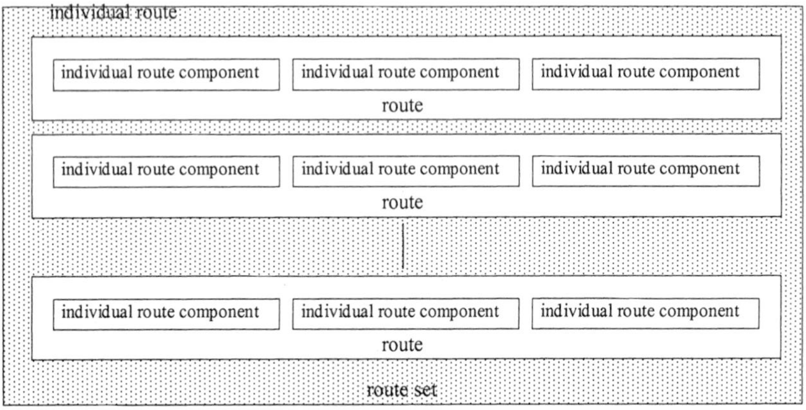

example return information for a route without protection (the route and the individual route
are the same in this case)

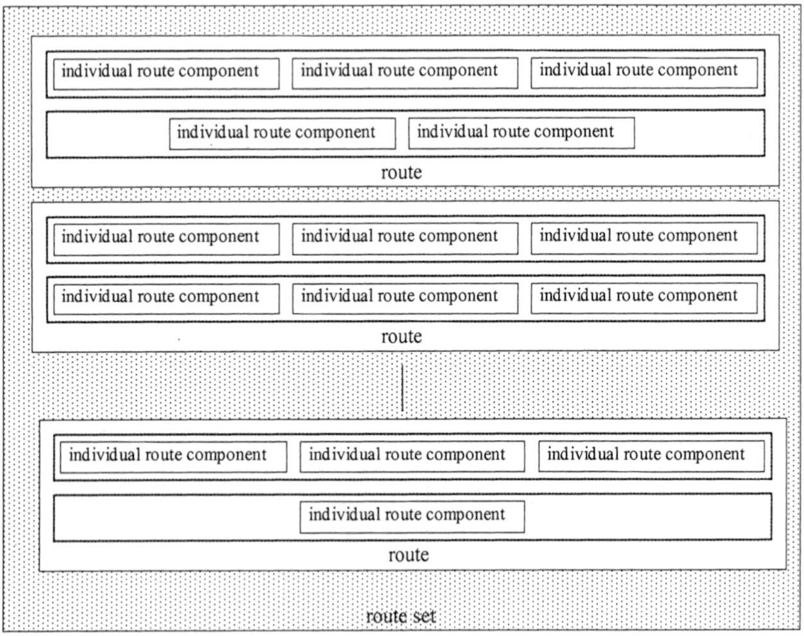

example return information for a route with 1+1 protection

Figure 6-11. Route set with route components

Figure 6-11 (a) above illustrate the individual routes and the components
that make up the route. Because there is no protection scheme supported, the
individual route equate to the route discovered by the request. In the 1+1
protection case there are two individual routes with possibly different

number of components each to form one route thus offering the protection. The number of components associated with an individual route depends on other considerations such as diversity, cost etc.

The route discovery interface for obtaining the routes is defined with the following input and output parameters based on the requirements. The input parameters are:

a) A and Z end corresponding to the two ends of the trail or SNC for which the route is being determined.

b) Routing conditions where technology specific conditions should be provided as discussed in SDH specific network model for this same function.

c) Direction of the route.

d) User identifier.

The output parameters when the request succeeds are:

a) Set of returned routes (including the route components).

b) Ends of the route.

The pre conditions that must be met before the request can be completed and the post conditions after identifying the route are specified in Recommendation G.854.16. One of the precondition for example is that the end points of route exist and available. When the route is identified the post conditions are the route components are free and meet the conditions supplied in the request.

If the route for the requested end points cannot be determined a number of exceptions may be generated. Examples of the errors are invalid ends of the route, requested conditions are not met by one or more route components, some of the route components are not free etc.

6.4.2.2 SDH Models

6.4.2.2.1 Sub-network Connection Management

At the network level technology specific additions are not always required because of the level of abstraction used in these definitions. It is the semantics of the properties and the values of the attributes that contain the necessary specialization.

6.4.2.2.1.1 Requirements

The requirements specified as part of generic SNC management are the same as for the generic models. The additional requirements for SDH, ATM and Dense Wave division multiplexing (DWDM) are specified in the TMF specification for multi-technology management.

a) For SDH different topologies are supported. The SNC type indicates the topology to be supported. The values are singleton, chain, ring, mesh etc.

b) Creation of an SNC may be requested with a specific protection level. The SNC created matches with the best available protection matching the requested effort.

c) The SNC may be requested with one of the following protected level: preemptible, unprotected, partially protected, fully protected and highly protected. For the best case of protection level there are no shared links and subnetwork connections (facilities in generic term) except for the origination and termination points. In SDH, dual ring architectures provide better survivability than what is available with only diverse routing.

d) Protection effort is specified in the SNC creation request to indicate that the protection level of the SNC that should be supported. The values are mandatory, same or better and same or worse.

e) Creation requests with routing constraints should be supported. In the case of SDH this may be specified using for example time slots, BLSR direction etc.

f) The SNC type determines the traffic flow through termination points at A and Z ends. The values for the SNC type are defined to be implicit or explicit. When the type is implicit, the possible values as follows: simple, add drop at A, add drop at Z, mesh, protection switched ring etc. For complex SNCs such as bi-directional rings, where the implicit type is not applicable the explicit value is used.

6.4.2.2.1.2 Analysis

To meet the requirements for SDH specific additional properties for SNC managed entity beyond those described for the generic model is included here.

a) Static Protection level with the values defined above. The protection level varies with the type of the subnetwork. A mesh arrangement for example is highly protected while a ring environment may not have the same high level of protection.

b) Protection effort is used to define whether SNC should support the requested level or better than the requested level or lower than what can be offered.

c) Based on the termination point forming the SNC, the role may be either primary or back up.

6.4.2.2.2 Route Discovery

As noted in the generic model section, discovering the route in terms of the route components connecting the end points is applicable to all technologies. The differences are in the technology dependent conditions that must be met in selecting the routes. The requirements and analysis subsections below are based on the muti technology modeling effort in TMF.

6.4.2.2.2.1 Requirements

In addition to the generic requirements, the following are relevant in the context of supporting SDH technology.

a) The route information provided for a protected SNC corresponding to an SDH layer should include normal and alternative paths.

b) The route information should be retrievable for any specified SNC.

c) If the SNC supports dynamic routing, only a single route with the connecting cross connects may be provided. This is because with dynamic routing there are multiple alternative paths and enumerating them or reserving them is not an efficient allocation of resources.

d) The route information should include only the resources that transfer that specific signal rate (VC 4 for example) of the SNC.

e) If rerouting on failure is not supported, a route discovery after the route setup should yield the same route. If rerouting is supported, a different route may be provided upon a route discovery request.

6.4.2.2.2.2 Analysis

The operational details and the input and out parameters defined in the generic model section meet all the requirements mentioned above for SDH. These requirements either result in specialization of the output parameters or specific values for the signal rate. For SDH, the structure of the route is defined in terms of a set of cross-connects set up within a number of NEs in a subnetwork.

In addition to the end points of the route, cross connects noted in the route structure are further composed of lists of A and Z end connection termination points, connection type and the direction of the cross-connect. The connection type assumes the same values defined earlier for the SNC type.

6.4.3 Functional models

The functional models have been developed to address multiple management areas. Examples are models for log management, event reporting, identifying the location of the managed entity etc. In most cases these models also support multiple technologies. Many of these models have been discussed in the past by several authors. These models may be found in Recommendation series X.73x, X.74x and X75x as well as few new services in Q.816.Two examples of functional models are discussed below. Log Service and multiple object selection service. The log service has been defined both Recommendation X.735 and Q.816. The second function offers a service that was part of the management protocol in most technologies including SDH and not considered as a function to be modeled in the initial designs. However with the introduction of new protocols this function has been explicitly modeled in Recommendation Q.816. The following shows various classes in the functional fragment.

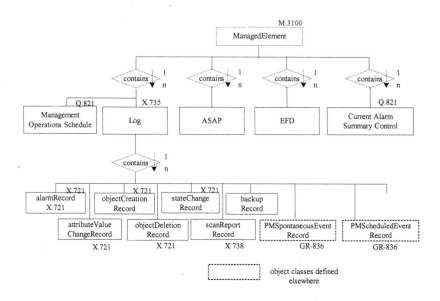

Figure 6-12. Functional Fragment classes

6.4.3.1 Log Service

The log service is used by multiple technologies including SDH and by different management function. For example when a failure alarm or performance threshold crossing or security violation event is generated it may be logged for persistent storage. The event may be logged either in the network element such as an SDH Add drop multiplexer or in an element management if for example the event is a network level e.g. link failure. The requirements are the same in both cases even though the event logged is different.

6.4.3.1.1 Requirements

The above needs give rise to the following requirements to be satisfied:

a) Storing evens persistently as log records must be supported.
b) The records to be stored should be selectable by the management systems. The selection criteria for storing records should be settable by the management system.
c) An external management system should be able to modify the criteria used in logging records;

d) The log containing the records should be able to generate events if for example the storage threshold allocated exceeds the specified value.

e) The external management system should be able to determine if changes to the characteristics of the log has been changed (either by external request or through internal operation of the managed system).

f) The external system should be able to determine if log records have been lost.

g) Support management system request to suspend and resume the logging action

h) When the log reaches its full capacity two possible options are supported – wrap where the new records will over write the existing records replacing the oldest record first, halt where no further records will be added and information is lost.

i) It should support retrieving and deleting log records from the log.

j) It should support creation and deletion of logs by the managed system

6.4.3.1.2   Analysis

Figure 6-12 from RecommendationQ.816 defines a functional model for the log service. The figure shows an event supplier and consumer model for event service based on which the log service has been defined. The supplier of the event is the managed system such as a network element and the event may be an alarm. Based on the quality of service that determines the priority of delivery of the event, the supplier sends the events to the channel. The even channel is a mechanism that is used to queue the events and the events are selected for queuing based on the filter criteria. There are three filters that may be used in this model. The filter on the input of the event channel determines the events that are queued in the event channel for reporting purpose. The filter on the side of the consumer determines which of the queued events are to be sent to which supplier. For the purpose of the log service a third filter is included. This filter is used to control the events that are moved to the persistent storage. In addition to the events being logged using the log filter, the model also notes that a data may be written to the log by applications that are not part of the event model.

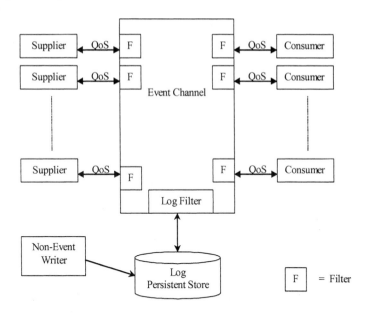

Figure 6-13. Model of Log Service

By manipulating the Log Filter, a managing system is able to control which events are logged and which aren't, in exactly the same way it is able to control which events are forwarded and which aren't. The only exception is the "Non-event Writer," which is an application that writes data directly to the log.

Table 6-13 Properties of Log Managed Entity

Property	Operations	Requirement Trace
Attributes		
Administrative state	read and write	g
Filter criteria	read and write	b,c
Maximum log size	read and write	d
Log full action	read and write	h,f
Capacity Alarm Threshold size	read and write	d,f
Events and parameters		
Threshold crossing event when the current log size reached	Name of the protecting unit, reason for protection	d
Creation and deletion	Name of the entry created or deleted, time stamp and values for attributes if creation event	j

Table 6-14. Creation of Log managed entity

Operation fields	Name	Description
Input Parameters	Administrative State	This parameter specifies the initial administrative state of the Log managed entity.
	Filtering Criteria	This parameter specifies the filtering criteria of the Log managed entity to record events.
	Max Log Size	This parameter specifies the max size for of the log in octets. An empty value indicates there is no limit.
	Log Full Action	This parameter indicates the action this log will take when it is full, which can be wrap or halt.
	Capacity Alarm Threshold	This parameter specifies the percentage point at which an event will be generated to indicate that a log full or log wrap condition is approaching.
Output Parameters	Log Id	This parameter specifies the name of the Log managed entity.
Return Value	-	Success indication
Exceptions raised	Invalid Parameter	One or more supplied parameter has an invalid value or not included in the definition
	Processing Error	An error was encountered while processing the request
	Communication Error	A communication error has occurred and it may be the result of an internal failure in the system or because of link failure.

Even though there is a appearance that the models shown in RecommendationX.735 and Recommendation Q.816 are different, this is not true. In X.735, the event processing cloud abstracts the details of what happens when an event is generated. That function is described in Q.816 by introducing event channel and the relationship to the filtering criteria. The publisher subscriber model of Q.816 includes that a consumer of the events registers with the even channel for the events, log being one of the subscribers. Thus the two models are not proposing differing views for the function but some of the internal processing within a system is explained in Q.816 compared to X.735.

Table 6-14 represents the properties of the log managed entity and how they meet the requirements above.

In addition to the attributes of the log managed entity and the notification, Table 6-14 defines the parameters included when a request is sent by a managing system to create the log entity. The response is one of the possible responses listed in the table. This operation meets the requirement "j" listed above.

6.4.3.2 Multiple Object selection service

An efficiency consideration in managing multiple entities of the same type or different is to be able to address more than one managed entity. Consider the example where a lock operation is to be performed on all the the protection units in a protection group. Instead of having to address each one individually and set the protection status to lock out mode, it is more efficient in terms of message exchanges to address all of the protection units and perform this request with one message. As noted above this feature to select multiple instances was included as part of the management protocol and thus was not considered in the common services set.

6.4.3.2.1   Requirements

The following is a set of requirements based on which the model in Recommendation Q.816 was developed. The efficiency in requesting management operations becomes important in large network elements that may support millions of entities to be managed. To optimize the number of interactions and thereby increase the efficiency, the following objectives are to be met by the information model.

a)   Ability to define a scope for selecting the managed entities where an operation is requested

b)   Further refinement of the scope using filtering criteria.

c)   The minimum set of operations that should be supported within the selected scope is the ability to read the values of attributes, modify the values and deletion of the selected objects.

6.4.3.2.2   Analysis

In order to define an information model, one of the assumptions made is how the managed entities are logically structured in the NE. The use of a hierarchical position of the entities using the aggregation concept defined in an object-oriented approach offers a simple way to analyze the requirements and model the common functions. However this approach may not be easily translatable to interactions with other approaches such as sending simple text messages or file transfers. The general concepts defined in the analysis however can be adapted when non-object oriented approaches are used.

To meet the requirement for selecting multiple objects within a specific context, the root of the search is defined. This is referred to as the base object and the scope for the operation starts at this root. Having defined the root, the next parameter is defining the scope of the search relative to the root. The scope should support different choices. These are the root object only, all objects included within a level such as two levels down from the root entity, all the entities up to the last level from the root (also known as the sub tree), entities at a specific level. The number of objects selected will be first based on the specified choice.

Further refinement of the selection of objects is possible by defining criteria referred to as "filter". This is achieved by specifying a constraint so that result of the evaluating the conditions is "true" or "false". When a managed entity from the first selection process satisfies the constraint it is selected for the operation. Thus the use of both the scope and filter parameter offers an effective mechanism to select a subset of a tree of objects and requires one request from the management system. The criteria can be complex Boolean expression verifying the values or presence of properties of the entity.

The multiple object operation server object is defined in the information model with a number of parameters for the read and write methods. The input parameters of these methods include the scope and filter fields defined above as well as the type of operation (for example read) and the properties to be read.

The methods also include the output parameters and depending on the requested operation they vary. A read operation of the properties returns values whereas a modify operation may only need to indicate whether the change was done successfully or not.

A number of error causes have been defined when the requested multiple object operation cannot be performed. Examples of these errors are invalid filter, the name of the root entity is invalid etc.

As discussed earlier, the information model for the multiple object selection service is not specific to a technology or management application. It is a generic or common functional model similar to the log service.

6.5. CONCLUSIONS

This focus of this chapter is on the information architecture of network management framework discussed in Chapter 5 with emphasis on monitoring and controlling networks using SDH technology. A number of information models have been developed in ITU and other public fora to support both generic and technology specific models. The models, to a large extent, have been specified using a specific design approach determined by the interface protocol. However it is now accepted in the industry that semantics of the management information is the key to future proof the models and offer evolution paths for future technologies and interface protocols. In alignment with this goal, information models are specified in this chapter in terms of requirements for managing the characteristics of the entities as well as the interactions between them. It is expected that standards groups will be using this approach going forward.

The chapter divides the information models for any technology into two broad categories; models required from the viewpoint of managing a single network element (referred to as network element view), and models required from the network view point of managing a network (referred to as network view). In the network element view the examples illustrate two cases: how refinements were carried out for SDH from generic models and how in some cases models were developed to support technology specific requirements. In addition to the two views, this chapter briefly introduces functional models that are both independent of the technology and applicable to multiple management applications. Even though implementations have modified the standard models based on reasons such as ease of implementation, time to market, complexity of definitions etc. these models (both generic and technology specific) form a foundation structure from which simpler models have been derived.

REFERENCES

[ITG805]   G.805, Generic Functional Architecture of Transport Networks, ITU-T, March, 2000

[ITG803]   G.803, Architectures of Transport Networks Based On the SDH, ITU-T, March, 2000

[ITG806]   G.806, Characteristics of Transport Equipment - Description Methodology and Generic Functionality, February 2004

[ITG774]   G.774 Management Information Model for the Network Element (NE) View

[ITG7741]   G.774.01, SDH Performance Monitoring Management Information Model for the NE View

[ITG7742]   G.774.02, SDH Configuration of the Payload Structure for the NE View

[ITG7743]   G.774.03, SDH management of Multiplex Section Protection Structure for the NE View

[ITG872]   G.872 Architecture of Optical Transport Networks, Nov, 2001, Amendment 1 Dec 2003

[ITG7744]   G.774.04, SDH management of the Subnetwork Connection for the NE View,

[ITG783]   G.783 Characteristics of SDH Multiplexing Equipment Functional Blocks, Feb, 2004

[ITG784]   G.784 SDH Management, ITU-T, July,1999

[ITG798]   G.798, Characteristics of optical transport network equipment functional blocks", January 2002, Amendment 1 June 2002

[ITG8080]   G.8080/Y.1304, Architecture for Automatically Switched Optical Network (ASON), ITU-T, 2001

[ITM3100]   M.3100, Generic Network Level Info Model ITU-T, 1995, Corrigendum-1, 1998, Amendment,1999

[ITM3120]   M.3120, CORBA Information Model

[TINA97]   Network Resource Information Model Specification, Version 3.0, TINA-C, 1997

[RAM99]   Fundamentals of Telecommunications Network Management, Lakshmi G. Raman

[AT247]   ANSI T1.247 v2.1-1996 Performance Management Functional Area Services and Information Model for Interfaces between Operations Systems and Network Elements

[AT231]   ANSI T1.231-1997, Digital hierarchy – Layer 1 in-service digital transmission performance monitoring

[OMG]   CORBA Telecom Log Service. Version 1.0, 2000

[Wei98]   "Connection Management for Multiwavelength Optical Networking", J.Y.Wei, et al, JSAC, Sept 1998

[ATT98]   AT&T/TMN TRANSPORT EMS to NMS Q3 Specification Oct 1998
               "Measurement Report File Format" of the 3G 32.104 PM specification

[IETF]   IETF RFC 2558, K. Tesink, March 1999

[ETSI]   ETSI EN 3000 417-7-1 Generic requirements of transport functionality of equipment, Part 7-1: Equipment management and auxiliary layer functions", October 2000

Chapter 7

MANAGEMENT TECHNOLOGIES
LANGUAGES AND PROTOCOLS

7.1. INTRODUCTION

Earlier chapters discussed various components, layers, and interconnection structures of the telecommunications networks, in particular transport networks. Next, telecommunication management network (TMN) architecture was discussed in Chapter 5 followed by functional information required to manage components, as well as aggregates such as subnetworks in Chapter 6. Continuing the management theme of the previous two chapters, this chapter will discuss protocols and technologies used to access and manage TMN entities such as NEF, EMF by various managing entities.

Usually, languages used to describe the structure of management information and the protocols used to access management information of an entity, such as a network element (NE), are treated as one[94]. This chapter deliberately treats these as two separate topics since there is no real dependency between them as will become apparent later. Also kept separate is the functional information (the content)[95]. These three pieces form part of information architecture of TMN as explained in Chapter 5 and illustrated in Figure 7-1. This figure also captures relation between the three topics and sets the context for this chapter.

---

[94] For instance, SNMP and Abstract Syntax Notation One (ASN.1) are usually described together

[95] For instance, CORBA IDL, ASN.1, or XML can be used to express management information of a DCS[95] NE but the functional information or the content itself will be same.

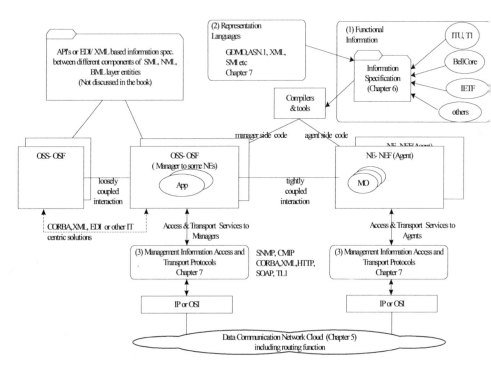

Figure 7-1. Context: Management Information, Languages and Protocols

The first topic was discussed in Chapter 6 and this chapter will discuss the last two. Specifically, this chapter will discuss XML[96] and CORBA as part of languages, TL1, HTTP, SNMP, CORBA, and CMIP as part of management information access protocols. The description of various protocols will be based on a normalized communication model consisting of protocol entity and users (applications). Besides, protocol neutral information modeling and access operations being defined in ITU-T using UML will be discussed.

The chapter also includes discussion of web services - an emerging technology that may have influence on OS-OS as well as OS-NE interfaces. The technologies associated with web services are: Web Services Description Language (WSDL); Universal Description, Discovery, and Integration (UDDI); Simple Object Access Protocol (SOAP)[97]. WSDL will be discussed in the languages section. Though SOAP is used to perform the equivalent of remote procedure calls supported by CORBA, CMIP and hence is in the category of access protocols, it will be discussed together with WSDL.

7.2. MANAGEMENT PROTOCOLS/LANGUAGES GALORE

There are a number of management protocols available to access management information. Similarly there are a number of languages available to describe structure of management information. Given this, there are three points worth highlighting.

First, Operation Support Systems (OSSs) from different carriers may use different protocols to interact with NEs. Even within a single carrier OSS environment, different OSS applications might use different protocols. For instance, a carrier's alarm management OSS application may use TL1, while performance monitoring OSS application uses SNMP. Besides, carriers are likely to stay with their existing OSS technologies (i.e. legacy) with exceptions. Given this, any new equipment or system to be introduced into a carrier network will be expected to support protocols that will be compatible with that carrier's TMN environment.

---

[96] See reasons for leaving out ASN.1 in §7.3

[97] Java Messaging Specification (JMS), part of J2EE platform, is another option. Applications, brokered by a JMS provider (commonly packaged with Application Servers), communicate using JMS send/receive and other advanced APIs.

Second, vendors are likely to design new products based on management technologies used in their exiting products. For instance, a vendor with OSI/GDMO technology in their existing product line is likely to design new products with that technology. Such an approach (i.e. using legacy infrastructure) enables vendors to derive benefit out of internal tools and processes associated with that technology.

Third, vendors are likely to support multiple protocols (TL1, SNMP, CORBA, etc) so that they can sell their products to different carriers. As mentioned before, a given carrier may require vendors to support one or more protocols that are compatible with that carrier's TMN environment.

Setting aside the issue of legacy on part of vendors and carriers, one can evaluate languages in terms of - features, flexibility, tools, and other metrics. Features such as inheritance can be very useful to schema designers to model generic as well as specific resources and such modeling increases specification re-use. Similarly, protocols can be evaluated in terms of features, flexibility, scalability, performance, etc. In some cases the need for protocol to be human-readable (as in the case of TL1) might influence the choice more than the protocol performance alone. Another important feature is support for asynchronous events and event channel management. In general, the rest of the chapter will try to highlight these aspects.

7.3.   INFORMATION SPECIFICATION LANGUAGES

This section will describe some of the commonly used languages to specify or declare management information. Such declaration is also called SMI –structure of management information or schema. A good understanding of ASN.1 is very useful for understanding many ITU-T information model standards and the following quote aptly emphasize its practical role as well:

> *"ASN.1 is a critical part of our daily lives; it's everywhere, but it works so well it's invisible!" - Olivier DUBUISSON.*

Despite its ubiquitous role, since ASN.1 is discussed elsewhere extensively, this chapter will focus on other languages. For readers interested in exploring ASN.1, Dubisson's book is an excellent resource.

7.3.1   XML

EXtensible Markup Language (XML)[W3C], though started as a mark up language in publishing world, has become a dominant technology of choice for information representation, exchange, and processing. As shown in Figure 7-2, XML, along with HTML and others, is derived from SGML (Standard Generalized Markup Language).

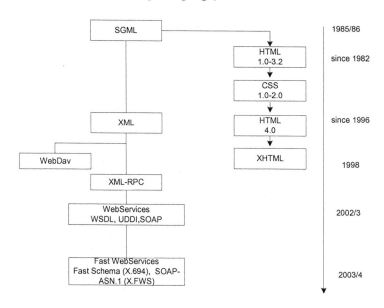

Figure 7-2. SGML derived technologies and time line

Table 7-1 lists the various technologies associated with XML:

Table 7-1. XML Technologies

XML/Web Technologies	Description
DOM	Document Object Model - a platform and language-independent interface that allows applications to dynamically access and manipulate the content, structure, and style of XML documents.
XPath	Language for addressing specific parts of an XML document.
XQuery	Query language for XML, designed to be broadly applicable across various types of XML data sources
SOAP	Simple Object Access Protocol – a lightweight protocol for exchanging structured and typed information of services and objects in a distributed environment using XML, and HTTP in a platform-independent manner.
WSDL	Web Service Definition Language
UDDI	Universal Description, Discovery, and Integration – a WS registry technology similar to X.500

XML allows, using markups, information exchanged between systems or between a system and a human user to be explicitly described. In other

words the information being exchanged is self-describing. This allows
humans, as well as automated systems, to understand the information more
easily. However, markups do increase verbosity of information being
exchanged. Any one who looked at an HTML source page might know how
difficult it is to get to the real content hidden behind myriads of HTML tags.

Each XML document contains a prolog before the start tag that begins
the document proper. The prolog can include:

- An optional XML declaration
- A document type declaration (DOCTYPE)
- Comments
- Processing Instructions (PI)
- White Space

A PI represents an instruction to the document processor and is of the
form `<?target instruction?>`. Any PI with target=*xml* represents and
indicates an XML document. The following lists a simple XML document
where `<?xml version="1.0"?>`, as part of prolog, indicates that it is an xml
verion 1.0 document

```
- <?xml version="1.0"?>
- <element1 attr1="aaa"  attr2="bbb" …>
       - …              ← content of element1
       - …
       - …
- </element1>
```

Each element in XML will have an opening and a closing markup[98]. The
opening markup consists of a tag and some attributes with each attribute
being a *name="value"* tuple. The end markup contains the same tag as
used in the opening markup. An element's start and end markups define the
scope of that element. With in the scope of an element, simple unmarked
data or other markup data may be included and such inclusion is recursive.
The allowed type of unmarked data or   child elements can be constrained
using either DTD or Schema, to be discussed shortly.

An XML document containing management information structure (i.e.
MIB) description (i.e. declaration) of a network element can be rather very
large. If such document is stored in one large physical file it can lead to

---

[98] A markup is a tag enclosed in brackets e.g. <xyz> for starting markup or </xyz> for ending
markup where xyz is the tag.

versioning or other file management related issues. Such MIBs can be easily managed by breaking them into sections or modules. The MIB then becomes a logical document consisting of one or more pieces with some or all of the individual pieces stored in separate files and Figure 7-3 Illustrates this.

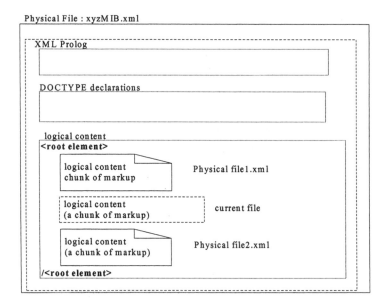

Figure 7-3. Logical and Physical structure of XML Documents

To give an example, one can structure an NE MIB such that it consists of PM (Chapter 6) related XML MIB definitions in one file, Fault/Alarm related definitions in another XML MIB file etc. To be able to include such files, containing chunks of markup, into a master file, XML provides an ENTITY referencing, a way to define a reference to a chunk of markup located in the current file or in another file. Use of ENTITY[99] to include chunk of mark up is similar to <include> directive in C, C++ or other programming languages.

Before presenting syntax for entity definitions, meta symbols used by EBNF[100] (Extended Backus Naur Form) and used throughout the chapter are listed in Table 7-2. EBNF is a meta language and notation commonly used in computers science to specify syntax of programming languages using meta symbols listed below with their meaning.

---

[99] ENITY referencing can also be used to alias constants and is similar to #define in C/C++.

[100] EBNF is more powerful than the plain BNF. It contains three meta symbols -? *, +. Any EBNF production can be converted to an equivalent set of BNF productions.

Table 7-2. EBNF Notation

Meta symbol	Meaning
+	Item or group of items preceding + occurs one or more times
?	Item or group of items preceding ? occurs at most one time
*	Item or group of items preceding * occurs zero or more times
::=	"is defined as"
()	Items within the brackets are a group and qualified further using ? , *, or +
|	Choice between left and right side items
{}	To express repeatability, indefinitely

The following listing contains EBNF productions for general entity (GE) and parameter entity (PE, to be discussed shortly).

```
;EBNF productions for Entity Declaration

EntityDecl::=    GEDecl | PEDecl
GEDecl    ::=    '<!ENTITY' Name   EntityDef   '>'
PEDecl    ::=    '<!ENTITY'  '%'   Name PEDef '>'
EntityDef ::=    EntityValue| (ExternalID NDataDecl?)
PEDef     ::=    EntityValue | ExternalID
ExternalID::=    'SYSTEM' URI
                 | 'PUBLIC' PUB_NAME_ID URI
NdataDecl   ::=    'NDATA' Name

;concrete syntax and examples

<!ENTITY [entity name] SYSTEM "URI">
<!ENTITY [entity name] PUBLIC "PUB_NAME_ID" "URI>
<!ENTITY [entity name] "Replacement Characters/Identifier" >

;Examples

<!ENTITY xyzPM SYSTEM "xyzPMMIB.xml">

---Example Referencing ----
<xyzNEMIB>
    &xyzPM ← including PM MIB definitions here
    … ← other definitions
</xyzNEMIB>
…
```

The SYSTEM keyword is used to specify an external markup located at the specified URI. One can use PUBLIC keyword, instead, to refer to a common or standard chunk of markup.

The embedded files must contain only markup and should not be full XML documents by themselves i.e. no <?xml ...?> or other PIs are allowed in them. Figure 7-4 captures the above discussion related to an NE MIB consisting of multiple functional MIBs.

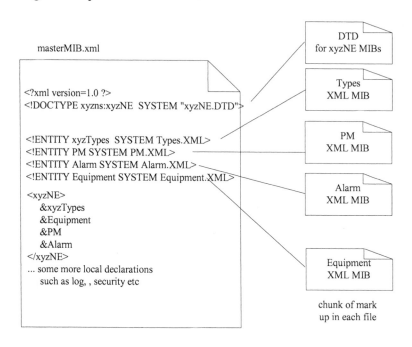

Figure 7-4. An Example XML document derived from multiple physical documents

The chunk of included markup must be balanced in the sense it must contain a whole number of elements, plus optional text or comments at the start or end. Some times it is useful to treat files containing a chunk of markup as valid XML documents in their own right. One way to accomplish this would be to have a separate file whose only purpose is to act as the document entity for a section of MIB as illustrated in the following example:

```
- File: pmDoc.xml
-
- <?xml version="1.0" ?>
- <!DOCTYPE xyzPM SYSTEM "xyzMIB.DTD" [
- <!ENTITY xyzPM SYSTEM "xyzPMMIB.xml"> ← include markup from
PMMIB.xml
- ]>
- &xyzPM;
-
```

Besides the above general aspects, there are three important features of XML: Namespaces, Document Type Definition (DTD), and XML Schema. A brief description of each follows.

7.3.1.1 Namespaces

An XML application can declare and use many types (user-defined or built-in). Therefore, to avoid type name clash, XML supports the concept of *namespace* [W3C] in which each type is associated (implicitly or explicitly) in the context of its name space. When using a type, the namespace to which it belongs is prefixed if necessary. The prefix together with the type name (called the local name) is referred to as QNAME in XML recommendations and can be used in document, element or attribute declarations. The following provides an example.

```
<?xml version="1.0" ?>
<transportNE xmlns:q3ml="http://www.xyz.com/q3ml"
    xmlns="www.xyz.com/xyzxsd">
    <q3ml:managedElement neType="sdhNE">
        ...
    </q3ml:NE>
    <operationalValue> ... </operationalValue>
</transportNE>
```

XML namespaces are defined using the name space attribute. In the above example, complex element, transportNE, specifies q3ml as its namespace. The value, www.xyz.com/q3ml, of the namespace attribute is a hint to the document processor to locate the schema associated with that namespace. The value need not actually point to a physical location. Any child element or attribute that is prefixed with q3ml will import the syntax from q3ml namespace. Any unqualified element or attribute in the scope of trasnportNE would be in the context of the default namespace. For instance, the unqualified element, operationalValue is from the default name space - xyzxsd.

7.3.1.2 DTD

Each XML document must follow two basic rules. The first rule is that an XML document must be *well–formed*. A well formed XML document is one that has every tag closed that is opened, and has no tags nested out of order.

The second rule is that an XML document must be *valid*. A *valid* document is one that conforms to a structure that is pre-defined using

Document Type Definition (DTD) notation. A DTD establishes a set of constraints for the content of an XML document. Each element in an XML document must meet the constraints specified for that element in the DTD. The pre-defined structure is called document template. Many XML DTD templates exist for use in banking, telecom, insurance, automobile, and other areas of information exchange and processing. Using DTD, one can specify allowed elements, allowed attributes, acceptable values for each attribute, nesting and occurrence of each element etc.

Each XML document can specify its type (i.e. the DTD to use) at the beginning using `<!DOCTYPE` declaration.

The EBNF productions for element declaration are given below along with some concrete examples.

```
- Elementdecl  ::= '<!ELEMENT' Name contentspec'>'
- contentspec  ::= 'EMPTY' | 'ANY' | Mixed | children
- children     ::= (choice | seq) ('?' | '*' | '+')?
- cp           ::= (Name | choice | seq) ('?' | '*' | '+')?
- choice       ::= '(' cp ('|' cp )+ ')'
- seq          ::= '(' cp (',' cp )* ')'
- mixed        ::= '(' '#PCDATA' ('|' Name)* ')*'
               -   | '(' '#PCDATA' ')'

- <!ELEMENT [element-name] [Element definition] >
- <!ELEMENT [element-name] ([Nested Element] [,Nested Element]…]) >
- <!ELEMENT [element-name] EMPTY >
- <!ATTLIST [element-name] [attribute-name] [attribute-type]
[modifier] [default-value] >
```

The various data types available in DTDs to specify attribute type are listed in Table 7-3.

Table 7-3. DTD Attribute types

DTD Data Types (used to specify attribute data type)	Value Description
CDATA	character data
(eval | eval ...)	An enumerated type with the values specified in the list
ID	unique id
IDREF	id of another element
IDREFS	a list of other ids
NMTOKEN	a valid XML Name
NMTOKENS	a list of valid XML Names
ENTITY	an entity
ENTITIES	a list of entities
NOTATION	name of a notation
Xml	predefined

The attribute modifier can be one of #REQUIRED, #IMPLIED, or #FIXED. An implied attribute is an optional attribute and can remain unspecified. A required attribute is a mandatory attribute and must be present in the element declaration. A fixed attribute is one with fixed value. Attributes can have default value specified after the modifier.

The following listing illustrates how GDMO managed object class template definition can be defined using DTD. The EBNF meta symbols (+,?, *, |, etc) have been explained earlier.

```
<!ELEMENT moClass
        (derivesFrom? | behaviour? | nameBinding?
        | packageRef? | package? | gdmoAttributeGroup* |
        gdmoAttribute*
        | action* | notification*)
>

<!ATTLIST gdmoManagedObjectclass
    name    NMTOKEN    #REQUIRED
>

<!ELEMENT derivesFrom (classRef+) >

<!ELEMENT package >
        (behaviour? | nameBinding?
        | attributeGroup* | attribute*
        | action* | notification*)
>


<!ELEMENT classRef    EMPTY>

<!ATTLIST classRef
```

```
-     name CDATA    #REQUIRED>
-
-   <!ELEMENT packageRef EMPTY>
-
-   <!ATTLIST packageRef
-     name        CDATA    #REQUIRED
-     conformance     (mandatory | Conditional| private) "private"
-     mandatoryAttributes CDATA   #IMPLIED>
-
-   <!ELEMENT nameBinding (parent+)>
-
-   -- NE MIBs contain many data types (sets, lists, structs etc)
-   -- but for brievity only two types are included
-   <!ELEMENT sval (CDATA)#REQUIRED>
-   <!ELEMENT ival (CDATA) #REQUIRED>
-
-   --
-   <!ELEMENT gdmoAttribute ((svalue|ival)>
-   <ATTLIST gdmoAttribute
-       - name NMTOKEN #REQUIRED
-       - >
-
-   <!ELEMENT gdmoAttributeGroup (gdmoAttribute+)>
-       - <!ELEMENT package
-       (gdmoAttributeGroup*|gdmoAttribute*|actions*|notification*)>
-
-   <!ELEMENT parent
-     (creationRule?,
-     deletionRule?,
-     modificationRule?,
-     RetrievalRule?)>
-
-   ...
-
-   <!ELEMENT xyzNE
-           (moClass+| packages* | attributeGroup* | attribute* )>
-   _____
-
```

The following is an instance document based on the example DTD listed above.

```
-   _____
-   <?xml version="1.0"?>
-   <!DOCTYPE xyzNE SYSTEM "xyzNEMIB.DTD">
-
-   <xyzNE>
-
-   <moClass name="managedElementR1">
-       - <derivesFrom> </derivesFrom>
-       - <attribute name="userLabel" value""/> </attribute>
-       - <package  name="administrativeStatePkg"/>
-           - <gdmoAttribute name="administrativeState">
-               - <svalue>unlocked</svalue>
-           - </gdmoAttribute>
```

```
-    </package>
-    <package  name="operationalStatePkg"/>
   -    <gdmoAttribute name="operationalState">
   -    <svalue>enabled</svalue>
   -    </gdmoAttribute>
-    </package>
-    <package  name="availabilityStatusPkg"/>
   -    <gdmoAttribute name="availabilityStatus">
   -    <svalue>available</svalue>
   -    </gdmoAttribute>
-    </package>
-
-    ...)
-  />
-  <moClass name="sdhNE">
   -    <derivesFrom> managedElementR1 </derivesFrom>
   -    <gdmoAttribute name="neType">
        -    <svalue> "DCS" </svalue>
   -    </gdmoAttribute>
   -    ...
-  />
-  </xyzNE>
-
```

A document's first element, appearing after the PIs, and `<!DOCTYPE` declaration is called the root element of the document. The root element's tag is used in `<!DOCTYPE` declaration to specify the document type along with any external DTD reference as per the syntax for document type declaration shown below

```
-  doctypedecl::= '<!DOCTYPE'
        -    rootElementTag  (ExternalDTD)?  ('[' internalDTD ']' )? '>'
-  ExternalDTD::= 'SYSTEM'  URI
                -    |'PUBLIC'  PUB_NAME_ID URI
-
```

The reference to external DTD can use either the SYSTEM or PUBLIC referencing format. External DTD referencing is similarly to `<!ENTITY` referencing explained earlier to include chunks of mark up from external documents. The URI specified in the SYSTEM external entity reference is the location of the external DTD. Alternatively, if one uses PUBLIC external entity reference then the XML document processor resolves the PUB_NAME_ID to an URI. If the document processor is not able to locate the DTD at the derived URI then it will use the URI specified after the PUB_NAME_ID.

Whether using SYSTEM or PUBLIC entity declaration format, the document processor will retrieve the external DTD and use it to validate the current document. In the above example, the root element is the xyzNE and uses the SYSTEM entity reference format.

In the above example, three packages were defined in managedElementR1. Other MOCs, such as terminationPoint, also use these and other common packages (see chapter 6). Instead of repeating such common package definitions in every MOC, one can can define them as stand alone packages. One can then provide reference to them using packageref element. However, in-memory representation of a MOC containing a packageref element must actually contain all the attributes of the referenced package. The sender and receiver have an implicit understanding with respect to the structure of serialized instance document versus the structure declared in the DTD. Note though that the above is merely an example to aid understanding of DTD for use in telecommunications and should not be construed as the only way to structure and model information.

ENTITY element introduced earlier has a variant called the *parameter entity* which can only be used in DTDs. It helps in reusing some commonly used definitions. The only difference in its declaration and use is an extra % symbol. When referencing a parameter entity a % is used instead of & (as done with replacement entity referencing). The following provides an example for parameter as well as for general entity referencing.

```
<!ENTITY  %severityChoices " indeterminate |
                            critical|
                            major |
                            minor|
                            warning|
                            cleared"
>
<!ENTITY tMrk "Trade Mark of XYZ Corporation">

<!ELEMENT perceivedSeverity EMPTY>
<!ATTLIST perceivedSeverity
          severity   (%severityChoices;)   #REQUIRED>
<xyzProd1>
  <tradeMrk> &tMrk  </TradeMrk>
  ...
</xyzProd1>
<xyzProd2>
  <tradeMrk> &tMrk  </TradeMrk>
  ...
</xyzProd2>
```

Though DTD notation allows concise declarations, the notation is different from the one used in XML documents. Therefore, people have to learn new syntax and rules to create DTDs. Besides; DTDs are not flexible enough due to lack of equivalence, derivation, and mapping to OO languages. XML Schema is the solution for all these problems.

7.3.2   Schemas

XML schema language [W3C] allows document/template designer to specify the structure of an XML document (also called instance document) and serves identical purpose as DTD. However, XML Schema language provides richer set of constructs enabling one to constrain instance documents in more ways and with more flexibility and ease than is possible with DTDs. However, unlike DTD, XML schema is just another XML document and therefore the template document looks just like an XML document. Besides, in DTDs each element's structure and constraints are included in its declaration. If another element needs to be declared with identical structure and constraints then the same specification need to be repeated in the new element's declaration. Therefore, a DTD can be too verbose and not concise and as a result prone to errors. On the other hand, XML schema does allow one to define such reusable specification fragments resulting in less verbiage. These are principles for having attribute groups and packages in GDMO - reuse of packages across multiple MO definitions.

Typical use of XML schema involves generating code in C++, Java or some other OO/programming language and the process is called *data binding* as shown in Figure 7-5.

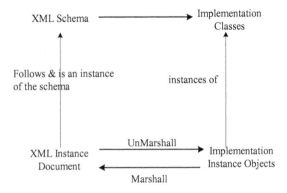

Figure 7-5. Schema data binding

Some of the simple schema types such as *int, long, double, string* may have natural mapping to the OO language types. Other complex types defined in a schema may have to be mapped to user-defined types constructed using chosen OO/programming language facilities.

A collection of XML schemas for use in telecommunications OAM&P is called tML – telecommunications markup language. Accordingly interfaces that support XML schema as per tML guidelines are referred to as tML interfaces. One should note, however, that tML is not a new language but only an application of XML schema. Framework and guidelines for defining XML schemas for various applications across X interfaces is described in ITU-T M.3030 standard. T1M1 document [Ree01] describes how some of the GDMO/ASN.1 constructs can be modeled using tML schema framework. Figure 7-6 illustrates, based on the above reference a hypothetical mapping. WSDL and SOAP specifics in the figure were added to take into account WS technologies that emerged after the tML was published by ITU-T. These specifics will be highlighted in § 7.3.2.3.

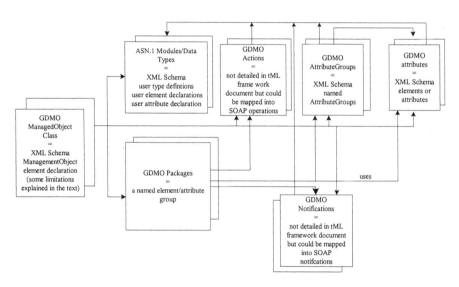

Figure 7-6. XML Schema elements for describing management information (similar to GDMO/ASN.1)

The mapping described in [[Ree01] is not yet standard but only a T1M1 contribution. The discussion on XML schema in this chapter derives from [M.3030] and [Ree01]. Also the rest of the discussion will use tML and XML schema interchangeably.

Since tML schema language is used to describe only the structure of management information and not the interaction aspects - actions, notifications etc, there is difference between tML based interfaces with those of CORBA or other paradigm based ones. Given current trend, it may be possible that XML version of GDMO, for use across Q interface, may become commonplace even if standardization of the same lags market place.

An XML schema contains:

- Definitions of:
  - User-defined complex types – with any child elements and attributes
  - User defined simple types – with no child element but constraints
- Declarations of:
  - Elements (of simple or complex type)
  - Element Groups
  - Attributes (of simple type), and
  - Attribute groups (of simple type)

7.3.2.1 User defined Types and Derivation

The XML schema has rich type system with more than 44 different built-in types shown in Figure 7-7. .

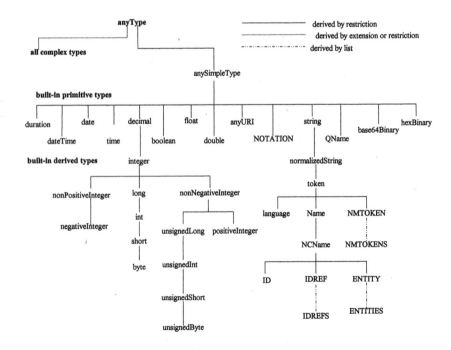

Figure 7-7. XML Schema built-in types

The built-in simple types are divided into `built-in primitive` and `built-in derived` type blocks. The `SimpleType` element allows definition of user defined simple data types. The `ComplexType` element allows definition of user defined complex (aggregate) data types. Both simple and complex user defined types can be used to specify an element's type. Element's attributes, however, can only be of simple type (native or derived). In addition, some built-in data types such as `ID`, `Key` etc can only be used for attribute types. Any element that contains attributes or other elements is considered as a complex type. For instance, `<sdhNE neType="DCS"/>` is a complex type even though there is no content inside the `sdhNE` element. The following listing contains syntax[101] for declaring user-defined complex and simple types. Examples in the rest of this section will explain the usage, at least the simple cases.

```
<complexType
   abstract = boolean : false
   base = QName
   block = #all or (possibly empty) subset of {extension,
restriction}
   content = elementOnly | empty | mixed | textOnly
   derivedBy = extension | restriction
   final = #all or (possibly empty) subset of {extension,
restriction}
   id = ID
   name = NCName>
   Content: (annotation? , (((minExclusive | minInclusive |
maxExclusive | maxInclusive | precision | scale | length | minLength |
maxLength | encoding | period | duration | enumeration | pattern)* |
(element | group | all | choice | sequence | any)*) , ((attribute |
attributeGroup)* , anyAttribute?)))
</complexType>


simpleType
   final = (#all | (list | union | restriction))
   id = ID
   name = NCName
   {any attributes with non-schema namespace . . .}>
   Content: (annotation?, (restriction | list | union))
</simpleType>

<restriction base="existing Type Name">
      [content:(annotation?, (simpleType?, constrainingFacet*)))]
</restriction>
```

---

[101] Refer to [W3C] for detailed explanation of the syntax.

The syntax for declaring element type (i.e. user defined markup/tag) is provided   in the following listing with usage examples provided in the rest of the section.

```
<element
       [abstract = boolean : false]
       [block = (#all | List of (extension | restriction |
       substitution))]
       [default = string]
       [final = (#all | List of (extension | restriction))
       [fixed = string ]
       [form = (qualified | unqualified)]
       [id = ID ]
       [maxOccurs = (nonNegativeInteger | unbounded)   : 1]
       [minOccurs = nonNegativeInteger : 1]
       [name = NCName ]
       [nillable = boolean : false]
       [ref = QName ]
       [substitutionGroup = QName ]
       [type = QName ]
       [{any attributes with non-schema namespace . . .}]>

       Content: (annotation?, ((simpleType | complexType)?, (unique
       | key | keyref)*))
</element>
```

Inside an element one can include, as part of element's content, simple or complex type data.

One can also define new types by deriving from existing types and there are three ways to do that:

– By deriving from an existing type using *restriction* or *extension*
– As *list* or *union* of existing simple types

Different kinds of constraints (i.e. restrictions) that can be applied on a data type are called *facets*. They allow one to constrain the value space of the original type. Some of the facets are listed in the   Table 7-4.

Table 7-4. XML schema's facets[102]

Facet	Description
enumeration	Specifies allowed set of values and thereby constrains a data type to the specified values
length	Number of units of length. Units of length depend on the data type. This value must be a nonNegativeInteger.
maxExclusive	Upper bound value (all values are less than this value). This value must be the same data type as the inherited data type.
maxInclusive	Maximum value. This value must be the same data type as the inherited data type.
maxLength	Maximum number of units of length. Units of length depend on the data type. This value must be a nonNegativeInteger.
minExclusive	Lower bound value (all values are greater than this value). This value must be the same data type as the inherited data type.
minInclusive	Minimum value. This value must be the same data type as the inherited data type
minLength	Maximum number of units of length. Units of length depend on the data type. This value must be a nonNegativeInteger.
pattern	Specific pattern that the data type's values must match. This constrains the data type to literals that match the specified pattern. The pattern value must be a regular expression.

The following provides some examples for deriving by restriction using the perceivedSeverity GDMO attribute from X.721 and its equivalent declaration in XML schema. The <union> element will be explained shortly.

```
; Examples to illustrate XML Schema facet

<simpleType name="sPerceivedSeverityType">
  <restriction base="string">
    <enumeration value="indeterminate"/>
    <enumeration value="critical"/>
    <enumeration value="major"/>
    <enumeration value="minor"/>
    <enumeration value="warning"/>
    <enumeration value="cleared"/>
  </restriction>
</simpleType>

<simpleType name="iPerceivedSeverityType">
  <restriction base="integer">
    <maxInclusive value="5">
    <minInclusive value="0">
```

[102] Borrowed from http://www.xmlme.com/ .

```
-   </restriction>
-   </simpleType>
-

-   <simpleType name="securityAlarmSeverityType">
-       <list  itemType=" iPerceivedSeverity"/>
-       <maxInclusive value="4">
-       <minInclusive value="0">
-   </simpleType>
-

-   <simpleType name="perceivedSeverityType">
-       <union
-           memberTypes= " sPerceivedSeverityType
            iPerceivedSeverityType"
-       />
-   </simpleType>
-
```

While `facets` are used to constrain value space of user defined simple
types, occurrence constraints allow constraining occurrence of child
elements. The purpose of a `facet` is to constrain the values that can occur
inside the element while the purpose of occurrence constraints is to constrain
the structure (and hence the content) of an element. The two occurrence
constraints are *minOccurs* and *maxOccurs*. Many examples presented
throughout this section will help explain their usage.

As part of derivation feature, XML Schema attempts to mimic object
oriented (OO) concepts. An existing type (user defined or native) can be
used as base to derive new type. Similarly, an existing element can be used
to declare new elements. Concepts such as 'abstract', 'virtual', and 'final' so
common in OO languages are available in the schema as well. The
following listing contains examples for 'abstract' and 'final'.

```
-
-   <!--no derivation -->
-   <simpleType  name="personName" final="#all">
-       <restriction base="string/">
-   </simpleType>
-

-   <!-- no instanantiation -->
-   <element name="person"  type="personName"  abstract="true" />
-

-   <!-- neither instanantiation nor derivation are allowed -->
-   <element name="person"
-       type="personName"  abstract="true" final="#all"/>
-

-   <!-- derivation through restriction -->
-   <simpleType  name="personName" final="restriction">
-       <restriction base="string/">
-   </simpleType>
-
```

Table 7-5. Some of XML Schema Elements and their description

Schema Element	Description
all	Allows the elements in a group to appear (or not appear) in any order in the containing element.
annotation	Defines an annotation.
any	Enables any element from the specified namespace(s) to appear in the containing *complexType, sequence, all,* or *choice* element.
anyAttribute	Enables any attribute from the specified namespace(s) to appear in the containing *complexType* element.
attribute	Declares an attribute in the context of an element or at a global lev.
attributeGroup	Groups a set of attribute declarations so that they can be incorporated as a group into complex type definitions.
choice	Allows one and only one of the elements contained in the group to be present within the containing element.
complexContent	Contains extensions or restrictions on a complex type that contains mixed content or elements only.
complexType	Defines a complex type, which determines the set of attributes and the content of an element.
documentation	Specifies information to be read or used by users within the annotation element.
element	Declares an element.
extension	Contains extensions on complexContent or simpleContent, which can also extend a complex type.
field	Specifies an XML Path Language (XPath) expression that specifies the value (or one of the values) used to define an identity constraint (unique, key, and keyref elements).
group	Groups a set of element declarations so that they can be incorporated as a group into complex type definitions.
import	Identifies a namespace whose schema components are referenced by the containing schema.
include	Includes the specified schema document in the target namespace of the containing schema.
key	Specifies that an attribute or element value (or set of values) must be a key within the specified scope.
keyref	Specifies that an attribute or element value (or set of values) correspond with those of the specified key or unique element.
list	Defines a simpleType element as a list of values of a specified data type.
notation	Contains the definition of a notation.
redefine	Allows simple and complex types, groups, and attribute groups that are obtained from external schema files to be redefined in the current schema.
restriction	Defines constraints on a simpleType, simpleContent, or complexContent definition.
schema	Contains the definition of a schema.
selector	Specifies an XPath expression that selects a set of elements for an identity constraint (unique, key, and keyref elements).
sequence	Requires the elements in the group to appear in the specified sequence within the containing element.
simpleContent	Contains either the extensions or restrictions on a complexType element with character data, or contains a simpleType element as content and contains no elements.
simpleType	Defines a simple type, which determines the constraints on and information about the values of attributes or elements with text-only content.
union	Defines a simpleType element as a collection of values from specified simple data types.
unique	Specifies that an attribute or element value (or a combination of attribute or element values) must be unique within the specified scope.

XML schema comes with many pre-existing tag/elements, many of which are listed in   Table 7-5. Note, however, that applications can still use these tags since every tag is unique within its own name space – for instance, xsd:annotation is different from myApp:annotation.

7.3.2.2 Aggregate Data Types

Schema allows user defined structured (i.e. aggregate) data types that combine one or more simple or complex types using - group, all, choice, and sequence elements. These elements are called *compositors* and the members they contain (i.e. compose) are called *particles*! Schema's *compositors* are similar to struct, choice, and set constructs available in ASN.1.

The compositor <sequence> is used to model an *ordered* set. The members (i.e. particles) of the set can be of simple or of complex type. The member can also be an aggregate type. Besides, any member of the sequence can be declared as optional. Also, tML schema supports ASN.1's 'sequence of' aggregate data type similar to ASN.1's sequence of type. The following provides usage example for the <sequence> compositor. The scenario involves adding a cross-connection by invoking create--connection action on an NE. The parameter to this request is a sequence of (i.e. a list of) connectionInformation elements. Each element contains an ordered sequence of – xcType, fromTP, toTP, xcName, userLabel etc (M.3100 ASN.1 definition is more generic and the following is a simplified version to aid discussion here).

```
- ; ASN.1 description
- _____
-
-
- ConnectionType ::= ENUMERATED (unidirectional(0), bidirectional(1)}
- ConnectionInformationItem ::= SEQUENCE {
        - connType ConnectionType,
        - fromTP ObjectInstance,
        - toTP   objectInstance,
        - userLabel UserLabel OPTIONAL
        - }
-
- ConnectionInformation ::= SEQUENCE OF
        - ConnectionInformationItem;
-
- ;----equivalent declaration using XML schema ----
-
- <simpleType name=" ConnectionTypeType">
  - <restriction base="integer">
    - <enumeration value="0"/>
    - <enumeration value="1"/>
  - </restriction>
- </simpleType>
- <complexType name="connectionInformationSequenceType">
```

```
-   <sequence maxOccurs="unbounded">
  -   <element name="value">
    -   <complexType>
        -   <sequence>
            -   <element name="connectionType"
                -   type="ConnectionTypeType"/>
            -   <element name="fromTP" type=ID/>
            -   <element name="toTP" type=ID/>
            -   <element name="userLabel" type=string"/>
          -   </sequence>
      -   </complexType>
    -   </element>
  -   </sequence>
-   </complexType>
-
-   <element name="connInformation"
    -   type="connectionInformationSequenceType"/>
-
-   -----example instance -----
-   <connInformation>
  -   <value>
    -   <connectionType>1</connectionType>
      -   <fromTP>1234/>
      -   <toTP>1235</>
      -   <userLabel>"XC 1"/>
    -   </connectionType>
  -   </value>
    -   <connectionType>1</connType>
      -   <fromTP>2234/>
      -   <toTP>2235</>
      -   <userLabel>"XC 2"/>
    -   </connectionType>
  -   </value>
-   </connInformation>
-
-
```

A type that aggregates multiple data types but whose content is not known apriori can be modeled as "ANY" data type. Though there is no explicit ANY data type in XML schema, tML proposes realization of similar semantics. The following example is based on tML:

```
-
-   <complexType name="ANY">
  -   <complexContent mixed="true">
    -   <restriction base="anyType">
        -   <complexType>
          -   <sequence>
              -   <any processContents="lax"
                  -   minOccurs="0" maxOccurs="unbounded"
                  -   namespace="##any"
                -   />
            -   </sequence>
```

```
-   </complexType>
  -   </restriction>
 -   </complexContent>
-  </complexType>
-
-
```

The compositor `<all>` is used to model a collection, the members of which can be any of the elements specified inside the compositor. It is similar to ASN.1's SET construct except that the members of `<all>` compositor can only be <element> and can't be another compositor.

The compositor `<choice>` is used to describe choice between several possible elements or groups of elements and models 'selection' or 'one of '.

As per tML, the semantics associated with ASN.1's *set* type, which models fixed, unordered list of items of distinct types (some of which may be declared optional), can be achieved using *schema's <sequence>* compositor. Similarly, tML suggests using *<sequence>* compositor with *maxOccurs* facet set to "unbounded" as equivalent to ASN.1's *'set of'* type, which models fixed list of items of single type. The following example provides an example for realizing ASN.1's 'set of' type. Also included are other ASN.1's declarations and their equivalent realization in schema.

```
-  ASN.1 declaration of currentProblemList from M.3100  and X.721
-
-
-  ProbableCause ::= CHOICE {
                     -  globalValue OBJECT IDENTIFIER,
                     -  localValue INTEGER}
-  AlarmStatus ::= SET OF INTEGER {
                     -  underRepair(0), critical(1), major(2),
                     minor(3), alarmOutstanding(4) }
-  CurrentProblem   ::=   SEQUENCE {
                     -  problem      [0]   ProbableCause,
                     -  alarmStatus  [1]   AlarmStatus}
-  CurrentProblemList   ::=   SET OF CurrentProblem
-
-  --one of communications alarms
-  aIS ProbableCause ::= localValue : 1
-  --one of equipment alarms
-  backplaneFailure  ProbableCause ::= localValue : 51
-
-
-  Equivalent declarion using XML schema and tML guidelines
-
-
-  <simpleType name="probableCauseLocalValueEnumType">
       -  <restriction base="Integer">
            -  <enumeration value="1">
                 -  ...
```

```
-          -   <enumeration value="51">
      -   </restriction>
-   </simpleType>
-   <complexType name="probableCauseType">
      -   <choice>
              -   <element name="globaleValue" type="ID"/>
              -   <element name="localValue"
              type="probableCauseLocalValueEnumType"/>
          -   </choice>
-   </complexType>
-
-

-   <complexType name=""currentProblemSequenceType">
      -   <sequence>
              -   <element name="problem" type="probableCauseType"/>
              -   <element name="alarmStatus" type=alarmStatusType"/>
          -   </sequence>
-   </complexType>
-

-   <complexType  name="currentProblemListSetType"
        -   <sequence maxOccurs="unbounded">
        -   <element name="value" type="currentProblemSequenceType"/>
-   </complexType>
-

-
```

All types declared in an XML Schema are created in the target name space specified at the beginning of a schema. The following listing contains an example XML Schema borrowed from T1X1 committee's work on tML.

```
-   <!-- XML Document tmlCommon.xsd declaring two types: Name and
Telephone in a schema >
-   <?xml version="1.0"?>
-   <schema name="tmlCommon.xsd"
xmlns="http://www.w3.org/1999/XMLSchema">
-   ..........
-   <element name= "Telephone" type="string"/> ← Telephone is a type
-   <element name= "Name" type="string"/> ← Name is a type
-   ..........
-   </schema>
-   <!-- XML Document tmlX790.xsd declaring a schema for Contract
(reuses some types from tmlCommon.xsd) >
-   <?xml version="1.0"?>
-   <schema name="tmlX790.xsd"
-           xmlns=http://www.w3.org/1999/XMLSchema    ← default namespace
-           xmlns:tmlCom="http://www.xxx.org/tml/tmlCommon.xsd">  ←
explicit namespace
-   ......
-   <element name="Contact">  ← Contract is a type
  -   <complexType>
    -   <sequence>
      -   <element ref="tmlCom:Name"/>
```

```
-   <element ref="tmlCom:Telephone"/>
-   <element name=" LocationAddress" type="string" />
-   <element name="Details" >
    -   <attribute name="status" default="notComplete">
        -   <simpleType base="string">
                -   <enumeration value="customerComplaint">
                -   <enumeration value="TechnicianDispatched">
                -   <enumeration value="ModemProblem">
                -   <enmueration value="customerPCProblem">
                -   <enumeration value="ServiceComplete">
        -   </simpleType>
    -   </attribute>
    -   </sequence>
-   </complexType>
-   </element>
-   ......
-   </schema>
-   <!--XML Document troubleReport.xml An instance document>
-   <?xml version="1.0"?>
-   <TroubleReport xmlns ="http://www.xxx.org/tml/tmlX790.xsd" >
        -   <Contract>    -- instance 1
                -   <Name>CompanyABC</Name>
                -   <Telephone>+1 888 555 1212</Telephone>
                <LocationAddress>...........</LocationAddress>
                -   <Details status="ServiceComplete"/>
        -   </Contract>
        -   <Contract>   -- instance 2
                -   <Name>CompanyXYZ</Name>
                -   <Telephone>+1 888 777 1212</Telephone>
                -   <LocationAddress>...........</LocationAddress>
                -   <Details status="ModemFault"/>
        -   </Contract>
-

-   </TroubleReport>
```

In the above example a schema (structure) is defined for the element 'Contract' in tmlX790 document. The TroubleReport contains instances (records) of the 'Contract' element. Syntax for elements without explicit namespace prefix is looked up in a default namespace. TroubleReport, the top-level element, specifies a *default namespace* using xmlns = http://www.xxx.org/tml/tmlX79.xsd. Therefore, syntax of Contract, Name, Telephone, and LocationAddress are looked up in the default namespace since they don't have explicit namespace prefix.

Table 7-6 provides illustrative comparison of attribute declaration in DTD and Schema.

Table 7-6. Illustrative comparison of attribute definition in DTD vs Schema

DTD	XML Schema

`-<!ATTLIST xyz`
`–     a CDATA`
`#REQUIRED>`
	`–  <element name="xyz">`
	`–   <complexType content="elementOnly">`
	`–    -- some child elements of xyz`
	`–     <attribute name="a" type="string" use="required"/>`
	`–   </complexType>`
	`–  <element>`

`–  <!ATTLIST xyz`
`–     a CDATA`
`#IMPLIED>`
	`–  <element name="xyz">`
	`–   <complexType content="elementOnly">`
	`–    -- some child elements of xyz`
	`–     <attribute name="a" type="string" use="optional"/>`
	`–   </complexType>`
	`–  <element>`

`–  <!ATTLIST xyz`
`–     a (x|y|z)`
`#REQUIRED>`
	`–  <element name="xyz">`
	`–    <complexType content="elementOnly">`
	`–     -- some child elements of xyz`
	`–      <attribute name="a">`
	`–       <simpleType base="string">`
	`–        <enumeration value="x"/>`
	`–        <enumeration value="y"/>`
	`–        <enumeration value="z"/>`
	`–       </simpleType>`
	`–      </attribute>`
	`–    </complexType>`
	`–  <element>`

Finally, before concluding this section, Table 7-7 provides tML guidelines for translating GDMO information model into XML schema model [Ree01] (not a standard though). Though the T1M1 contribution suggests structural mapping, the suggestion of mapping GDMO actions, notifications to SOAP is author's own.

As listed in Table 7-7, many of the GDMO facilities are not mapped into tML - instead they are annotated. The next section will revisit this mapping and explain how GDMO's actions and notifications can be realized using WSDL/SOAP facilities.

Table 7-7. GDMO to tML Schema mapping

GDMO/ASN.1 concepts/items	Mapped to
Managed Object Classes (MOC)	Schema type (containing only the GDMO attributes as elements)
Packages (attributes)	Name element group with the package name used as the name for the group
Packages (Operations)	Annotation
Attributes (in a package or in a MOC)	Elements
GET, REPLACE, ADD, REMOVE semantics	No equivalent
Attribute Groups	Named Groups
Actions	Annotations
Notifications	Annotations
Behaviours	Annotation
NameBinding	Annotation
ASN.1 modules	tML target name spaces
ASN.1 types	schema's equivalents - simple or complex type or elements

7.3.2.3 Web Services/SOAP

XML can be applied to Q3 interface in two ways; first the GDMO/ASN.1 information model can be equivalently described using XML. Second, XML can be used to access and manipulate management information in place of CMIP or CORBA. Though newer developments overshadow this, an effort to apply XML to Q3 interface was first attributed to Q3ML [Q3ML]. Currently, a new technology that is popular in the IT industry and called - web services (WS) [W3C] has the potential to be useful for OS-OS and OS-NE interfaces. The various pieces of which are in Figure 7-8.

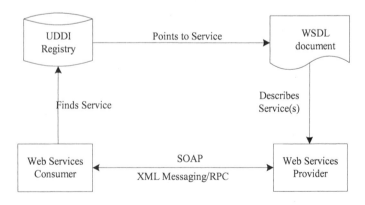

Figure 7-8. Web Services Architecture

One should note that currently a number of technologies exist, such as CORBA/GIOP, Java, EDI, XML, .NET etc. Not withstanding such existing technologies, IT industry has found favour with WS. Given its potential, particularly as an OSS technology, the next few paragraphs will describe basic concepts of WS.

As illustrated in the figure, WS consists of - WSDL, SOAP, and UDDI. Web services are searched and published using public-directory service. Currently the web-services directory standard is the UDDI. It is different from from other direcories in that there are no documents or specifications stored but only references to them are stored. For more on UDDI refer to [Eri02]. while a brief discussion of SOAP and WSDL follows.

7.3.2.3.1 SOAP

SOAP [W3C] is the transport protocol of WS and consists of four parts:

1. A message envelope consisting of a body and header
2. Transport binding framework with default HTTP binding.
3. Serialization framework – plain XML documents or based on a set of encoding rules for expressing instances of application-defined data types.
4. SOAP RPC for remote procedure calls.

WS/XML protocol layers and SOAP message envelope structure, consisting of a header and body, is illustrated in Figure 7-9.

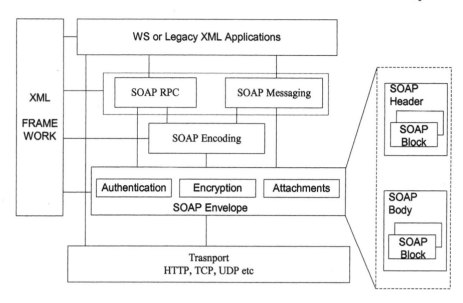

Figure 7-9. SOAP Layers and Message Structure

The SOAP header, if present, in SOAP message may contain one or more SOAP blocks. The SOAP blocks in the header are called header blocks and each of them contains instructions on how to process the message. The instructions include routing and delivery settings, authentication and authorization information, and transaction contexts. The header element and its children can have any or all of these attributes - encodingStyle, mustunderstand, role, and relay. The encodingStyle defines rules used to serialize parts of a SOAP message and its format (i.e. the data type) is Uri. The format of role attribute is also an URI. Allowed values and their meaning (where prefix = http://www.w3.org/2003/05/soap-envelope/role) are listed in Table 7-8

Table 7-8. SOAP role attribute values

Value	Description
[Prefix]/next	Each SOAP intermediary and the ultimate SOAP receiver MUST act in this role.
[Prefix]/none	SOAP nodes MUST NOT act in this role.
[Prefix]/ultimateReceiver	The ultimate receiver MUST act in this role.

The following provides an example for envelope containing a header and body. The header contains one soap block

```
-  <?xml version="1.0"?>
-  <s:Envelope xmlns:s ="http://www.w3.org/2003/05/soap-envelope"
s:encodingStyle="http://www.w3.org/2003/05/soap-encoding">
      -  <s:Header>
         -  <t:transaction
            -  xmlns:t="http://example.org/2001/06/tx"
            -  s:mustUnderstand="true" > 5
         -  </t:transaction>
      -  </s:Header>
      -  <s:body>
         -  ...
         -  </s:body>
-  </s:Envelope
-
```

The SOAP body contains actual message to be delivered and processed by the ultimate SOAP receiver that is providing a given WS service. It may also, like the SOAP body, contain one or more SOAP blocks. The child elements of <body> can use any number of attributes and of type including SOAP name space qualified attribute - encodingStyle.

One of the SOAP blocks in the body can be a SOAP fault block to be explained shortly.

The structure of SOAP response message is similar to that of request message. It may contain header or simply the body. The response may contain failure (i.e. fault) information and the following listing provides an example:

```
-  <?xml version="1.0"?>
-  <s:Envelope xmlns:s ="http://www.w3.org/2003/05/soap-envelope">
      -  <s:body>
      -  <s:Fault> ← SOAP block
            -  <s:Code> ... </s:faultCode>
            -  <s:Value> ... </s: Value>
            -  <s:Subcode> ... </s: Subcode>
               -  <s:Value> ... </s: Value>
            -  </s:Subcode >
      -  </s:Fault>
      -  </s:body>
-  </s:Envelope
-
```

A SOAP message is originated by a source SOAP node and consumed by one or more SOAP consumer nodes. However, the message may traverse one or more intermediate SOAP nodes called the 'actors'. Only source node can decide the path to be traversed by a given SOAP message. The path information is included with the message itself. The mechanism is similar to source routing available in IP protocol. Besides, since SOAP message contains multiple components, each component can be specifically targeted to a particular receiver (actor) by adding "*target*" attribute as shown in the following example.

```
- <?xml version="1.0"?>
- <s:Envelope xmlns:s = "http://www.w3.org/2003/05/soap-envelope" >
  - <s:Body>
    - <s:Header>
        - <x:validate actor="uri of the validator" >
        - </x:validate>
    - </s:Header>
    - <xyz:addUser  xmlns:xyz="some uri" >
    - </xyz:addUser>
  - </s:Body>
- </s: Envelope>
-
```

SOAP specifies data models for representing programmatic data types. This includes rules for mapping compound data structures, array types, and reference types. The specification contains encoding rules for serializing values that conform to the SOAP data models to XML for transfer over wire. With respect to compound data structures, the approach taken is reasonably straightforward; all data is serialized as elements, and the name of any given element matches the name of the data field in the programmatic type. For example, given the following Java class,

```
- class Person
- {
-     String name;
-     float age;
- }
```

The name and age fields would be serialized as elements <name> and <age> respectively. Both elements would be unqualified, that is, their namespace name would be empty. In cases where the name of a field would not be a legal XML name   a mapping algorithm is also specified in the XML encoding specification.

The mapping of reference types is more complicated. It involves serializing the instance and marking it with an unqualified attribute whose local name is id. All other references to that instance are then serialized as empty elements with an unqualified attribute whose local name is href. The value of the href is a URI that references the relevant serialized instance via its id attribute.

SOAP encodings have been controversial and a shift away from SOAP encodings might be inevitable.   Not withstanding the controversies, the SOAP encodings might be useful for non-XML and legacy applications.

A compound value represents aggregation of two or more accessors. There are two compound value types – *arrays* and *structs*.   A *struct* compound value type is one in which each member accessor has different name. An *array* compound value type is one in which the accessors have the same name.

SOAP is a simple lightweight coding of HTTP POST messages and their replies. It may suffice to some applications, but it is clear that it will not offer an efficient coding in all cases.

The messaging model of SOAP consists of :

1. Request/response - synchronous
2. One way - asynchronous
3. Multicast

SOAP supports a tighly coupled communication scheme based on RPC. The method and parameter information is exchanged using XML. SOAP also supports very loosely couple communication using regular XML messaging. SOAP also allows exchange of non-XML data as attachments to a SOAP message. For instance, one can retrieve non-XML if one were to retrieve logs on an NE that is usually text one can use use this mechanism. The SOAP attachment specification uses MIME encoding which is widely used. However, SOAP also specifies an alternate mechanism called DIME.

A Web Service is described using WSDL – a language based on XML and XML Schema. There are 6 parts to describing service in WSDL :

1. Definitions (of data types)
2. Messages (the parameters used in various operations)
3. Operations
4. Port types (the interfaces)
5. Bindings (transport and data format bindings)
6. Location (of the service i.e. the server information)

Concpets associated with WSDL and SOAP are illustrated in the following example. The example is based on an hypothetical NE that provides OAM&P services using XML across a Q3WS (Q3 Web Services Interfaces). To keep it simple, only the services supported by NE to setup, delete or retrieve SONET/SDH cross-connects will be discussed. Other services, based on various fabrics discussed in Chapter 6, can be similarly described using WSDL.

The various name space prefixes used in the example are listed Table 7-9 to avoid cluttering the example:

Table 7-9. name spaces used in the example

Prefix	ID/URI
- q3xsd	- schemas.xyz.com/q3scd
- q3ws	- schemas.xyz.com/q3ws
- soap	- www.w3.org/2003/06/wsdl/soap12
- Wsdl	- www.w3.org/2003/06/wsdl
- xsd	- www.w3.org/2001/XMLSchema

The `wsdl:definitions` specifies `q3ws` (the current or target name space) as default name space for use with unqualified types.

```
-
- -----------WSDL description------------
- <?xml version="1.0"?>
-
- <!-- wsdl:definitions defines services offered at Q3WS interface -->
- <wsdl:definitions name="q3WssdhNE"
    - targetNamespace="schemas.xyz.com/q3ws"
    - xmlns:q3ws="… "
    - xmlns:q3xsd="…"
    - xmlns:soap="…"
    - xmlns:wsdl="…"
    - xmlns:xsd="…"
    - xmlns = "schemas.xyz.com/q3ws"
- /">
-
    - <!1-- wsdl:types -->
    -
    - <wsdl:types>
    -
        - <xsd:schema targetNamespace="schemas.xyz.com/q3ws">
        -
            - <xsd:element name="CrossConnectRequestParamsType">
            - <xsd:complexType>
                - <xsd:sequence>
                    - <xsd:element name="fromTP" type="uri"/>
                    - <xsd:element name="toTP" type="uri"/>
                    <xsd:element name="connectionType"
                                 type="connectionTypeType"/>
                - </xsd:sequence>
            - </xsd:complexType>
            - </xsd:element>
            -
            - <xsd:element name="CrossConnectResponseParamsType">
            - <xsd:complexType>
                - <xsd:element name="crossConnectionStatus"
                  type="string"/>
            - </xsd:complexType>
            - </xsd:element>
            -
            - <xsd:element name="CrossConnectFaultParamsType">
            - <xsd:complexType>
                - <xsd:all>
```

```
        - <xsd:element name="errorMessage" type="string"/>
      - </xsd:all>
    - </xsd:complexType>
  - </xsd:element>
-
- </xsd:schema>
-
- </wsdl:types>
-
- <!2-- wsdl:messages -->
-
- <wsdl:message name="CrossConnectRequestMsg">
  - <wsdl:part name="body"
    - element="q3ws:CrossConnectRequestParamsType"/>
- </wsdl:message>
-
-
- <wsdl:message name="CrossConnectResponseParamsMsg">
  - <wsdl:part name="body"
    - element="q3ws:CrossConnectResponseParamsType"/>
- </wsdl:message>
-
- ...
-
- <!4-- wsdl:portType describes (in and out params)
    - for each operation -->
    -
- <wsdl:portType name="CrossConnectPortType">
  - <!3-- wsdl operations -->
  - <wsdl:operation name="CreateCrossConnectOp">
    - <wsdl:input message="q3ws:CrossConnectRequestMsg"/>
    - <wsdl:output message="q3ws:CrossConnectResponseMsg"/>
    - <wsdl:fault message="q3ws:CrossConnectFaultMsg"/>
  - </wsdl:operation>
  - ...
- </wsdl:portType>
-
- <!5 bind to a message transfer protocol>
- <wsdl:binding name="Q3WS-SOAP"
        - type="q3ws:CrossConnectPortType">
  - <soap:binding styleDefault="document"
        - protocol="
        http://www.w3.org/2003/05/soap/bindings/HTTP"/>
  - <wsdl:operation name="CreateCrossConnectOp">
    - <soap:operation soapAction="http://www.xyz.com/fabric"/>
    - <wsdl:input>
      - <soap:body encodingStyle="document"
            - namespace="schemas.xyz.com/q3ws"/>
      </wsdl:input>
    - <wsdl:output>
      - <soap:body encodingStyle="document"
            - namespace="schemas.xyz.com/q3ws"/>
```

```
    -   </wsdl:output>
    -   <wsdl:fault>
       -   <soap:body encodingStyle="document"
                  -   namespace="schemas.xyz.com/q3ws"/>
       -   </wsdl:fault>
   -   </wsdl:operation>
   -   ...
  -  </wsdl:binding>
  -
  -   <!6-- declare the service  -->
  -   <wsdl:service name="Q3WS-SOAP">
  -     <wsdl:documentation>
         -   NE1 Q3WS management interface
       -   </wsdl:documentation>
  -
  -      <!-- connect service to the binding "Q3WS-SOAP" -->
  -      <wsdl:port name="CrossConnectPort"
  -                 binding="q3ws:Q3WS-SOAP">
          -   <!-- give the binding a network address -->
          -   <soap:address location="http://NE1.xyz.com/Q3WS"/>
       -   </wsdl:port>
       -   ...
  -   </wsdl:service>
  -
-  </wsdl:definitions>
```

The `wsdl:binding` element has a child element - `soap:binding` or `http:binding`. The `soap:binding` child element takes any or all of these attributes- `protocol`, `styleDefault`, `nameSpaceDefault`, `encodingStyleDefault`. The `protocol` attribute specifies the transport protocol used for SOAP envelopes. Generally, it is HTTP and correspondingly the attribute value is as shwon in the example above is : `//www.w3.org/2003/05/soap/bindings/HTTP`. The `styleDefault` attribute (of type `xsd:string`) has a value of either `document` or `rpc`. The value of this attribute is set to `document` if the SOAP message contains plain XML document and set to `rpc` if SOAP-RPC encoding is used.

Other elements that can appear inside `wsdl:binding` include `soap:module` and `wsdl:operation`. The `wsdl:operation` element of step **5** in turn contains an optional child element- `soap:operation`, which provides binding information for the `wsdl:operation`. The binding information is specified as attributes of `soap:operation` element and there are two of them - `style` and `soapAction`. The `style` attribute is of type `xsd:string` and indicates whether the operation is serialized using document-oriented or RPC style. The value of `style`, if specified, overwrites the encoding style specified at an outer scope. Next, `soapAction` attribute, of type `xsd:anyURI`, specifies the value for HTTP `SOAPAction` header   (HTTP headers discussed in §7.4.2). The `wsdl:operation` elements inside the `wsdl:binding` element of step **4** are the same ones as declared in **3**. However, the purpose in step **4** is to declare information about the body of the SOAP request/response message associated with an operation. The request message is specified using `wsdl:input` element and the response message using `wsdl:input`.

The following SOAP request and response illustrates usage of the WSDL cross-connect service.

```
— ------------SOAP request------------
— <s:Envelope xmlns:s ="http://www.w3.org/2003/05/soap-envelope"
s:encodingStyle="http://www.w3.org/2003/05/soap-encoding">
—    <s:Body>
—      <q3ws:CrossConnect xmlns:q3ws="http://www.xyz.com/q3ws">
—         <fromTP>/bay1/shelf1/slot3/port1/timeslot4</fromTP>
—         <toTP>/bay1/shelf1/slot3/port2/timeslot4</toTP>
—         <connectionType>STS-3c</connectionType>
—         ...
—      </q3ws:CrossConnect>
—    </s:Body>
— </s:Envelope>
—
— -----------SOAP response------------
— <s:Envelope xmlns:s ="http://www.w3.org/2003/05/soap-envelope"
s:encodingStyle="http://www.w3.org/2003/05/soap-encoding">
—    <s:Body>
—      <q3ws:CrossConnectResponse xmlns:q3ws="http://www.xyz.com/q3ws">
—        <crossConnectionStatus> </crossConnectionStatus>
—      </q3ws:CrossConnectResponse>
—    </s:Body>
— </s:Envelope>
—
```

WSDL also allows binding directly to HTTP using either GET or POST method and the followign is an exmaple:

```
— <!5 bind to a message transfer protocol>
— <wsdl:binding name="Q3WS-HTTP" type="q3ws:CrossConnectPortType">
—   <http:binding verbDefault="GET"/>
—     <wsdl:operation name="CreateCrossConnect">
—       <http:operation location="q3ws/fabric/>
—         <input>
—         ...
—         </input>
—         <output>
—         ...
—         </output>
—       </http:operation>
—     </wsdl:operation>
—     ...
—   </http:binding>
```

```
-  </wsdl:binding>
-
```

Earlier, as part of XML schema discussion, in Figure 7-6, tML suggested mapping from GDMO→XML Schema was presented. Discussion of WSDL/SOAP specifics shown in that figure was deferred to this section. Now that reader has basic understanding of WSDL it will be easy to expand on these specific now. A GDMO package can be equivalently described using `wsdl:definitions` and `wsdl:portType`. A GDMO MOC would then consist of either `WSDL:definitions` or simply a schema element type that contains one or more `wsdl:definitions`. The following listing provides an example.

```
-
-
-  <!-- assume generic NE types are in xsd, and all actions and their
input,output, fault parameters are in q3ws files -->
-
-  <q3xsd:gdmoAttribute xmlns:q3xsd ="…"
   -  name="…" type="…">
-  </q3xsd:gdmoAttribute>
-
-  <!-- use portType defined in WSDL file -->
-  <q3xsd:gdmoAction xmlns:q3xsd ="…"
   -  xmnls:q3wsdefs ="…" name="…" type="xyzPortType…">
   -
-  </q3xsd:gdmoAction>
-
-  <!-- map GDMO notifications to SOAP notifications -->
-  <q3xsd:gdmoNotification xmlns:q3xsd ="…"
   -  xmnls:q3wsdefs ="…" name="…" type="xyzPortType…">
   -
-  </q3xsd:gdmoNotification>
-
-
-  <q3xsd:gdmoPackage xmlns:q3xsd ="…" name="…">
   -  <!-- one or more attributes -->
   -  <gdmoAttribute> … </gdmoAttribute>
   -  <!-- one or more actions -->
   -  <gdmoAction> … <gdmoAction>
   -  <!-- one or more notifications -->
   -  <gdmoNotification> .. </gdmoNotification>
-  </q3xsd:gdmoPackage >
-
-
-
-  <q3xsd:gdmoMOC xmlns:q3xsd ="…" xmnls:q3wsdefs ="…">
   -  <!-- one or more attributes -->
   -  <gdmoAttribute> … </gdmoAttribute>
   -  <!-- one or more packages -->
   -  <gdmoPackage> … </gdmoPackage>
   -  <!-- one or more actions -->
   -  <gdmoAction> … <gdmoAction>
```

```
-    <!-- one or more notifications -->
-    <gdmoNotification> .. </gdmoNotification>
-  </q3xsd:gdmoMOC>
```

It is important to understand that management information and operations on that information are two different things. What WSDL deals with is the second part – operations to access and modify management information – directly or indirectly. Naturally, the structure of management information associated with technology specific NE (sdhNE for instance) can be fully described using XML Schema constructs explained earlier. The WSDL then declares operations, with input and output parameters, re-using some of the simple and complex types described in the XML schema for that NE type. There are also mappings between WSDL and CORBA [OMG04] [OMG03]

Webservices is promising to be 'the next wave' in IT technology space, with more standards work and software solutions needed in the areas of :

- WS-Notification framework
- WS-Transaction framework
- WS-Security framework

The last one, namely WS-Security, seems to have a closure. The Organization for the Advancement of Structured Information Standards (OASIS) has ratified WS-Security (WSS) 1.0 [WSS04]. This standard has the support of many software companies (Microsoft, Sun, IBM, BEA, and many others). The WSS proposes SOAP extensions to provide message confidentiality and integrity. Besides, WSS accommodates a variety of security models and encryption technologies. It is expected that WSS, being an open standard, will usher in broad range of WS products - WS management solutions, XML firewalls, Identity management solutions, etc. Microsoft's .NET and Sun's J2EE platforms natively support these standards. In telecom, if NEs use WS architecture (i.e support WS as Q3 interface) then supporting WSS would be a challenge, as many NE vendors don't use either of these platforms. Typically, NE vendors use embedded operating systems like Wind River's VxWorks or their own in-house operating systems. Recently, Microsoft embedded versions of NT/2000/XP are being used as platforms in such applications as IP-PBXs, and transport NEs. It may be possible for such devices to support WSS easily as would be for devices based on versions of Linux OS. Alternatively, NE vendors can add WS blade on gateway network elements. This will allow them to support WS interfaces at Q3 but keep their legacy and valuable NE software implementations intact. Such blades are easy to add as many vendors' chassis are based on standard back planes such as cPCI and emerging telecom backplanes such as AdvancedTCA. For these backplanes, NT/2000/XP or Linux blades can be easily added from the market place.

7.3.2.3.2   Fast Web Services (FWS)

Started initially by Sun Microsystems, efforts are underway in ITU-T to standardize Fast Web Services. The motivation for making WS faster came from bandwidth limited applications (wireless) and performance sensitive applications (High Performance Computing -HPC). Bandwidth sensitivity is a result of verbosity of character-based XML messages. Performance sensitivity is a result of having to process (using DOM or SAX XML document processors) character-mapped XML messages.   In general character-based XML affects in three areas:

– Transmission - bandwidth sensitivity
    Character-mapped XML SOAP messages
– Persistency - storage sensitivity
    Storing and retrieving XML data using RDBMS or native file systems
– Processing - memory and CPU sensitivity
    Parsing, validation, binding, or transformation

The efforts to solve WS performance issues is commonly called   'Fast Web Services (FWS)' and considered as alternative to WS but retaining the architectural principles of WS. Particularly, *"Fast Web Services is the term applied to the use of ASN.1 to provide message exchanges based on a SOAP envelope and WSDL specification of services that can have a higher transaction-processing rate and less bandwidth requirements than use of a character-based XML representation* [Dub04].

The difference between WS and FWS is in the use of ASN.1 as XML Schema notation and use of PER (or other compact encoding rules) for transfer. An FWS node can still support legacy character-mapped XML transfer mode as illustrated in Figure 7-10.

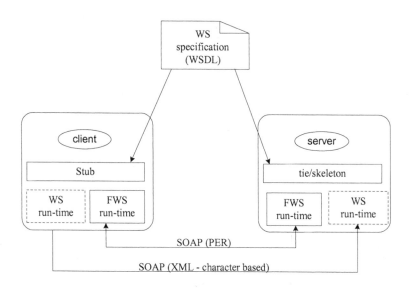

Figure 7-10. Dual-Mode WS stack

The mapping of XML schema to ASN.1 is standardized in X.694. The FWS run-time converts from XML to fast schema using X.694 rules. However, this may require the FWS node to spare processing power for such conversion. An alternative is being proposed in X.finf, which specifies a static conversion of whole XML document to ASN.1. Both approaches produce ASN.1 output, which is transferred over wire using binary encoding (BER or PER).

Some of the Binary XML proposals being debated are listed in Table 7-10 below:

Table 7-10. Some Binary XML Formats [Mic]

Format	Technique
ASN.1	Schema-based compression
BXML	Direct Encoding
FastSchema	ASN.1 based compression (X.694)
XBIS	XML Binary Information Specification
X.finf	Static mapping of whole XML document into ASN.1 for transfer using PER/BER

There is strong support from wireless industry to develop a Binary XML format. Such interest can be understood based on bandwidth constraints of wireless applications. Binary XML, as a standard or embraced by industry at large, should also help OS-OS (and OS-NE) XML interfaces.

7.4. INFORMATION ACCESS AND TRANSPORT PROTOCOLS

This section will present some of the important information access and transport protocols. The discussion here will be limited to higher level transport protocols such as HTTP, SNMP, and CMIP. Lower level transport protocols such as TCP, UDP etc are not discussed here since there is enough information available elsewhere. Also excluded from discussion are dynamic routing protocols such as OSPF. However, some discussion related to routing, albeit in separate context, can be found in Control Plane (Chapter 9).

7.4.1   Transaction Language one (TL1)

TL1, a text-based man-machine language, is the most widely used protocol to access transport network elements in North America. The interest in providing text based man-machine language   for use by craft personnel started in 70's.   Unlike CMIP and SNMP, the overriding theme in the development of man-machine languages is the need for humans to be able read the protocol messages The ITU-T addressed this requirement in its Z.300 series of standards on user interfaces in telecommunications. In 1984, Bellcore, in its role as standards-setter for the Bell regional holding companies, decided to specify a standard, based on Z.300 series, a man-machine language for controlling network elements. It was to be called transaction language 1 (TL1). While GR-831 (1996) is the core document that details the grammar of TL1, various other Bellcore/Telcordia generic requirements documents detail TL1 messages for operations functions, such as alarm surveillance, performance monitoring, testing and provisioning of transport and switching network elements. Figure 7-11 illustrates communication model of TL1.

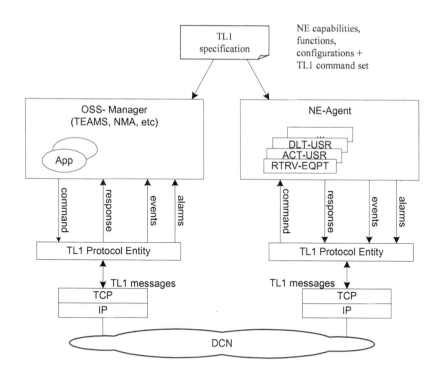

Figure 7-11. TL1 Communication Model

There are three types of TL1 messages:

- Input or command messages (also called request messages)
- Output/Response messages
- Autonomous messages

The general format of a TL1 input message is shown below:

```
- command_code:[staging parameters]:[general block]:[one or more
parameter blocks];
```

The *command code* usually consists of a verb and one or two modifiers. For instance, one may use RTRV-EQPT-ALL to retrieve all equipment of an NE with RTRV as the command verb, EQPT the first modifier, and ALL the second modifier.

The *staging* block consists of three parameter blocks delimited by *semicolon* (;). The first block contains TID (Target Identifier) parameter while the second block contains one or more AID (Access Identifier) parameters. Together, first and second block describe the location or address (i.e. the stage) where the command verb conducts its operations.

The TID (Target Identifier) is an ID assigned to Network Element. Service providers obtain TIDs from Telcordia for each NE in the network. The TID is typically from 1 to 20 characters in length. To construct TID all alphanumeric characters except spaces can be used. However, TID values also are not case-sensitive. In terms of locating the NE with a given TID in the DCN, a standardized address resolution protocol called TARP (TID ARP) is available which resolves TID to OSI[103] address. Many vendors have implemented TARP that resolves TID to an IP address. TL1 message may be sent directly to an NE   or routed through, one or more intermediate NEs. If an NE is the only possible target (over a given connection) then TID may be omitted from the message. Otherwise, a TID value must be specified in a command message.

The AID identifies physical equipment or resource that is the target (i.e. the stage) for a command. In the case of autonomous (alarm or event) message, the AID describes the source of the message. The AID, usually consisting of upto 20 characters, may be functional or physical or a common language equipment identifier (CLEI).   The physical AIDs are used to identify such entities as circuits and equipment. For instance, B1-CH2-CP1 may be used to identify the first circuit pack of the second chassis of the first bay while B1-CH2-CP1-P2 may be used to identify the second port on that circuit pack.   AID can also be used to identify entries in NE database – such as Performance Monitoring (PM) or log record entries.   When a group of equipment or resources are to be accessed then grouping (&) and ranging (&&) operators may be used as the following examples illustrates.

```
-   --use of group operator to specify CPs 5&7
-   RTRV-EQPT:NE1:B1-CH1-CP5&B1-CH1-CP7;
-
-   --alternative way of using group operator
-   RTRV-EQPT:NE1:B1-CH1-CP5&-CP7;
-
-   --use of ranging operator to specify  CP1 to CP10
-   RTRV-EQPT:NE1:B1-CH1-CP1&&-CP10;
-
-   --alternative way of using ranging operator
-   RTRV-EQPT:NE1:B1-CH1-CP1&&B1-CH1-CP10;
-
```

---

[103] For NSAP/OSI addressing information/formats, see chapter 9

The CLEI codes used as AIDs are obtained from Telcordia by equipment vendors. CLEI codes identify vendor specific functional equipment and consist of ten alphanumeric characters split into 5 sub-fields as shown in Table 7-11.

Table 7-11. CLEI structure

Sub Element	Element Name	No. of Characters	Element Codes
1	Family	2 [1-2]	Alphanumeric
2	SubFamily	2 [3-4]	Alphanumeric
3	Features	3 [507]]	Alphanumeric
4	Reference	1 [8]	Alphanumeric
5	Complemental	2 [9-10]	Alpha

The first seven characters present a concise summary of an equipment entity's circuit or transport capabilities, e.g. functional, electrical, bandwidth, etc. The eighth character denotes the reference source used for coding the item, and the last two characters denote manufacturing vintage or version, and other complemental information. For more details on the usage of these sub fields refer to Telcordia CLEI documentation.

TID and AID, along with CTAG (correlation tag), are part of the *staging block*. The CTAG is an alphanumeric reference containing up to 6 ASCII characters. The CTAG is used for correlating response(s) with commands. The Operations System (OS) assigns a value for inclusion in the input message and the NE in the response message for that command copies the same value.

Finally, parameter blocks of TL1 messages contain parameter values. Each parameter block is delimited by colon (:) with parameters within a parameter block separated by comma (,). There are two types of parameter blocks – named and positional. The named parameter block contains one or more parameters using <name=value> format. The positional parameter block contains only parameter values delimited by *comma* (,) with the name of each parameter implicitly understood by sender and receiver. The following illustrate some examples for the two types of parameter blocks. The first is a template describing a facility alarm from a hypothetical NE. The location parameter in the alarm refers to near end or far end failures (refer to Ch3 and 6 for description of far end and near end faults). Next, two instances of this alarm template are provided – both relate to OC-48 facility alarm. The difference between the two is that one uses named parameter blocks and the other uses positional parameters. Reader may note that within a message (input or output), there can be positional and named parameter blocks. Just that within a parameter block one can't use mixed format.

In the context of transport network elements, setting up a cross connection is a common operation that involves specifying two termination points (TP) − a `fromTP` and a `toTP` (in case of bi-directional cross-connection one simply uses tp1 and tp2 without prefix `from` or `to`). The two TPs involved are specified using two AIDs in a TL1 command as the following examples illustrates.

```
-- one template to create cross-connection where
-- pst (primary status) = IS or OOS
ED-CRS:[TID]:<fromTP>,<toTP>: [CTAG]::[<pst>];
--one instance of the above to create xc between
-- (CH1,CP3,Port1,Time Slot 1) and (CH1,CP1,Port2,Time Slot 3)
ED-CRS:NE1:STS1-B1-CH1-CP3-P1-S1,STS1-B1-CH1-CP1-P2-S3:R123::IS;

-- another template where cct(cross connect type=1WAY or 2WAY
ED-CRS-<STS_PATH>:[TID]:<fromTP>,<dstTP>:[CTAG]::<cct>::;
--one istance to create XC between
--(Port 4,Time Slot 6) and (Port 7,Time Slot 12)
ED-CRS-STS3C:NE1:STS-6-4,STS-12-7:123::2WAY::;
```

The following listing contains a command and its response. In the staging block of the command, NE1 is the TID, CH1-CP1-P1 is the AID, and R123 is the CTAG. The general block is empty and the parameter block contains one positional parameter with value 2WAY.

```
--command to retrieve all cross-connects of port1 on
--circuit pack 10 in  chassis 1 of NE1----
RTRV-CRS:NE1:CH1-CP1-P1:R123::2WAY;

--Response
--assuming there are only two cross-connects ----

<cr> <lf>
<lf>
NE1^2004-03-10^16:30:00 <cr> <lf>
M^^R123 COMPLD <cr> <lf>
^^^"STS1-CH1-CP1-P1-S1,STS1-CH1-CP10-P1-
S1:2WAY:MONTYPE=NORMAL:PST=OOS,SST=SGEO" <cr> <lf>
^^^"STS3-CH1-CP1-P1-S4,STS3-CH1-CP10-P1-
S4:2WAY:MONTYPE=NORMAL:PST=OOS,SST=SGEO" <cr> <lf>;
```

The following provides a template for alarm followed by two equipment alarms.

```
Equipment Alarm Template   (see Ch6 for various TMN alarms)
```

```
–
–   SID DATE TIME
–   ** ATAG REPT ALM EQPT
–
"<aid>:<ntfncde>,<condtype>,<srveff>,<ocrdat>,<ocrtm>:[<condescr>]null
[<rtag>]"
–   ;
–
```

The various fields of alarm template are listed in Table 7-12

Table 7-12. Alarm Template parameters

Template parameter	Description
Ntfncde	Notification code
Condtype	Alarm condition type
Srveff	Indicates the effect of the alarm condition on service: SA – effecting, NSA – not effecting
Ocrdat	Alarm occurrence date
Ocrtm	Alarm occurrence time
Rtag	Reference tag to be used in all alarm messages associated with the same root cause

The following listing contains two equipment alarms based on the above template:

```
–
–       NE1 04-03-14 11-30-00
–   ** 001 REPT ALM EQPT
–       "CH1-CP1:MJ,EQPT,SA,2004-03-14,11-30-00:\"EQUIPMENT
FAILURE\",ATAG1"
–   ;
–       NE1 04-03-14 11-31-00
–   *^ 002 REPT ALM COMM
–       "CH1-CP1:MI,COMM,NSA,2004-03-14,11-31-00:\"DCC COMMUNICATION
FAILURE\",ATAG2"
–   ;
–
```

The following is a template for facility alarm (vendors can specify their own formats and this is just an example template)

```
–
–
–
–       SID DATE TIME
–   ** ATAG REPT ALM <FacilityType>
–
"<aid>:<ntfncde>,<condtype>,<srveff>,<ocrdat>,<ocrtm>,[<locn>],[<Drn>]
:[<conddescr>],[<rtag>]"
–   ;
```

The following provides two examples based on the above template. In the first alarm, all parameter blocks use name=value format. In the second alarm, all blocks use positional format.

```
   NE1 04-03-19 08-00-02
** 001 REPT ALM OC48
   "B1-CH1-CP1-P1:NTFNCDE=MJ,CONDTYPE=AIS-L,SRVEFF=SA,OCRDAT=2004-
03-19,OCRTM=08-00-00,LOCN=NEND,DIR=RX:\"LINE ALARM INDICATION
SIGNAL\",ATAG1"
;
   NE1 04-03-19 08-00-02
** 001 REPT ALM OC48
   "B1-CH1-CP1-P1:MJ,AIS-L,SA,2004-03-19,08-00-00,NEND,RX:\"LINE
ALARM INDICATION SIGNAL\",ATAG1"
;
```

The following lists segments of TL1 language specification from GR-831 and using Backus-Nauer Form (BNF) meta-language (predecessor to EBNF and explained earlier in §7.3.1). The character ^ is used to represent space (i.e it is not the literal ^)

```
input message::= <cmd code>: [<tid>]:[<aid>]:<ctag>:[<general
block>] (:<payload block>)*;
general block   ::= <nil>
                  | <pos def gen block>
                  | <name def gen block>

payload block   ::= <nil>
                  | <pos def param, (<pos def param>)*
                  | <name def param, (<name def param>)*>


output message ::= <ackMsg>
                  |<responseMsg>
                  |<autonomousMsg>

responseMsg    ::= <header> <response id> [<text block>]
<terminator>
autonomousMsg  ::= <header> <auto id> [<text block>] <terminator>
ackMsg         ::= <ack code>^<ctag> <cr> <lf>
header         ::= <cr> <lf> <lf>^^^<sid>^<yeaar>-<month>-
                  <day>^<hour>:<minute>:<second>
response id    ::= <cr> <lf>M^^<ctag>^<completion code>
text block     ::= (
                  ( <cr> <lf>^^^<unquoted line>)
                  | ( <cr> <lf>^^^<quoted line>)
```

```
           - | ( <cr> <lf>^^^<comment>)
           - )*
-  autoid::= <cr> <lf> <almcde>^<atag>^<verb>[^modifier>[^modifier>]]
   -
   -  _____
```

Except for the comment line, rest of the response message is parsable. The form of the terminator is `<cr><lf>` followed by either *semicolon* (;) or the *greater than* character.

Table 7-13 lists acknowledgement, completion and alarm codes used in TL1 messages.

Table 7-13. TL1 (a) Ack, (b) Alarm and (c) Completion Codes

Ack Codes	
Code	Meaning
IP	In-Progress
OK	OK
PF	Partial

(a)

Alarm Codes	
Code	Meaning
*C	Critical Alarm
**	Major Alarm
*^	Minor Alarm
A^	Non-alarm message

(b)

Completion Codes	
Code	Meaning
COMPLD	Command completed successfully
DENY	Command denied
PRTL	Command completed partially
DELAY	Command queued
RTRV	Response to an input retrieve command

(c)

Though TL1 is a simple protocol there is no standard command set. Each vendor specifies their own set of commands with associated syntax and semantics. The following tables list some example TL1 commands and serve to highlight the different functional aspects of managing a NE.

Table 7-14. TL1 commands-1

Equipment Related	Description
– ENT-EQPT –	– Add equipment such as chassis, circuit pack, plug-in
– ED-EQPT	– Edit information of an existing equipment
– DLT-EQPT	– Delete equipment such as circuit pack
– RTRV-EQPT	– Retrieve list of equipment of an NE
– RESET-EQPT –	– Reset equipment for instance -reset a given circuit pack
– REPT-ALM-EQPT	– Equipment alarm report
– REPT-EVT-EQPT	– Equipment event report - for instance AVC or creation or deletion report
– RTRV-ALM-EQPT	– Retrieve equipment alarms
Facility Related	
– REPT RMV	– Report removal
REPT RST	– Report reset
RMV-<PortType>	– Remove a facility
RST-<PortType>	– Reset a facity
– REPT ALM <FacilityType>	– Ffacility Alarm Reports
– RTRV-ALM- <FacilityType>	– Retrieve facilitiy alarms
– REPT EVT <FacilityType>	– Facility Event Reports
– RTRV EVT <FacilityType>	– Retrieve facilitiy alarms
Cross-Connect Related	
– DLT-CRS-<TPType>	– Delete a Cross-Connect
– ED-CRS-<TPType>	– Edit an existing Cross-connect
– RTRV-CRS-<PortType> –	– Retrieve Cross connect(s) by port AID and additional criteria specified in the command parameter block
– RTRV-CRS-<TPType>	– Retried Cross connect(s) by termination point
Alarm/Event related	
– ALW-MSG-ALM	– Allows autonomous alarm reports/messaes
ALW-MSG-EVT	– Allows autonomous event reports/messaes
INH-MSG-ALM	– Inhibit Reporting of Alarms
INH-MSG-EVT	– Inhibit Events of Events
REPT ALM COM	– Communication Alarm reports
REPT EVT COM	– Communication Event Reports
RTRV-ALM-ALL	– Retrieve all Alarms

Table 7-15. TL1 Commands –2

NE System Management	Description
– ACT-USER	– Activate User
CANC-USER	– Cancel user account
ENT-NETYPE	– Enter NE/System forrmation
ALW-MSG-ALL	– Allow NE to publish events/alarms
– INH-MSG-ALL	– Inhibit all messages from the NE
RTRV-HDR	– Retrieve Header - null command
RTRV-NETYPE	– Get NE/System information
SET-SID	– Set NE TID
Security Management	
DLT-USER-SECU	– Delete User
ED-PID	– Edit User password
ED-USER-SECU	– Edit User security information
ENT-USER-SECU	– Add/Enter User security informatin
REPT ALM SECU	– Security Alarm report
RTRV-ALM-SECU	– Retrieve all security alarms
RTRV-USER-SECU	– Retrieve User security information
Log Related	
– REPT DBCHG	– Retrieve DBCHG/AVC Events
RTRV-ALM-LOG	– Retrieve Alarm Log
RTRV-DBCHG-LOG	– Retrieve DBCHG Log
RTRV-EVENT-LOG	– Retrieve Event Log

The process of getting Telcordia OSS systems to support a particular vendor's command set (representing equipment capabilities, features, services) is called OSMINE and is described next.

7.4.1.1 OSMINE Process [Jackie Orr[104]]

OSMINE[105] (Operations Systems Modification for the Integration of Network Elements) is a process used by Telcordia to integrate Network Elements into Telcordia's OSS environment. NE vendors (called the NEPs - N Network Equipment Providers) contract with Telcordia for the OSMINE related work. Carriers also pay software maintenance fee to Telcordia to upgrade to the necessary OSS software to manage a newly introduced network element or technology. Because Telcordia OSS solutions are widely used by operators, getting OSMINE integration can enhance the prospects of a product, particularly in the RBOC segment.

---

[104] Jacqueline Orr - currently Director, Program Management at CIENA Corporation. Ms. Orr has spent more than 25 years with major equipment vendors defining and developing products for the Telecommunications Industry; with extensive experience in delivery of products to the Service Provider market.

[105] OSMINE and all other products mentioned and described in this section are trademarks of Telecordia Technologies.

OSMINE is a multi step process. The first step involves NEP filling out information about a new product or technology using a template called the NEIR (Network Element Information Request). Telcordia, using the NEIR as a guideline to introduce them to the NEP's new product or technology, proceeds with an estimation of the scope of work that will be required to integrate the defined NE into the various required OSS. Once both Telcordia and the NEP agree upon the products that will require integration and the configuration scenarios that will need to be supported, the necessary work to integrate the vendor's product/technology can begin. The NE will be modeled as SONET, DWDM, ADM, Hybrid or whatever is appropriate to the technology involved and the application for which the NE will be used. Some of the NE types and their associated functions and applications supported by Telcordia OSS products are listed in the following. Many SONET/SDH NE functions and NE applications were discussed in Chapter 1 and 3 discussed.

Table 7-16. Some Telcordia supported NE types, configurations, and applications

Supported NE Type	Configurations	Functions associated with the NE	Network application	Service Applications
DCS (W-DCS, B-DCS)		Grooming Add/Drop Bridge & Roll (Facility Rolling) Broadcast PM Test Access Remote Configuration	Gateway Hubbing Meshe Ring Ring-Interconnection	Bandwidth Management Switched private networks Unbundling
ADM (Hardwired/TSA or Programmable-TSI)	Terminal ADMs Intermediate ADMs Hubbed ADM	Add/Drop Ring Protection – UPSR, BLSR Drop & Continue Liner APS protection Equipment Protection	Linear ADM chains ADM rings Point to Point ,	Bandwidth Management

Telcordia adds support in the OSS for new technologies/applications, as service providers adopt/deploy them to rollout new services. Since products based on a given new technology will be available from multiple NE vendors, Telcordia determines the baseline changes that will be required and spreads the cost of OSS changes to integrate the new technology through a program called GFDS – Generic Feature Development Services (GFDS). NEP's that choose to participate in funding the development offered by a specific GFDS are call the OFG (Original Funding Group) and can get rebates from Telcordia as other NEP's buy into the capabilities of the GFDS post the funding period. Some of the technologies that have been done or are in progress include OC192, OC768, VCAT, Hitless rearrangement using LCAS and so on. Members of the funding group are the NEP's and Telcordia. The Service Providers may also choose to participate.

Each member of the funding group participates in the review of the development proposal from Telcordia and has an opportunity to ensure that the generic support being developed by Telcordia meets their needs as much as practical. However, even for a common technology, each NEP may have special implementation needs requiring changes to the OSS over and above what was done through GFDS. In such cases, any additional customization of Telcordia OSS suite specific to an NEP is done through the regular OSMINE process in stage 3 / stage 4 development.

For companies that will be targeting the licensee's of Telcordia OSS's as potential customers, it is highly recommended that they engage with Telcordia as the earliest possible date to gain as guidance in the implementation decisions they will make with regard to TL1 and Alarms. A network complimentary implementation of these functionalities can significantly reduce the cost and time required to achieve OSMINE integration. A company can begin engagement with Telcordia starting with a negotiating a master agreement. Next steps would be to engage in a joint review of product design documents, especially with consideration for design of TL1 messages and the maintenance reporting on the system. These design reviews can be arranged with Telcordia for a reasonable fee and can result in significant savings when it is time for the company to engage in OSMINE integration. Figure 7-12 shows various products associated with Telcordia.

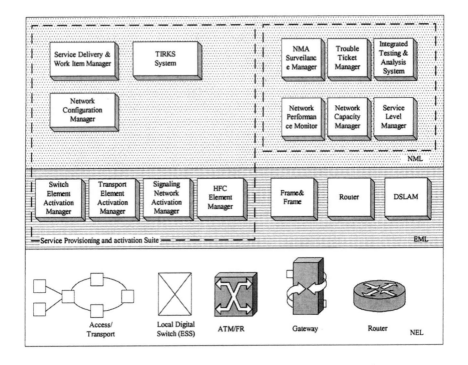

Figure 7-12. Telcordia OSS Suite

7.4.1.2 NMA

NMA (Network Management and Analysis) system is a network monitoring and surveillance operations support system. It is used to monitor the network, administer, analyze, and resolve trouble tickets for switching and transmission facilities. The system also monitors the environmental conditions that may affect operations. The data collected and analyzed by NMA algorithms provide the ability to view the total network in real-time, monitor, and control network elements. NMA assists in pinpointing network failures to the element at fault, minimizing truck rolls and service outages

7.4.1.3 TIRKS:Inventory Management

TIRKS (Trunks Inventory Record Keeping System) is an integrated system that supports the total network provisioning process consisting of:
- Inventory planning, provisioning and management of equipment and facilities
- Order flow & automated circuit design

– Preparation and distribution of Work Order

TIRKS system supports provisioning process related to special service circuits, message trunks, and carrier circuits. TIRKS software supports a full range of transmission technologies- SONET self-healing rings and sophisticated SONET configurations; digital circuit hierarchy (DS0, DS1, DS3); analog voice circuits; and European digital hierarchy standards (SDH). All of the network planning starts with TIRKS and other "downstream" systems gain their data from TIRKS in order to support flowthrough.

It tracks progress of service and work orders, performs end-to-end circuit design and maintains accurate inventory of facilities and equipment (current and pending), as well as their circuit assignments. It supports interactive queries and customized reports on work orders, quality measurements, etc. It supports interfaces to number of other Telcordia OSS products as shown in Figure 7-13[106]. The numbers in the figures indicate the order flow sequence as described in the Telecordia document.

---

[106] Borrowed from BD-TIRKS-TOVVEN-PG, March, 2003, Issue 1, © Telecordia technologies

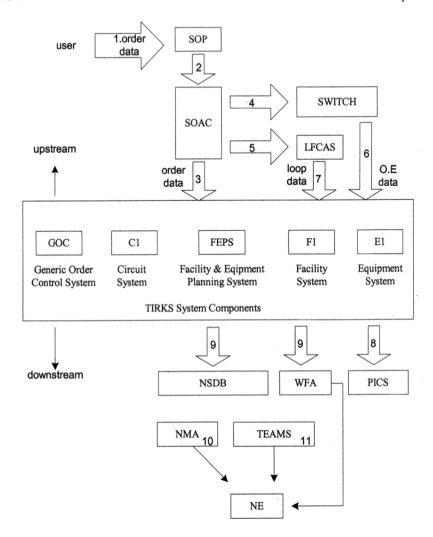

Figure 7-13 TIRKS's System Components and workflow

The various components of TIRKS work together to provide the necessary elements for circuit design and network planning. The GOC system inputs, tracks and manages orders related to special service, message trunk and carrier system circuit orders. The C1 system manages circuit assignment, automated design and distribution of special service, message trunk and carrier system circuits. The FEP system is used to aid planning and allocation of facility and equipment to meet orders from GOC. The F1 system provisions, selects, assigns and manages inventory records for cable, fiber, D/WDM, and other such items. The E1 system is used to provision, select, assign and manage Termination, Signaling, Transport, Multiplexing, ADM, DCS, Hybrid and switching equipment records.

Some of the other products mentioned in Figure 7-13 are briefly described in Table 7-17. For full description as well as latest information reader is advised to consult Telcordia. Some Telcordia OSS products/solutions that are related to an NE in a managing role or otherwise are listed in the following table (they are called heritage OSS solutions and Telcordia also has what are called NG OSS solutions)

Table 7-17. Telcordia OSS products (related to TIRKS)

Product	Description
SOP	Service Order Processor - any of various order formatting systems used by the RBOCs to input service orders
SOAC	Service Order and Analysis Control
SWITCH	System that assigns line-side central office switching equipment inventory for POTS, DS0, and other high capacity circuits
LFACS	Loop Facilities Assignment and Control System – inventory management and assignment of subscriber access facilities across– copper-based loops, DLCs, DSLAMs, and PONs. Maintains records of all outside plant facilities – cables (copper and fiber), DLC facilities, fiber-in-the-loop systems, terminal, and customer service locations.
WFA	System that tracks the flow of circuit intallation and maitenance operations.
PICS	Plug-In Inventory Control System
NSDB	Network and Services Database - system that stores in-effect and pending service records and acts as the interface between TIRKS and the downstream OSSs
TEAMS	Transport Element Activation Manager System- an EML layer entity - manages multiple NEs (of same or different type - ADM, DCS, etc) and from single or multiple vendors. NE provisioning and service cross-connections are accomplished using TEAMS

It is opined that on an average North American operators are 40% more efficient than their counter parts elsewhere. It is also believed that the main reason for their efficiency is implementation and usage of robust OSS systems (mainly Telcordia OSS suite), through all phases of network operation. Therefore, Telcordia OSS solutions occupy an important role in transport network management. TL1 is therefore an essential part of any transport network element. However, at the time of writing, Telcordia is adding XML interfaces to some of their products (such as migrating NSDB and WFA/C system interfaces from FCIF (Flexible Computer Interchange Format) to XML message format).

TL1 has been an integral part of North American networks. It remains to bee seen, as converged networks take shape and new software technologies such as Web Services mature, how TL1 and Telcordia solutions will evolve. Given TL1's widespread use in RBOC (and to some extent in IXC) transport networks, the information presented here should help the reader achieve some rudimentary understanding of TL1.

7.4.2   Hypertext Transfer Protocol (HTTP)

HTTP is a state-less text based client-server request/response protocol. A client wishing to access a resource on a server contacts that server and specifies the action to be performed on the resource. The action is called *method* and Table 7-18 lists some commonly used HTTP/1.1 methods.

Table 7-18. HTTP/1.1 Methods

HTTP Request Method	Description
GET	Retrieve a resource
HEAD	Retrieve metadata
POST	Similar to GET
PUT	Upload file
DELETE	Delete a resource
OPTIONS	Query Server Options
TRACE	Trace proxy chain

POST is similar to GET. But, in addition to providing the URI of the resource, the message also contains a number of variables and values the collection of which is called the *'payload.'* Typically, GET is used when the intent is to get some resource and the POST is used when the request can result in modification of one or more resources on the website. Figure 7-14 shows typical HTTP request/response transaction.

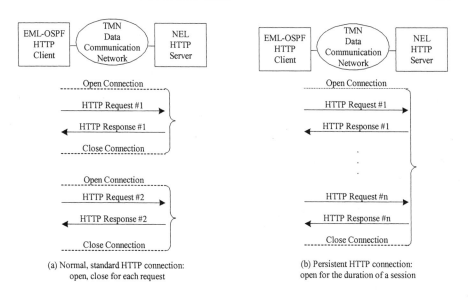

Figure 7-14. HTTP Communication Model

HTTP communication model is simple; open connection, send request, receive response and close connection. However, since TCP connection setup takes some time, such simple process is not an efficient way for getting multiple resources on a server. Persistent connections, introduced in HTTP/1.1[107], are an efficient way for those applications that need to get multiple pieces of information from servers. In Figure 7-14 we illustrate the persistent connection model of HTTP where the TCP connection is *'kept alive'* and re-used across multiple requests. The concept of persistent session used in LDAPv3, a directory access protocol, is similar in principle to HTTP's persistent connections,. However, LDAPv3's persistency is state based and used to retrieve a large query response in chunks.

Once persistent connections became available it was natural to expect a change to 'one request at a time' communication model. The change is in the form of pipelining of requests in which client can send new requests without necessarily waiting for response for previous requests. However, it is simpler to stick with 'one request at a time' model for majority of applications.

The resource on which a requested *method* is to be performed is specified using *Uniform Resource Identifier* or *Locator* (URI/URL). The structure of URI is similar to a file path and allows hierarchical addressing and naming

---

[107] A connection is persistent by default in HTTP/1.1. If connection needs to be closed after current request, client can indicate that by using *connection: close* header.

of resources. For instance, a circuit pack managed object on an NE (with name NE10) may be referred to with an URI of /NE10/Bay1/Shelf2/Slot3/cp. The number of components in a URI implies the depth of a particular object starting from the first component in the URI. For instance, with the previous example, cp MO is at a depth of four relative to the NE10 object.

The first part of the following listing shows HTTP request and response message formats while the second part provides an example. Although it allows binary content to be exchanged, HTTP is predominantly used to exchange text-based information. Each request consists of a request line, one or more headers and a body. Response follows similar structure.

```
- HTTP Request  Line    - <method>  < uri> HTTP/<version>
- Request Headers        - … general headers
-                        - … request specific headers
-                        - … entity headers (optional)
                         -              ← Empty line
- Request body           - … additional information/parameters suitable
- (if present)           for the method and the uri
                         -
- HTTP       Response    - HTTP/<version> <status code>  reason-line
Line                     - … general headers
- Response Headers       - … response specific headers
-                        - … entity headers (optional)
                         -                ←Empty line


- HTTP Request  Line    - GET  /foo HTTP/1.1
- Headers                - Connection: Keep-Alive      ← general
-                        headers
-                        - Date:xxxx
                         - Accept: text/html,text/plain ← request
                         headers
                         - From: gnu@somecompany.com
                         - If-Modified-Since: xxxx
                         - Content-Length: xxx          ← entity
                         headers
                         -               ← Empty line
- Request body           - …
- (if present)           -
- HTTP Response          - HTTP/1.1 200  OK
Line                     - Date: xxx                    ← general
- Headers                headers
-                        - Content-type: text/html      ← response
                         headers
                         - Content-Length: xxx          ← entity
                         headers
                         -              ←Empty line
- Response  body         - …. Response body
- (if present)
```

Table 7-19 lists HTTP response *status codes*.

Table 7-19. HTTP response status codes

HTTP Response Status Code	Category	Description
1xx	Informational	Indicates the request has not been refused and is being processed
2xx	Successful	Request received and processed successfully
3xx	Redirection	Used to redirect the client to a different location (URL)
4xx	Client Error	Indicated an error in the request due to the request originator
5xx	Server Error	Indicates an error on the request processor i.e the server

Table 7-20 lists some of the headers and their type. The header *from* can be used to send information about the user on behalf of whom the client machine is sending the request. The header, *SOAPAction* is used by SOAP to indicate a particular service (i.e application) on the server to which the incoming HTTP message containing SOAP request should be delivered. The value for this header element is declared in the binding section of WSDL description of the service as explained earlier.

Table 7-20. HTTP Headers

Header Name	General	Request	Response	Entity
Accept		√		
Allow				√
Connection	√			
Content-Base				√
Content-Encoding				√
Content-Location				√
Content-MD5				√
Content-Type				√
Date	√			
From		√		
Host		√		
If-Modified-Since		√		
Last-Modified				√
Server			√	
SOAPAction		√		
Transfer-Encoding	√			
WWW-Authenticate			√	

Another important feature o HTTP/1.1 is virtual hosting in which server can manage more than one domain name for an IP address. This helps, though not so common, NE that need to host multiple agents.

To support asynchronous notification, HTTP server (the agent) has to push events to one or more HTTP clients (managers). However, the push model doesn't fit with the request/response pull model of HTTP. To solve this, there are two techniques available. Both techniques still use request/response model except that the response is a never-ending reply.

The first technique is used in WebDAV [Gol99] and relies on manager sending a request to the agent asking for notifications. As part of the response, the agent replies with multi-status code indicating that the response contains more than one reply. Each reply can contain information about one event.

The second technique is based on Transfer Chunk Encoding (TE) used by servers to segment large response into small chunks. A HTTP response contains Content-Length header indicating the size of the response. If Chunk encoding is used then, instead of using Content-Length header, the server will use Transfer-Encoding header. Each chunk is a normal HTTP response and the response body contains the chunk length as first data. The last chunk contains '0' in the response body and indicates the end of the response. An event can be sent in one chunked response or split across multiple chunks. The chunk management is delegated to transport layer with the higher layers. There should never be chunk length of '0' on an event channel realized using TE method unless the intention is to close the channel.

In both schemes, the OSS needs to send request to receive events from the NE. The request must contain all   information required to setup the event channel (information such as filter, object lists etc). The agent responds to the request and starts sending events using multi-status response or Chunk encoding technique.

HTTPS is HTTP over Secure Socket Layer (SSL). The SSL protocol provides encryption of data exchanged between clients and servers offers privacy (confidentiality) as well as protection against manipulation (integrity).   Secure access to NEs or to OSSs has become extremely important in recent times due to a number of reasons. Other security protocols include RADIUS, TACAS. User security management includes user account and privilege (or access control lists - ACLs) management. The account and privilege information may be stored on a centralized external policy/access control server running such systems as - Microsoft's Active Directory or other solutions such as RADIUS (or TACACS). The RADIUS (or TACACs) server keeps userid/password information. When user logs into an NE, the NE (or OS) contacts a centralized RADIUS server to authenticate the user. This is different from SSL/ Transport Layer Security (TLS).

7.4.3 SNMP

Simple Network Management Protocol (SNMP) is commonly used to manage IP centric devices such as Ethernet switches, IP routers etc. However, it is also used in telecommunications environment and so it is important that we describe it here. It is based on agent-manager paradigm as illustrated in Figure 7-15.

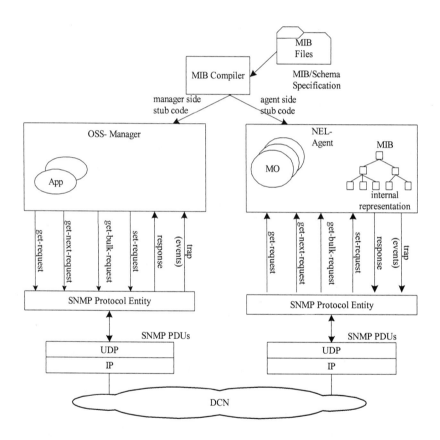

Figure 7-15. SNMP Manager-Agent Communication Model

Table 7-21 provides a brief summary of SNMP messages. Messages *get-requet, get-next-request, and set-request* are generated by manager applications. The *response* message *is* generated by agents in response to a set or get request message from managers

Table 7-21. SNMP Messages

Message	Description	Direction
get-request	Retrieves a value from a specific variable	
get-next-request	Retrieves the value following the named variable; this operation is often used to retrieve variables from within a table. With this operation, an SNMP manager does not need to know the exact variable name. The SNMP manager searches sequentially to find the needed variable from within the MIB.	Manager→Agent
get-bulk-request	to get multiple objects , typically from tables	Manager→Agent
get-response	The reply to a get-request, get-next-request, get-bulk-request, or set-request	Agent→Manager
set-request	To set values of objects. Similar to CMIP set operation.	Manager→Agent
trap	An unsolicited message sent by an SNMP agent to an SNMP manager indicating that an event has occurred	Agent→Manager

SNMP typically uses UDP[108] to send and receive messages. The agent generally uses UDP port 161 to receive requests while managers listen for traps i.e events on port 162. The maximum length of SNMP PDU is 1472 (taking into account 20 byte IP header, 8 byte UDP header and 1500 MTU on Ethernet LANs). The PDUs themselves are expressed using ASN.1. The following lists a part of SNMP PDU specification while Figure 7-16 illustrates the structure of the PDU.

```
— max-bindings ::= 2147483647
— VarBind ::= {
        — name ObjectName,  ← OBJECT IDENTIFIER
        — choice {
            — value          objectSyntax, ← any of SNMP data types
            — unspecified    NULL,          -- tag 5
            — noSuchOBject   [0]  IMPLICIT NULL, -- tag 0
            — noSuchInstance [1]  IMPLICIT NULL, -- tag 1
            — endOfMibView   [2]  IMPLICIT NULL  -- tag 2
        — }
— VarBindList ::= SEQUENCE (SIZE (0..max-bindings)) of VarBind
    —
— ErrorStatus ::=
        — INTEGER {
```

---

[108] Because of datagram nature of UDP, there is no notion of session in SNMP; managers can request management operations on agents at any time without trying to establish an association as in CMIP or HTTP.

```
          –   noError(0),
          –   tooBig(1),
          –   noSuchName(2),
          –   badValue(3),
          –   …
          –   noCreation(11),
          –   …
          –   }
–
–   PDU ::=
        –   SEQUENCE {
              –   request-id            Integer32,
              –   error-status          ErrorStatus
              –   error-index           INTEGER (0..max-bindings),
              –   variable-bindings     VarBindList
              –   }
```

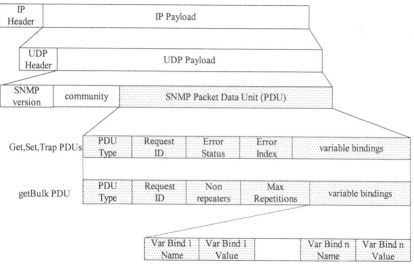

Figure 7-16. SNMP PDU structure

Management application wishing to perform a management operation on an NE constructs a request with a list of name-value pairs (NV). Each name is an Object identifier (OID) of a management variable/object on the NE The OIDs are registered in a management information tree (MIT). Each object/variable can be of standard or enterprise specific one as illustrated in Figure 7-17. In this figure, an entry is registered for CompanyX - a fictitious one to illustrate the fact that each vendor defines their product MIBs using OIDs registered under their own specific branch off of the global registration tree.

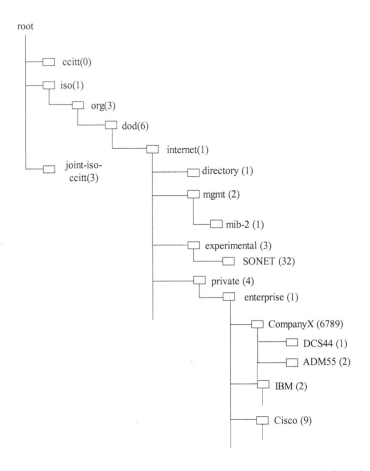

Figure 7-17. Management Information Tree (MIT) – Internet and private branches

Each entry in the *varBindList* is encoding using Basic Encoding Rules (BER) [25], a simple and widely available encoding technique illustrated in Figure 7-18. The BER encoding is based on *<Type, Length, Value>* (TLV) structure where:

- Type/Tag: indicates ASN.1 type; class, tag and whether the type is primitive or constructed.
- Length: indicates the length in octets of the ensuing value field
- Value: contains the value of the variable as a string of octets

There are three ways to encode ASN.1 values:

- Primitive Definite Length Encoding (pDL)
- Constructed Defined Length Encoding (cDL)
- Constructed Indefinite Length Encoding (cIL)

All three encodings use TLV structure. The size of the type field can be one or more octets. First two bits of the first octet encode class information. The third bit encodes the primitive or constructed type information. The remaining five bits of the first octet encode a tag number that distinguishes one data type from another with in the designated class. For types whose tag number is greater than or equal to 31, those five bits contain the binary value '11111', and the actual tag number is contained in the last seven bits of one or more additional octets. The first bit of each additional octet is set to '1', except for the last octet, in which it is set to '0'. Table 7-22 lists some example encodings.

Table 7-22. BER Encoding Examples

– valueX ::= {285_{10} }	– Tag=2	– L=2	– V=285
	– Integer	–	
– TaggedIntVar ::= [99] INTEGER	– T=99,	– L=3	– V=
	– Tagged Type		– [TLV,T=2,L=2,V=285]
– valY taggedIntVar ::= {285}			

pDL: The primitive, definite length encoding method can be used for simple types and types derived from simple types by implicit tagging. The class can be any of the four classes. The P/C bit is zero to indicate that it is primitive encoding. This encoding method requires that the length of the value be known in advance.

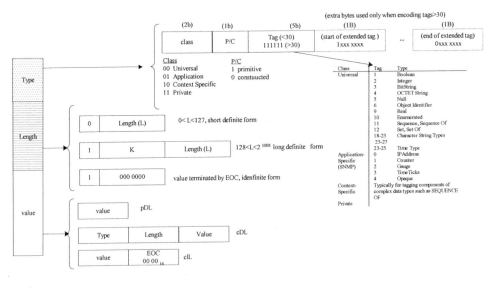

Figure 7-18. ASN.1 BER TLV formats

cDL: This method of encoding can be used for simple string types, structured types (SEQUENCE, SEQUENCE OF, SET, SET OF), types derived from simple string types as well as from structured types by implicit tagging, and from any type defined by explicit tagging. As in pDL, the class can be any of the four classes. The P/C bit is 1 to indicate that it is constructed encoding. Both pDL and cDL encoding methods requires that the length of the value be known in advance.

cIL: this method of encoding is identical to cDL except that the length of the value field is not required to be known in advance and instead the end of value field is indicated by a two-octet End-Of-Contents (EOC) field with a value of 00_{16}.

Other encoding techniques such as "Packet Encoding Rules" (PER) and "Lightweight Encoding Rules" (LER) also exist that allow quick encoding and decoding. There is also XML Encoding Rules (XER) to encode ASN.1 values as XML. For instance, the variable valY in Table 7-22 will be encoded as <valY>285</valY> using XER encoding. Obviously, XER will not be bandwidth efficient.

Once all the data for an SNMP request is ready, the protocol entity adds version, community name and encodes the request, which includes encoding NV list as well as PDU header information using BER.

Besides, basic get-request, SNMP also supports `get-bulkRequest`. To describe its use, let us consider a hypothetical example consisting of retrieving a list of interfaces on of an NE. In a transport network element, interfaces are identified (ifID), usually, using "`bay.rack.shelf.slot.port`" scheme as illustrated with *portEntry* in example. We use similar structure for *ifEntry* but, for simplicity, use just "`shelf.slot.port`", Figure 7-19 shows the MIB view.

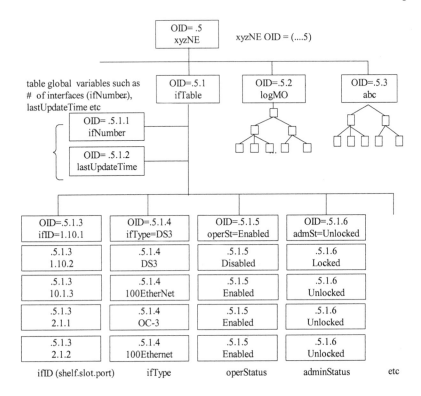

Figure 7-19. Example to illustrate Bulk Request

For retrieving small and simple tables, getRequest and getNextRequest commands can be sufficient. However, when there are hundreds of interfaces, as can be common with transport DCS (c.f. Chapter 1 and 2), retrieving all rows can be slow. Get-bulkRequest is meant to help with such large retrieval operations. The following listing shows the sequence of steps involved in retrieving all four rows of the above table: (Reader may recall that a particular object on an agent is identified by *object TypeID.instanceName*)

```
—  Manager→agent
        —  GetBulkRequest [non-repeaters=2, max-repetitions=2]
                —  (ifNumber, lastUpdateTime, ifID, ifType, operStatus,
                    adminiStatus )

        —

—  Agent→manager
    —  Response (
        —  (ifNumber.0="4"),
        —  (lastUpdateTime.0="123456"),
        —  (ifID.1.10.1=1.10.1"),
        —  (ifType.1.10.1="DS3"),
```

```
       - (operStatus.1.10.1="Enabled"),
       - (adminiStatus.1.10.1="Unlocked")
       - (ifID.1.10.2="1.10.2"),
       - (ifType.1.10.2="DS3"),
       - (operStatus.1.10.2="Disabled"),
       - (adminiStatus.1.10.2="Locked")
       - )
```
- Manager→Agent
    - GetBulkRequest [non-repeaters=1, max-repetitions=2]
        - (lastUpdateTime, ifID.1.10.3,ifType.1.10.3,operStatus.1.10.3, adminiStatus.1.10.3)
-
- Agent→manager
    - Response (
        - (lastUpdateTime.0="123466"),
        - (ifID.1.10.3="1.10.3"),
        - (ifType.1.10.3="DS3"),
        - (operStatus.1.10.3="Enabled"),
        - (adminiStatus.1.10.3="Unlocked"),
        - (ifID.2.1.1="2.1.1"),
        - (ifType.2.1.1="OC-3"),
        - (operStatus2.1.1="Enabled"),
        - (adminiStatus.2.1.1="UnLocked")
        -)
- Manager→Agent
    - GetBulkRequest [non-repeaters=1, max-repetitions=2]
        - (lastUpdateTime, ifID.2.1.2,ifType.2.1.2,operStatus.2.1.2, adminiStatus.2.1.2)
- Agent→Response
    - Response (
    - (lastUpdateTime.0="123476"),
    - (ifID.2.1.2="2.1.2"),
    - (ifType.2.1.2="100Ethernet"),
    - (operStatus2.1.2="Enabled"),
    - (adminiStatus.2.1.2="UnLocked"),
    - (logMo.1="…."), ← indication to the manager that the table fetch is done.
    -)
-
-

In the above example, there are no more table entries after 2.1.2. A request to get two more interface entries starting from 2.1.2 will return just one more table entry and the second variable returned in the response will be a lexicographical successor to 2.1.2, which will be the logMO object. This should indicate to the manager that it has reached the end of table and hence should stop. If it is the end of MIB the agent may respond with "endofMibView".

The following lists specification of bulkRequestPDU, the structure of which was already illustrated in Figure 7-16.

```
- BulkRequestPDU {
    - requestId Integer32,
    - non-repeaters INTEGER (0.. max-bindings),
    - max-repetetions INTEGER (0.. max-bindings),
    - variable-bidning VarBindList
    - }
```

Even with get-bulkRequest, retrieving large amount of information from an NE can be very expensive. Typically OSS, when it connects first time to an NE, will retrieve all management information stored on the NE. This first step is usually called discovery and involves discovering object instances on the NE (different from network discovery). Because of protocol overhead and relative weakness in bulk-data transfer in basic request/response mechanism, such initial discovery part can incur excessive latency unless special techniques are used. For instance, vendors use FTP-based solution where the MIB data is retrieved and stored in a file on the device, and then transferred to the manager via FTP. It is a three-step solution. In the first step the manager writes the data (i.e list of variables or tree root OID) to be fetched into a vendor specific MIB. In the second step, the manager writes to another vendor specific MIB specifying the FTP address. In the third step, the SNMP agent FTPs the requested data to the specified FTP address. Similar technique is used in TL1 implementations as well. A TL1 command (for instance, RTRV-ALL) requesting all management information on an NE is issued. Along with such command an FTP address is specified. Later, when the NE agent has prepared the response file, it uploads it to the specified address.

The most important thing for telecommunications environment is the event or alarm reports emitted by agents to inform one or more managers of some event of significance. Some of these events may be related to changes occurring within the device and some to internal or external faults. The set of all notifications that can be emitted by a device will be known to manager entities apriori from the SMI (the schema) associated with that device. In SNMP such asynchronous notifications are called traps. With traps, unlike manager→agent request/response interaction, an agent doesn't know where to send them. Therefore an explicit trap destination IP address is configured into each agent. Some agent implementations may support multiple destinations so that multiple managers can all receive same copy of event reports.

The events of most interest to telecommunications management are equipment and link or facility faults. Whatever may be the nature of events, each event report carries some common information as explained in Ch6. In the case of SNMP, the common information consists of sysUpTime and snmpTrapOID. Besides these two pieces of information, each event report may contain   values of zero or more objects as the following example illustrates.

```
- TrapPDU {
     - requestId Integer32,
     - error-status ErrorStatus,  ← errorStatus listed earlier
     - sysUpTimeOID      ← var-1 name
     - sysUpTimevalue    ← var-1 value
     - snmpTrapOID       ← var-2 name
     - snmpTrapValue     ← var-2 value
     - variable-bidning varBindList ← var-3 … var-n
     - }
-
- snmpTrapOID OBJECT-TYPE
     - SYNTAX      OBJECT IDENTIFIER
     - MAX-ACCESS accessible-for-notify
     - STATUS      current
     - DESCRIPTION
           - "The authoritative identification of the
           notification currently being
           - sent. This variable occurs as the second varbind in
           every
           - SNMPv2-Trap-PDU and InformRequest-PDU."
           - ::= { snmpTrap 1 }
```

The notification object is indicated using the snmpTrapOID variable in the trap PDU. Vendors typically define a large number of notification objects i.e. even types to model many different aspects of transport network equipment; notifications for NE configuration changes, faults and any events occurring in the NE that might of significance to OSS. Earlier we provided xyzCriticalAlarmOnReport as an example.

This concludes description of SNMP as a management information access protocol. The discussion has focused on simple protocol aspects and is based primarily   on SNMPv2. Issues such as security, use of SNMP over TCP and other transport protocols; enhancements brought out in SNMPv3 have all been skipped. The basic SNMP protocol mechanisms will be invariant and hence the discussion here should suffice. One can describe SNMP as a fine-grained management information access protocol because it relies on the fact that every variable has an address (OID). There is no concept of higher-level objects. For example, there is no way to specify a request to get a managed object and all its attributes. Also, ASN.1 used to describe SNMP MIBs has not no support for object inheritance. Java, Corba, CMIP, WebDAV all support better of way of providing such functions. However, despite lacking such advanced features, SNMP is widely used because of its simplicity. Besides, many of today's network management applications don't need such advanced features. However, newer generation management applications (such as those to be built for Grid computing, Active Networks, etc) may require object-oriented abstractions to be part of protocol. In that context, it is worth noting that SNMP is an old and simple protocol that lacks advanced capabilities.

7.4.4   CORBA

Common Object Request Broker Architecture (CORBA) is a middleware technology used to allow applications to interact with each other independent of (1) the platforms they are running on and (2) the languages used to build them. This is similar to the philosophy of Java. However, to some extent the emergence of Java, and later XML, overshadowed CORBA. However, Java and XML are complimentary technologies to CORBA. Work on CORBA started around 1990 and picked up momentum[109] in the years following by as object orientated approaches (OOA) to software development became a highly discussed activity though the interest has waned a bit by around 2000.

Figure 7-20 illustrates CORBA communication model based on client-server paradigm, which is similar to TMN's manager-agent paradigm with manager being a client and agent being a server.

---

[109] As evidenced by lot of papers at *Object-Oriented Programming Systems, Languages and Applications (OOPSLA)* and other conferences during 90-94.

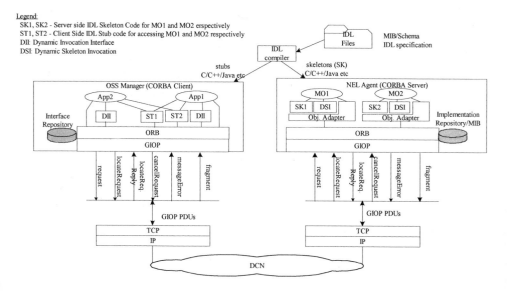

Figure 7-20. CORBA Manager-Agent Communication Model

The server in a CORBA framework hosts one or ore more objects that offer services/interfaces for invocation by remote clients. The invocation request can be to get/set some attributes of a serving object. Alternatively the request can be to activate a method supported by a serving object.

Figure 7-21 illustrates CORBA architecture as standardized by the Object Management Group (OMG), the organization responsible for all aspects related to CORBA.

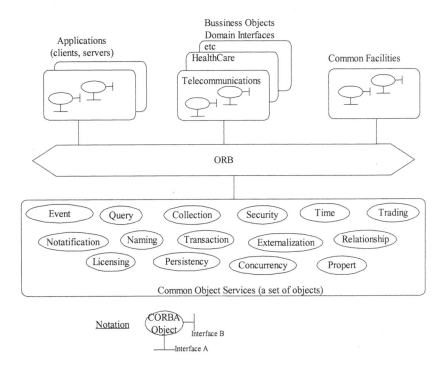

Figure 7-21. OMG CORBA Architecture

First generation of CORBA technology focused on defining infrastructure for object oriented distributed computing with the simple goal of remote method invocations. Currently CORBA standard not only contains basic protocol elements to support request/response but also Common Object Services (COSS), Common Facilities (CF), and Domain Interfaces (DI). Domain interfaces (previously called Vertical facilities) are industry and business specific and include such domains as health care, telecommunications, etc. Both application objects and standard CORBA objects (Naming, Event etc) are accessed through the Object Request Broker (ORB). Table 7-23 lists standardized object services.

Table 7-23. CORBA Services

Service	Description	conformance
Collection	Facilities for grouping objects into lists, queues, sets etc	Mandatory
Query	Facilities for querying collections of objects in a declarative manner	Optional
Concurrency	Facilities to allow concurrent access to shared objects	Optional (trasactional lock support optional)
Transaction	Flat and nested transactions on methods calls over multiple objects	Optional
Event	Facilities for asynchronous communication through events	Mandatory (typed events optional)
Notification	Advanced facilities for event based asynchronous communication	Mandatory (trasactional support and repository optional)
Externalization	Facilities for marshalling and unmarshalling of objects	Mandatory
Life Cycle	Facilities for creation, deletion, copying and moving objects	Optional
Licensing	Facilities for attaching a license to an object	Optional
Naming	Facilities for systemwide name of objects	Mandatory
Property	Facilities for associating (attribute, value) pairs with objects	Mandatory
Trading	Facilities to publish and find the services an object has to offer	Mandatory (proxy, link, and admin interfaces optional)
Persistence	Facilities for persistently storing objects (i.e its state)	
Relationship	Facilities for expressing relationships between objects	
Security	Mechanisms for secure channels, authorization, and auditing	
Time	Provide the current time within specified error margins	

In CORBA, interfaces, consisting of data and methods, are specified using Interface Definition Language (IDL). A set of interfaces may be realized by a run-time object/application. Roughly, one can conceive of an IDL interface that contains attributes, actions as corresponding to any of the GDMO MOCs from M3100 fragments discussed in the previous except for notifications. The GDMO notification needs to be handled separately in CORBA and will be explained as part of CORBA event service.

Figure 7-22(a) shows basic data types supported in CORBA.

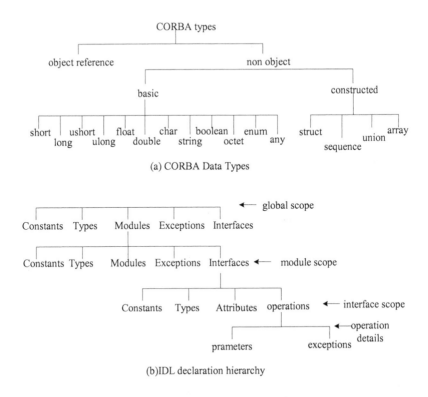

(a) CORBA Data Types

(b)IDL declaration hierarchy

Figure 7-22. CORBA data types and Interface Repository Containment Hierarchy

The IDL data types, interfaces, and exceptions can be defined at global, module level as shown in Figure 7-22 (b). Module construct allows a naming context and a way to qualify reference to a type or interface. For instance, `moduleA.interfaceA` is different from `moduleB.interfaceA`.

As shown in Figure 7-22, an object (or interface) IDL specification is given to tools that produce client side stub (ST) code in regular languages, such as C, C++, Java, etc. They also generate server side skeleton (SK) code. The server application object implementing/hosting an interface can be implemented in one language while the client application that accesses that interface is implemented in another language. CORBA allows such decoupling.

With similar objective as in GDMO, and XML Schema, CORBA IDL allows specifications to be housed in a *module* as illustrated here. An *interface* refers to a package of *methods* (actions), *attributes*, and *exception*. Unlike GDMO in which packages can't be inherited, CORBA does allow interface inheritance. A run time application or object can implement one or more *interfaces*.

```
-   ;declarations at application's global scope
-   [<type declarations> ;]
-   [<constant declarations>;]
-   [<exception declarations>;]
-   [<interface declaration>;
-   [<module declarations> ];
-
-   ;skeleton module declaration syntax
-   module <identifer> {
        - [<type declarations> ;]
        - [<constant declarations>;]
        - [<exception declarations>;]
        - [interface <identifier> [: <inheritance>] {
          - [<type declarations> ;]
          - [<constant declarations>;]
          - [<attribute declarations>;]
          - [<exception declarations>;]
          - [<op_type> <identifier> (<parameters)>]
                - [raises exception]  [context]
          - …
          -
        - }  ← end of interface
        - … more interfaces
        -

        -
        - }  ← end of module definition
-   ;
-   ;example module delcaration
-   module itut_m3120 {
      - IMPORTS
            - /* Types imported from OMG standard */
            - typedef CosNaming::Name NameType;
            -
            - /*Types imported from itut_x780*/
            - typedef itut_x780::AdministrativeStateType
                            -   AdministrativeStateType;
      - interface ManagedElement: itut_x780::ManagedObject {
        -

          AlarmStatusType alarmStatusGet ()
                -   raises (itut_x780::ApplicationError);
-
        - CurrentProblemSetType currentProblemListGet ()
-           raises (itut_x780::ApplicationError);
-
          AdministrativeStateType administrativeStateGet ()
-           raises (itut_x780::ApplicationError);
-
          void administrativeStateSet
-             (in AdministrativeStateType administrativeState)
-             raises (itut_x780::ApplicationError);
-
          OperationalStateType operationalStateGet ()
```

```
–            raises (itut_x780::ApplicationError);
–
–        UsageStateType usageStateGet ()
–            raises (itut_x780::ApplicationError);
–
–        AlarmSeverityAssignmentProfileNameType
–        alarmSeverityAssignmentProfilePointerGet ()
–            raises (itut_x780::ApplicationError,
–                NOalarmSeverityAssignmentPointerPackage);
–
–            ...
–
–        }; // interface ManagedElement
   – .... More interfaces
 – }// end of module m3120
```

The skeleton `itu_module3120` IDL module listing is borrowed and adapted from M.3120 (which is a CORBA version of M.3100 GDMO information model discussed in Chapter 6). The adapted listing's purpose is to illustrate IDL module syntax usage. The `managedElement` CORBA interface supports get/set of administrative, usage, operational state and alarm status attributes. Besides, it provides alarm severity assignment profile (ASAP) that can be get/set as well. There are other details inside this interface but omitted for brevity. Reader may consult M.3120 for full and exact listing.

CORBA supports Dynamic Interface Invocation (DII) in which clients that don't have stub code statically linked into them can, by consulting an Interface Repository (IR), perform operations on server objects. The IR stores information using containment similar to IDL declaration containment shown in Figure 7-22 (b). Another use of DII will be discussed as part of typed events.

7.4.4.1 CORBA Event Service

CORBA Event Service (ES) [OMG01] provides a framework for asynchronous event notifications. Suppliers and consumers of events are separated using Event Channel (EC). Suppliers send events to the channel, which, if necessary, buffers and forwards them on to one or more consumers. The ES defines two modes for event channel; *push* and *pull*. A consumer can use either model to receive events. Similarly, suppliers may use either model to publish events. A single event channel may support both push and pull model and all   Figure 7-23 illustrates the above discussion and the various EC objects.

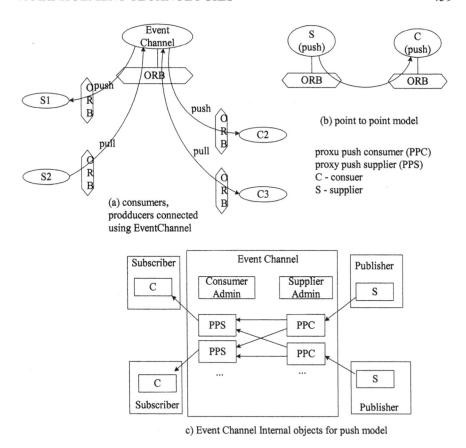

Figure 7-23. CORBA Event Channel

All the administrative work of maintaining lists of consumers and suppliers, handling of requests to subscribe or unsubscribe to the channel is delegated to the EC. Registration of suppliers and consumers is a two-step process. A publisher application (such as NE) first obtains a proxy-consumer (PC) from the EC then connects to the PC by supplying it with the object reference of itself. Similarly, an event consumer application (such as an OSS) first obtains a proxy-supplier (PS) from the EC and then connects to the PS by supplying it with the object reference of itself as the example in Figure-21 (c) illustrates. The reason for such a two-step process, as per OMG specification, is that it allows event channels to be connected in daisy-chain fashion or what can be described as *event channel composition*. The purpose of such chaining is, of course, application dependent. In TMN, such an advanced use may be applicable to EML and WSF as shown in Figure 7-24.

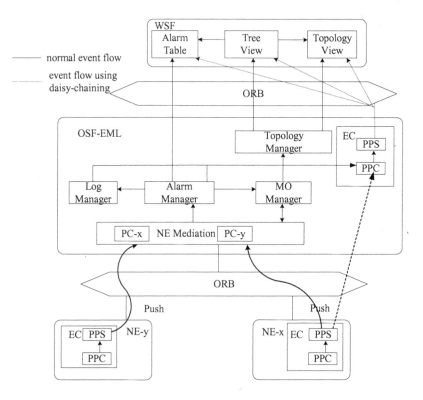

Figure 7-24. CORBA Event Channel Daisy Chaining

Table 7-24. CORBA Notification Service

Features of Notification Service (NS)		Comment/description
typed or structured events	Typed events now part of ES as well. Structured events are special in NS.	Requires consumers and suppliers to use structured or typed events – strong-type systems are always prefereable from software quality perspective.
Filtering	Using Filter Constraint language	Filter installed at consumer or supplier end of EC
QoS	Reliability	Event Reliability best-effort or guaranteed
		Connection Reliability transient or persistent
	Event Queue Management	Maximum queue size
		Delivery policy
		Discard policy
	Individual Event Management	start time
		time out
		priority
	Event Batching	

A typical deployment will have one or more WSF entities connect to a OSF-EML that manages a group of NEs. In such deployment, events from NEs as well as those generated inside of OSF-EML can all be published through the two ECs setup in daisy chain configuration.

It is commonly recognized, that the Event Service has severe deficiencies and therefore many extensions were proposed and standardized as part of CORBA Notification Service (NS) [Har97]. The extensions, listed in Table 7-24, support filtering, typed events (to be explained), and quality of service with respect to event delivery.

In push model, consumers implement a generic call back method "push" which takes *Any* data type parameter. To support typed events, suppliers and consumers need to use a common interface specification that declares one or more typed events. A typed event is nothing but a regular method with semantics of *'oneway'* with no *'out'* and *'return'* parameters. The type of the event includes the name of the method (aka event name) plus the method signature consisting of one or more arguments (aka event parameters). The following listing contains notification interface from M.3120. All of the notification operations defined in this interface pass a number of parameters, some of which are common.

```
- interface Notifications {
-     void equipmentAlarm (
-           in ExternalTimeType              eventTime,
-           in NameType               source,
-           in ObjectClassType            sourceClass,
-           in NotifIDType              notificationIdentifier,
-           in CorrelatedNotificationSetType
      correlatedNotifications,
-           in AdditionalTextType             additionalText,
-           in ProbableCauseType          probableCause,
-           in SpecificProblemSetType       specificProblems,
-           in PerceivedSeverityType         perceivedSeverity,
-           in BooleanTypeOpt          backedUpStatus,
-           in NameType             backUpObject,
-           in BooleanTypeOpt        alarmEffectOnService,
-           in SuspectObjectSetType       suspectObjectList
-           ...
-     );
-     ...
- }; // end of Notifications interface
```

The other typed notification methods are similar to the one above and Table 7-25 lists them (from M.3120):

Table 7-25. NE level typed CORBA Notifications, M.3120

1.Attribute Value Change	9.Physical Violation
2.Communications Alarm	10.Processing Error Alarm
3.Environmental Alarm	11.Quality of Service Alarm
4.Equipment Alarm	12.Relationship Change
5.Integrity Violation	13.Security Violation
6.Object Creation	14.State Change
7.Object Deletion	15.Time Domain Violation
8.Operational Violation	

Assuming push model, to deliver an event, both typed and untyped, EC invokes a matching method of consumer object's notification interface. To invoke consumers, since it will not have stub code for consumer's application notification interfaces (such as the TMN notification interface of M.3120), the EC builds a CORBA request dynamically (i.e it uses DII) by consulting an Interface Repository (IR). Similarly, since EC's *TypedProxyPushConsumer* (*tPPC*) doesn't implement any application notification interface, event suppliers use DSI to invoke operations on the *tPPC*. Similarly, since *TypedProxyPushSupplier* (*tPPS*) doesn't have skeleton code related to application events, it supports DII. Figure 7-25 (a) illustrates the above description. Essentially, in typed event model, the consumer is the CORBA server and the supplier is the CORBA client.

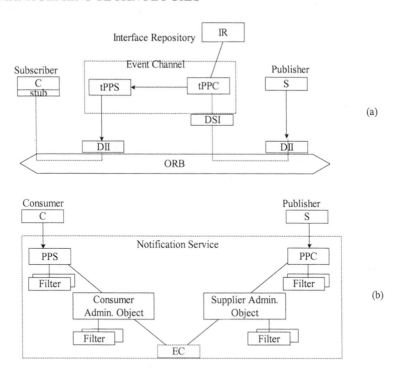

Figure 7-25. CORAB (a) typed event flow (b) NS filters

The application event suppliers (such as NEs) connect to tPPC and the application consumers (such as OSSs) connect to *tPPS* using administrative interfaces of EC.   Table 7-26 provides a simple example where a supplier (an NE) publishes, using push model, and equipment alarm to EC, which in turn sends it to the consumer (an OSSApp) by invoking a method on the consumer. The example also includes NE publishing an *untyped* event using *Any* type data parameter and the EC invoking *push* method on the consumer.

```
-
-   // common IDL specification for all NE level typed events/alarms
-   interface NEAlarms {
    -   void equipAlarm (in arg1, in arg2, … )
    -   void commsAlarm (in arg1, arg2, ….)
    -   …
-   }
-   // consumer object/application specification
-   OSSApp: CosTypedEventComm::TypedPushConsumer, NEAlarms {
    -   …
    -   }
-
```

Table 7-26. Illustration of typed Event

Consumer (OSSApp)	EC tPPS	Supplier (NE) tPPC	comment
– Connect to tPPS	–	–	– connecto to tPPC
–	–	–	–
– OSSApp.**equipAlarm** (…)	– ←	– ←	– tPPC.**equipAlarm** (…)
– ;process the event	–		
–			
–	–	–	– Any x
–	–	–	– 1. marshal event
–	–	–	– parameters into x
–	–	–	–
– OSSApp.**push** (Any y)	– ←	– ←	– 2. tPPC.**push** (Any x)
–			
– 1. unmarshall event			
– parameters from Y			
–			
– 2.process the event			

The benefit of typed events is that it offers filtering based on event type. However, one should note that there is no filtering based on event parameters. Such filtering is supported by installing filters and is not supported in ES but in NS. CORBA event channel can also be extended to provide customizable filtering services using hook objects [Tom99]. Filters can be supplier or consumer specific as illustrated in Figure 7-23.

For issues associated with performance penalty of using typed events, as well as advantages of structured events refer to [Pri01a],[Pri01b].

7.4.4.2 CORBA Invocation Styles

Table 7-27 lists different CORBA invocation styles available for clients to invoke operations on a server.

Table 7-27. CORBA remote operation invocation styles

Invocation Style	Failure Semantics	Description
Call	At-most-once	Caller blocks until a response is receiver or an exception is raised
One-way	Best effort delivery	Caller continues immediately without waiting for any response from the server
Deferred synchronous	At-most-once	Caller continues immediately can later block until response is delivered.
AMI call back		Call back model for asynchronous invocation
AMI Polling		Poll for response

The synchronous (the default) method involves client blocking until response is received from server or an exception is raised. Using the asynchronous invocation style, specified using the IDL keyword *oneway,* allows clients to invoke remote operations without waiting for response. The server responds by doing another simple oneway call back on the client. For this to work, the client, when it invokes a *oneway* remote operation must supply reference to a call back operation to the server. Oneway invocations have best effort semantics in the sense ORBs don't guarantee that the request gets delivered to the server object. However, this style makes the server and client appear as peers to each other from communication perspective. From service perspective, server is still providing some services to its clients.

The Asynchronous Method Invocation (AMI) style was introduced into CORBA to provide true non-blocking remote invocation as one-way had some shortcomings[110]. AMI allows a client to invoke methods on server objects without blocking for response. Within AMI two types are possible – call back and polling. The call back model involves an upcall from ORB to client when response is received from the server. The polling style involves the client checking for response periodically.

It is possible to integrate object-oriented interactions into OS to provide maximum benefits to application developers. Apertos[111], a message oriented OS developed by Sony, took CORBA approach one-step further and integrated object interactions as part of the OS and provided such semantics as Call (synchronous), Send (with oneway semantics), SendWithContinuation (oneway or AMI call/poll semantics), KickContinuation (response to client who used SendWithContinuation).

7.4.4.3 GIOP

GIOP is the **G**eneral **I**nter-**O**RB **P**rotocol standardized by OMG, which enables end-to-end interoperability between ORBs from different vendors. It is similar to SNMP interoperability where SNMP protocol entities of agent and manager can be from different vendors.   ORB is not tied to any transport protocol though the requirements and assumptions are primarily based on TCP. Accordingly GIOP can be transported over TCP, SCTP and any other that meets byte-oriented, stream based, reliable delivery requirements.

---

[110] Oneway invocation can   block a client if IIOP/TCP is used by ORB to deliver the request. On the other hand, if UDP is used then there is no guarantee that the message would be delivered.

[111] In Apertos, the equivalent of IOR was is an OID and clients invoke methods on local or remote objects transparently. Resolution of OID to a location is done using Name Server.

Table 7-28. GIIOP Messages

GIIOP Messages	Comment	Direction
Request	To invoke operation on a remote object	Client→server
Reply	Response to Request	Server→Client
LocateRequest	Query current location of an object	Client→server
Locate abstractReply	Response to location query	Server→client
CancelRequest	Used to notify servers that a client is no longer interested in response to a previous request	Client→server

There are four aspects to GIOP:
- GIIOP Messages
- Common Data Representation
- Transport Protocol assumptions
- Object References

Table 7-28 lists a set of GIOP messages   and Figure 7-26 illustrates the structure of some of the messages.

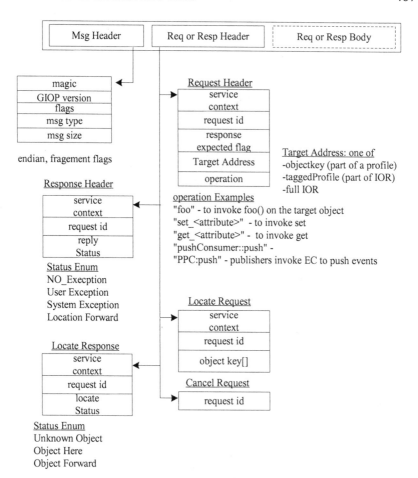

Figure 7-26. GIOP Message Structure

GIOP uses Common Data Representation (CDR) for encoding CORBA messages. CDR is a transfer syntax that maps various IDL data types into a sequence of encoded octets. The mapping is commonly called marshalling or serialization and is done so that the machines involved in communication can be either of little   or big endian type.

GIOP realized using TCP/IP as the transport protocol is called the Internet Inter-ORB (IIOP) protocol. IIOP is a high-level protocol that takes care of many of the services associated with the levels above the transport layer, including data translation, memory buffer management, dead-locks and communication management. It is also responsible for directing requests to the correct object instance within an ORB.

7.4.4.4 GIOP Object References

In CORBA/GIOP, object references are specified using a language-neutral technique called the Interoperable Object Reference (IOR). The destination of each GIOP message is specified using IOR. Each IOR consists of a type-id identifying the destination object type (e.g., M3100::logInterface), and one or more opaque tagged profiles. Each tagged profile contains protocol specific representation of an object's location i.e. address. One of the tagged profiles standardized is the IIOP profile. Figure 7-27 illustrates the structure of GIOP IOR and that of IIOP profile.

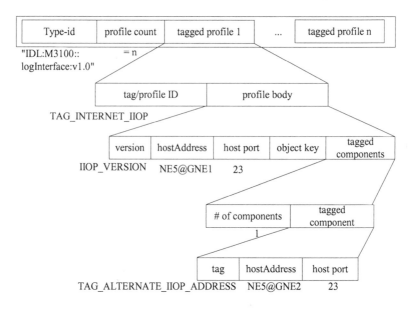

Figure 7-27. IOR, and IIOP Profile structures

Each IIOP profile contains a version to identify IIOP version, object's host IP address and port number, an object key, and one or more tagged components. The ORB sending a GIOP message uses information stored in the tagged components. The receiving ORB uses just the object key to locate one of many local objects to which a received GIOP message should be delivered. When a CORBA object comes to life it will instantiate its IOR (hence the object key) and supply the same to its clients somehow i.e clients wishing to send messages to a target server object must know that server object's IOR *apriori*. In the client's machine, a table can be created to map IORs to proxy names that are easier to use. The client application can access the object using the IOR, which masks the client application's ORB implementation from the server ORB implementation.

Figure 7-27 also provides an example of fail-over. The alternate address will be tried in the event that a client can't communicate to the server at the primary address.

Initially, CORBA was supposed to enable a world of objects that communicate and live together forming object web world much like the WWW. With that said, the following are some interesting and simple analogies:

IDL is to CORBA what HTML is to the WWW
GIOP/IIOP is to CORBA what HTTP is to the WWW
IOR is to CORBA what URL is to the WWW

As mentioned earlier, SNMP lacks of support for object-oriented abstractions. While CORBA overcomes the shortcomings of SNPMP with such abstractions as inheritance, managed object (interface) etc, it still needs richer set of abstractions to qualify as true distributed computing object-oriented technology[112]. However it is already meeting requirements of many of today's TMN and IT applications where the need is not so much on advanced object-oriented distributed capabilities but of seamless integration between heterogeneous systems[113], particularly before the advent of Java.

7.4.5 CMISE/CMIP

Common Management Information Service Element and its associated Common Management Information Protocol (CMISE/CMIP) is commonly associated with TMN and vice versa. Except for WebDAV/XML (and lately WS[114]), CMISE/CMIP is the most powerful access protocol of all discussed thus far. CMISE/CMIP's, the communication model of which is illustrated in Figure 7-28, power comes from its ability to support *scoping, filtering, linked responses*, and *event forwarding descriminator (EFD)*.

---

[112] Object oriented distributed computing is an interesting field and, reader can refer to [Agh93] for more.

[113] Heterogeneous systems are defined as systems with different implementation/platform technologies. For instance a C++ application on Unix is heterogeneous with a C++ application on MS-Windows. A Java application on a Unix platform is homogeneous with Java application on MS-Windows platform as Java hides the platform details.

[114] Including FWS

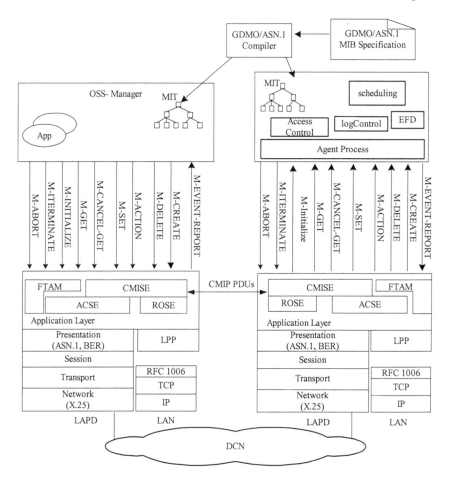

Figure 7-28. CMISE/CMIP Communication Model[115]

The *association control service element* (ACSE) is used to establish association between two entities wishing to exchange management information. One can equate the process of setting up association with setting up of HTTP session- consisting of TCP connection setup, validation of user credentials using HTTP authentication etc. ACSE is an OSI application layer entity that helps applications establish and maintain association. Two ACSE's interact using ACSE PDUs (A-PDUs).

---

[115] LPP – Lightweight Presentation Protocol is part of CMOT architecture, which provides details of CMISE/CMIP over IP networks using either TCP or UDP.

The *remote operations service element* (ROSE) provides remote invocation services using request/response primitives. ROSE services are analogous to CORBA or remote procedure call (RPC) mechanisms for invoking operations. Like ACSE, ROSE is also an OSI application layer entity. Two ROSE entities interact by exchanging ROSE PDUs (R-PDUs). The conversion of ROSE service requests from users into ROSE PDUs is done in ROSE protocol module (ROPM).

The *common management information service element* (CMISE) provides services to applications (such as OSFs) to invoke management operations on another application entity (such as NEF). The various CMISE services, which can be invoked in confirmed or nonconfirmed mode, are listed in Table 7-29. A management operation invoked in confirmed mode will have a response from the remote end while a command invoked in non-confirmed mode will not.

Table 7-29. CMIP Services

Services	Mode
M-GET	confirmed
M-SET	Confirmed / unconfirmed
M-CREATE	Confirmed
M-DELETE	Confirmed
M-CANCEL-GET	Confirmed/ unconfirmed
M-EVENT-REPORT	Confirmed/ unconfirmed
M-ACTION	Confirmed/unconfirmed

Two CMISE providers interact using common management information protocol (CMIP). The CMISE service requests from users are converted into CMIP messages/PDUs in CMIP Protocol Module (CMIPPM). For instance, M-GET service request from user is converted into m-Get CMIP PDU. Each CMIP PDU is then converted into a ROSE service. Collectively CMIPPM is the service provider with the above set of service primitives to users and informing them of incoming requests, responses, and event reports using *indication primitives.* Figure 7-29 details CMIPPM.

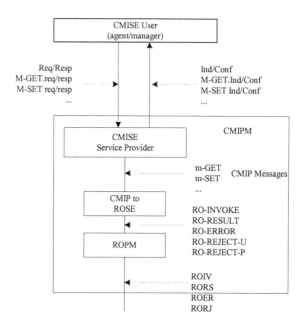

Figure 7-29. CMIPM - CMIP Protocol Module

Each CMISE/CMIP message contains invoke-ID, operation code, MOC[116], MI, and operation specific information as shown in Figure 7-30

---

[116] Managed Object Class (MOC), c.f. Chapter 6

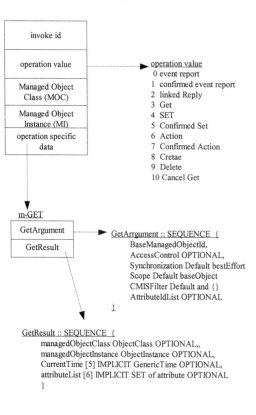

invoke id
operation value
Managed Object Class (MOC)
Managed Object Instance (MI)
operation specific data

operation value
0 event report
1 confirmed event report
2 linked Reply
3 Get
4 SET
5 Confirmed Set
6 Action
7 Confirmed Action
8 Cretae
9 Delete
10 Cancel Get

m-GET

GetArgument
GetResult

GetArrgument :: SEQUENCE {
    BaseManagedObjectId,
    AccessControl OPTIONAL,
    Synchronization Default bestEffort
    Scope Default baseObject
    CMISFilter Default and { }
    AttributeIdList OPTIONAL
}

GetResult :: SEQUENCE {
    managedObjectClass ObjectClass OPTIONAL,,
    managedObjectInstance ObjectInstance OPTIONAL,
    CurrentTime [5] IMPLICIT GenericTime OPTIONAL,
    attributeList [6] IMPLICIT SET of attribute OPTIONAL
}

Figure 7-30. CMIP Message Format

M-GET service is used to retrieve values from one or more managed objects. It is a confirmed service that requires response from the remote end. A pending *get request* can be cancelled by invoking M-CANCEL-GET service. The attributes to be retrieved can be specified or by default all attributes of selected objects will be returned. The candidate objects are specified using *scope* and *filter* parameters. Response to a *scoped get request* can be potentially large and is required to be broken into small chunks. Such chunks are sent using *linked response* mechanism. Each *linked response* contains the invoke-id from the request as well as a linked-id parameter. This parameter is set to '0' to indicate start of linked response in the first response PDU, to '1' in the intermediates PDUs, to '2' in the last PDU. A simple GET response as well as linked GET response are illustrated in Figure 7-31   (a) and (b).

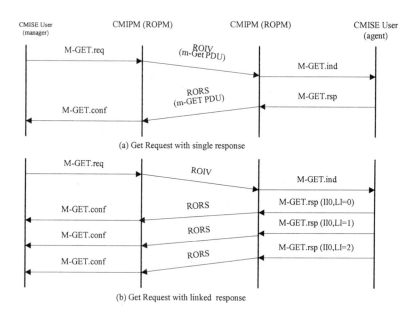

Figure 7-31. M-GET operation and corresponding CMIP Messages

7.4.5.1 Scoping

Scoping selects one or more candidate MOs for management operations such as Action, Get etc. The candidate objects can be all or some subset of objects of a subtree rooted at a given base MO. The base MO specified in scoped operations is the reference point for scoping and the MO is said to be at level zero. There are four levels of scoping as follows:

Single object: The base MO alone is selected. This is the default action for scoping. In Figure7-26, if the base object is the NE MO, then the get/set/action management operation is applied on NE MO only.

Single level: The nth-level subordinates can be selected. For example, if n is the first level in Figure 7-26, the objects selected are log (1), fabric (f100), log (2)

N Levels: base object plus subordinates up to the nth level. In Figure 7-26, if n=1 then the candidate objects will be NE (xyz, the base MO), log (1), fabric (f100), log (2) .

Whole sub tree: the candidate objects include all objects of a subtree to depth and breadth of infinity below a base MO. In Figure 7-32, if we take NE (xyz) as the base MO then the subtree consists of all objects below the NE object.

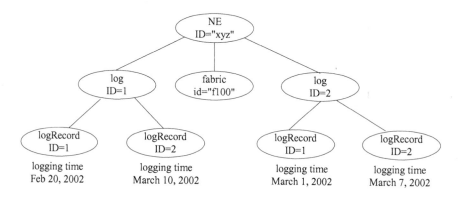

Figure 7-32. CMIP Scoping

7.4.5.2 Filtering

While scoping selects all objects in a subtree, sometimes it is desirable to filter and select a subset of objects in the subtree. The criterion for filtering is expressed in terms of assertions about the presence of attributes or values of attributes called attribute value assertions (AVA). The AVAs are expressed using matching rules or operators: equality, greaterOrEqual, lessOrEqual, substrings, subsetOf, supersetOf, and nonNullSetIntersection. If a filter involves more than one assertion, the assertions can be grouped using and, or, and not. Filters can be arbitrarily complex, as one can nest filters within filters. The following lists a skeleton of CMIS Filter specification.

```
- CMISFilter ::= CHOICE {
-      item [8]  FilterItem,
-      and  [9]  IMPLICIT SET OF CMISFilter,
-      or   [10] IMPLICIT SET OF CMISFilter,
-      not  [11] IMPLICIT SET OF CMISFilter
-      }
- FilterItem CHOICE {
-      equality       [0] IMPLICIT Attribute,
-      substrings     [1] IMPLICIT SEQUENCE OF CHOICE { …},
-      greaterOrEqual [2] IMPLICIT Attribute,
-      lessOrEqual    [3] IMPLICIT Attribute,
-      present        [4] IMPLICIT AttributeId,
-      subsetOf       [5] IMPLICIT Attribute,
```

```
    –  supersetOf          [6] IMPLICIT Attribute,
    –  nonNullSetIntersection   [7] IMPLICIT Attribute
    –  }
```

Using the example MIT from Figure 7-26, to retrieve log records after Feb 20 the following filter (semi-formal) can be used:

```
–  SampleFilter CMISFilter::= {and  {
       –  item equality {
              –  attributeId managedObjectClass,
              –  attributeValue logRecord
       –  },
       –  item  equality {
              –  attributeId loggingTime,
              –  attributeValue 20020220
              –  }
–  }
–
–
```

It is interesting to note that filtering supported by directory access protocol, LDAP is similar, and perhaps borrowed, from CMIP.

7.4.6   Interworking

By now reader can see that there are many options or paradigm (CMIP, SNMP, CORBA, XML etc) available to implement TMN entities (NELs, OSFs). Such availability results in proliferation of different paradigms and management domains. A management domain is defined as consisting of one or more TMN entities that interact among themselves using one and only one paradigm. Figure 7-33 shows a multi-domain environment. In general, one can consider the following as possible interworking scenarios:

- GDMO ←→ CORBA
- GDMO ← → Internet (SNMP)
- CORBA ← → XML or WS.
- Internet (SNMP) ←→ XML or WS
- GDMO← → XML or WS

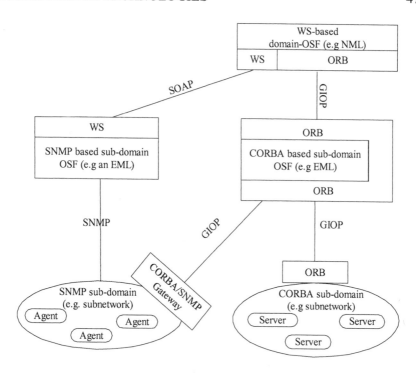

Figure 7-33. Interoperability of CORBA TMN entity with (SNMP | CMIP) based entities or domains

For domain-to-domain interaction two different approaches were explored as part of an effort called Joint Inter-Domain Management (JIDM) [OPG00]. Internet Management Coexistence (IIMC) is another initiative that conceived gateway functional element for interworking OSI and SNMP domains. In general, translators/gateways can be built as Adapter service on top of OSS or on one or all of the NEs. Alternatively, one can build stand-alone gateway entities as illustrated in Figure 7-34. A more recent discuss can be found in

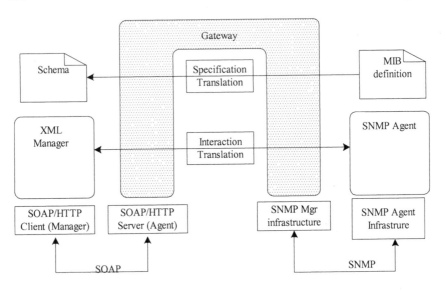

Figure 7-34. XML/SNMP   Gateway

No matter where the translation function is placed, there are two parts to such function - information translation and protocol or interaction translation. The information translation can be static or dynamic as discussed below:

Static Translation (ST) of Information: With JIDM's work on static translation, there are two parts to translation. The first part deals with translation of data types. JIDM specification specifies translation algorithms to map ASN.1 types to IDL types. In translating from ASN.1 to IDL, information such as default values, allowed values, tagged types, constrained types, and subtypes are lost. In ASN.1, such information is specified using comments and can't be translated into any implementation code. Even if it can, CORBA doesn't deal with how the data type is used but only with how it is coded and transferred.

The second part of JIDM static translation deals with   translation of GDMO managed object classes (MOCs) into   CORBA interfaces. The translation is simple but has the disadvantage that when there is change in the information model the applications (managers, agents) need to be rebuilt (i.e., code-churn) requiring re-installation (commonly called upgrade). Such a process means   Opex for network providers and R&D expense for the vendors. Dynamic translation, to be discussed shortly, addresses this issue.

For GDMO← → Internet (SNMP) static translation, two specifications are available - [NMF30] and [NMF26]. The translation is static and operates at syntactic level only. It appears that practical deployments involved existing SNMP domain managed by an OSI manager using the gateway/translation functional element [Hee98]

For CORBA← →WS, two OMG specifications [OMG04] exist and describe how to map WSDL to IDL conversion and vice versa.

For SNMP (ASN.1) ←→XML (DTD or Schema) static translation, [Yoo] offers algorithms and suggestions. For SNMP ←→ WS, static translation may be done using X.694.

Dynamic Translation (DT) of Information: As mentioned earlier, static translation can require restarting manager, agent or both. Therefore, JIDM specification, using advanced capabilities of CORBA, specified dynamic translation. These include dynamic invocation interface (DII) for clients (i.e. OSF and others in client role) and dynamic skeleton interface (DSI) for servers (i.e. NEF and others in server role). With this approach, servers can add new MOs, make changes to existing MOs, or in general modify the structure of management information without requiring that clients be re-built or be re-installed. This is made possible by having a gateway element that can intercept each request from clients and build a dynamic invocation request and send it the server. However, the usefulness of dynamic translation and gateway element architecture is not clear. Besides, there is performance impact, since the gateway has to translate each request and build dynamic request instead of using static stubs.

Interaction or protocol translation (IT): This involves supporting request/response as well as any filtering support for the various possibilities listed earlier. Both JIDM and IIMC include details on GDMO←→ CORBA and GDMO←→SNMP protocol conversions respectively. For SNMP ←→ WS interaction translation, currently there is no known standards work on going. Such dynamic translation involves mapping SNMP's requests/responses to SOAP requests/responses. Similarly, SNMP's trap must be mapped somehow to SOAP's notifications. For CORBA ← WS, the OMG specifications listed earlier address the interaction translation part. For the remaining combinations, author is not aware of any work on going at the time of writing.

In the context of management inter-working, it is useful to mention multi-domain control plane being standardized by G.ASON (Ch9). Such control plane can support TMN or obviate the need for management gateways. However, only limited set of functions, such as network topology, resource discovery, and service provisioning can be accomplished by control plane. Equipment provisioning and management (in general functions associated with FCAPS), typical of regular management interfaces, still need to be supported. Therefore, multi-domain environments with different management paradigms may still require gateway architecture as described above unless efforts are made in enhancing control plane standards to support equipment provisioning and management.

7.5   CONCLUSIONS

In this chapter we presented an overview, and to some extent a condensed one, of XML, ASN.1, CORBA, SNMP, HTTP, CMISE/CMIP. In the languages section, we discussed ASN.1, including Basic Encoding Rules (BER). We also explained how to specify aggregate types (such as tables) and how to identify columnar variable of single and multi-dimensional tables.   As part of XML, we discussed three things– name spaces, DTD, and Schema. Specifically, we explained how templates and objects are specified using DTD and Schema. We also highlighted object-oriented nature of schema constructs. Such constructs are useful to define abstract, final, and restricted types.   In the protocols section, using a normalized communication model, we provided a brief description of SNMP, HTTP, CORBA, and CMIP.

Throughout the chapter, the separation between functional information and technologies and protocols required for storing, processing, and accessing of such information is highlighted. The choice of communication model and hence protocols and technologies is a matter of paradigm preference and comfort for vendors and service providers. In TMN environments, associated with transport networks, OSI based manager and agent is the dominant paradigm. This paradigm treats manager and agent as closely coupled systems. Such tight coupling provides well-defined behaviour and consistency. On the other hand, such tight coupling   forces OSSs to upgrade each time a new device is added to network or an existing device is upgraded with different schema. Such process can be expensive. Migration towards 'loosely coupled' systems requires some maturity in technologies and increased use of dynamic schema, object discovery, modularization etc.   XML and dynamic-CORBA are ideal candidates for modeling such 'loosely coupled' interactions. The same argument applies to

TL1. Being text-based protocol (a TL1 message is an XML document with hidden! markup) it inherently makes the OSS and the NEs as loosely coupled entities.

REFERENCES AND FURTHER READING

Agh93   Research Directions in Concurrent Object-Oriented Programming, Gul Agha (Ed), MIT Press, 1993

ASN102
- ITU-T Recommendation X.680 (2002) | ISO/IEC 8824-1:2002, Information technology – Abstract Syntax Notation One (ASN.1): Specification of basic notation.
- ITU-T Recommendation X.681 (2002) | ISO/IEC 8824-2:2002, Information technology – Abstract Syntax Notation One (ASN.1): Information object specification.
- ITU-T Recommendation X.682 (2002) | ISO/IEC 8824-3:2002, Information technology – Abstract Syntax Notation One (ASN.1): Constraint specification. †
- ITU-T Recommendation X.683 (2002) | ISO/IEC 8824-4:2002, Information technology – Abstract Syntax Notation One (ASN.1): Parameterization of ASN.1 specifications. †
- ITU-T Recommendation X.690 (2002) | ISO/IEC 8825-1:2002, Information technology – ASN.1 encoding Rules: Specification of Basic Encoding Rules (BER), Canonical Encoding Rules (CER), and Distinguished Encoding Rules (DER).
- ITU-T Recommendation X.691 (2002) | ISO/IEC 8825-2:2002, Information technology – ASN.1 encoding rules: Specification of Packed Encoding Rules (PER).
- ITU-T Recommendation X.692 (2002) | ISO/IEC 8825-3:2002, Information technology – ASN.1 encoding rules: Encoding Control Notation (ECN) for ASN.1. †
- ITU-T Recommendation X.693 (2002) | ISO/IEC 8825-4:2002, Information technology – ASN.1 encoding rules: Specification of XML Encoding Rules (XER).
- ITU-T Recommendation X.693 (2002) / Amd.1 (2003) | ISO/IEC 8825-4:2002/Amd.1: 2003, Information technology – ASN.1 encoding rules: XML Encoding Rules (XER) – Amendment 1: XER Encoding Instructions and EXTENDED-XER.

ASN104
- ITU-T Recommendation X.694 (2004) | ISO/IEC 8825-5:2004, Information technology – ASN.1 encoding rules: Mapping W3C XML Schema Definitions into ASN.1.
- ITU-T Recommendation X.finf (2004) | ISO/IEC ???:2005, Information technology – ASN.1 encoding rules: Mapping W3C XML Schema Definitions into ASN.1.

BCR   Operations Application Messages - Language for Operations Application

	Messages, GR-831, BellCore
Cas96	Introduction to Community based SNMPv2, J. Case et al, Internet RFC 1901, IETF, Jan 1996
Cha01	Efficient Transfer of SNMP Bulk Data, S. Chandragiri, Internet Draft, IETF, April 2001
Dub01	Dubuisson, O., ASN.1 - Communication Between Heterogeneous Systems Morgan Kaufmann, 2001
Dub04	Dubuisson, O., ASN.1: A Powerful Schema Notation for XML and Fast Web Services, ITU-T SG17 Contribution, 2004
Fen01	Dynamic Evolution of Network Management Software by Software Hot-Swapping, Feng, N., et al, IEEE/IFIP, 2001
Fie99	"Hypertext Transfer Protocol, HTTP/1.1," R. Fielding, Internet RFC 2616, IETF, June 1999
Fran99	J. Franks et al, "HTTP Authentication: Basic and Digest Access Authentication," Internet RFC 2617, IETF, June 1999
Gol99	Y. Goland et al, "Hypertext Extensions for Distributed Authoring-WEBDAV,", Internet RFC 2518, IETF, Feb 1999
Har97	T.H. Harrison et al, "The design and performance of a real time CORBA event services," OOPSLA, 97
Heg98	H., Hegering et al, Integrated Management of Networked Systems Concepts, Architectures, and Their Operational Application, Morgan-Kaufmann, 1998
Hop02	Understanding the WevDav protocol, Hopman, A., Microsoft, 1999
IIOP	"IIOP: Internet Inter-ORB Protocol," http://panuganty.tripod.com/articles/iiop.htm
Joh03	Larmouth, *J., ASN.1 is getting sexy again! or ASN.1, XML, and Fast Web Services*, ETSI MTS#37, Sophia Antipolis, France, Oct. 14-16, 2003
Lev03	Leventhal, M., Lemoine, E. ,and Williams,S., Binary Showdown, XML-Journal, Dec. 1, 2003
Lia	"Evaluation of SCTP as a Transport Mechanism for CORBA GIOP Messages," Y.W Liang, et al
Mcc00d	The Interfaces Group MIB, K. McCloghrie et al, Internet RFC 2863, IETF, June 2000
Mcc91d	Management Information Base for Network Management, of TCP/IP-based internets: MIB-II, K. McCloghrie, Internet RFC 1213, IETF March 1991
Mcc99a	Structure of Management Information for SNMPv2, K. McCloghrie et al, Internet RFC 2578, IETF, April 1999
Mcc99b	Textual Conventions for SNMPv2, K. McCloghrie et al, Internet RFC 2579, IETF April 1999
Mcc99c	Conformance Statements for SNMPv2, K. McCloghrie et al RFC 2580, IETF April 1999
McL	Java and XML, Bret McLaughlin, Oreilly
Met03	XML Data Binding Simplifying XML Processing, Eldon Metz, Allen Brookes, Dr. Dobbs Journal, March 2003
Mun03	Mundy D.P., Chadwick D. D, Smith A., "Comparing the Performance of Abstract Syntax Notation One (ASN.1) vs eXtensible Markup Language (XML), Terena Networking Conference, Zagreb, May 2003
OMG01	CORBA Event Service v1.0, OMG, March 2001

OMG02	CORBA Notification Service v1.0.1, OMG, Aug 2002
OMG03	CORBA to WSDL/SOAP Interworking, v1.0, OMG, 2003
OMG04	WSDL-SOAP to CORBA Interworking, v1.0, OMG, 2004
OPG00	Open Group, Inter-Domain Management: Specification and Interaction Translation, C.802, Jan 2000.
OSS03	*ASN.1, a schema language for XML*, White paper by OSS Nokalva, Oct. 2003
Pre02	Transport Mappings for the Simple Network Management Protocol (SNMP), R. Presuhn (Ed), Internet 3417, Dec 2002
Pre02	Management Information Base for SNMPv2, R. Presuhn et al, Internet RFC 3416, IETF, Dec 2002
Pri01a	Open Fusion J2EE & CORBA Services Notification Service White Paper, Prism Tech, May 2001
Pri01b	Open Fusion J2EE & CORBA Services Notification Service Performance Evaluation White Paper, Prism Tech, August 2001
Q3M00	Q3ML – XML-Based Management Interface for OSS, Lever Wang, Contribution T1M1.5/2000-204, 2000
Ree01	Proposed Standard Recommendation using GDMO/ASN.1 mappings to XML schemas and Applied to Configuration Audit Support Function (X.792), Raymond E. Reeves, T1M1.5/2001-090 R2
RFC1904	Version 2 of the Protocol Operations for the Simple Network Management Protocol (SNMP), RFC 1904
RFC2496	Definitions of Managed Objects for the DS3/E3 for the DS3/E3 Interface Type, D. Fowler (Ed), Internet RFC 2496, IETF Jan 1999
Ric97	Presenting XML, Richard Light, Sams.Net, 1997
San03	Sandoz, P., et al, Fast Web Services, Sun Technical Paper, Aug 2003
Sla99	Enterprise CORBA, Dirk Slama et al, Prentice Hall PTR, 1999
Ste02	"WebDAV and Apache," Greg Stein, Apache Conf 2002
Sub00	Network Management Principle and Practice, Mani Subramian, Addison Wesley, 2000
Tes99	Definitions of Managed Objects for the SONET/SDH Interface Type, K. Tesink, Internet RFC 2558, IETF March 1999
TML02	ITU-T Recommendation M.3030 (2002), Telecommunications Markup Language (tML)
Tom99	Tomono, M., "An Event Notification Framework based on Java and CORBA," Proc. IEEE IM, pp. 563 - 576, May 1999
Udu99	Udupa, D. K., TMN Telecommunications Management Network, McGraw-Hill, 1999
W3CSOAP	– *SOAP Version 1.2 Part 1: Messaging Framework, W3C Recommendation, http://www.w3.org/TR/2003/REC-soap12-part1.*
	– *SOAP Version 1.2 Part 2: Adjuncts, W3C Recommendation, http://www.w3.org/TR/2003/REC-soap12-part2.*
W3CXML	– *Extensible Markup Language (XML) 1.1, W3C Recommendation, Feb 2004, http://www.w3.org/TR/2004/REC-xml11-20040204.*
	– *Namespaces in XML 1.1, W3C Recommendation, Feb, 2004, http://www.w3.org/TR/2004/REC-xml-names11-20040204/*
	– W3C XML Information Set: 2001, XML *Information Set, W3C Recommendation,http://www.w3.org/TR/2001/REC-xml-infoset-20011024/.*

W3C XML Schema: 2001, *XML Schema Part 1: Structures, W3C Recommendation,http://www.w3.org/TR/2001/REC-xmlschema-1-20010502/.*

W3C XML Schema:2001, *XML Schema Part 2: Datatypes, W3C Recommendation,http://www.w3.org/TR/2001/REC-xmlschema-2-20010502/.*

Whi01   White Ed, Requirements for an XML based Q interface for ease of integration and fast introduction of new telecommunication services, T1M1/2001-113, 2001

WSS04   Web Services Security v1.0 (WS-Security 2004), OASIS, 2004

X.208   Syntax Notation, CCITT X.208/ISO 8824

X.290   Basic Encoding Rule CCITT X.209/ISO 8825

Yok92   Yokote, Y., The Apertos Reflective Operating System: The Concept and its Implementation. OOPSLA 1992: 414-434

Yoo02   Yoon-Jung Oh, Hong-Taek Ju, Mi-Jung Choi, James Won-Ki Hong: Interaction Translation Methods for XML/SNMP Gateway. DSOM 2002: 54-65

[Tor04]   Integrating SNMP Agents with XML-Based Management Systems, Torsten Klie and Frank Strauβ, IEEE Comm Magazine, July 2004, 76-83

Chapter 8

TRANSPORT NETWORK SURVIVABILITY
Present and Future Options

8.1. INTRODUCTION

The ability of a system to continue to provide services in the presence of failures internal to itself is traditionally the art of high availability system design. For many decades the emphasis in telecommunications was on designing high availability *equipment elements* and routing services over such elements in a non-redundant way. But in the last decade, driven by fiber-optics as the preferred—but vulnerable—physical medium we have seen the advent of *survivable networks* as a specialized new area of network technology and network planning. A survivable network has abilities to continue providing services in the face of either internally arising, or externally inflicted failures. Survivability itself is only a qualitative term referring to the overall ability to carry on providing service in the face of such failures. Survivability can thus be derived through many measures ranging from armoring of cables at the physical layer all the way up to multi-carrier traffic splitting at the business level. A *self-healing network* refers to the specific vision of a transport network that has a basic topology, a planned distribution of spare capacity, and an autonomous mechanism embedded in the network that makes it immune to any single failure via sub-second[117] automatic reconfiguration to ensure service continuity.

---

[117] Of the schemes to be discussed, ranging from automatic protection switching, to rings, to mesh preplanned protection, and adaptive dynamic mesh restoration, restoration times are

Self-healing properties are essential for all future transport networks. The user networks that we perceive individually, such as the Internet, the phone network, banking networks, travel networks, and so on, all ride on a common transport network infrastructure. This is especially true and ever increasing with the trend toward "converged" networks where all applications and services are converted into IP packet flows and routed over DWDM optical transport. The ability of transport networks to recover from failures is thus crucial to commerce and society[118]. The vision of transport networks as self-healing systems—established in the communications field since about 1987—is so compelling that it has also recently been picked up as a catch-phrase to describe major new goals set for computing systems as well [Frye01]. Self-healing computing systems share the same motivation and importance as the drive towards self-healing transport networks.

In general, all layers of a network need certain self-healing capabilities to address faults arising at their own layer. When this is done successfully, higher layers will never be aware of the failures that actually occurred at the lower layers. Obviously therefore, the transport layer, which is just above the passive physical layer is fundamentally important to invest with self-healing capabilities. The most common type of physical-layer failure is cable damage arising from natural or man-made causes; trench digging, construction work, craftsperson errors, ship anchors, sabotage, tress falls, earthquakes, rodents, fires, floods, etc. The sheer mileage of fiber-optic cable now deployed in ducts, direct-buried underground, or on overhead pole-lines, is so large that cable-cuts dominate all other sources of externally-imposed network failures. Cable-cutting events occur virtually every few days in extensive networks with 50,000 or more route-miles of fiber. The surprisingly high rate of cable cuts, despite many measures for the physical protection of cables, is evidenced in the industry by the black humor of referring to backhoe equipment as "Universal Cable Locators."

It is surprising how frequent failures are. Snow reported [Snow00] that since 1992 there have been about 16 outages *per month* in the United States alone that each affected over 30,000 users. And many interesting (even bizarre) reports of cable cuts and their impact can be found daily on the Internet. A sampling on just one day in 2004 yields stories of ship anchors, train derailments, and the more typical cable dig-up events as well [QSIN04] [Tham04] [Wear04] [Bowm04] [McAu04]. Physical failure of node

---

characteristically around 50-60 ms, 150-200 ms, 200-300 ms, respectively, up to perhaps 1.5 seconds at most for the lowest priority path in the last scheme mentioned.

[118] Carriers are required to file network outage reports with regulating authorities when failures affecting over 30,000 users occur. The Network Reliability Council's website (www.nric.org) provides interesting reports on network reliability.

infrastructures (by fire or power loss) is far less frequent, although software-related crashes of routers within nodes are a growing concern. These are, however, outside the current scope.[119] The good news is that cable cuts can be *efficiently* and completely protected against in next-generation transport networks. By efficiently, we mean through methods that can ensure 100% restorability against any single span failure with *considerably less than 100% redundancy* in terms of equipment and capacity resources that have to be set aside for survivability. The same methods can also be applied or adapted to provide 100% survivability to the paths that transit a failed node.

8.1.1 Outline and Scope

The plan of the chapter is to explain the key survivability concepts and techniques used in currently deployed transport networks and then to introduce some of the main alternatives and key ideas for transport networks of the future. Our initial focus is thus on existing methods such as APS, rings and centrally controlled mesh restoration, which are widely deployed in existing networks. Following that we look at some promising "next generation" technologies not yet deployed in practice. Examples of some of the leading options for next-generation transport networks include distributed GMPLS mesh restoration, self-organizing mesh restoration, self-planning mesh protection, *p*-cycles, and shared-backup path protection. We conclude with a selection of views on ways that next-generation network (NGN) transport will differ from current practice in requirements, technology, and concepts, including new directions in transport network planning, evolution, and business strategies. This "past, then future" orientation results in the roadmap for the chapter shown in Figure 8-1. Along the way, wherever possible, we try to address common misconceptions that are relevant to the topic.

The emphasis is on qualitative explanations of the basic mechanisms, concepts, issues, and principles for achieving survivable transport. Where

---

[119] Other than in regards to software reliability issues, which are under study as a separate field of research, failure of the nodes of the transport network itself is thousands of times less frequent than failures of the transmission between nodes. Node recovery is also a categorically different problem than recovery from transmission failures. The latter *can be* 100% restored by network-level re-routing but the former cannot because there is no network-level action that can restore paths sourcing or sinking at a failed node. Fire extinguishers, backup generators, physical site security, and so on, are much more crucial to achieve physical node survivability. And equipment redundancy, spares inventory, and high-availability system design are the primary measures to address nodal element survivability. Our present scope is accordingly limited to survivability that can be derived through network-level re-routing principles only.

the reader might like to delve more deeply into the related mathematics and
network design or optimization problems, we try to give key references.

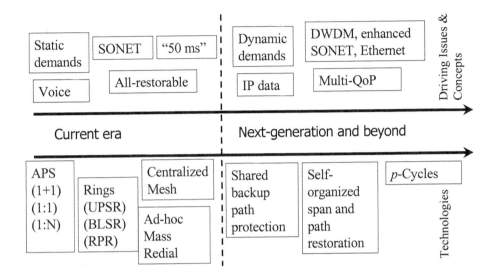

Figure 8-1. Environments and methods for current and next-generation transport survivability

At the top of Figure 8-1 we identify some of the main paradigms and
issues that characterize the transport solutions currently used. Until recently
voice was the dominant and most important payload—and the money-maker
for network operators. The demand requirements were not difficult to
forecast accurately as input to network planning, and business growth was
unrelenting—incumbent carriers didn't necessarily need to focus on
efficiency in transport to be profitable and competitive. However, by the
mid 1990s, survivability had reached crisis proportions and any solution that
seemed simple and easy to deploy was urgently needed. SONET rings filled
the bill despite later experience that showed how inefficient and inflexible
multi-ring network solutions really are. SONET was the main payload
multiplexing and transport signal technology and, although sometimes
contested, network operators believed that they had to have equal
survivability measures in place for all signal flows and that 50 ms was the
requirement for restoration times.   In this era, the main methods for
survivability were varieties of APS and ring technologies (to be defined),
although centrally controlled mesh schemes also saw use by some advanced
carriers.

Amongst other differences in the coming era for transport networking, is
the prospect of less easily forecast and more "on-demand" types of demand

patterns. IP data (including Voice over IP) becomes the dominant payload type and extensive wavelength division multiplexing is used in conjunction with SONET formats that are enhanced with better flexibility for IP data payloads. Ethernet-framed LAN signals also find their way into use as WAN transport formats at 1 Gb/s and 10 Gb/s. In addition, more pragmatic views about restoration time requirements will prevail and transport networks will begin offering a range of Quality of Protection (QoP) options, not a single "50 ms" restoration class for all. In this environment the transport network can offer ultra-high availability (withstanding *dual* failures) to some services, while other services require only single-failure protection in or even best- efforts restoration only. All of the new concepts for survivability listed in Figure 8-1 emphasize inherent capacity efficiency and flexibility both in terms of easy adaptation or reconfiguration to shifting demand patterns and in offering different survivability options to different services within a single transport network. With this setting of the stage, let us now start in with APS and ring-based transport.

8.2. AUTOMATIC PROTECTION SWITCHING (APS)

The simplest class of mechanism for survivability is automatic protection switching (APS). These schemes involve reserving a protection channel (dedicated or shared) with the same capacity as the channel or facility to be protected. Different APS techniques are characterized by the following criteria:

- The topology, either linear or ring,
- Whether the protection channel carries a backup copy of the traffic permanently or only when requested for protection,
- Whether the protection channel is shared among working channels that may potentially need protection,
- Whether both directions of transmission switch (bi-directional switching) to protection channels when a failure occurs in one direction or only the affected direction switches (unidirectional switching), and
- Whether the network automatically reverts traffic back to the working channels after they have been repaired/restored (revertive switching) or continues to use the protection channel after the repair/restoration (non-revertive switching).

In so-called "linear protection" the entity to be protected follows a point-to-point route for that layer of the network where the end-nodes of the protected path segment are different. "Linear" simply distinguishes this type of APS scheme from ring schemes where the protected path closes on itself at the end-nodes. SONET/SDH APS is implemented at the Line/Multiplex Section layer. Ring protection makes use of having an alternative path around the ring in the opposite direction as the one affected by the failure. The standards for ring and APS techniques covered in this chapter are still evolving. SONET linear and ring APS is covered in ANSI T1.105.01 [SONET00] and ITU-T Recommendation G.841 covers linear and ring protection for SDH networks [ITU96] [ITU98]. The ITU-T has recently begun a project to define protection in a general set of recommendations that are independent of the underlying transport technology. The first of these is G.808.1 [ITU03] for linear networks. G.808.2 will similarly cover ring networks. The protection recommendations that will be developed for G.709 OTN networks (and later revisions of G.841) will then be able to reference these generic protection recommendations for common concepts rather than repeating them in each technology-specific recommendation.

8.2.1 1+1 and 1:N SONET APS

In any APS scheme the network element (NE) that detects the fault condition also initiates the protection switching action and is referred to as the tail-end node. The other node is referred to as the head-end node. Note that these definitions apply separately to each direction of transmission. The head-end node's main task is to electrically split the affected working signal (i.e., make an electrical duplicate copy of the signal) and feed this bridged signal into the standby protection channel (while continuing to feed the working channel as well). This is referred to as the head-end bridge function. In 1+1 APS, the bridge is always present. In 1:1 or 1:N the tail-end signals upstream to request the bridge upon failure. In SONET/SDH, this signaling information is communicated in the K1 and K2 overhead bytes and these two bytes constitute the SONET/SDH APS signaling channel. G.709 OTN has also reserved overhead bytes for implementing an APS signaling channel. The typical criteria for initiating APS are: [120]

- Detection of a failure (e.g., loss of signal (LOS), loss of framing (LOF)),
- Signal Fail (excessive bit error rate (BER)),
- Degraded Signal (relatively high BER), or
- Externally initiated commands from the craft or OSS (e.g., manual switches or forced switches).

These criteria form a hierarchy of priorities when multiple channels or conditions compete for access to the protection resources. Forced switches

---

[120] A protection lockout command also allows the craftsperson to prevent traffic from being switched onto the protection channels. Protection lockout typically has the highest priority.

are the highest priority, and would typically be performed when the craft needs to lockout a facility during maintenance or upgrades when other fault or failure conditions could be affecting other working channels. Failures are the next highest priority. In practice, high BER is treated as a failure. Manual switches are those initiated for maintenance purposes when the craftsperson wants the network to still automatically protect itself in the event of a real problem. Manual switches have the lowest priority.

In addition to being used by the tail-end to request signal bridging actions from the head-end NE, the APS signaling channel is also used to communicate status information between the head- and tail-ends. In SONET/SDH, the K1 byte primarily carries the tail-end requests and/or status, and the K2 byte communicates the head-end status. It should be noted that the APS communication is conducted over the protection channel rather than over the working channel. Because the system cannot rely on the working channel for communication when it has failed, the receiver ignores the APS signaling on the working channel. SONET/SDH standards also include a protection switch Exerciser mode for purposes of testing the protection control functionality. The Exerciser function causes a protection switch action to be taken except for the final action of bridging or switching the actual traffic. This is performed routinely as a background maintenance activity to avoid the implications of a "silent failure" in the redundant standby equipment and protection switching logic. All steps of protection switching are operated except the tail-end transfer function, because that is the one step that would inevitably put a 'hit' on the operating signal.

1+1 protection switching is by far the simplest type of APS. In 1+1 there is a permanent head-end bridge of the working channel onto the protection channel. Typically, in SONET/SDH the channel here is an entire OC-n rate line signal. As illustrated in Figure 8-2(b), when the receiving NE detects that the traffic arriving from the protection channel is healthier than the traffic from the working channel, it switches to taking its traffic from the protection channel. 1+1 protection can be either unidirectional or bi-directional. In uni-directional 1+1 protection switching there is no need for the tail-end to communicate with the head-end (i.e., the tail-end simply chooses the best signal without informing the head-end). On the other hand, in bi-directional 1+1 protection switching, the tail-end needs to inform the head-end so that the head-end can initiate protection switching steps for the other direction of transmission that result in the head-end also taking its traffic from the protection channel. In *non-revertive* operation, the distinction between the working and protection channels disappears, except for the default starting state of the network. In 1:N APS, a single protection or standby line-rate transmission system is shared by N working systems as illustrated in Figure 8-2(c).

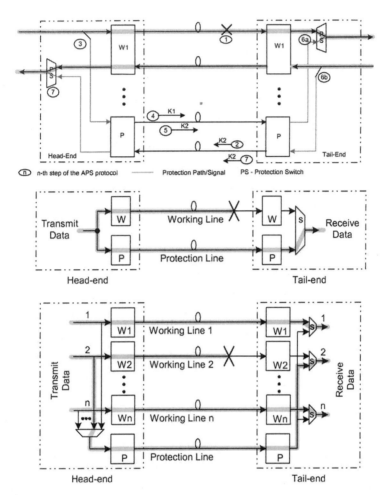

Figure 8-2. Linear APS: (a) Basic protocol steps, (b) 1+1, (c) 1:n example

To support the shared access to a standby system, no head-end bridge is established until failure occurs. Signaling is then needed to specify the requested channel and request the head-end bridge. To this end, SONET/SDH standards define the K1 and K2 byte APS channel definitions in Table 8-1. 1:N APS greatly enhances system availability against failures of single fibers, transmitters or receivers—anything that affects only one working channel at a time. Following a failure of a working system, the tail-end informs the head-end that it needs its working-line signal to be bridged into the protection line. To do this the tail end originates K1 with a reference to the working line number and the fault type (SD, SF etc). The head-end NE then checks the current status of the protection system and the priority of the conditions (K1 bits 1-4) for each working system requesting protection.

Table 8-1. SONET/SDH Byte K1 and K2 definitions for linear APS (from ANSI T1.105.01)

Byte K1			Byte K2		
	1111	Lockout of protection	Bits 1-4		These bits shall indicate the number of the channel that is bridged onto protection unless channel 0 is received on bits 5–8 of byte K1, when they shall be set to 0000.
	1110	Forced Switch			
	1101	Signal fail – high priority (not used in 1 + 1)			
	1100	Signal fail – low priority			
	1011	Signal degrade – high priority (not used in 1 + 1)			
Bits 1-4	1010	Signal degrade – low priority	Bit 5	1	Provisioned for 1:n mode
	1001	(not used)		0	Provisioned for 1 + 1 mode
	1000	Manual switch			
	0111	(not used)		111	AIS-L
	0110	Wait-to-restore (revertive only)			
	0101	(not used)		110	RDI-L
	0100	Exerciser		101	Provisioned for bi-directional switching
	0011	(not used)	Bits 6-8		
	0010	Reverse request (bi-directional only)		100	Provisioned for unidirectional switching
	0001	Do not revert (nonrevertive only)			
	0000	No request		011	Reserved for future use for other protection switching operations, e.g., nested switching.
Bits 5-8	Bits 5-8 are only used in 1:n protection where they communicate the number of the working channel for which request is issued. 0 Null Channel (indicates protection channel) 1-14 Working Channels/Lines 15 Extra Traffic Channel			010	
				001	
				000	

NOTES –
1. The Lockout of Protection switch priority uses bits 5–8 = 0000).
2. For Signal Fail and Signal Degrade only, bit 4 indicates the priority assigned to the working channel requesting switch action.
3. The Exerciser function may not exist for certain protection switching systems.
4. Reverse Request assumes the priority of the request to which it is responding.

As the K1 byte encoding in Table 8-1 illustrates, the maximum number of working channels is 14 plus one protection channel. The K1 byte encodes channel number zero to indicate protection line and fifteen to indicate extra traffic. Because 1:N protection has a protection line that is normally idle, it can be used to carry Extra Traffic.

Obviously 1:N protection is much more capacity-efficient than 1+1. The limitation is, however, that when applied to several fibers or wavelengths to form a 1:N APS system on the same cable, there is no survivability against cable cuts. There is thus an excellent internal system availability enhancement but no means of network restoration unless all working channels and the protection system follow disjoint physical routes. 1:N protection has therefore, not seen much use in practice. Its real significance, however, has turned out to be the simple extension of 1:N SONET linear APS into the SONET BLSR ring configuration where the signaling for channel number is adapted to apply to node number instead, in a closed ring configuration which then withstands cable cuts as well as single-channel failures.

8.2.2   Subnetwork Connection Protection (SNCP)

Subnetwork connection protection (SNCP) can be thought of as 1+1 APS applied end-to-end over an entire network or subnetwork, and implemented at the tributary signal level, as opposed to the whole OC-n line-rate signal level. For example, it could be used over the entire customer path (trail) from ingress to egress from the SONE/SDH network. This type of user-level 1+1 APS is a commonly employed technique in Europe, where the line-rate ring protection schemes that SONET/SDH can also support and UPSR are rarely used. The concept is identical to 1+1 APS but it would be implemented at, say, an STM-1 (STS-3c) tributary signal level, end-to-end over an entire network/subnetwork. This is in contrast to 1+1 APS applied at the entire OC-192 line rate of a specific transmission system facility, and routing customer tributary signals over the protected facility. At the egress of the subnetwork on which an SNCP arrangement is established, a selector evaluates the two signals that it receives over the two paths and chooses the best signal.

To avoid unnecessary or spurious protection switching in the presence of bit errors on both paths, a switch will typically only occur when the quality of the alternate path exceeds that of the current working path by some threshold (e.g., an order of magnitude better BER). SNCP allows protection of multiple subnetworks along an end-to-end path.   The UPSR ring protection mechanism (to follow) can be considered as a special case of SNCP. North American SONET networks typically use UPSR for access and BLSR for transport rather than SNCP.[121]

---

[121]   SDH has an overhead channel reserved for "trail" protection. SNCP can take place at any of the subnetwork boundaries, while trail protection takes place only at the trail termination points. Otherwise the two are conceptually the same.

8.2.3 Survivable Rings

Ring-based protection schemes have enjoyed enormous popularity, especially in North America for several reasons: Technically rings are enhancements of APS technology, and were thus relatively easily developed and standardized based on extensions to the SONET APS signaling protocol. Rings are thus the closest technology to prior generations of transmission systems that used 1+1 and 1:N APS and were thus relatively easy to develop and quick to gain acceptance by the transmission engineering community in the telcos. Secondly, rings address both the need for single-channel protection switching and by virtue of the ring topology, also protect against cable cuts. Rings also provide a system design that collects demands together to exploit the economy-of-scale in transmission technology (for instance an OC-192 ring may cost only two to three times what an OC-48 system at ¼ the capacity costs)[122]. In addition rings became available just when the need in industry reached crisis proportions after several spectacular cable cuts in the 1990s. At the time, rings were also perceived as being much simpler to understand and operate than distributed mesh-restorable schemes. Although more efficient mesh-based survivability architectures were then under study for SONET, rings filled the void as the first reasonable, standardized, and available solution for transport survivability and hence were rapidly and extensively deployed. A conservative estimate is that from 1990 to 2004, more than 100,000 SONET ring ADM terminals (both BLSR and UPSR and over the full range of OC-n rates) were deployed in North America alone. Rings are therefore fundamentally important in transport survivability and will be present as a legacy technology for a long time. One drawback with rings is that despite the impression of simplicity when considering a single ring, practical networks employing *multiple* interconnected rings have turned out to be extremely complex to design, operate, and grow, and to be quite inefficient in overall capacity usage. They are also relatively inflexible to changes in demand pattern. These are some of the reasons for growing interest in the mesh-based alternatives that follow.

There are two main categories of survivable ring. Path switched rings perform their protection at the path level, and line switched rings perform their protection at the line level (equivalently the "multiplexed section" level in ITU/SDH terminology). In principle a ring can support either unidirectional or bidirectional routing as illustrated in Figure 8-3, and either

---

[122] A reviewer of this chapter advised that in practice carriers aim for four times capacity at 2.5 times the cost, for a 40% reduction in cost per unit capacity for the larger OC-n system.

type of protection action (path or line). In practice, however, only two of the four possibilities have been developed as SONET rings. The SONET Unidirectional Path Switched Ring (UPSR) uses unidirectional routing and path level protection. The SONET Bidirectional Line Switched Rings (BLSR) and Multiplexed Section Shared Protection Ring (MS-SPRings) in SDH use bi-directional routing and line-level loopback protection. UPSRs are inherently "two-fiber" structures but BLSRs are defined in both "two-fiber" and "four-fiber" variants, to be explained shortly.

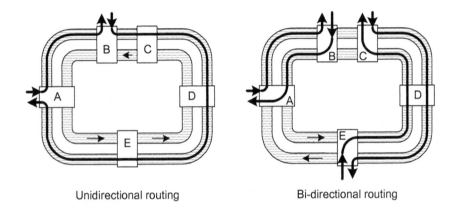

Unidirectional routing   Bi-directional routing

Figure 8-3. Ring routing examples

Consider the example of traffic exchanged between two ring nodes A and B in Figure 8-3 to see how ring routing affects the protection switching. In unidirectional rings, the working path from A to B and the path from B to A are routed in the same direction around the ring (clockwise in this example). The A to B and B to A connections typically occupy the same time slot on all spans of the ring. For bidirectional rings, the same channels can potentially be used for different inter-node communications in different parts of the ring. In the bidirectional ring example, nodes A and B communicate with each other directly over counter-propagating channels via the shortest route between the on the ring. This allows nodes C and E to communicate over the same time-slots on another portion of the ring. Such reuse of wavelengths or timeslots on different portions of the ring is known as spatial reuse and can provide a great increase in ring capacity efficiency.

A disadvantage of the unidirectional routing is that it does not allow spatial reuse of time slots on the ring. Spatial reuse is only useful, however, if the demand is distributed in a general way between pairs of nodes on the ring. Commonly in access networks all of the ring traffic originates from, and is destined to a single node, which is the hub node accessing the wider outside world and core network. In this case there is no efficiency benefit to bidirectional routing because the way flows all add up in this case, the cross-sections of the spans next to the hub node wind up requiring the same line capacity in a BLSR as in a UPSR for the same application. This is why access ring networks, which are typically arranged as a collection of remote nodes that are all logically connected to the same central office node, are virtually always based on UPSR rings.

For long haul networks, the individual facilities are typically long and expensive, which makes spatial reuse economically attractive. Hence, most backbone long haul networks use BLSR. For metro transport networks, however, the facility costs are relatively lower and BLSR equipment has typically been somewhat more expensive than UPSR equipment. As a result, BLSRs are deployed most often in long-haul networks and core metro networks, while UPSRs are the most common in access/aggregation networks.

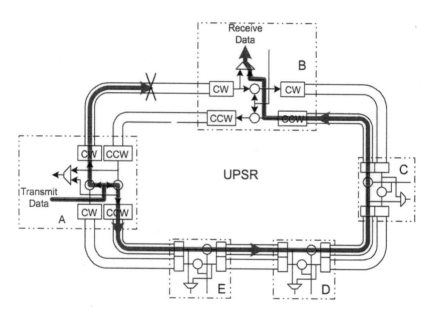

Figure 8-4. UPSR illustration

8.2.3.1 SONET Unidirectional Path Switched Rings (UPSR)

A UPSR is illustrated in Figure 8-4. Each UPSR node bridges traffic that enters (is "added") at its site onto channels in both ring directions. The receiving node then compares the quality of signal received from the two paths around the ring and chooses the best path based on measures such as BER, framing loss, or signal level. UPSRs are typically non-revertive, so other than an initial provisioning default, the two paths are not typically designated as working and protection paths. Notice that this is effectively a collection of tributary-level 1+1 APS arrangements, or equivalently SNCP setups, within the ring subnetwork. As with SNCP, any path being dropped from the ring (i.e., at its egress point) is monitored. If a BER threshold, synch loss, or signal loss alarms arise, a tail-end transfer switch to the alternate path will be performed.

The main virtue of the UPSR is its simplicity. Protection switching decisions are made locally on a per-path basis, by each receiving node independently, so there is no need for APS signaling between the switching node and the corresponding head-end node. UPSR technology is simple and low-cost, especially for the access network application. Imagine the cost of separate (even non-redundant) point-to-point access links at, say, the OC-3 rate, compared to the same set of accessing nodes sharing a single ring-structured OC-48. This is the economy of scale effect mentioned above. First cost, space, power, inventory and maintenance implications correlate strongly with the simple *number* of transmitter and receiver circuit packs required, and only secondarily to the actual speed of those circuit packs. So a UPSR using a single pair of OC-48 optics cards at each site can be considerably more economic than the equivalent capacity obtained with lower-speed cards. Outside the access network, however, the UPSR is notoriously inefficient in its use of capacity. Because each signal carried in a UPSR appears everywhere in each span of the ring, it follows fairly directly that the UPSR line rate must equal or exceed the sum of all the individual demands it serves. This is the penalty associated with not supporting spatial reuse of capacity as in BLSRs.

Another practical design challenge with UPSRs is that the receiving NE requires circuitry to process BER and other monitoring functions continually to examine both paths and to compare the status of these paths on every individual tributary signal that egresses the ring at its site. This can turn out to require a surprisingly large amount of CPU capability and related power consumption and heat dissipation, especially if every UPSR NE is designed to possibly terminate (drop) all tributaries from the ring. The "hub" node on a fully loaded OC-12 VT accessed UPSR would have to monitor the performance of 672 VT1.5 paths, including calculating and comparing the BER under degraded signal conditions. In contrast, in the BLSR, only aggregate states of the entire line-signal need be monitored to activate protection switching.

8.2.3.2 SONET Bi-directional Line Switched Rings (BLSR) and SDH Multiplexed Section Shared Protection Rings (MS-SPRings)

Line switched rings in general seek to provide greater capacity efficiency and to avoid the complexity of looking at each individual path for both the working and protection channels on higher rate systems (i.e. ≥622 Mb/s). By switching at the line level, all paths carried over that line are protected simultaneously. This in turn allows the receiving node to only provide monitoring and termination circuits for the paths that it receives over the working line. While unidirectional line switched rings have been proposed, carriers have typically preferred to include the bi-directional routing capability into line switched rings. Figure 8-5 illustrates a four node BLSR. In a typical configuration, half the bandwidth in each direction around the ring is reserved for working traffic and half is reserved for protection traffic with a fixed one-to-one correspondence between working and a protection channels. (An exception to this is noted below for non-preemptible unprotected traffic.) In a two-fiber BLSR, the working traffic uses half the STS paths for working traffic and half for protection. In a four-fiber BLSR there is a working fiber and a corresponding protection fiber in each ring direction, with the working fiber dedicated to working traffic and the protection fiber to protection.

There are relative advantages of four-fiber (4F) and two-fiber (2F) BLSRs. The obvious difference is in capacity and numbers of fibers, receivers, and so on. A 4F OC-n BLSR yields a full OC-n of bidirectional working capacity. A 2F OC-n BLSR yields only half the channels of the OC-n rate as working channels. The other half are set aside for matching protection. On the other hand a 2F BLSR uses half the fibers and transceiver circuit packs, etc. Less obviously, the 4F BLSR has theoretically higher service availability because it can support normal 1:1 APS switching within a span as well as loopback-type BLSR ring action (to follow). For more analysis of the relative merits of 2F and 4F BLSRs Wu's book is recommended [Wu92].

Following a node or span failure, the nodes adjacent to the fault will detect the condition[123] and initiate the ring APS protocol. Each node detecting a fault sends an APS request to the node to which it was connected in the direction of the fault (i.e., the node that is sending it the failed signal). The case of a bi-directional fiber cut between nodes C and D is illustrated in Figure 8-5. Here, both nodes C and D detect the failed signal. Node C sends D a bridge request with an indication of the nature of the failure (e.g., signal

---

[123] Fault detection is based on line level measures such as LOS, etc., explained in Chapter 3.

fail), and node D likewise sends C a bridge request. The byte K1 and K2 definitions for this APS communication are shown in Table 8-2.

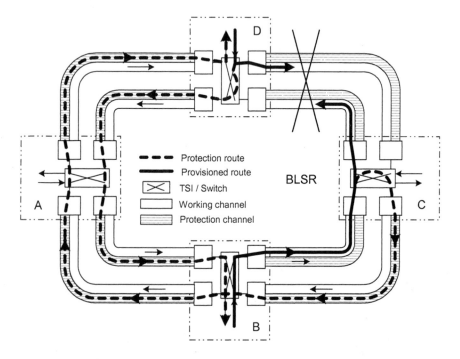

Figure 8-5. Two-fiber BLSR illustration (with cable cut fault)

When nodes A and B see these APS requests, they allow the APS channel to pass through transparently. When C and D see each other's requests, they loop back the signals that they were transmitting on the C-D span so that they go around the ring in the opposite direction over the protection channels. This loopback is known as a ring bridge (i.e., a bridge of the affected working channels onto to the ring protection channels— analogous to the head-end bridge in APS). At the same time they do a ring switch (i.e., the APS tail end transfer function) to substitute the output of the protection channels for the failed working channels. They then update their outgoing K2 bytes to indicate that they are in the bridged and switched state. If only one direction is affected, then only one node will need to perform the ring switch.

In Figure 8-5 the working and protection channel pair can either be STS-1/STS-*N*c (VC-4/VC-4*N*c) channels in the same fiber on a two-fiber ring, or they can be separate fibers of a four-fiber ring. In the case of a four-fiber ring, individual span failures between nodes can be protected through a 1:1 type of span switch. Span switch requests are signaled over the affected link between the nodes, which is referred to as the short path. Since it would be possible to support multiple simultaneous span switches around the ring, span switches are given higher priority than ring switches. The existence of a span switch is signaled around the ring (i.e., on the long path) by the switching nodes so that a node needing to request a ring bridge will defer to the span switch and not signal its ring bridge request. The interaction of span and ring switching on four-fiber rings, and the desire for consistent, predictable behavior, added a large degree of complexity to the BLSR protocol. For long haul carriers, the added protection capabilities of the four-fiber rings have made them worthwhile. For networks covering less geographical area, the relative simplicity of the two-fiber rings has sometimes made them more attractive. In some circumstances, it has even been advantageous to deploy two separate, overlaid two-fiber rings instead of a single four-fiber ring.

Table 8-2 – *Byte K1 and K2 definitions for a SONET BLSR (and SDH MS-SPRing)*

Byte K1			BYTE K2		
Bits 1-4	1111	Lockout of Protection [Span] or Signal Fail [Protection] (LP-S)	Bits 1-4	Source node ID is set to the node's own ID	
	1110	Forced Switch [Span] (FS-S)			
	1101	Forced Switch [Ring] (FS-R)			
	1100	Signal Fail [Span] (SF-S)	Bit 5	0	Short path code (S)
	1011	Signal Fail [Ring] (SF-R)		1	Long path code (L)
	1010	Signal Degrade [Protection] (SD-P)			
	1001	Signal Degrade [Span] (SD-S)			
	1000	Signal Degrade [Ring] (SD-R)		111	AIS-L
	0111	Manual Switch [Span] (MS-S)		110	RDI-L
	0110	Manual Switch [Ring] (MS-R)		101	Reserved for future use
	0101	Wait-To-Restore (WTR)	Bits 6-8	100	Reserved for future use
	0100	Exerciser [Span] (EXER-S)		011	Extra Traffic (ET) on Protection Channel
	0011	Exerciser [Ring} (EXER-R)			
	0010	Reverse Request [Span] (RR-S)		010	Bridged & Switched
	0001	Reverse Request [Ring] (RR-R)		001	Bridged (Br)
	0000	No Request (NR)		000	Idle
Bits 5-8		The Destination Node ID is set to the value of the ID of the node for which that K1 byte is destined. The Destination Node ID is always that of an adjacent node (except for default APS bytes).			

NOTE – Reverse Request assumes the priority of the bridge request to which it is responding.

A BLSR can also recover much of the ring traffic in the event of either a node failure, or even multiple span failures. Consider the case of the node D failure in Figure 8-6 where D was in intermediate node for a circuit between B and C. The failure is detected by nodes A and C, which will both respond by sending bridge requests to node D in the opposite direction around the ring. When A and C see each other's bridge requests to D, they know this means that a failure has removed D from the ring. As a result, A and C will then perform ring switches.

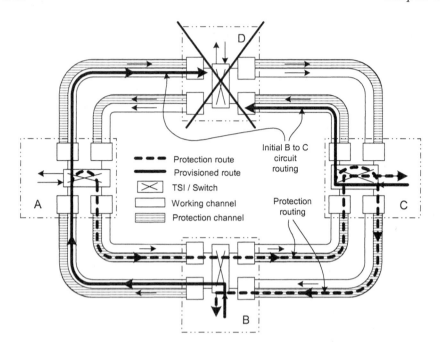

Figure 8-6. Two-fiber BLSR illustration with node failure

These ring switches due to node failures, however, have changed the connectivity of the ring in a manner that could potentially result in misconnected traffic, as illustrated in Figure 8-7. In this example, Circuits 1 and 2 use the same channel (time slot) in different portions of the ring. Due to the ring switches, the traffic from A to B is now dropped by D, and the traffic from D to B is now dropped at A. This unacceptable situation was handled in the SONET/SDH standard in the following manner. Each ring node has one database that shows the ring connectivity (i.e., the sequence of node ID numbers around the ring) and a second database that shows the connectivity of each STS-1/STS-Nc (VC-4/VC-4Nc) channel on each fiber. (This is an extra step in provisioning that must be performed whenever a path is set up through a BLSR.) When a node sending a ring bridge request receives a ring bridge request from another node, it examines the destination node ID in that request. For a span failure, this destination ID will be its own ID. For the case of a node failure (or multiple failures that have isolated a section of the ring) the destination node ID can be compared to the ring connectivity table to determine which node (or nodes) is missing from the ring. The node will then perform the ring bridge and switch, and simultaneously squelch all STS-1/STS-Nc (VC-4/VC-4Nc) channels that were added or dropped at the failed node(s). The other ring node sending a ring bridge request will perform the same functions. This squelching action, which is accomplished by sending AIS-P in these channels, prevents the misconnection.

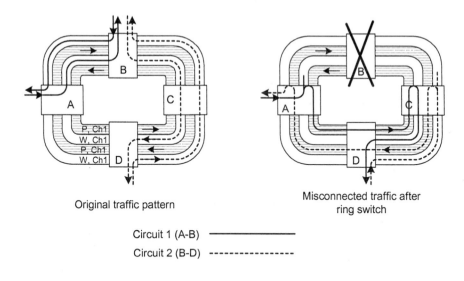

Original traffic pattern

Misconnected traffic after ring switch

Circuit 1 (A-B) ——————————

Circuit 2 (B-D) ------------------

Figure 8-7. Potential traffic misconnection problem that a BLSR must handle

Handling potential misconnections becomes more complicated if the ring adds and drops VTs (VC-1/2s). For VT access, it is not desirable to squelch all the VTs in an STS-1 just because a subset of these VTs is added/dropped at the failed node(s). The method adopted for SONET/SDH handles the VT squelching in the following manner. In its initial stages, the protection of the VT accessed ring proceeds in exactly the same manner as an STS-1 accessed ring, including STS squelching. Next, however, once a switching node receives a bridged and switched indication from the other switching node (in byte K2), it will unsquelch those STS-1s that are VT accessed. Once the squelching is removed, it is the responsibility of the nodes dropping those VTs to squelch them at their drop output. To implement this additional VT functionality, the nodes on the ring need additional database information. Each ring node, whether it adds/drops VTs or not, needs to keep track of which STS-1s on the ring are VT accessed so that the node will know which STS-1s to unsquelch if it is a ring switching node. Each node that drops VTs needs to keep track of the other nodes that are the source of the VTs that it drops. Whether a ring node is a ring switching node or an intermediate node that passes through the K1/K2 bytes and protection channels, it will see the K1 ring bridge requests from the two switching nodes. By comparing this pair of crossing K1 bytes, the intermediate nodes can also determine which node(s) have been removed from the ring and use their VT connectivity information to determine which of its dropped VTs originated on the failed node(s). Each drop node will then squelch any of its dropped VTs from the failed node(s) by inserting AIS into the dropped signal.

Because the BLSR uses dedicated protection channels, it is possible to use these channels to carry "extra traffic" also known as standby-line access. This allows low-priority traffic to use the protection channels when they are not needed for protection. When the protection channels are required for the restoration of their associated working channels, the extra traffic is removed (preempted). Carriers typically sell extra traffic bandwidth at a discounted rate, which gives the carrier additional revenue from the otherwise unused bandwidth, and allows cheaper bandwidth for the subscriber with the understanding that the availability is not guaranteed. One additional interesting wrinkle in BLSRs is the ability to support non-preemptible, unprotected traffic (NUT). When a channel that would normally be a working channel is assigned to carry NUT, the corresponding protection channel may or may not also be assigned to carry NUT. When a failure affects the NUT-bearing channels, the NUT will be lost. A failure of the working channels, however, will not affect a NUT bearing channel that would have been a protection channel in non-NUT applications. In other words, a failure on the ring will not result in the traffic on this NUT channel being preempted like a protection channel carrying Extra Traffic would have been. One of the main applications for NUT is to carry traffic that is protected by some other means, such as data traffic protected by a Layer 3 restoration mechanism.

While UPSR implementations can be taxed to process all tributary status information continually, one of the challenges in designing BLSR nodal equipment is that each node must quickly compare the destination node ID of a received ring bridge request with its ring topology databases, and then, if a node is missing from the ring the switching node needs to squelch the affected channels. In a four-fiber BLSR, it must also evaluate whether the failures should be handled through span or ring switching. The ring bridge loopback and squelch functions are typically performed in the cross-connect circuit of that NE. This squelching-speed challenge for the BLSR comes when a ring failure occurs that removes a node or whole portion of the ring.

As mentioned in Chapter 2 and Chapter 6, the OIF (Optical Interworking Forum) has recently begun work on a chip-to-chip interface between framer circuits and switch fabric circuits. This interface is called TFI-5 (TDM Fabric to Framer Interface). The TFI-5 interface, which could be confined to a printed circuit board or go over a system backplane, contains overhead channels that can be used to simplify protection implementations. The idea is that the evaluation of the signals (at either line or path level) can be performed by the framer circuits and communicated over the TFI overhead bytes to the switch fabric. A state machine associated with the switch fabric chips can then determine how to reconfigure the switch fabric to implement the required protection switch action. This approach can be used for linear protection, UPSR, or BLSR and allows all the decision processing to take place in the unit that contains the TSI fabric, and distributes the processing load by having the line cards determine the fault or degradation condition.

8.2.4 Interconnected Rings

Ring protection can provide a building block for creating larger protected networks by interconnecting rings. For example, a long haul backbone ring can have multiple metro rings subtended from it. Also, a cascade of smaller rings allows more protection capability than a single large ring. A desirable property of ring-based networking in general is that rings are closed subsystems; problems in one ring can be handled in that ring and not require (or cause) actions to be taken in an adjacent ring. The failure of a node associated with the interconnection between the rings is, however, not so clear-cut and needs some special consideration.

In general, it is desirable to have rings interconnect at two geographically distinct locations so that a failure at one interconnection point can be protected by the other interconnection point. Dual-interconnected rings are illustrated in Figure 8-8. Because there are two points at which rings are inter-connected, this scheme is commonly called dual-ring-interconnect (DRI) or dual homing (DH)[124,125]. It is also known as the 'matched node' arrangement by some vendors. A drop-and-continue function is typically used in which the interconnecting traffic is dropped at one of the interconnection nodes and then continues around the ring to also be dropped at the other interconnection node. UPSRs and BLSRs differ in how they select the best interconnection signal.

In a UPSR (Figure 8-8.a), each interconnection node performs a path switch function on the traffic that it sends to the other ring so that the best path is chosen by the same principles discussed above for path switching. A UPSR receiving traffic from another ring behaves somewhat differently than a normal UPSR. Rather than an interconnection node adding the traffic in both directions around the ring, one interconnection node adds the traffic in one direction and the other adds it in the opposite direction. The destination drop node then performs the same type of path switch function as a normal UPSR drop node.

---

[124] "Dual homing" is a BellCore term while T1 and ITU use dual-ring interconnect to refer to the same architecture for ring-to-ring interconnection.

[125] DH also refers to the survivable PSTN architecture where a remote access concentrator or small CO connects to two hub COs for protection against failure of either hub CO.

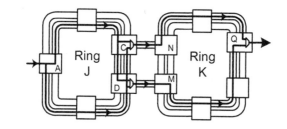

a) Dual node interconnection of UPSRs

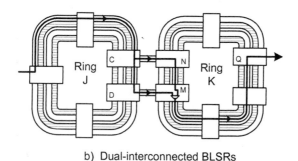

b) Dual-interconnected BLSRs
with service selectors

Figure 8-8. Drop-and-continue and selection functions in interconnection of BLSRs
(only one traffic direction shown)

For BLSRs (Figure 8-8(b)), the line switching eliminates the need to perform path switching on the traffic dropped from the feeding ring to the receiving ring. The receiving ring, however, must choose the best signal coming from the two interconnection nodes. Looking at Figure 8-8(b), consider the example of a failure of interconnection node C in feeding ring J. This failure would mean that node M in receiving ring K would be receiving good traffic from its connection to node D and failed traffic from its connection on the ring to node N. The selection process at node M is referred to a service selection and is essentially the same as path switching except that the selection is made on add/pass-through signals rather than a dropped signal. The example of Figure 8-8(b) shows what is called opposite routing in which the signals in one ring are routed around the "opposite" side of that ring as the signals are routed in the other ring. Same side routing, which is also allowed, would place the service selection function in node N in Figure 8-8(b). Other variations on BLSR interconnection involve the methods used for carrying the drop-and-continue traffic between the interconnections nodes on a ring. UPSRs and BLSRs can be interconnected to each other using a combination of the interconnection techniques of Figure 8-8(a) and (b).

Two general techniques are shown in Figure 8-9 for connecting the interconnection nodes of one ring with those of the other. For either technique, the physical connection can be a SONET/SDH or asynchronous/PDH signal. In the first interconnection technique, the logical connections between the nodes in the rings are made directly at the same level in both rings (e.g., VT1.5, STS-1, DS1, E1, or DS3). In the second interconnection method (Figure 8-9 b), there is an intermediate NE between the interconnecting ring nodes that performs a multiplex function. This multiplex function is typically used to add traffic from other locations. For example, if ring W was a VT-accessed ring (e.g., from a metro network) and ring Z was an STS accessed ring (e.g., from a long haul network), the intermediate multiplexer or cross-connect would take VTs from ring W and VTs from other sources and multiplex them into a STS-1 that it hands off to ring Z.

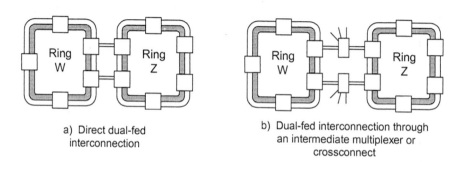

a) Direct dual-fed interconnection

b) Dual-fed interconnection through an intermediate multiplexer or crossconnect

Figure 8-9. Ring interconnection examples

The intermediate NE thus allows more efficient fill of the STS-1s in ring Z. (Clearly, the same external traffic would need to be added at both intermediate multiplexing points. This traffic could come, for example from a different ring.) When the destination drop node is in a STS accessed UPSR, this example of Figure 8-9 (b) leads to an interesting problem. A failure of one of ring W's interconnecting nodes would mean that the STS-1 formed by the intermediate NE would comprise failed VTs in the time slots associated with ring W traffic. Meanwhile, the STS-1 handed to ring Z through the other interconnection node would be comprised of good VTs from ring W due to ring W's protection switching around the failed node. Since ring Z does not process VTs, it would have no way of knowing that the constituent VTs of one STS-1 is good while those of the alternative STS-1 are failed. Hence, the drop path switch in UPSR Z could drop a STS-1 with bad payload when a STS-1 with good payload was available. It is clearly not desirable to require a STS drop node to have the circuitry to examine every constituent VT Path. The alternative was to define a Payload Defect Indicator (PDI) that is generated and used in the following manner. An intermediate NE is required to keep track of how many incoming tributary signals are bad. This number is encoded into the SONET/SDH C2 byte (see Table 4 of Chapter 3). When the STS UPSR drop node is receiving two good STS-1s, it examines the C2 bytes of each to determine the relative health of their payloads. The STS-1 containing the lowest number of failed constituent tributaries is then chosen as the dropped signal.

Two points warrant closing emphasis, however. One is to keep in mind that dual-ring interconnect is a *per-path* decision to treat the signal this way. The two rings involved need to topologically share two so-called "matched nodes" to enable the possibility of DRI, but DRI itself is not a whole-ring to whole-ring relationship or concept. In other words it is not a pair of rings that that are DRI connected as a whole, or not, but rather each path transiting ring to ring is or is not set up with DRI. It is at the level of each individual path that the decision can be made to provision it with a single geographical ring-to-ring transition point,[126] or to set up a drop-and-continue arrangement for DRI. Secondly, if readers imagine that DRI setups, especially the drop-and-continue aspect, tend to rather quickly exhaust the span between the matched nodes, they are quite right. The premature exhaustion of provisionable capacity between the matched nodes is one of the issues with DRI. For this reason, variations such as "drop and continue on protection" have been proposed where protection capacity is used for the continue signals. If an intra-ring switch occurs, the continue signal may be temporarily bumped off. Otherwise if an inter-ring transition failure occurs, the continue signal is intact, but without consuming working capacity. There is a theoretical availability reduction, but in practice this may be less important than extending the provisioning life of the rings. It is also possible the move the drop signal over to the other ring and provide its continue function in the other ring itself but again this is not without some loss of theoretical availability.

Further discussion of strategies for dual-ring interconnection, including detailed availability analysis and resource implications of a variety of ring interconnect strategies can be found in [Grov99] [Grov99b] [Flan90] [Nor96] [ToNe94] [DDH97]. A good general reference for more details on rings and APS technology is Wu's book [Wu92].

---

[126] The "cross-office" (in-building) wiring to go from an ADM drop on ring X over to the add interface on an ADM for ring Y, will often be via an OC-48 or higher rate 1+1 or 1:N APS protected transmission system with "short-reach" optics. So DRI really is a measure to protect against the possible loss of an entire building (presumably via fire, power loss, sabotage), not primarily just the cross-office cabling.

8.2.5 Resilient Packet Ring Protection (IEEE 802.17)

A Resilient Packet Ring (RPR) [RPR04] is like a packet-oriented BLSR. It makes use of the ring topology to provide survivability. It also uses bi-directional routing of working flows and the principle of looping back to the protection ring (called wrapping in RPR) at the edge of the failure. The main difference is that an RPR uses the physical SONET signal capacity between sites in a static point-to-point manner and does both working and protection routing at the MAC frame level rather than at the level of the transport line signal. RPR supports two protection method options, referred to as the "wrapping" and "steering" modes, both implemented at the MAC layer rather than the physical layer. The wrapping mode was modeled after the SONET BLSR and the steering mode was modeled after the SONET UPSR.

Each RPR node sends periodic messages in both directions of the ring to its neighbors so that a node will be able to detect when a link has failed. In one of RPR's protection options, when a link failure is detected, the node(s) detecting the failure take the traffic going in the direction of the failure and place that traffic on the fiber in the opposite direction. This is equivalent to the loopback operation in a BLSR, but is called wrapping in the RPR context. It allows MAC-layer packet frames to reach their destination in spite of the failed link, just as loopback on protection does in a BLSR. A "passed source" bit is toggled when a frame passes its source station to prevent a frame from being wrapped continuously by a pair of wrap nodes.

However, the RPR does not include a 100% dedicated redundant protection channel, so in the wrap mode the apparent capacity of the ring between most node pairs is effectively halved. This means that if the pre-failure node-to-node packet flows happened to be fully utilizing available bandwidth up to the limits of acceptance delay and packet loss, then in the wrap state the available capacity is "100% oversubscribed." This worst case scenario obviously would severely impact most traffic flows. But in practice the pre-failure utilization would rarely be that high and when or if severe congestion or delay resulted from the wrapped state, the applications using the RPR will by nature back off on their attempted throughputs. Some packets may be lost due to both the original failure and then due to overload but as each node detects the overload condition, the fairness protocol that controls packet insertion at each node will throttle the traffic to match the reduced ring capacity. This throttling removes the overload condition so that a stable state is reached in which no additional packet loss occurs. Thus RPR provides a capacity-efficient BLSR-like structure that is well suited to data applications.

But RPR also has a second way of improving performance following entry to a failure state. The second protection option is to have the NE detecting the failure send messages to all the other NEs so that they can actively redirect (switch) their packets away from the failure. This option is referred to as the *steer* protection mode, and is logically equivalent to path switching at the tributary level in a UPSR (except that the alternate direction signal feed has no traffic flow on it until needed).

Finally, RPRs can be configured in a compatible way with existing BLSRs or UPSRs. For instance an RPR may be logically established by forming a ring using STS-3c tributaries of, say, an OC-192 SONET BLSR or UPSR. RPR protection holds off until the SONET protection has time to complete its operation within the ring. If SONET ring protection operates, then the RPR does nothing, and experiences essentially no degradation whatsoever.[127] But if the SONET ring fails to operate for some reason, or, perhaps more likely, the RPR is established over unprotected (or even preemptible "extra traffic") SONET channels, then RPR protection will kick in and retain a still-operating, but perhaps degraded, form of connectivity between its nodes.

8.2.5.1 Partial Protection Using the Link Capacity Adjustment Scheme (LCAS)

LCAS [ITU01] works with virtually concatenated channels, as described in Chapter 4, to create and manage physically diverse routings for component signals of concatenated SONET pipes. For data-centric applications, this provides the option of provisioning a data pipe between IP traffic source/sink nodes which, in the event of a failure, appears (to the service-layer) end-nodes to have simply fallen back to being a pipe of lesser bandwidth, but with no service interruption. A virtually concatenated channel is one in which a number of paths are combined to form a larger channel such that the individual constituent paths (i.e., the members of the virtually concatenated group (VCG)) can take different routes through the network. The virtual concatenation sink node performs the realignment of the members so that the payload data can be extracted from the VCG channel. A carrier can provision the VCG such that the members are intentionally routed along diverse routes. When a failure on one of the routes removes some of the members, LCAS signaling from the VCG sink tells the

---

[127] Physical-layer propagation delays in the recovered-ring state may be increased, but this is usually milliseconds or less, and is a *constant* delay. It is usually trivial compared to delays associated with packet queuing. The two delay issues are sometimes confused, however. In a wrapped RPR, more queuing nodes are encountered. In a looped-back BLSR underlying an RPR, all RPR flows see the same number of RPR nodes on their routes. The general point is that a change in the physical layer propagation time is not the same as an increase queuing delay in a packet service layer.

VCG source which members have failed.[128] The LCAS source then stops using the failed members and falls back to using only the healthy members.

This is illustrated in Figure 8-10, which builds on the prior discussion of LCAS in Chapter 4. In Figure 8-10(a) the three physically diverse LCAS paths that were established in Figure 4-1 are shown, along with the allocation of STS-1 constituent flows to the three paths of the LCAS group. In the pre-failure state, the service seen by the end nodes is equivalent to an STS-5c (hence the five SQ numbers). Fully disjoint routes will obviously be a common preference where possible, but in general as shown, there can still be efficiencies in bandwidth-use by defining multiple diverse routes that are not mutually disjoint. (For example here, the downward diversion that SQ0 takes for part of its route, before re-joining SQ1, could be warranted if the link that SQ0 takes alone is restricted in available bandwidth.)

In Figure 8-10(b) a failure occurs on a span that is common to two of the three paths forming the LCAS VCG. At this point, two options exist. The failed members can be logically removed from the VCG through a provisioning configuration change, or, the VCG can continue to exist in its initial logical form but using only the bandwidth of the surviving path until the failure is cleared. In the second scenario, when the failure clears, the sink will see signaling indications that the source is still not placing data in these members (although those paths are now noted to be working again). When the source receives an OK message for the restored members, it will resume placing data in them.

Although this is not protection or restoration of the prefailure state as seen by the application flows, in the sense of the other protection techniques in this chapter, this form of response to failure can be highly attractive for cost-sensitive data applications. LCAS allows use of unprotected links with a fallback to lower bandwidth in the event of a failure, rather than suffering a complete loss of the connection, or having to pay for full protection. Thus LCAS gives a form of graceful degradation rather than a complete and sudden outage that could arise from failure of a single unprotected service path. Also note, however, that nothing prevents one or more of the constituent LCAS member paths from itself being routed *with* conventional protection. This would allow one to design services that exploit extra bandwidth when available (by adding LCAS members), with fall back upon failure but only to a certain point—strongly 'drawing the line' at a minimum LCAS capacity represented by the protected-service group members.

---

[128] See Chapter 4 for the signaling details.

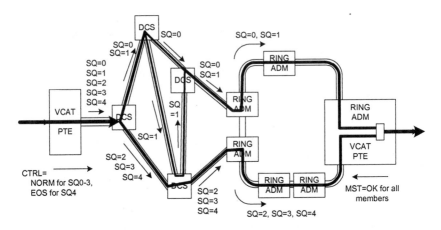

a) Initial, provisioned state of the VCG with diverse routing of members

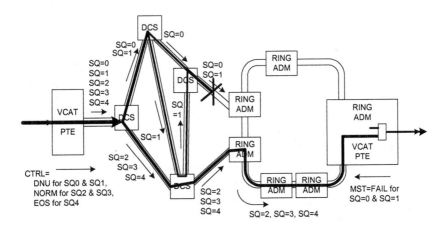

b) State of the VCG in the response to the failure on a link affecting two members

Figure 8-10. LCAS fallback to reduced bandwidth due to failure affecting two out of three members of the VCG.

8.2.6 Ring-based Network Design

As part of the planning phase of new network builds, network operators perform identification of sites as well as estimation of static traffic demand between sites. The planning phase may also take into consideration demand growth between existing sites or potential new sites (i.e. forecast). All such information is used in selecting optimum placement of nodes using ring or mesh topology. At the time of SONET/SDH introduction many studies were done to find optimum designs in terms of minimizing number of rings, ADM cost, inter-ring traffic and other parameters. Basic principles of ring-based network design are well explained in [Flan90]. A survey of planning tools and related algorithms used for automated ring network design is available in [MoGr98].

Access, Metro, and long-haul ring designs may have different constraints. For instance, a new access ring should be at least two-connected in the sense it should be connected at two distinct nodes of an existing ring. If the protection used on the ring is BLSR then further a constraint is that there can only be sixteen nodes in the ring. Considering such constraints, the problem of access ring design involves four inter-related sub-problems:

1. Determining the number of rings to use,
2. Determining the membership of the rings (which nodes are on each ring),
3. Determining the order of the sites with the rings, and
4. Determining the routing of demands through the ring set (the "ring loading" sub-problem).

Automated planning tools take a number of parameters into account and output the number, sizing and placement of rings required as well as the assignment of ADMs to rings and the routing of demands through rings. They also allow minimization of inter-ring traffic. Most of the optimization problems are difficult to solve to optimality and so heuristics are available. An overall strategy that has proven to work well is embodied in a system called "RingBuilder" [GrSl95].[129] In a first phase, an iterative process places rings one at a time in a "greedy" way, based on how well loaded and cost effective each individual ring candidate is within the environment of demands remaining unserved at each stage in the design. The process considers the total amount of demand-distance served by each ring candidate relative to the detailed cost of actually deploying each prospective candidate ring. The set of candidate ring systems can include UPSR, BLSR, and 1+1 APS structures, each at a variety of available OC-n rates. The resulting design is a completely detailed, near-optimum, ring-based network design. Often these designs are suitable for comparative or longer range planning studies, without further improvements. But the capital costs can be so significant that if the design is for a real project that is to be purchased and built, a few more hours of staff and computer time is a trivial expense given the further savings possible. In this case, the initial RingBuilder design can be followed by a tabu-search phase that seeks cost improvements in the

---

[129] Subsequently developed for commercial use as part of VPISystems TransportMaker™ network planning software.

overall design by repeatedly testing to find rearrangements in the routing assignments that can lead to deletion of a ring from the design [MoGr01]. The basic method is one known in the field of tabu search as 'strategic oscillation' between feasible (but sub-optimal) solutions and solutions with lower than optimal cost but which are infeasible. In the ring network context, the tactic is first to simply delete some ring entirely from the design and see if the affected demands can be "mopped up" somehow by re-routing through the other rings. Generally this will not be entirely possible, so a ring (or rings) is added back by the basic Ringbuilder approach of phase one, to restore feasibility. Eventually as this process explores different combinations of rings and routing plans, it discovers complete solutions that use fewer rings in total, which is usually a major cost saving.

It can be extremely worthwhile for network operators to employ design optimization software not just computer-mechanized manual planning systems. Depending on the company, however, it is surprising to see manual planning of ring-based networks still occurring in medium sized network operators. A human should always be in the loop to guide and sanity check the results, and design intuition can always be valuable, but ring-based network design (even just choosing the routings for a set of demands to go through one ring) is a strongly combinatorial problem. Given the economic pressure on network operators these days, adoption of modern optimization methods is almost crucial to remain cost-competitive. As a case in point to underline the potential benefits one of the authors was involved in a test-case of the RingBuilder software. A ring-based design had been manually developed over a six-week period by a team of experienced ring network planners, for a 30-node network at a capital cost of over $200M. The same topology, demands, and ring system options and costs were provided as a challenge case to RingBuilder. In a minute on a laptop computer, the system found a design costing $30M less. About 10 minutes later the tabu search phase found another design improvement worth $5M in saving.

On the proverbial "bottom line," cost savings are indistinguishable from added revenues and so should be reaped where possible—network planning and optimization software can contribute to business success in this way. In fact, this is an important aspect of the NGN transport environment in general: as in the airline industry, it will be the adopters and users of state of the art optimization, planning, and decision support, applied to all aspects of their business, that are more likely to remain viable. Birkan et al. [BiKe03] independently voice the same encouragement of network operators to adopt and benefit from formal methods of operations research. More on ring-network planning can also be found in [CDS95] [ChS97] [DDH95].

8.3. MESH-BASED NETWORK SURVIVABILITY

The main competitor to rings is mesh-based restorable networking. The vision of self-healing mesh transport networks was initially laid out in 1987 for SONET DCS-based transport networks [Grov87] [GrVe90] [Grov97]. This spawned a period of intense study of distributed restoration algorithms (DRAs) by network operators (notably MCI, Telecom Canada, AT&T, and Bellcore) on methods for deploying and operating a mesh-restorable network. Initially, the motivation for mesh-based survivability was capacity efficiency and—through a suitable DRA— the attainment of "split second" restoration times to beat the "call dropping threshold" of the switched telephony network without the delays and database dependencies associated with centralized control.

By the mid-1990s the concepts had been validated and developed for implementation on SONET digital cross-connects switching typically at OC-48 granularity. About the same time, however, OC-192 ring-based transport and optical fiber amplifiers significantly further cut the per-channel cost of transport. This extended the life of ring-based transport for many operators. Especially in sparsely connected networks, OC-192 ADMs were more economic than establishment of the corresponding cross-connect node infrastructures to deploy mesh-based operations. As it turns out the standardization of rings also took most of the 1990s to complete, with corresponding effort on standardizing distributed mesh restoration being delayed. Thus, the ideas of mesh restoration remain valid but the mesh alternative has been delayed first by crisis proportions of survivability in the early '90s which called for the first-available solution (extension of APS systems into rings), by the cost-discontinuity represented by OC-192 ring technology, and by a lack of effort so far on mesh-restoration standards.

Recently, however, the pendulum has swung to greater appreciation not just the efficiency of mesh-based methods, but also the greater flexibility, ease of growth, and inherent multi-service abilities it provides. New "data-centric" entrants to telecommunications, and Internet organizations such as the IETF and CANARIE, have challenged the status quo. Regarding speed, [Mok00] and [Scha01] have both argued that the requirement of 50 ms restoration time is a "telecom myth" with no basis in actual service requirements. Similarly, the assertion that mesh restoration is too complicated to consider in practice is increasingly countered by familiarity and comfort with distributed routing processes used in the Internet, and adaptive call routing processes [Ash98] that are even more complex and autonomous and have been in use for over a decade now. Thus, mesh-based network survivability is a fairly long-standing vision, but it remains one of the most promising alternatives for next-generation transport networks.

Centrally controlled mesh-restoration schemes have been implemented in a few large-scale networks such as AT&T, Wiltel, MCI and British Telecom. AT&Ts FASTAR system is the most publicized centrally controlled scheme. More recently, efforts at intelligent optical cross-connect development by companies like Ciena and Tellium have enabled AT&T to develop its own 100+ node mesh-based survivable transport network. Operational details tend to be closely held but industry contacts advise that several other mesh-based networks of 30 to 100 nodes are now operating based on variants of both span- and path-oriented provisioning, protection, and restoration. A key enabler to access the efficiencies of mesh-based networking is the development of GMPLS standards for the optical control plane, discussed in Chapter 9. In future it is possible that truly self-organizing approaches may also be adopted in which there is no requirement for nodes to have current database of global network state, but in which local interactions result in emergent large-scale behavior that effectively copes with either failures or traffic shifts.

8.3.1   Mesh Protection or Mesh Restoration?

Although for discussion we will speak simply of mesh *restoration* schemes, there is a conceptual difference between mesh protection or restoration. "Restoration" generally refers to a survivability method where the required replacement paths are fully determined in real-time upon failure. An advantage of this is that the recovery pattern can be adaptive to the actual network state at the failure time, although it may take somewhat longer than when a pre-planned set of switching reactions are already decided in each node. The latter is what is termed protection. An important point is, however, that any restoration scheme always enables a corresponding pre-planned protection scheme because "dry-runs" the dynamic adaptive restoration process can generate the pre-planned reactions. In [Grov94] (and [Grov97]) it is explained how even a fairly slow DRA, or centrally computed control, can be used to continually generate and update distributed protection pre-plans in the network nodes.

8.3.2   Protection *and* Restoration: The 1FP-2FR Strategy

One of the most promising strategies for providing ultra-high availability for priority services is to have both protection and restoration mechanisms embedded in a transport network. This enables a self-planning "first failure protection – second failure restoration" (1FP-2FR) capability. This strategy relies on an embedded DRA to continually generate protection plans during the normal operating times between failures. Then upon a first failure, the fast pre-planned protection reactions are ready. A restoration time under 150 ms would be typical of what can be achieved in this pre-planned protection state. The second part of the strategy is based on recent findings that under adaptive restoration re-routing in a mesh network after a single failure, the level of restorability to a *second* failure is almost always still a significant fraction of the working capacity lost in the second failure. Often it is over 50% in networks that were optimally designed only for an assurance of full *single* failure restorability. (See [ClGr02] [ClGr00] or Chapter 8 in [Grov03b]). This means that if the same DRA which was previously being used to generate distributed pre-plans is now executed in real time in the face of the dual-failure state, it is virtually bound to come up with yet further restoration paths to employ. The second-failure recovery level would generally take longer than the pre-planned first-failure reaction, and will generally not provide for complete restoration. But if the second-failure restoration paths are then preferentially allocated to higher-priority service paths they will enjoy a level of service availability that is incredibly high— better even than with 1+1 dedicated APS because even 1+1 APS can only withstand one failure.

Under 1FP-2FR, a certain number of services can be guaranteed to withstand all single *and* all dual failures.[130] This results in almost ring-like speed of reaction to single failures, but with far less spare capacity, while also providing a slower but still extremely fast recovery from dual-failures for premium services. It is easily demonstrated that even if seconds elapse for recovery to the second failure situation, the unavailability impact on services is infinitesimal. (We will later show (Section 8.6) that what impacts availability in a survivable network is not the speed of any successful restoration reaction, but the likelihood of a second failure being restored or not.) Thus, in a mesh-based survivable network, protection and restoration are closely inter-related and it is highly advantageous to employ both. In contrast ring and APS survivability schemes provide protection reactions to a first failure, but a second service-affecting failure on the same structures leads directly to outage (on the order of half the physical MTTR on average).

There are two basic approaches within mesh restoration: *span* and *path* restoration. Both of these schemes reflect the basic idea of what is meant by "mesh-based" survivability. This is to exploit the diversity and connectivity of the physical-layer graph to achieve efficiency through shared use of

---

[130] The only exception is if the service path is routed through one or more degree-2 nodes. Usually, however, a priority path can be routed to avoid such nodes. End-node failure of the path itself is also excluded as this is not a restorable failure under any network restoration scheme. (See [ClGr02].)

standby capacity over non-simultaneous failure scenarios. In all mesh-restoration schemes to date, working paths which are not directly affected by the failure are not affected in any way by the restoration response of the network. In other words an unaffected path undergoes no hits or re-routing. Only failed paths are re-routed for their survivability. This is, however, something that may be re-thought in future under increasing business competition. A rearrangeable or even preemptible service class offering could suit some customers, especially in a data-centric environment where application throttling and IP-layer rerouting options also exist.

8.3.3   Span and Path Restoration

Span restoration is also variously been referred to as local, link, or patch restoration. Span restoration can be thought of as the mesh-based scheme that corresponds to the BLSR ring in that the re-routing of affected paths occurs between the immediate end-nodes of the failed span and may imply a loop-back in the end-to-end routing of a path while in the restored state. It is as if the path simply takes a detour around the failed span. The detour replaces the failed span, and routing continues as it previously did end-to-end away from the end-nodes of the failed span. In an optical network with a mixture of optically transparent ("o-o-o") and electro-optical ("o-e-o") cross-connects, it may be that the failure span is defined between the two nearest o-e-o cross-connects because it is easier to detect failure and rapidly re-route affected signals between nodes where electronic access to carrier signals is available for fault detection and signaling insertion.

Span restorable networks offer a desirable combination of properties. Simple and exact theories for optimal spare capacity planning exist (See [HeBy95] or Chapter 5 in [Grov03b].) The amount of spare capacity required to provide 100% restorability is much lower (typically factors of 2 to 3 times) than required for the same ring-based protection. The length, hop-count, and transmission properties of the set of replacement paths for each failure scenario can also be completely controlled and pre-determined. Because the reaction to each failure is localized, restoration can also be very fast. Propagation delays for restoration-related signaling, and the number of nodes needing to make new cross-connections in real time are both minimized.

The main difference with *path restoration,* also called end-to-end restoration, is that each pair of path end-nodes that is affected by the failure goes about simultaneously undertaking means that will lead to a new end-to-end replacement for each of their affected paths. This makes the scope of the recovery process much more widespread, allowing even greater potential efficiencies in the sharing of spare resources than with span restoration. On the other hand, to realize these efficiencies, and to ensure predictable outcomes, path restoration requires sophisticated overall coordination of the interactions between the individual path formations concurrently being undertaken between all affected end-node pairs. This requirement is met in some schemes by pre-planning backup routes that dovetail with each other in advance (such as SBPP to follow) or through self-organizational principles that dynamically recognize and coordinate multiple pathset formation to avoid destructive mutual interactions (such as the SHNp to follow). Another approach (which we call "ad-hoc" path restoration) is a solely luck-of-the-draw approach based on the picture of GMPLS-based reprovisioning of a single failed path as a restoration mechanism. The flaw is that what happens nicely for a single path failure does not scale with any predictability to situations where a entire cable is cut and the signaling and capacity contention for thousands of such independent reprovisioning attempts are blindly unleashed without mutual coordination.

Figure 8-11 contrasts the two basic ideas of span and path restoration. It shows two demands from A to C and one demand from E to B. The fiber links between nodes are shown in dotted lines and may be hidden behind demands. Following the HG span cut, three demands are affected. In span restoration, these demands are re-routed around the failure point; one of the A-C demands is re-routed over HD and DG; the second A-C demand is rerouted over spans HF and FG; the single E -B demand is re-routed over spans HF, FB, and BG. Even though B is the ultimate destination for the E-B demand, as a result of span restoration it traverses BG span twice.

This is what is called a loopback and it arises unavoidably, as span restoration doesn't know (indeed the point not to have to know) the destination of each demand being restored. Sometimes much is made about the possibility of such loopbacks or "backhauls" arising. Although diagrams of this effect seem to catch the eye and appear quite anomalous, every BLSR protection action involves a guaranteed loop-back that is identical in nature and is inherent to how a BLSR works [131]. In span restoration such loopbacks are harmless as long as the transmission properties of the restoration paths are adequately planned, just as they have to be for the reverse path of a

---

[131] One exception to this is in trans-oceanic BLSR applications (TLSRs). TLSRs have modifications to intercept the normal BLSR loopback and drop the signal the first time it passes a shore station, rather than from its normal drop point that could include a loopback path that again crosses part or almost all of the ocean. Note, however, that as with the loopback issue in a span restorable mesh, there is no capacity savings associated with loopback elimination if all other ring failure scenarios also still have to be protected.

BLSR as well. They are furthermore not even wasteful, despite first appearances, because what is used as loop-back capacity in one failure is almost always otherwise used and needed spare capacity for other failures to use without loopback in their re-routing paths. The point is that the spare capacity in an span-restorable mesh network is allocated to support all the planned failure scenarios, through shared reconfiguration of the common pool of spare capacity. What may look like inefficient routing and capacity use under one failure scenario, will be revealed to be quite efficient under other failure scenarios. See Chapter 6 in [Grov03b] for more on this, including the related but false notion that detecting and eliminating loop-backs in span-restorable routing would lead to spare capacity reduction.

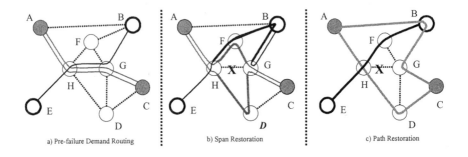

Figure 8-11. Contrasting span and path restoration (from [IrGr00])

Figure 8-11 (c) shows in contrast a possible path restoration reaction for the same span failure. Following the HG span cut, path end-nodes A, C, E and B are alerted by the loss of signal or the appearance of an AIS payload in the signal.[132] Subsequently, to restore demands between A and C, either A or C can initiate restoration. Similarly, E or B can take responsibility to restore E-B demand. The resultant paths are also different and (usually) require less total spare capacity for their formation compared to span restoration. Path restoration is more complex to implement as more nodes

---

[132] Note that end-node alerting happens on its own. The fault does not need to be first located and then explicitly signaled to the end-nodes. If the service fails the end-nodes detect it themselves by BER or loss of signal. However, for the most theoretically efficient path restoration, it is advantageous to know the location of the failed span so that unaffected working capacity up to and away from the break can be exploited in the recovery pattern. Otherwise path restoration is forced to use only fully disjoint paths from the failed working paths. One way to indirectly achieve this is for the nodes adjacent to the failure to generate AIS signal with overhead bytes asserting the immediate release of working channels of affected paths upstream and downstream of the failure.

are involved (A, C, E, B in the above example) and the allocation of available spare capacity on spans to the multiple simultaneously required replacement path sets is strictly a "hard" optimization problem. Even with more than the theoretical minimum of spare capacity present the problem of mutual coordination between simultaneous end-to-end path replacement efforts must be carefully solved (or planned for in advance) or the target restoration level is not assured of being achieved. Ways of assuring this are to employ central control and computation, or to use a path DRA which is capable of solving this mutual-capacity allocation problem so that the overall recovery closely matches a solution to what is called the *capacitated multi-commodity flow* problem. Pre-planned sharing arrangements can also inherently take such coordination considerations into effect when working paths are first provisioned.

The multiple *commodities* that arise in the context of path restoration are the distinct flows between different end-node pairs that are all simultaneously being re-routed between different end-nodes. In contrast, span restoration inherently involves a single-commodity *maximum flow* (MF) recovery problem. The difference in complexity is great. Say a span bearing 100 working channels is cut. In the span restoration framework the problem is equivalent to finding 100 replacement paths between the same two end-nodes. If the spare capacity to support a flow of 100 units through the surviving network between the two ends is present, then finding a set of reroutes that do so is solvable with a simple k-shortest paths or maximum-flow routing algorithm. Under path restoration the same failure may manifest itself as 100 distinct end-node pairs having to simultaneously re-provision one end-to-end path through the surviving network, in the presence of 99 other simultaneously occurring attempts. The damage of 100 units may could alternately affect 50 end-node pairs with two paths each, and so on. Each of these scenarios is a multi-commodity, finite-capacity, routing problem. The key difference is that span restoration presents a single flow problem. Multiple routs may have to be used, but there is no interference in doing this from other simultaneous capacity-using processes. Under path restoration, multiple rerouting subproblems are effectively triggered concurrently and are strongly coupled together under the finite capacity present. If there is no mechanism for coordination whatsoever (as in 'ad hoc' restoration) the overall outcome can be completely unpredictable.

Suitably implemented, however, true path restoration techniques offer some other advantages over span restoration. These are primarily advantages in the reaction to a node failure or to complicated multiple failures of the physical plant and advantages in terms of offering end-customers direct control over protection path planning (under SBPP). Another advantage arises in the context of an all-optical network when rapid fault-detection is considered. Generally, fast fault detection is more easily done at locations where the signal is accessed electronically. In a fully optically transparent network, this strongly suggests that failure detection be done only at path end-nodes, in conjunction with end-to-end path protection using fully disjoint backup and working path routes so that the actual location of the failure does not need to be known immediately. On the other hand, recent advances in theory and mechanisms for span restoration are also addressing node and multiple span failures (c.f. [Grov03b]) and simplified end-customer control of protected path provisioning (c.f. [Grov04]).

8.3.4 Shared-Backup Path Protection

In the most efficient form of path restoration, the response is specific to each failure that occurs. Two different span failures may affect the same individual path and its end-to-end route in the restored state can differ for each recovery pattern. The importance of this is that the absolute minimum of spare capacity (maximum capacity efficiency) is achievable if the restoration pathsets are failure-specific in this way. However, it is possible to consider failure independent path restoration, which has several practical advantages (and the difference in overall capacity requirements is not usually more than ~10%). In this context "failure independent" means that no matter where on a path a failure occurs, the path will be switched over to a single pre-determined end-to-end backup route. Any failure evokes switchover to the same pre-planned end-to-end alternate path. Obviously for this one backup route to serve for all possible failures along the path, it must have no spans (optionally also no nodes) in common with the initial working path. This is the basic approach of what we call shared backup path protection (SBPP).

SBPP can be thought of as arranging a set of 1+1 APS diverse protection paths but keeping the spare links on the backup paths unconnected and shared with other 1+1 setups, with connection occurring as needed upon failure. SBPP has the desirable properties of being a shared spare capacity scheme, so it is efficient. It is also an end-to-end scheme so it is amenable to end-user (or application) control for establishment of both primary and backup paths. (This is seen as being desirable to some, while in other contexts there seems no reason to burden the end-user or application with the task of establishing protection. In the later view protection can be just a service attribute guaranteed by the carrier.) But perhaps the biggest advantage is seen for optical networks where fault sectionalization is thought to be slow or difficult. In that case, dynamic operation is simplified because the switchover is always only to the one pre-planned backup path and activation is completely controlled by the end-node. No matter where the failure occurred, when the end-nodes see the alarm, they just trigger the switchover. The SBPP scheme is also quite amenable to GMPLS-type path provisioning with survivability arrangements made at provisioning time in the service layer itself. The problem of finding efficient fully disjoint or span-disjoint backup routes for each primary route is treated in [Bhan99].

The key ideas for routing under SBPP are that one tries to route the working path over the shortest or least cost path over the graph. This is called the "primary" path. Usually, but not always, there will be one or more possible backup routes between the same end nodes of the primary path. To be eligible as a backup route, a route has to have no nodes or spans in common with the route of the primary path and no spans or nodes in common with any other primary path whose backup route has any spans in common with the route being considered. Together these considerations ensure that when a primary path fails (under any single failure scenario):

(a) No span or node along its backup route is simultaneously affected. This means it will be possible to assemble a backup path along that route if sufficient spare channels have been pre-planned. This can be called the self-disjointness requirement of the backup route, and is the fairly obvious condition for survivability.

(b) No other primary path that is affected by the same failure has a pre-planned backup path that assumes the use of the same spare channel(s) on any span of the first primaries backup route. This is referred to as the failure disjointess requirement and is not actually needed to enable survivability, but *is* required to enable the efficient *sharing* of spare channels over different backup routes. This more complicated set of considerations basically makes sure that if primary A and B are both pre-planned to use a certain spare channel on span X in their backup routes, then there is no (single) failure where primaries A and B would ever both need that spare channel at the same time. (If they did, then two spare channels must be provided on span X to be used simultaneously, instead of one spare channel only, which can be shared by A and B over failures that do not happen simultaneously.)

As a last consideration, the preferred choice for the backup route (assuming there are several possibilities) is the one that requires the fewest *new* channels of spare capacity to be placed (or committed from the available capacity). In other words one tries to chose a backup route on which a backup path can be formed for primary A which, to the greatest extent possible, assumes the use only of spare channels already associated with other primaries that have no "shared risk" in common with primary A.

For example in Figure 8-12, the backup paths (0-4-8-12) and (1-4-8-11) may share a spare channel on the span (4-8) because their corresponding primary paths have no common-cause failure scenarios (and so would not ever have a simultaneous need to use the shared backup channel). Technically, such primary paths are said to have no Shared Risk Link Groups (SRLG) in common [Raja03] [OkNo00]. As a result, one unit of spare capacity is saved on span (4-8) relative to dedicated 1+1 APS for the same two primary paths. Depending on network details, it has been found that if at most three to five such sharing relationships or "claims" are allowed to be established by diverse primaries on each spare channel, the theoretical minimum total investment in spare capacity can be closely approached.

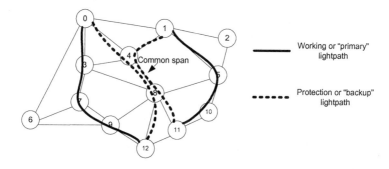

Figure 8-12. Illustration of 1+1 APS and SBPP provisioning methods

An implication of spare capacity sharing is that although backup routes are preplanned and known in advance, the individual backup lightpath channels to form a protection path still have to be seized, cross-connected, and confirmed in real time when a failure occurs. (With 1+1 APS, or with *p*-cycles to follow, protection paths are already both known and fully pre-connected before failure.)

Something to be aware of under SBPP also is that when the shortest route is taken for a primary path, there may sometimes be no disjoint route remaining on which to plan a backup path. In other words a disjoint path is infeasible if the shortest route is taken first. In such cases the shortest path traverses the graph in a way that leaves no disjoint route options. The basic phenomenon is known as the "trap topology" first studied in the context of restoration routing in [DuGr94]. The routing trap can typically be overcome in practice by simple trial and error on different choices of primary routes, or avoided entirely in the first instance by using a least cost cycle-finding algorithm (rather than iterating on shortest paths). [133]

The strictly minimum cost solution for routing new SBPP demand is based on finding the cycle on the graph that contains both desired end-nodes and has least total cost of *new* channels required. The total cost determination must reflect that new channels are needed on every span of the primary route, but that the backup can share already commissioned spare channels associated with other (already placed) primary paths that have no shared risk elements with the new primary. Thus a strictly optimal SBPP routing decision may involve a primary path choice that is longer than the shortest path if it makes an efficient (well-shared) disjoint backup path possible. It can thus also be appreciated that optimized global or incremental planning of SBPP-based networks is actually quite complex. For more advanced treatments on SBPP readers are referred to Chapter 6 in [Grov03b] or [KiKo00] [KaSa94] [LiWa02] and the references therein. We will later return to SBPP with a discussion of issues and alternatives for dynamic demand provisioning.

The SBPP restoration mechanism requires a signaling phase from each tail-end switch to confirm availability of the backup route and to seize and cross-connect capacity to activate the backup path. SBPP is thus not quite the same kind of "protection" scheme as in 1+1 APS or rings. It would more accurately be called a type of pre-planned restoration scheme in that although the backup *route* is uniquely pre-defined, all spare channels needed to form the backup *path* have to be seized and cross-connected upon the failure in real time.

Under global optimization or good incremental routing heuristics, SBPP is generally slightly more capacity efficient than span restoration, and slightly less efficient than failure-specific path restoration. Typically, however, the range of variation from the most theoretically efficient form of path restoration, through to SBPP, and on up to span restoration is less than 10% in terms of total capacity requirements. One study recently found that when nodal equipment costs and modularity effects are included in an optimized cost (not just capacity) comparison of these schemes, there is almost no difference at all in total *cost* [OrWe03]. This strongly suggests that for next-generation mesh-based survivable networks, the choice between mesh schemes should be based primarily on operational and management considerations because virtually any shared-mesh scheme (or a scheme with

---

[133] This is equivalent to solving a min-cost network flow problem where there is only one channel on each span and the required flow quantity is two. In more general mesh re-routing contexts, any method that can find a pathset with capacity equal to the maximum flow quantity between end nodes will also not succumb to the trap topology effect.

similar capacity efficiency such as *p*-cycles) can be expected to be about the same in terms of cost. Operational issues such as flexibility, forecast tolerance, and scalability are then where the bigger cost differences really lie in the choice of technology.

8.3.5 "Ad-Hoc" Path Restoration

GMPLS [GMP01] [KoRe00] [OkNo00] mechanizes the process of path provisioning with protocols for collecting global link resource information and commanding connections through a network from any originating node. With the advent of GMPLS there is growing popularity to the idea of just using GMPLS to re-provision paths when needed upon a failure. The notion is that when a disaster such as a cable cut occurs all affected end-node pairs just "redial their path." This risks providing a false sense of security, however. Let us explain the concern.

Most of the time in most networks, it will be quite likely that a single isolated re-provisioning attempt can find a replacement path on-demand for a single failed path. Just as when a single phone call is accidentally dropped, the re-dial attempt is very likely to succeed. The network as a whole is still essentially stable and intact in these circumstances and so GMPLS can then reliably replace the single failed path with the next-shortest path through the graph of unused capacity. So everything is fine for a single isolated path failure and this is a useful extra functionality from GMPLS; it gives a simple on-line mechanism to cope with single isolated path failures such as may typically arise, for example, from a single interface card failure on a router.

The confounding issue arises when this single-path replacement idea is extrapolated to be the basis of a network's entire response to a cable cut. When a cable is cut (and this is the most frequent form of physical layer failure), an arbitrarily large number of service paths, involving a large fraction of all end-node pairs, fail simultaneously. In this case there are two sources of uncertainty about the overall outcome if every affected node-pair then pursues an independent GMPLS path re-provisioning attempt for each path failure that affects it. First is the message-handling behavior of nodes and protocols under the mass onset of concurrent, semi-synchronized, distributed path re-establishment attempts occurring at the same time that OSPF-TE type of link-state update information is being disseminated about the failure. The complete dynamics of the mass concurrent onset of all the relevant protocol instances, and the message handling and buffering ability of nodes coping with the suddenly huge messaging onset is fundamentally difficult to model. Constituent protocol instances of GMPLS, primarily OSPF-TE for state update and CR-LDP for path resource seizure will be running on the network simultaneously in large numbers. Opportunities for deleterious interaction between simultaneous CR-LDP instances seem particularly worrisome: one instance has reserved a wavelength on a span and is propagating back to complete its path, but it fails in back-propagation due to other instances of the CR-LDP being initiated. Meanwhile another path attempt may fail elsewhere because of the reservation on the first resource mentioned, which is then freed. Such failed CR-LDP attempts result in reattempted connections between the host nodes, compounding the dynamic deadlocks, fall backs, and re-attempts.

There is also the fundamental issue of "mutual-capacity" allocation that affects the restoration level that can be achieved. The key issue in reliable formation of efficient restoration pathsets is the ability to coordinate individual restoration paths so that the way limited spare capacity on all spans is allocated supports the maximal overall recovery level. This is why simple mass-redial for restoration is a risky proposition. When studied experimentally in [KaGr03] it was found that achieved restoration levels could be as low as 20% or as high as 100% of what was theoretically possible within the spare capacity present, just depending on the random dynamics of how the individual path re-dial attempts interact with one another in seizing spare capacity. It has been shown that in general there is no assurance about the overall recovery level or pattern of recovery that will arise if the only response to a cable cut is a mass independent redial attempt by every end node. This is a fundamental difference from all the other schemes discussed for protection or restoration, which by design can give explicit assurances of target levels of restorability.

The view that GMPLS also provides a restoration mechanism is causing some misunderstanding in the industry that all forms of restoration are similarly unassured. This seems to be based on an assumption that any restoration scheme must be equivalent to a mass of individual blind re-attempts. However, from what has just been covered on SBPP, and will follow on SHNp, this is untrue. SBPP delivers assured outcomes by virtue of its pre-failure coordination of spare channel sharing over the backups of failure disjoint primary paths. SHNp delivers assured outcomes by virtue of a built-in mechanism for distributed capacity coordination that ensures maximally efficient allocation of available spare capacities to the set of affect end-node pairs. It is only *ad-hoc* restoration by mass uncoordinated redial that is unassured in its outcomes because it embodies no coordination of spare capacity amongst simultaneous end-node pair restoration attempts.

Having thus given a caution about relying solely on GMPLS reprovisioning for restoration, there is a context where its use can make sense, however. This is in the 1FP-2FR strategy previously mentioned, where an assured protection mechanism is present, along with capacity adequate for 100% restorability against any single failure. In this case if we want to assure the highest availability for a relatively small number of premium service paths, then it does not really matter too much whether a GMPLS mass reattempt is a perfectly assured method to derive the second failure restorability. The philosophy is to attempt the second-failure "redial" only for the priority paths, assuming there are considerably fewer of them than the affected demands of all classes. This last-ditch attempt to save the day for priority paths in the face of a second failure may not fare too badly. (But it is still not the same as a 1FP-2FR strategy involving predictable restoration mechanisms for which the outcome of all dual-failure scenarios can be pre-determined as the basis for strict dual-failure restorability guarantees for some customers based solely on the capacity state of the network following the first-failure.)

8.4.   APPROACHES FOR CONTROL OF MESH PROTECTION AND RESTORATION

8.4.1   Centrally Controlled Mesh Restoration: FASTAR

AT&T's Fast Automatic Restoration (FASTAR) is one of the early-centralized restoration implementations, introduced into its network in 1992. [Chao94] and [ChDo91] describe the architecture and operation of FASTAR and [AmCw00] gives an excellent treatment of the corresponding network spare capacity planning process to ensure target levels of restoration against all single failures using FASTAR. As illustrated in Figure 8-13, reaction to a fault occurs at three different levels. The rerouting of voice traffic is done by Real-Time Network Routing (RTNR) and implemented by Voice Switches (VoSW). This is triggered if FASTAR was not able to re-route affected DS3s following a fiber optic failure. FASTAR itself is triggered if underlying SONET protection failed or was not in place. FASTAR works by managing the reconfiguration of up to 250 DS-3 cross-connect systems. Figure 8-13 is based on the basic description of FASTAR from references above with added illustration of some specific further considerations about spare capacity described in [AmCw00]. The changes involve connecting SONET/SDH APS protection capacity to the cross-connects where feasible and economical. What this means is on a SONET 1:1 APS span both its working and protection OC-N link capacities are accessible by the cross-connect. The APS protection bandwidth can then be used either for restoration rerouting for other span failures or used directly on the same span for 1:1 like protection switching for the corresponding working OC-N link.

FASTAR uses a Restoration Node Controller (RNC) at each CO and a Restoration and Provisioning Application (RAPID) at the NOC. The RNC collects alarms from the co-located cross-connects (DACS) and lightwave terminal equipment (LTE). RNCs also verify new paths. RAPID, once it knows of a fault and computes an alternate path, will send commands to the DACS to reconfigure cross connections as needed. To verify that affected services have been restored, RAPID also sends test commands to remote RNCs. They in turn tell their local DACS to bridge a DS3 test signal generator into the path to be tested.

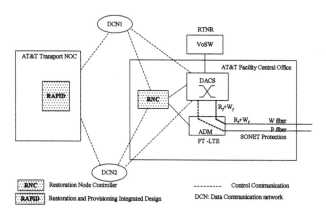

Figure 8-13. FASTAR: AT&Ts centrally controlled network restoration scheme

The restoration speed of FASTAR reportedly varies from 2 seconds to 5 minutes to restore more than 90% of affected DS3s. FASTAR also includes a priority system to decide what to recover first. The CCS#7 signaling network and FAA air traffic control get high priority as well as trunking for 1-800 call services. The regular switched voice network gets lower priority, because the central office switches (4ESSs) have their own recovery capabilities. Private line DS3 services also get a priority, as well as DS3s that on aggregate contain many customer DS1s that are assigned priority.

Before FASTAR is triggered, SONET/SDH protection, if available on a span, is allowed to provide the restoration. The data communications network itself to control FASTAR obviously has to be of extremely high availability and independent as much as possible from the links it protects. For this reason FASTAR control communications relies on complete redundancy in the form of a land-line data network as well as satellite communication to all FASTAR nodes. Reportedly, FASTAR was used to good effect (as well as other measures) to quickly restore a basic level of telecommunications services following the Sept 11, 2001 event.

8.4.2   Internet-style Control: GMPLS

The growth and pervasiveness of the Internet marks it as one of the most successful technical developments in history. Not surprisingly, therefore, concepts used in the Internet are being adopted and extended for the control of optical transport networks as well. One of the keys to the pervasiveness of Internet applications is the seemingly transparent way in which any logical connection that we desire is routed through the Internet. In this section we will have a quick overview of these principles and then see briefly how they have been adapted and extended for use in distributed control of routing and protection in transport networks. Chapter 9 is devoted in its entirety to the various protocols and standards for network control. Here, we need to just set down enough background on GMPLS to see how it can be used for signaling and control in provisioning services in optical networks with explicit survivability arrangements made at the same time, and activating the protection when needed

Under IETF development, several de-facto standards are primarily used in conjunction with the SBPP principle for end-to-end path protection [KoRe03a] [KoRe03b] [KoLa00] [LiWa02]. The same protocols for network state orientation and signaling can in principle also be applied to the realization of span or path-restorable mesh networks as well. A basic concept of Internet-style control is the aspect of periodic state dissemination amongst nodes and the requirement of synchronized global network state databases in every node for these schemes to operate. This is the key difference from a latter paradigm that we will consider—that of truly self-organizational behavior.

The overall idea of adapting Internet protocols to control transport networks, both for service provisioning and for activation of protection paths can be summarized as follows.

(i)   Every node maintains a database of the network state (link topology and currently available capacity resources).

(ii)   Each time a node becomes aware of (or causes) a change in the state of a link incident upon itself, it broadcasts an update to all other nodes about this change (called a Link State Advertisement or LSA).

(iii)   The accumulation and integration of LSAs is what allows nodes to maintain the database in (i).

(iv)   Whenever a new path is required, the controlling end-nodes solve a least-cost routing problem using the global database.

(v)   The desired path is connected through the network through a series of "label switching" commands forwarded along the path. This is based on a set of signaling standards called GMPLS.

(vi) For SBPP service setups, the same global database of network state is used to solve for SBPP primary and backup routes. The primary path is activated and channels on the backup path are earmarked for protection of the corresponding primary both using GMPLS signaling.

(vii) When a primary path fails, the end-nodes of the path detect the failure and use GMPLS signaling to seize, cross-connect and activate the reserved spare channels on the backup route to form the backup path.

Note that this whole approach is essentially equivalent to centralized control in that it assumes and relies upon a database of global state information to solve the required routing and reconfiguration problems. An instance of the same routing algorithm that would be solved in a centrally controlled architecture is solved instead in each responsible node.

GMPLS unifies and extends traditional Internet routing and control protocols to control most all types of networks, including IP networks, ATM networks and the circuit-oriented networks like SONET/SDH and wavelength routed WDM networks [Man03]. Before GMPLS, each network layer has its own independent control protocol suite. For example, ATM has its own UNI and PNNI. SONET/SDH has its own TNM-based control system; and wavelength routed networks have so far often been controlled by private software packages based on SNMP. With GMPLS, however, the same protocol suite can control routing and connection-making in the packet layer, the TDM layer, the wavelength layer, and ultimately in the fiber layer.

The process of network state database synchronization is based on the original Internet OSPF protocol but with extensions to support so-called "traffic engineering" (TE) [KoRe03b]. Basic Internet protocols historically were only concerned with logical link topology. TE extends them to include details of the capacities and features available on each link in the network so that route computation can take QoS issues and protection arrangements into consideration. For SBPP, the TE database is used to compute a working route as well as a protection route, which is required to be link, node, or SRLG-disjoint from the working route and suitably coordinated with sharing arrangements for protection already established for other working paths.

The signaling functions of GMPLS establish, modify, query, and release service connections using the component protocols LMP, OSPF-TE and RSVP-TE or CR-LDP. In following discussions, we assume that RSVP-TE is the signaling protocol. Given a desired route, the process of signaling to actually establish a path on the route consists of two steps involving PATH and RESV messages. The PATH message goes forward requesting the desired resources along the route from the source to destination node. Once the PATH message has reached the destination node, that node can determine whether the requested route is available or suitable for the new connection. Based on this decision, the destination node can respond back the source node declining the request, or acknowledge the request proceed to cause creation of the desired connection through RESV messages relayed back to the source over the same route as the PATH message took.

Once a connection is established, it is recorded in a connection database at each node that stores all the information related to the connections that start or end at the node. This can include the hop information of the connection (which can be a sequence of link IDs), protection priority of the connection (which can be best effort, spare capacity sharing protection, or dedicated protection), capacity or bandwidth of the connection, etc. When a connection finishes its service, the source node will trigger a release process by sending a release message to the destination node. After receiving the release message, the destination node will send back a confirmation message to confirm the success of the release. All the network resources used by the old connection are released for future use and a new round of LSA message flooding updates all other nodal databases.

GMPLS route computations and signaling assume a synchronized network state database. This is not always the case, however, because it takes time for the LSA flooding process to synchronize the network state databases. Thus, before the database is fully updated, the local node may find a route that has enough network resources, but actually not enough resources under the real network state. As such, when the signaling process tries to set up the connection, it is sometimes declined either because there is not enough resources on some spans or there are but resource contention causes blocking. Contention is possible when multiple connection requests arrive at different nodes at nearly the same times. The nodes compute the routes for the new connections independently. If coincidentally, the routes use the same network resources, then contention arises. To resolve such contention, one proposal is to compare the source node IP addresses of all the connection attempts involved and to select the one with the largest IP address to use the competed resources. Other metrics such as the service priorities of connections, the required bandwidth of connections, etc., may also be used for the comparison.

8.4.2.1 GMPLS Control of SBPP

Under SBPP, a centralized algorithm running with the global-view database in the originating node finds the working and protection routes, then signals the establishment of the two paths using the PATH message and

RESV messages above. In the "concurrent" strategy two signaling sessions are initiated at once—one for the working path and the other for the protection path. If both succeed, then the survivable service is provisioned; otherwise, the provisioning process fails. The sequential strategy establishes the working and protection paths one by one. Only when the working path is established successfully, can the protection path be set up. This reduces signaling activity if a path cannot be provisioned. Two network-update processes follow a successful service provisioning. One is to update the local connection database, and the other is to update the network state database and connection databases in other nodes including spare capacity sharing information, by issuing "TE" type LSAs. Because SBPP also needs to consider the spare capacity sharing relationship when searching for a protection route, the update of connection database needs to be carried out globally in the network.

Figure 8-14 summarizes the network state database and the connection database needed to operate a survivable network based on GMPLS-controlled SBPP. The link state database includes (for every network link): 1) deployed link capacity, 2) used link capacity, 3) working capacity and spare capacity on the link, and 4) the list of end-to-end paths currently with a spare capacity sharing relationship on each spare channel of that link. The SBPP connection database stores (for every connection): 1) names of the two end nodes, 2) capacity of the connection, and 3) the routes of the working and protection paths.

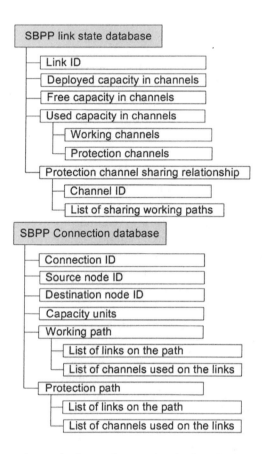

Figure 8-14. Network state database and connection database for GMPLS-based SBPP
service provisioning (adapted from [ShGr04]).

With this information available in each node, and with the use of GMPLS
signaling protocols, any node can establish and tear down protected and
unprotected service connections. When a failure does occur, the end nodes of
affected paths use the same PATH-RESV signaling process, now along the
backup route, however, to activate the pre-planned backup path.

8.4.3   Self-organizing Control

Although initially studied in depth in the early 1990s, truly low-level
self-organization of sets of mutually coordinated paths through the transport
network is a basic technology that remains to be attention for next-
generation transport networking. By "self-organizing" we refer to
mechanisms that operate without *any* database of network state, with small
amounts of code and with no assumption of a global view of the rerouting
problem.   Self-organization is about simple rules for local interactions

between nodes that results in desired emergent behavior at the network scale. This is quite different from routing in today's Internet. Although routing in the Internet (including uses of MPLS and GMPLS) is often referred to as "distributed," in important ways it is not. Under these protocols every node participates in a distributed database synchronization exercise, following which routing problems are solved by conventional centralized algorithms running on a local copy of the database of global network state. Experience has shown that this is relatively frail, inefficient, and poor-scaling solution mainly due to its heavy database and software dependences. So far, however, it has been imported wholly into IETF standards for operation and control of the transport network, including protection and restoration.

The original vision and methods for distributed self-organizing restoration and path-provisioning [Grov97] are much simpler, elegant, and robust. The self-organizing approach is characterized by small event-driven finite state machines (such as those that implement the SONET APS protocol), implemented in kilobytes of firmware, interacting through channel-associated signaling that is no more complex than the K1-K2 overhead bytes that SONET uses. In contrast, the conventional approaches to network control have been described as "the software mountain"[134] characterized by hundreds of thousands of lines of conventional source code and databases. In contrast, the nodes of a self-organizing self-healing network store no database at all. The environment and configuration in the physical layer of the graph at the time of a failure *is* the database within which the self-organizing protocol executes.

In the self-organizing approach simple low-level interactions between nodes have a desired *emergent effect*. For restoration, the emergent effect is the spontaneous formation, like crystallization, of the required path-set as a *pattern of interconnected spare channels*. The problem is not solved through explicit route-finding algorithms, but through the execution of a low-level multi-way interaction (more like playing a game) that has the *side-effect* of forming restoration path sets. It is more like the spontaneous formation of a crystal, once triggered by the precipitating event. The result is uniquely defined by the initial conditions of the network and the failure that triggers it off. Once triggered, the process leads inexorably and spontaneously to creation of the desired restoration pathset by independently made decisions at each node.

This kind of highly parallel interaction between nodes via simple channel-associated signaling bytes is dramatically faster and more autonomous than nodes interacting through queued messaging over packet

---

[134] The term "software mountain" is used in the editor's introduction to the *IEEE Proceedings Special Issue on Communications in the 21st Century*, where [Grov97] appears.

data networks formats, using conventional software in reliance on up-to-date global network state. Although the self-organizing approach offers desirable properties, it is such a different paradigm that it has been hard for it to catch on. Even after a decade of validations and implementation by several groups, many in the industry continue believe (or at least assume without question) that restoration is actually impossible without centralized control or databases of global network state on which to run explicit routing programs. The self-organizing effect operates on such simple (but indirect) principles that a common reaction is still that it is impossible. Recently, however, with continued growth of cost and complexity associated with conventional software and database-intensive approaches, the pendulum of interest has started swinging back to searching for truly autonomic approaches. So the time is right to revisit the concepts of self-organizing formation of restoration pathsets for future transport networks.

The basic mechanism was called the SHN protocol[135] (for *Self Healing Network*). The SHN protocol effects spontaneous formation of restoration path sets between the end-nodes of a failed span. It may also be used in "Capacity Scavenging" mode to find maximal or desired sets of feasible paths through the existing unused capacity of a network between any two nodes. Other applications include background audit of the network restorability state and autonomous background generation and continual updating of span-protecting pre-plans. The basic SHN for span restoration was later extended into a version that self-organizes "multi-commodity" pathsets for path restoration, called SHNp, [IrGr00] and also adapted into a version that self-organizes network protecting *p*-cycles in response to any given environment of working capacity requirements [GrSt98]. (*p*-Cycles follow.)

8.4.3.1 How the SHN Protocol Works[136]

Let us start with an overview statement that will give a general initial picture and also touch on key concepts to which we will then devote more in-depth explanation. The SHN is based on an event-driven finite state machine (FSM) with three main states: *Sender*, *Chooser* and *Tandem*. Sender and Chooser nodes are adjacent to the fault and together referred to as the "custodial" nodes. Briefly, the Sender initiates the self-organizing process through multi-index forward flooding and later uses the pattern of paths formed to restore the working paths affected by the failure. The Chooser responds to the first appearances of the forward flooding process on

---

[135] "SHN" is a trademark registered to TRLabs. (www.trlabs.ca)
[136] This description is a shortened and adapted version of what appeared in [Grov97].

each index and initiates the reverse-linking process, which creates paths out of the dynamically evolving forward flooding index trees rooted back at the Sender. The Chooser also receives live traffic substituted by the Sender over the paths created and performs its end of the traffic substitution phase. Other nodes (if involved) follow the Tandem node rules, which mediate the competition amongst forward flooding indexes for propagation and also collapses successful index trees into paths through reverse-linking.

Interaction between nodes occurs through *statelets*. Statelets can initially be thought of as the mesh equivalent of the K1-K2 bytes employed for event-driven operation of the SONET APS or BLSR protocols. They are channel-associated signaling bytes with static contents that are simply repeated in every outgoing frame of the corresponding carrier signal until an event at a node alters their contents. It is important to understand that statelets are *not* data communications packet messages between processors. They are semi-static tags of state information physically bound to the associated carrier signal. Statelets can be implemented either in SONET overheads[137] or in the generous new overhead fields provided within the recently "digital wrapper" scheme for digitally transparent lightwave applications.

The basic SHN statelet contains the following fields: (Other fields can be added for provisioning, network-audit, and self-pre-planning applications but are omitted from this overview. See [Grov97].) The four basic fields are:

1. **Node-pair label**: This tags all statelets with a common label identifying the failure to which they pertain. For span restoration the fault label is the two names of nodes adjacent to the span failure. This label is a primary basis for concurrently handling multiple faults. Each custodial node places its name first in this field when it emits a statelet. From this, intermediate nodes can tell a forward flooding statelet from a reverse-linking statelet.

2. **Repeat count**: Every Tandem node increments the hop count on statelets forwarded on by it. A hop limit sets the greatest logical distance from the Sender that statelets may propagate and hence sets a limit to the restoration path length. Any statelet arriving at a Chooser or Tandem with hop count greater than the repeat limit (RL)

---

[137] The K1-K2 bytes themselves are candidates for statelet-bearing purposes in a mesh-restorable network because they would no longer be needed to perform ring or APS functions. Ring-mesh hybrids are still accommodated: if channels are designated as part of a ring, APS, or mesh-protected subnetwork then a generalized restoration implementation knows which "language" pertains to the K1-K2 bytes on each channel. Existing K1-K2 syntax applies for ring and APS, and a new syntax would be standardized for K1-K2 self-healing mesh statelet functions. (Alternately the statelet functions could also be based on yet-unallocated Sonet overheads or digital wrapper overheads).

is ignored. The RL can be constant over a network or it can vary in regions of a network by setting the Sender initial repeat value (IRV) on a node-by-node basis.

3. **Node ID** (NID): In any statelet emitted by a node, that nodes' name is included so that adjacent nodes have a record of the logical span on which the statelet arrived and so that channel-to-span associations are always apparent from external data, not from internally stored data. Node names are not appended to a list, simply written into the dedicated last node ID field. In contrast to the node-pair label which is static, the NID field changes at every node.

4. **Index:** A unique index number is assigned to each statelet emitted by a Sender node, which triggers the pathset formation process. (The pattern of initial Sender flooding is detailed below). The indexing is not repeated in each span but runs sequentially over all statelets initiated by the Sender. Every statelet subsequently created by a Tandem or Chooser node bears the same index value as the incoming statelet to which it has a causal relationship (called the precursor relationship, below).

Figure 8-15 shows these fields within the overall structure of a statelet, in addition to checksum and mode fields. The mode field is like an op-code that permits a variety of different applications to follow. The main path-forming effect occurs through the interaction of Mode 0 or forward-flooding and reverse-linking statelets. As paths are formed, a Mode 1 statelet on each manages path mapping and traffic substitution. The RA bit asserts a "reverse alarm" indication in the event of a unidirectional failure on the failed span.

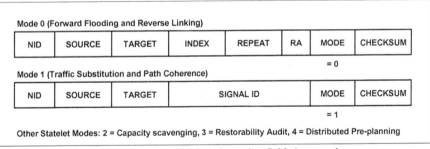

Figure 8-15. The semi-static information fields in a statelet

The overall operation of the SHN can now be described. Although we speak in terms of activation, forward flooding, reverse-linking and traffic substitution processes, these are not strictly sequential phases. For simplicity we describe these processes as if they were separate phases but they actually occur in a parallel asynchronous manner.

Before any failure, each node applies a *null statelet* to the overhead bytes on all working and spare channels leaving its site. The null statelet contains only one non-null field, the nodes' name in the NID field. This allows the SHN task in adjacent nodes to identify all channels in the same logical span, without requiring or maintaining nodal data tables. Changes in the span composition before a failure are therefore self-updating.[138]

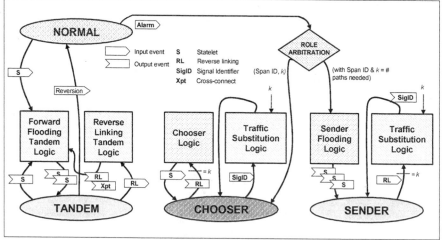

Figure 8-16. Structure of the Finite State Machine (FSM) that implements the Selfhealing Network (SHN) protocol.

In the case of a single span failure, the SHN task is initially in the Normal (suspended) state at all nodes. The onset of a signal fail or other alarm in a working port at a cross-connect invokes the local SHN task at that node. (The SHN task can be thought of as just an interrupt handler for two specific events at a node: an alarm arrival or a non-null statelet arrival). Activation in multiple failure cases or randomly dispersed individual fibre break times on the same cable are discussed in [Grov97] and references therein.) The first task when the SHN protocol is activated by a local failure is Sender-Chooser arbitration. Figure 8-16 portrays the overall structure of the finite state machine that implements the SHN protocol and will be used for reference in the following description of operation.

[138] This technique, originally part of the SHN protocol has been adopted as part of the Link Management Protocol (LMP) developed by the IETF (See Chapter 9 for LMP).

Sender-Chooser Arbitration: The SHN task (invoked by the alarm at one or both) custodial nodes reads NID in the last valid pre-failure statelet latched in the failure-affected ports, learning the identity of the node(s) to which connectivity has been lost. Each custodial node then performs an ordinal-rank test on the remote node name (NID) relative to its own to determine whether to act as Sender or Chooser. The outcome of the test is arbitrary but it guarantees that one node will adopt the Sender role and the other will become Chooser. To assure both custodial nodes are alerted in the event of a possible unidirectional failure a far-end failure indication state is also asserted immediately on each outgoing channel in the span on which the receiving alarmed ports arose locally.

Sender Flooding: Immediately after role arbitration, the Sender applies an active statelet with a unique index number to each available spare channel at its site, up to a maximum of *min* (w_{fail}, s_i) channels in each span, where w_{fail} is the number of working channels on the failure span and s_i is the number of spare channels on the ith span at the Sender. If there are surviving spare channels on the failure span itself, they are included in the Sender flooding pattern (resulting in partial or complete APS-like restoration over the same span as the failure channel(s) as part of the restoration reaction). The SHN thereby inherently emulates 1:N APS on every span in the event of single-channel failures. We proceed, however, to describe operation as if the failure is a complete cable cut. Note that the entire protocol proceeds upon, and only ultimately affects, the configuration of the spare capacity in the network. The only interactions with working capacity are to observe the number of failed working paths, and, as the restoration pathset is formed, to cross-connect failed working signals into those paths.

The Sender's initial statelets propagate at line-transmission speed and appear as receive statelet (RS) events in ports on neighboring cross-connects. This invokes the SHN task in those nodes in the Tandem state. At this stage the Tandem node rules affect selective forwarding of incoming statelets on spare channels at that node location. Each statelet that is forwarded by a Tandem node repeats the fault label (node-name pair) and index fields unchanged. The repeat count is incremented relative to the corresponding incoming statelet, and the node overwrites the NID field with its own name. As more statelets arrive, the rules for the Tandem node effect a form of competition for forwarding because the number of outgoing spare channels is limited, and only one index can be on a channel at any time. These Tandem node rules indirectly cause the desired path-set formation result. The Tandem node rules are given separate attention below. Eventually, through Tandem node forwarding, one or more of the forward flooding statelets arrives at the Chooser node (if not, the topology is fundamentally unrestorable). This triggers a reverse-linking sequence.

Reverse-Linking: When the first statelet on a given index arrives at the Chooser, the Chooser initiates a reverse-linking process that traces through a sequence of forwarding associations at Tandem nodes, back to the Sender. The path traced back follows the sequence of *precursor relationships* defined at each node for that index in forward flooding and it collapses the family of statelets for that index onto only one path between the Sender and Chooser. To initiate reverse-linking, the Chooser applies a *complementary statelet* (to be defined) on the transmit side of the port that has the received the first statelet bearing a specific index. The Chooser emits no other statelets and only reacts to the first arrival of a statelet with each index. When the reverse-linking statelet arrives at a Tandem node, it propagates the reverse-linking statelet in the direction of the precursor for that index, suspends all other statelets leaving the node with the given index, and revises the remaining statelet forwarding pattern as detailed below. The Tandem node also then requests the local host cross-connect to make a cross-connection between the ports through which the reverse-linking process traveled. Propagation of the reverse-linking statelet event itself, however, does *not* wait for completion of the cross-connection. By the time the reverse-linking process on the given index traces back to the Sender, the only statelets that remain in the network for that index act as a bi-directional *holding thread* for the individual path that has just been formed. At any time later, if either Sender or Chooser cancel their holding statelet, the path is released end to end. This enables a simple automated reversion option if desired. At this point in the SHN, the Sender knows without doubt that the local port on which a reverse-linking statelet just arrived is the local end of one complete path to the Chooser and immediately selects one of the affected working signals for substitution into the restoration path that just became available. Traffic substitution does not wait until the entire SHN execution is complete. As soon as a reverse-linking sequence reaches the Sender, an affected demand is mapped onto it by the Sender. Thus, one can easily map high priority demand units to the first available paths and accommodate traffic classes of various priorities. A property of the SHN is that reverse-linking arrival at a Sender *is* full confirmation that a path exists. No separate additional confirmation phase (sometimes called a "third phase" in other DRAs) is actually needed or part of the SHN. [139]

---

[139] The SHN does not use, nor does it require, a final explicit path-tracing confirmatory phase. The arrival of a reverse-linking statelet at the Sender cannot have occurred unless the path is complete and has its other end-node at the Chooser. Commitment of resources to form a path happens during the reverse-linking phase for each path not, as in other protocols, through a final third phase which is necessary to activate and confirm a path which may fail to have formed due to contention effects that are not present in the SHN.

Tandem Node Rules: Two promised definitions can now be made and are needed for describing the Tandem node.

Precursor: For each index value present amongst statelets incoming to a node, the port with the lowest repeat count for that index is defined as *the precursor for that index.* If more than one statelet has the same lowest repeat count for that index, the precursor will be the one that permits the greatest satisfaction of the target broadcast pattern (to follow). If both criteria are equally satisfied, then the lowest numbered port will be the precursor. The precursor for an index is always the root of the forwarding pattern for that index within the node.

Complement: A complement condition exists in a port when a forward flooding statelet is matched by a reverse-linking statelet having the same index and node pair (fault) label, and a valid repeat count. With these definitions in place, we can now specify the Tandem node rules as follows. These are:

(i) First, keep a list of ports where precursor statelets are presently found. Order this list firstly by increasing repeat count and secondarily (i.e, amongst equal repeat count groups) by increasing number of the port where they appear. (The secondary ordering is arbitrary but makes the detailed construction of the path-set that results exactly repeatable for the same failure scenario.)

(ii) Any time a new incoming statelet appears (or an incoming statelet changes), determine if the new statelet is a better precursor for its index according to the definition above. If so, remove the old precursor from the list in (i) and insert the new precursor in its properly sorted position. For each index present at the node, always *try* to provide one outgoing statelet on the same index on any one spare channel in each span other than the span in which the precursor for that index currently lies. This is called the basic forwarding pattern. It is important to realize that statelets each "occupy" the spare channel on which they are placed so in general it will not always be possible to fully satisfy the basic forwarding pattern for every index present.

(iii) When it is not simultaneously possible for all precursors to enjoy one instance of a statelet with its index in each other span, always effect a composite forwarding pattern that is *equivalent to*[140] the steps of erasing all outgoing statelets then, working in order from lowest to highest repeat count

---

[140] Implementations of the Tandem node rules can vary from the way the *equivalent pattern* is described here, as long as the state that results is always equivalent to the result of what is described. In particular, temporary removal of all statelets to recreate the entire pattern is not actually desirable or necessary in practice. This is, however, the easiest way to specify how to compute the new pattern. Only statelets that do change as a result should then be removed and reasserted in their new form.

in the precursor list, asserting the target forwarding pattern as fully as possible for that precursor, without taking over any already occupied channels at each stage in the list.

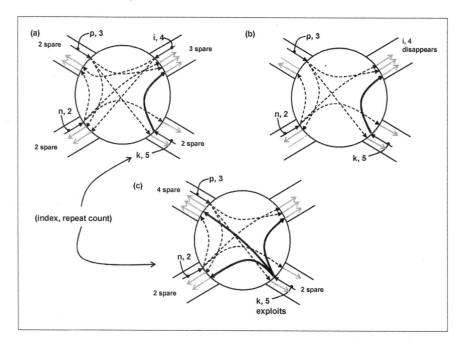

Figure 8-17. Example of the game-like rules of the SHN Tandem node which self-organize restoration pathsets: (a) four incoming statelets and their forwarding patterns. Index *k* is not fully satisfied. (b) the precursor for index *i* disappears (as may happen from reverse linking elsewhere). (c) Index *k*'s ranking now allows it to exploit available outgoing ports for a full forwarding pattern.

A reverse-linking event at a Tandem node is detected when a complement state arises in a port as the result of a new incoming statelet in that port. In this case the complement-forming statelet is copied to the outgoing side of the port where the precursor of the corresponding forward flooding statelet is located. This creates a complement condition in that port as well thus propagating the complement condition. The Tandem node then also sets the status of both ports to "working" (or equivalent) status and internally commands a cross-connection between the two complement ports. When reverse-linking on an index happens at a Tandem node, all other outgoing statelets on that same index (which are not part of the complement-pair) are cancelled. The basic forwarding pattern is then re-asserted amongst all remaining "unsatisfied" indexes at the node. This exploits the collapse of the forwarding tree for the reverse-linked index to benefit other indices that may still by vying for complete forwarding at the node. Once reverse-linked, any subsequent appearance of a statelet bearing the reverse-linked index is ignored at the node. Figure 8-17 gives an example of how the low-level rules interact at a Tandem node. Other detailed examples of applying the basic forwarding pattern for indexes by rank order of precursor repeat count can be found in [Grov97].

The interaction between nodes resulting from the forwarding rules, combined with the channel-releasing effects of reverse-linking, result in the following frequent low-level effects at Tandem nodes during self healing process.[141] (a) The precursor location for an index shifts (a lower repeat count appears in another port for an existing index). In this case the broadcast pattern for that index is re-rooted onto the new precursor port and the composite forwarding pattern is adjusted for consistency with the overall state rules above. (b) A distinct new index appears at the node: In this case the composite broadcast pattern is adjusted for consistency with the above rules, fitting the new index into the composite pattern on the basis of its repeat count and rooting its individual broadcast pattern on the single incoming instance of this index as its precursor. (c) A precursor disappears: In this case, if there is one or more other incoming statelets for the index of the precursor that disappeared, the rules will adopt the best of its kind as the new precursor. The forwarding pattern will be re-rooted to the new precursor, and the forwarding of that index will be revised in light of its new precursor location and integrated into a new composite pattern consistent with the rules above. If there are no other statelets present with that index to use as a new precursor, then all outgoing statelets for that index are removed and the composite forwarding pattern is revised to exploit the newly free outgoing channels to the benefit of any non-satisfied precursors. This is the type of example shown in Figure 8-17 above. (d) An index family collapses from reverse-linking at this node leaving one or more free outgoing channels. In this case, the composite forwarding pattern is reviewed, resulting in the possible extension of the pattern for one or more indices not previously enjoying a full target pattern of forwarding at the node.

8.4.3.2 The Emergent Network-Level Behavior

It is the nature of a self-organizing process that large-scale or "emergent" behavior that arises out of the low-level rules for interaction is not usually obvious. Let us try therefore to give an intuitive appreciation of how the

---

[141] Note that what follows from here is not a further specification of SHN rules themselves. The rules are complete as of the paragraph above. Rather, these are behaviors that can be observed at the level of the Tandem node as a whole arising from assertion of the low-level statelet-interaction rules above.

low-level rules above self-organize restoration path-sets at the network scale. First, each index family initially expands from the Sender, building a tree of precursor and forwarding relationships through those nodes. This tree is not regular or the same for all indexes, however. The form they evolve into depends on which nodes and spans each index is able to obtain forwarding onto under the Tandem rules. Through these rules each either expanding (or collapsing) index tree is interacting with all others concurrently also seeking to form a path. As soon as any index reaches the Chooser, its forwarding tree begins collapsing through reverse-linking onto a single path that re-traces the precursor sequence for that index. Freed channels arising from reverse-linking are immediately incorporated into updated forwarding patterns, expanding unsatisfied index families. If and when these still-expanding index trees also reach the Chooser, they too collapse under reverse-linking onto a single path releasing channels for other indices, and so on.

The emergent behavior is thus a rapid, completely self-paced, multi-lateral expansion of index trees in contention with others, punctuated by reverse-linking collapse of some index trees and further expansion of other index trees as a result. Without reverse-linking the process would result in a stable set of trees rooted at the Sender node. But index collapse advances other index trees; which in turn expand, succeed, and then also collapse. This all occurs asynchronously over the network as a whole limited only by nodal processing delays and link transfer times. The overall behavior is neither sequential nor wholly parallel in operation: it exploits the parallelism of the network to the greatest extent possible while and where the Tandem rules encounter no resource contention in forward flooding. Where contention arises, resources are allocated to always augment the index trees containing the shortest progression from the Sender at each incoming direction at the Tandem nodes. This may temporarily block one index tree in certain network regions, while the same tree grows in others.

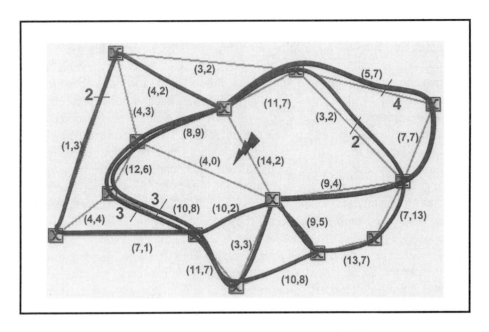

Figure 8-18. Example of a pathset self-organized by the SHN protocol.

The whole process automatically halts if 100% restoration is achieved, because the Sender then suspends any surplus initial statelets and all other indices have at that point been collapsed onto paths by reverse-linking. If the spare capacity of the network is not adequate for 100% restoration, the process stops evolving when no further reverse-linkings occur. In this case a Sender time-out suspends any uncomplemented initial Sender statelets, evaporating index trees that were not reverse-linked. In either case the final state of the network is one in which only complementary statelet pairs persist, each now tracing (and holding) the full length of one restoration path through the pool of spare channels on the graph, between Sender and Chooser nodes. The spare channels used now have working channel status (i.e., the post-restoration network state is immediately up-to-date should a subsequent cut arise). Figure 8-18 conveys through an actual example of SHN behavior, the sophistication and efficiency of the pathsets that it forms spontaneously, from available spare capacity, once triggered. The network shown has 66% redundancy and only the theoretically minimal amount of spare capacity to permit 100% restoration of the failure span shown. The restoration pathset is formed and cross-connected spontaneously as a single pattern of 14 paths formed over five distinct routes. The pathset is equivalent to a max-flow solution in the spare capacity present. This is a relatively complex example, chosen primarily to emphasize the capability and convey the basic concept. But the mechanism and corresponding spare capacity design of the network can both also be employed in a way that no restoration path is greater than some maximum desired length or in a way so that no more than a desired number of distinct routes are ever used.

8.4.3.3 Applications of the SHN Protocol

Extensive experimental studies (for example in [DuGr94] [GrVe90] and other studies summarized in [Grov94]) have confirmed that the path-sets that are spontaneously formed by this self-organizational process are essentially equivalent (i.e., over 99.9 percent of trials) to optimal centrally computed maximum-flow path-sets between the Sender and Chooser nodes with global knowledge of network state. As such, the SHN mechanism is actually a generalized tool for spontaneous path-set formation or discovery in networks between any two nodes and can be used in applications other than restoration. "Capacity Scavenging" is a name given to its use in automated path provisioning and when running in a pre-failure mode to conduct "mock" restoration trials, it can also be used for self-organized generation of distributed restoration pre-plans which can be rapidly activated in real-time upon failure [Grov94]. In fact the technique of Distributed Pre-Planning (DPP) using the SHN protocol in this way, really constitutes the first overall approach through which the capacity-efficiency of a "shared mesh" restoration scheme could potentially be combined with the same kind of speed that rings offer. The shared mesh efficiency arises because spare channels are not dedicated. The speed arises because all that has to be done in real-time upon a failure is to flood out the alarm information, which propagates as rapidly over the hops of a mesh as it does over the hops of a ring. Each node then immediately activates it pre-planned cross-connections to deploy its part of the overall pre-planned restoration path-set for the respective failure. In fact where signaling times dominate over cross-connection times, a DPP-based shared mesh restoration scheme actually stands to be faster than a BLSR ring in many cases because the ring may have up to 15 hops of signaling dissemination, but all of the nodes participating in forming mesh restoration path-sets are inevitably within 5 or 6 hops (perhaps 9 at the most) from the failure span.

In a practical implementation the SHN mechanism would operate autonomously in reaction to working channel failures or on a triggered basis for a variety of other applications built upon this path and path-set forming utility being embedded in the network. Following a restoration reaction, however, each node that participated in the event would subsequently notify the network operations centre (NOC) that it has an after-action report to send. The NOC collects the after-action reports and integrates them into a global picture of the fault and the recovery pattern that has been deployed spontaneously. At this stage the NOC has all its ordinary options and authority. The process acts autonomously for only a few hundred milliseconds and even then it operates only on spare channels. It can also be invoked in mock-failure trial mode at any time to observe the restoration patterns that would be formed. In fact, if the network operator is still uncomfortable with letting the process act autonomously, they can still use it only in trial mode simply to generate (and frequently update) protection pre-plans in each node. These can even be inspected and approved by the operator before enabling their use. Reversion after physical repair can be handled conventionally by NOC-to-OXC commands, explicitly returning each rerouted path to normal, or it can be automated by telling either the Sender or Chooser to simply "let go of" the holding statelets. (A temporary local record of the traffic substitutions made at Sender and Chooser nodes remains to tell them which ports to switch the affected traffic back into.)

8.4.3.4 Misconceptions about Distributed Restoration

Following the initial work on the SHN, other schemes were proposed: FITNESS, REACT, Two-Prong are a few. The basic ideas of the SHN are the same in these other schemes, but details vary to achieve different properties such as using packet message-based signaling, or finding paths with the largest concatenated OC-n replacement bandwidths as single units. A comparative discussion of various DRA proposals from the 1990s can be found in [Grov94]. A key point to address about DRAs in general, and especially the SHN, is the persistent assumption that DRAs will be too slow for use in real-time restoration and that dynamic restoration of any kind is always a "best efforts" approach that is not assured to give specific restoration guarantees. The latter misconception has been reinforced by advocates of the completely ad-hoc mass re-dial approach to restoration, which *is* without any assurance as to its outcome. The assumption seems to be that any scheme not involving preplanned solutions is inherently un-assured. However, the key to offering restoration guarantees is for the DRA has certain built-in performance properties characterized by *Path Number Efficiency (PNE)*. Restoration levels can then be guaranteed as a property of the spare-capacity design of the network. If PNE=1 it means that the DRA is capable of finding the maximum possible number of restoration paths out of any given environment of spare capacity. For example, in Figure 8-18, 14 paths is the limit that any method can achieve in the given spare capacity environment, so for that one failure scenario, *i*. the $PNE(i) = 1$. A DRA has an overall PNE=1 when it always produces the maximum number of possible paths, or a number equivalent to some reference algorithm. In this case off-line determinations (or self-audit via background mock-failure trials with the DRA) can both determine that restoration is assured in advance for all intended working paths on each span they cross because the restoration outcome is uniquely predictable from the spare capacity design alone.

Let us now talk about the speed achievable with a DRA, especially where the concern is about large-capacity failures. FITNESS uses explicit interprocessor messages between nodes and aggregates restoration flows as much as possible. For instance, if 36 working STS-1s fail, FITNESS will try to recover all of them as a group over one OC-36-capable restoration route, if possible. If not, its first iteration will fail and in another iteration it may discover instead an OC-24 route, followed with another iteration that discovers an OC-12 route. The motivation for aggregation is speed. But as with other DRAs as well, this is required primarily because the DRA is iterative—it finds its restoration routes one-at-a time, successively, not concurrently as a single overall path set. In contrast, the SHN is a single-step highly parallel-acting protocol that inherently discovers and exploits all routes simultaneously *in parallel.* What is often not realized about the SHN is that it *does not iterate at all.* Paths are *not* found one after the other in sequence. The SHN is triggered only once with the full magnitude of the restoration path number required and forms the complete required path-set in parallel, as a single pattern. As a result, the SHN can find all required or all feasible paths at the finest resolution desired, in a single parallel pattern-forming process on the network, making the speed of suitable implementations of the SHN essentially independent of the number of restoration paths required. Thus for example, compared to the example above explained how FITNESS would eventually find both the OC-12 and OC-24-capable routes after perhaps three iterations. The SHN finds both routes, and fully exploits their available capacity, as well as any other routes that may be involved, in a single pattern-forming phase.

The reason that FITNESS, REACT, Two-Pong and others all *iterate* to build up the required or desired path-set is that they do not embody an internal solution to the problem of coordinating capacity allocation amongst a set of *simultaneously* formed distinct restoration paths. The SHN solves this problem through the unique self-organizational effect of its Tandem node rules. Without some such solution to this problem, using resources to restore demand early in the process can potentially block restoration of another demand, especially as in FITNESS where the attempt is being made to find the maximal contiguous bandwidth blocks. Properties like this can lead to an undesirable requirement that the restoration mechanism proceeds to select routes in the same order used in the planning stage, and this does not help advance the adoption of DRAs. It is not guaranteed that restoring the largest bandwidth first always leads to the best possible use of capacity.

One proposal to put all DRAs on a common framework for performance assessment is based on the PNE concept. This links the theoretically maximal number of restoration paths that can be formed in a network to the spare capacity distribution in the network and experimentally validates a DRA relative to the achievable maximum. A DRA has a PNE of 100% only if it can be demonstrated that it finds or forms pathsets that are equivalent to the maximum flow that is theoretically possible within the spare capacity that is present in the network. If a DRA is tested and qualified as having a PNE of 100%, then all aspects of specifically having to place spare capacity on the network in amounts in excess of the theoretical minimum and/or distributed in particular ways so as to "guide" the DRA to correct operation are removed. If the embedded DRA is known to have PNE=1, then the network planner need only solve for and provide any distribution of the theoretical minimum spare capacity required—the DRA will find it and fully use it to achieve 100% restoration, by virtue of its PNE=1. Extensive testing of the SHNs PNE was conducted (in studies such as [Grov97] [DuGr94] and in extensive unpublished testing in industry evaluations), and its PNE is well-validated to be 100% (against a reference criteria of the perfect k-shortest paths replacement pathsets and well within 1% of the optimum max-flow solutions). Other DRAs were not tested this way, however, and required excess capacity to realize 100% restorability. This, in addition to the iterative nature of other DRAs in part led to the impression that DRAs in general would be slow and unassured in their outcomes. But with the passage of time, the simplicity and robustness of the maximally parallel self-organizational nature of the SHN does suggest its reconsideration for possible uses in future transport networks.

Perhaps because DRAs, especially the SHN protocol, are based on the idea of simple low-level rules leading to desired emergent behavior at the network scale (rather than explicit signaling and re-routing computations), there seems to have been important misunderstandings about them. One of the most common misunderstandings is to think of statelets as if they were conventional inter-processor message data packets. Another is to assume that the forward flooding process involves an exponential growth of flooding messages. Actually, statelet forwarding occurs in a controlled and limited way under the Tandem node rules and is strictly limited by the number of available spare channels on each span. Individually or together, these misimpressions lead to the presumption of high "messaging loads" with the SHN protocol. A quick look again at the Tandem node rules above easily shows that the forwarding is not exponential. But more important is to appreciate the ways in which distributed parallel state-driven interaction via statelets differs from serialized processing of queued inter-processor messages.

In classical inter-processor messaging, such as over the SONET DCC channel as an example, all information must pass through a serial communication channel between nodes and communication protocol stacks at both ends to handle re-transmission for errors and recovery from packet loss. This also requires buffers of adequate size, and protocol details for flow control. These considerations often do make it difficult to meet the speed objectives for restoration with any kind of distributed packet communications method to solve the problem. In contrast, the state-based interaction introduced in the SHN is potentially very fast because no queuing or communications protocols are involved, the spatial position of links on which signaling arises also encodes information relevant to the problem, and nodal interactions are highly parallel in nature. All of the dynamically changing network state needed at any time during the SHN execution is found in the statelets on the links of the network itself. The protocol action is always defined by the current state of links surrounding the node, not by previous messaging histories. In effect, the network surrounding a node turns into an active memory space within which the protocol executes. Nodes are directly coupled in this particular shared-memory sense. The protocol execution at any one node automatically updates the memory space within which adjacent nodes are concurrently executing, without any explicit communication between processors.

Also, with statelets, state changes propagate in the physical layer, are detected in hardware, and can be processed on an interrupt-driven basis by an event-driven, finite state machine (FSM) implementation of the DRA. The process does not even need to involve the main CPU of the cross-connect node. The cross-connect commands that the protocol deduces should be made in the local matrix can be directly coupled into the matrix controller. This is somewhat analogous to the way a peripheral device may be granted direct memory access (DMA) to a certain memory range without involving the CPU. Similarly, the SHN protocol task can be allowed direct access to the matrix and limited to requesting cross-connections only between spare channels (and substitution of failed working ports into restoration path ports). In processing statelets the SHN protocol requires no receive "message" buffers: a new statelet *overwrites*, rather than follows, any previous statelet. The host OS has no communications handling overhead to support the DRA; it views the protocol simply as an interrupt handler for statelet change events. And the port registers of the host cross-connect that receive (and transmit) statelets from (to) incoming (outgoing) signals *are* the memory space within which the SHN executes. Figure 8-19 illustrates.

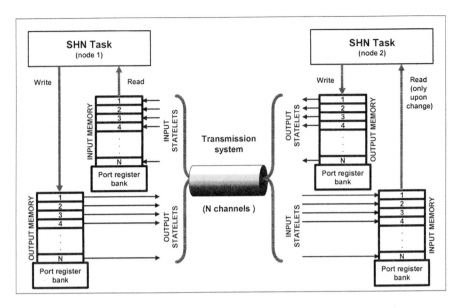

Figure 8-19. Functional model of how the SHN executes directly within a memory space that is coupled to its neighbor nodes comprised of the array of channel-associated statelets received from and transmitted onto each spare channel at the cross-connect.

Certainly if each change in a statelet is counted as a "message," then this paradigm for distributed interaction may definitely generate more "messages" than other means. But each spatially distinct state change is not a complete message to be processed in the usual sense. If warranted, signal grouping or bundling can also be applied to reduce the volume of state-changes to be processed at each DCS. In a bundled implementation a lightweight skeleton of spare channels is designated as a set of statelet bearing channels. Whatever is worked out for the cross-connection of those channels in an SHN event is then asserted identically for associated groups of other non-signaling spare channels. Scaling to an arbitrarily deep capacity environment is thus possible. Statelets are also robust to errors without requiring any buffering and ACK/NACK confirmations and message re-transmission. Simple triplication persistence testing is effective. In other words a statelet change event is simply not recognized until it has been repeated on three successive frames. If even more protection is desired, a checksum can be embedded in the statelet format to validate state changes on the first appearance, and waiting for the repetition only when a checksum fails. Finally, as mentioned, the best way to think about statelets is that they are like the "mesh equivalent" of the SONET K1-K2 bytes that drive the ring switching FSM protocol—but they instead mediate a distributed parallel formation of an entire path-set between any two nodes in a network.

Another misconception is that a DRA's speed depends on network size (total number of nodes). The presumption seems to arise from the notion that in a larger network, it takes more time for the flooding process to reach to the farthest edges of the network. But the key thing is that a process such as the SHN acts in a highly parallel way within the vicinity of the failure itself. It doesn't actually matter what size the complete network is because the reaction is always developed and only involves a selection of nodes within a pre-defined hop-limit from the failure. The network could be infinite in extent. While capacity depth and degree of connectivity can affect speed in more subtle ways, the total size of the network itself is actually irrelevant. By capacity depth we mean the sheer number of working channels to be restored in a typical failure. In a perfectly parallel implementation (where there are no significant delays in the SHN protocol processor or hardware itself), there are again logically no speed implications at all, because everything is happening perfectly in parallel. In practice, however, if the volumes of statelet change events to process become a source of significant delay, then the technique of bundling can be used. Bundling in effect scales down the magnitude of the capacity's involved in the problem by designating every n^{th} channel as a live statelet-bearing channel and applying the same cross-connection decisions that result for that channel to all n channels in the bundle. Finally, and somewhat related, is the common practice of using linear additive expressions to calculate restoration times. With this kind of highly parallel self organizing process, the path-set is formed simultaneously as a single pattern. Paths are not formed one after the other in a purely sequential way. Nor is each individual path formed as a direct succession of steps each with a fixed delay on each hop. It is therefore misleading to assume that restoration of 100 working paths will take ten times longer than for ten paths. Because of the parallelism involved, implementations using fast firmware (or even hardware) for execution of the SHN finite state machine, and/or bundling, with optical cross-connects that operate with microsecond switch times may take only a few percent longer to find the 100-path restoration pathset than that for ten paths. In fact in the limit of low cross-connect switching times, the time for SHN pathset deployment can be strictly independent of either network size or failure size because the SHN can be used in its DPP (pre-failure mock trial) mode to establish and continually maintain up-to-date action lists for each failure, in each node. All that has to be done in real time is failure announcement to nodes within a given radius of the failure. With a simple hop-limited flood announcement via statelets, the time for activation is thus low and constant regardless of capacity affected or overall network size.[142]

---

[142] When you throw a stone in a pond, the time for the wave to reach a certain distance away

8.4.3.5 Extension to Self-organizing Path Restoration: SHNp

Span restoration involves an inherently simpler form of restoration re-routing problem than path restoration, because the total replacement flow in span restoration has a single pair of end-nodes. Under path restoration the same total number of replacement paths has to be found but now they are distributed over a set of different end-nodes, making path restoration inherently a multi-commodity simultaneous flow-routing problem. Theoretically therefore, path restoration ideally requires solving multi-commodity maximum-flow (MCMF) routing problems (with upper limits on each commodity required to be routed). In this section, we briefly outline the central ideas of extensions that have been developed for the SHN that enable it to work as a *path*-restoration DRA. To avoid confusion, the path oriented version is called SHNp ("p" for path). Path restoration is an important extension because a path-DRA inherently copes better with a node failure or multiple span failures than does a span-DRA. In addition, it can provide a highly desirable form of maximal overall recovery against unplanned failure scenarios of arbitrarily severe damage such as in a "9/11" type of situation. With space in mind, this section is limited only to outlining some of the basic principles by which the SHN is extended to path restoration. More details can be found in [IrGr00] and references therein.

First, for path restoration the statelet format is extended to also include the Origin-Destination node name pairs of affected paths. The other main difference is an Interference Number (IN) field that is used in accumulating interference measures associated with potential restoration paths for different O-D pairs. Following either a span or node failure, surviving nodes adjacent to the failure substitute an AIS signal in the path. The AIS indications propagate back to the end-nodes of the affected O-D pairs, activating the path DRA in each affected end-node. A Sender-Chooser role arbitration is performed by such nodes for each destination to which it has lost demands. The Sender then creates initial statelets identifying the respective O-D pair and enumerating all such primary statelets with locally unique index numbers. The primary aim at this stage is to place one statelet on a spare link in each span connected to the Sender for each unit of lost demand for each distinct O-D pair of which this node is a Sender. This is effected through a concept of "internal" statelet arrivals. The Sender also creates initial interference values for all primary statelets at this stage. A key difference in the path restoration concept is that for any given failure, a node may be required to function as Sender-Chooser for its *own* affected paths, while

---

from its impact point is the same as if you throw a stone into the ocean. The time for the wave to propagate to the *edge* of the ocean is much greater, but that isn't what matters.

simultaneously functioning as a Tandem node to assist restoration of *other* O-D pairs affected by the same failure. This is why the nodes own need for restoration paths are handled by the concept of internal statelet arrivals. With this framework, all statelet handling by the Tandem node functionality can be unified by referring the node's own needs for restoration paths as a Sender/Chooser to a notionally collocated client end-node. The internally arriving statelets then come from this imaginary co-located client node which has actually sustained the end-node loss of affected paths with this destination or origin node.

Primary statelets thus propagate from each Sender to adjacent nodes. Such nodes may already be in a Sender or Chooser state for their own lost demands, or idle. In either case, an instance of the Tandem node protocol is created in these nodes. The Tandem node rules manage the contention amongst incoming statelets for subsequent re-transmission on the spare links available at that node based on incoming INs, secondarily by repeat count, and a final tie-breaking rule. It is the action of the Tandem node logic in rationalizing which statelets get forwarded where that affects an approximate solution to the mutual capacity constraint problem. INs are updated for each outgoing statelet based on the total local contention of statelets vying for propagation into each local span. Statelets initiated from the node in its Sender role are integrated into the same overall competition, having the advantage, however, of being internal arrivals with an IN of zero.

The Chooser responds with a reverse-linking (RL) statelet to any incoming statelet identifying an O-D pair to which it has lost demands and not yet responded. This triggers a reverse-linking process conceptually similar to that of the SHN, but with additional considerations that allow a reverse-linking process to legitimately fail en-route. In such circumstances, the failed RL statelets are rapidly cancelled and re-initiated by the Chooser. The reverse-linking process is, however, inherently locked in and firmly reserved once it reaches the Sender. Consequently, like the SHN, this is a "two-phase" protocol (forward flooding, reverse linking) that does not need an explicit third phase for final path confirmation. Each RL process causes collapse of the remnant statelet family, the freed link resources are exploited by other statelet families under the Tandem rules, and so on until a complete arrangement of mutually feasible, near-maximal replacement path-sets between O-D pairs is established.

In a Sender, one internally arriving statelet is created per span terminated at this node for each lost demand unit. These primary statelets bear the O-D pair name, an index number, and, while inside the node, an IN and Repeat value of zero. Records are kept in the node indicating the number of lost and restored demand units for each respective O-D pair. A return to idle follows after this initial 'Sender' work. This occurs for each distinct O-D pair of which this node is a Sender. All internal arrival statelets created as part of the Sender function are mapped onto available outgoing spare links and assigned an outgoing IN by the same logic that applies to competition between all statelets as described in the Tandem logic. The only other Sender function is, later in the process, to recognize a reverse-linking statelet when it arrives on a given combination of O-D pair and index. This arrival marks the discovery and locking-in of a guaranteed path for one demand unit that is available to a desired destination. The Sender then has to select and substitute a live traffic signal into the restoration path and check if restoration for this O-D pair is now complete, and if so, remove the corresponding internal arrival statelet from the outgoing link competition.

The Chooser has the simplest function. After initializing the number of lost and restored demands, it waits for the arrival of a forward flooding statelet (i.e. a statelet with the RL bit not set) from an O-D pair for which this node is the Chooser. If this statelet has an index to which this node (i.e. the Chooser) has not yet responded, and restoration for this O-D pair is not complete, the Chooser responds with a reverse-linking statelet. The RL statelet is created by setting the RL bit, and copying the O-D and index information from the incoming statelet to the outgoing statelet. Two events can subsequently occur from the Chooser's viewpoint: either the next event on the respective port is the arrival of a substituted live-traffic signal (which implies that reverse linking was successful and that the Sender has used this port for restoration), or, a new statelet arrives cancelling the prior reverse linking attempt. In the latter case, the Chooser resumes the watch for any unresponded statelet from that (O-D, index) family, and if one is immediately present, moves the root of the reserve-linking attempt to that port.

As in the SHN, the Tandem node rules are the key to the DRA's path forming efficiency. In SHNp they must effect a distributed resolution of the contention amongst prospective restoration paths for the finite capacity on each span. The Tandem node rules reuse the precursor and complement-port concepts in forward flooding and reverse-linking from the SHN, but the rules differ in the following main regards: (i) the main metric of competition amongst statelets is IN, not repeat count, (ii) IN updating is not a simple incrementing of values in retransmitted statelets as it is for repeat count, and (iii) nodes acting as a Tandem for some O-D pairs may also be Senders and Choosers for their own affected O-D pairs. In addition, the competition now has to be mediated over all O-D index families, whereas the SHN effectively involves only one O-D pair. First, all of the external and internally arriving statelets at a node are identified, and the precursors for each (O-D, index) family identified by finding the member of that family with the lowest incoming interference number (ties are broken by repeat count, then by index number if needed). By definition every internally arriving statelet attains precursor status. At this point the whole roster of demand for outgoing spare links is identified and a calculation is done of the total number of free links that would be required in each span to forward all valid statelets on that span. If this number is equal or less than the number of spare links physically present on that span, the span is said to have an IN of zero. If the difference between the total number of precursor statelets which aspire to be forwarded on a span, less the number of spare links present on that span, is a positive number at a particular instant in time, the span is assigned an IN equal to this difference.

The competition for actual forwarding onto the available spare links in each span is by ranking all precursors by order of increasing (incoming) IN (resolving ties by repeat count, etc.). With the complete ranking established, each precursor is visited and a statelet from that precursor's family forwarded on every span (except the span containing the precursor) capable by its spare link resources of sustaining one statelet. The forwarding pattern for each precursor is thus fully or partially satisfied depending on its rank and spatial relationship to other spans. Obviously, and importantly, internal arrival statelets do well in this process, "arriving" as they do with zero INs. And in general, some lower ranked statelets may do better than higher ranked ones, because a side effect of the rules is that where precursors arrive spatially (i.e., in different ports) at the node can also tend to promote or inhibit its competition over other statelets. One does not, therefore, stop proceeding down this list at the first entry for which no forwarding is possible. Lower ranked entries may still find forwarding opportunities. Finally, each outgoing statelet has the IN of the span it is forwarded into added to the incoming IN of its precursor statelet.

When the desired forwarding pattern of every statelet on the list is served to its fullest extent possible, in rank order, the node state is complete and current. Any new statelet that arrives has to be assessed to see if it warrants take-over of an existing precursor role at the node, or if it represents a new (O-D, index) family. If either of these events occur, or if a reverse-linking event occurs, or if a precursor statelet disappears (because of reverse-linking elsewhere), the ranking and forwarding rules are re-applied to keep the outgoing statelet pattern consistent with the nodal forwarding rules. The net effect is a distributed approximation to the centralized calculation of a multi-commodity maximum-flow solution. The "mutual capacity" issue which makes good solutions to path restoration routing problems hard is addressed by the strategy of statelet forwarding based on minimum Interference Numbers. By preferentially forwarding low IN statelets and accumulating interference numbers, a set of mutually cooperative and "courteous" routing choices is achieved within the available spare capacities.

8.4.3.6 On the Significance of Self-organizing Path Restoration to "9/11"-type of Events

Unprecedented recent acts of deliberate sabotage give a new importance to the approach of self-organizing recovery from failure. A feature of a truly self-organizing restoration process such as the SHNp is that it does not require pre-plans or any kind of comprehensive network-state database. In the traditional context for network survivability, the main focus is on single paths or single span failures as the most common failure scenarios. Single path failures arise from naturally occurring interface card or other electrical or optical component failures.   Span failures arise from cable-cutting accidents or shark bites, etc. Schemes using centralized control or involving distributed database synchronization such as GMPLS / SBPP above, are generally suitable solutions for these expected types of failures. But in a case of arbitrarily severe disaster, the SHNp protocol is of special interest because it has the ability to knit back together as many affected paths as possible without *any* preplanning or state databases. This gives it a special relevance in the face of arbitrary attack scenarios such as 9/11 tends to make us think about. Most existing design measures and mechanisms for ensuring survivability assume the potential failure scenarios are known and specifiable in advance (say individual spans and nodes, and possibly a few specific combinations of spans where these are linked by common-cause failure elements such as a bridge crossing).   In normal contexts working paths that are unaffected by the failure also remain untouched by the restoration process so as to minimize disruption to the network's customers. However, when dealing with catastrophic failures, the greatest concern will be for critical (i.e. emergency) communication services. All non-critical demands of lower priority can be considered preemptible, and any pre-failure traffic, including that of critical systems, would be allowed to be rerouted as needed. Network-wide reconfiguration with the goal of maximizing the post-failure survivability of critical traffic would be the ultimate goal and the SHNp is particularly well suited to this type of global reconfiguration. In the face or arbitrary disaster, the goal would also not be to achieve 100% restoration, but instead to achieve the maximum feasible restoration for critical services. This is an area of new research, specifically aimed at embedding a form of last-resort mechanism that would kick-in only upon recognition of major, un-recovered failure situations, and weave back an almost theoretically maximum pattern of recovery of designated lifeline services in the face of whatever actual set of demands are affected and whatever environment of surviving capacity and nodes exists.

8.5.   P-CYCLES: BRIDGING THE RING-MESH DICHOTOMY

If you were to ask someone involved with transport network architecture over the last decade "what are the best attributes of ring and mesh approaches," one would most likely hear that with rings it is the speed, structural simplicity, and the determinism of the rerouting process. On the other hand the best features of the mesh-based approach are the greater spare capacity efficiency and the flexibility of handling growth, changes, and multiple classes of survivability. Rather remarkably, *p*-cycles, which were only recently proposed [GrSt98], combine all these advantages in a single scheme for survivability. As such they are under increasing study as one of the most attractive new technology options for carriers in the next generation of transport networks.

p-Cycles are in some ways like BLSR rings, but with support for the protection of *straddling* span failures as well as the usual protection of spans on the ring itself. An important property is that *p*-cycles are fully pre-planned and pre-connected so that when a failure happens, only the two end-nodes of the span do any real-time switching. Unlike more generalized span or path restoration, no switching actions are required at any intermediate network nodes. This property is only found elsewhere in 1+1 APS or rings. *p*-Cycles thus have an inherent speed advantage—they inherit BLSR ring-like speed.

Less obviously, however, it has been found that by admitting the protection of straddling span failures, 100% restorable networks can be designed with essentially the same capacity efficiency as a span-restorable *mesh* network, which can be three to six times more capacity-efficient than ring-based networks. Another important form of efficiency is that *p*-cycles are "spare capacity-only" structures that (unlike rings) do not constrain the routing of working paths to coincide with the layout of the cycles themselves. Working paths are free to take the shortest routes between their end-points on the graph.

8.5.1   Span-protecting *p*-Cycles

Figure 8-20 illustrates the operation of basic *p*-cycles. A single unit-capacity *p*-cycle, as in Figure 8-20(a), is a closed path composed of one spare channel on each span around its cycle. When a failure occurs on a span covered by the cycle, the *p*-cycle provides one protection path for the failed span, as shown on Figure 8-20(b). In this aspect, *p*-cycles operate like a unit-capacity BLSR. But *p*-cycles also protect so-called "straddling spans"–spans that have end nodes on the cycle but are not themselves on the cycle, as shown in Figure 8-20(c). Significantly, because the *p*-cycle itself remains intact when a straddling span fails, it provides *two* protection paths for each straddling span failure scenario and straddling spans themselves require *no* spare capacity. This apparently minor difference actually has a great impact on the capacity requirements of *p*-cycle protection. When optimized for spare capacity, *p*-cycle network designs are often exactly as efficient (or within a few percent) of the capacity efficiency of a similarly optimized span-restorable mesh network. In fact it has been shown theoretically that *p*-cycles are the most efficient pre-configured pattern possible for network protection [StGr00]. If hosted on OXCs they have the added advantage that they can be created and updated on a unit-capacity basis without the inherent modularity present in rings. These combined attributes have made *p*-cycles an option that is of considerable recent interest and study.

One remarkable property is found with a particular type of *p*-cycle, one formed on a Hamiltonian cycle in the graph. A network is said to be Hamiltonian if a single cycle exists which visits all nodes exactly once. If a network is Hamiltonian (and contains suitable working capacity still requiring protection), it permits formation of *p*-cycles that are especially efficient in terms of the ratio of working capacity protected to the amount of protection capacity used to form the corresponding *p*-cycle [SaGr04]. The high efficiency arises because, for an investment of N unit-hops of protection capacity, in a network of S links, N + 2(S-N) units of working capacity can be protected, where N is the number of nodes (and hence, S-N is the number of links that have a straddling relationship to the Hamiltonian *p*-cycle). For example, a 25 node, 50 link network could contain a *p*-cycle that is only 33% redundant. In contrast, a ring protects N hops of working capacity with N hops of protection.

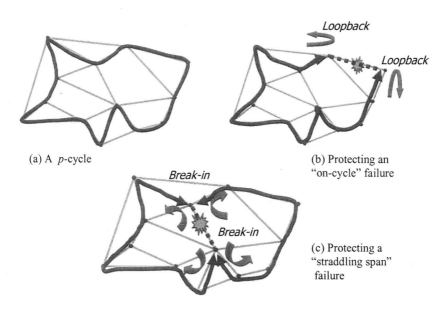

(a) A *p*-cycle

(b) Protecting an "on-cycle" failure

(c) Protecting a "straddling span" failure

Figure 8-20. The basic concept of *p*-cycles

This combination of attractive properties—ring-like speed coupled with mesh-like protection efficiency and unconstrained mesh routing for working paths—is not found in any other single scheme. *p*-Cycles have correspondingly been gaining much interest recently and are likely to be adopted in NGN transport at both optical and MPLS layers. Chapter 10 in [Grov03b] gives a more extensive overview of the concept and its literature to date. In addition a web-forum for the latest information on *p*-cycles has also been established [Sack04]. Some highlights of the work to date includes and studies on self-organization of the *p*-cycle sets, application of *p*-cycles to the MPLS/IP layer [StGr00], application to DWDM networking [ScGr02] [ScSc03] and studies on joint optimization of working paths and spare capacity [GrDo02].

8.5.2 Node-encircling *p*-Cycles

Node-encircling *p*-cycles [StGr00] [Grov03b] efficiently provide for the protection of transiting flows through a failed node. For a span failure, it is intrinsically possible (and is often the goal) to achieve 100% restoration of affected demands. But when a node fails, no method of network-level reconfiguration or re-routing can restore the originating or terminating demands at the failed node. Thus "node restoration" is really a misnomer. It is only pre-failure transiting flows through a node that can be restored by any type of network-level response. Source-sink flows at the failed node itself are inherently unrestorable by any type of network-level re-routing. Therefore span restoration and protection of transiting flows upon a node failure have inherently different target recovery goals. Perhaps counter-intuitively (because node failures can be viewed as equivalent to multiple simultaneous span failures) it turns out that 100% restoration of transiting flows affected by a node failure can often be feasible within the spare capacity required only for 100% span-failure restorability. The reason is that node recovery can only hope to address *transiting* traffic flows. The source-sink flows associated with the failed node itself are in effect (and of necessity) removed from the network recovery problem. For this reason it is often found that "100% node restoration" is easily achieved within spare capacity provided initially only with the intent of providing span restorability.

Node-encircling p-cycles provide an efficient strategy for protecting such transiting flows at every node, especially when applied in the MPLS layer to protect against router failure. In this context a set of virtual MPLS-defined node-encircling p-cycles can complement an underlying set of physical-layer span-protecting p-cycles using the same physical capacity. A node-encircling p-cycle provides an alternate path amongst all of the nodes that are adjacent to the failed node. To do this a node-encircling p-cycle must contain all of the nodes that were adjacent to the failed node, but not the failed node itself. This constitutes a kind of "perimeter fence" which is assured to be intersected at ingress and egress by all (transiting) flows that may be affected by the given node's failure.

A p-cycle, which protectively encircles a node, must be constructed within the sub-graph that results when the protected node is itself removed from the network. It may or may not be possible to form a simple cycle within the resulting sub-network. There is, however, always a logically encircling cycle construct for each node in any two-connected (pre-failure) graph. Chapter 10 in [Grov03b] provides details of the real time re-routing to and from node encircling p-cycles and how node failures are distinguished from span failures.

8.5.3   Path-segment or "Flow Protecting" p-Cycles

A third type of p-cycle provides a path-oriented correspondent to basic p-cycles, which are strictly span protecting. This is somewhat analogous to span and path mesh-restorable networks. Flow p-cycles generalize the protected straddling spans to the protection of contiguous flows of demand over path segments that may contain multiple spans and nodes. The protected path segments do not have to be end-to-end, but can be any segment of the whole path taken by a demand. In addition any bundle of demands that share the same segment of spans and nodes along their route can be treated as a protected capacity group, deriving protection from a correspondingly sized *p*-cycle. Flow *p*-cycles yield even somewhat greater capacity efficiency than regular *p*-cycles. Selective use of just one or a few flow *p*-cycles in conjunction with other protection schemes may also be attractive. One such use is to support transparent optical transport of express flows through a regional network. Another is to use a flow *p*-cycle around the perimeter of an autonomous system domain to provide a single unified scheme for protecting all flows that completely transit the domain. The idea of using both types of *p*-cycles in a network is presented in [ShGr03] [ShGr03b] as well as the related design and operational theory for flow *p*-cycles.

8.5.4 Efficiency of *p*-Cycles and *p*-Cycle Network Designs

Unlike other approaches to survivable network design, where the decision variables typically pertain to restoration path choices or flow assignments to eligible routes, *p*-cycles lend themselves rather handily to quite compact and useful measures of the potential or actual efficiency of individual *p*-cycles in different network and demand contexts. In contrast it is hard to identify useful general efficiency measures for individual candidate backup paths or eligible restoration routes. Such compact and characteristic efficiency measures enhance intuitive understanding about what determines a "good" *p*-cycle and also helps explain effects such as why an optimum solution may tend to contain large cycles, or even Hamiltonians. These efficiency measures can fairly easily be built upon to realize some quite simple and effective greedy heuristics, at least for the basic *p*-cycle design problem. In [GrDo02] (subsequently developed further in [Grov03b]), the *a priori efficiency* (AE) of a cycle as a candidate to be a *p*-cycle is defined as:

$$AE(k) \equiv \sum_{\forall i \in S} x_{i,k} \bigg/ \sum_{\forall j \in S} \delta_{j,p} \cdot c_j .$$

This is simply the ratio of the total number of protection relationships for individual working channels that cycle *k* can provide, relative to the cost of constructing a unit-capacity copy of a *p*-cycle on candidate cycle *k*. It is called the "*a priori* efficiency" because its main use has so far been to pre-select "high merit" candidate cycles large instances of the *p*-cycle spare capacity allocation problem. In this context it reflects the best potential efficiency achievable by the cycle if all its straddlers are fully loaded with two working channels each (but this may not always be the case in a complete design solution). When the actual working channel quantities (protected by the cycle as actually employed) are known, a closely related "demand-weighted efficiency" measure is defined [Grov03b].

The capacity efficiency of 100% span-restorable *p*-cycle designs can be remarkably attractive in practice and in a special case of semi-homogenous networks (probably most applicable for whole-fiber level protection) can actually *reach* the long-recognized lower bound on redundancy of *1/(d-1)* (see [Grov94]), where *d* is the network average nodal degree. A sample design follows to illustrate. Figure 8-21 shows a 13-node, 23-span test network, with *d*= 3.54. All span costs are 1 and the working channel counts on each span in Figure 8-21(a) are those that arise from least-hop routing of one demand unit between each node-pair. There are seven individual *p*-cycles in the solution and five unique cycles employed and the design is 100% span-restorable with a logical (channel count) redundancy of 85/158 = 53.8%. In practice this is a highly efficient design solution that could be used for protection at the lightwave-channel level.

But it is also possible with *p*-cycles to reach as little as 39.4% redundancy, which *is* the lower bound for span-protection if a single Hamiltonian cycle is used as a *p*-cycle in which all straddling spans are "fully loaded." Consider that in any such case we will have the ability to protect

$$\sum_{\forall j \in S} w_j = N + 2(S - N) = N + 2\left(\frac{N\overline{d}}{2} - N\right) = N(\overline{d} - 1)$$

working channels, thus making the overall redundancy exactly equal to

$$\sum_{\forall j \in S} s_j \bigg/ \sum_{\forall j \in S} w_j = \frac{N}{N(\overline{d} - 1)} = \frac{1}{(\overline{d} - 1)}.$$

Clearly for a scheme that retains ring-like switching speed and structural simplicity and determinism, these are remarkably motivating properties. Heretofore redundancies as low as 40% or so for 100% span survivability were only ever even approachable through end-to-end path restoration technologies that are far more complex to design and operate and much slower-acting in practice. It is important to understand that a span may be protected by more than one *p*-cycle. The total working traffic on a span, when it fails, may be re-routed over several distinct *p*-cycles that protect that span. The working traffic on a span may also consist of flows of varying granularities and be protected by *p*-cycles of the corresponding capacities.

p-Cycles also allow flexibility and adaptivity to growth and changing patterns of demand when cross-connects are used to create and manage *p*-cycles. Existing *p*-cycles can be recalculated and installed to cover new spans or new traffic patterns as they occur. The pre-configuration itself is a non-real time activity occurring in the spare-capacity only. Once a *p*-cycle is created by cross-connecting spare channels, restoration of traffic on spans protected by associated *p*-cycle is real-time.

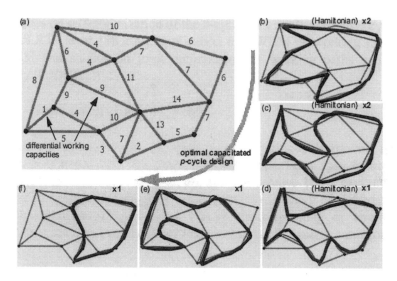

Figure 8-21. Understanding how a set of *p*-cycles efficiently protects a network [SaGr04]

It is also not necessary that all *p*-cycles in the network be created at once. One can create and change *p*-cycles as and when working paths are provisioned. At the time of provisioning a new working circuit (demand) between two nodes, a search can be made for existing *p*-cycles that cover spans traversed by the new circuit. If there exists a *p*-cycle with same bandwidth as the circuit and can be shared then there is no need to create a new *p*-cycle. If not, a new *p*-cycle needs to be created. If there is enough spare bandwidth along the route of an existing *p*-cycle then a new *p*-cycle is found. If there is no existing *p*-cycle or there is not enough spare bandwidth left to create a new *p*-cycle, a *p*-cycle discovery process can be initiated. If a *p*-cycle could not found then a warning and alert can be issued to prompt operations that an upgrade of network is necessary and such upgrade will go through the planning process. In fact as an ongoing matters of operations, the remain margin or service life of existing in spare capacity in the face of growth can be always monitored so that physical upgrades in capacity are planned as needed without ever encountering an actual exhaust scenario. As part of network upgrade, when a new link or a node is added to network existing *p*-cycles may need to be re-configured. In such situation, an incremental approach can be used where one *p*-cycle at a time is re-configured. Current research is looking at a number of highly flexible and simplifying operational strategies to accompany cross-connect managed *p*-cycles in terms of incremental reconfiguration for growth, post-failure reconfiguration to maximize secondary-failure restorability [ScGr04] (or to support the 1FP-2FR strategy for ultra-high availability services), and to support simplified automated provisioning for dynamic randomly arriving protected service path requirements [ShGr05].

8.6. OTHER ASPECTS OF NEXT-GENERATION TRANSPORT

In this section we close the chapter by offering further views on some of the most significant ways that we believe survivability issues will be handled differently in next generation transport networks. A few of the main future directions have already been developed; the swing to mesh-based networking principles, the adoption of "IP-centric" distributed control, the prevalence of the SBPP technique for dynamic provisioning of protected services, the eventual reconsideration of truly self-organizing network approaches, and the deployment of *p*-cycle based survivability solutions. Let us now touch on several other dimensions and issues, a bit more briefly.

8.6.1 Liberation from the "50 ms Myth"

How fast should protection or restoration be? A common answer is to the effect of "50 ms, of course—but don't ask me why." Some vendors and traditional network operators have long positioned 50 ms as an absolute necessity. But this is coming under much scrutiny and skepticism from a more data-centric standpoint. 50 ms is actually not a requirement but a specification of what was achievable in early automatic protection switching (APS) systems. There is essentially no linkage to what modern voice, data, leased line or video services actually can tolerate as interruption times without noticeable impact. In next generation transport networks, the restoration time requirements will be more rationally based on what services truly require. Several different restoration or protection times will be provided based on customer willingness to pay. What is at stake is significant because as long as 50 ms is insisted upon, 1+1 APS systems and UPSR rings are practically the only options. But these are massively brute-force solutions for survivability. Highly efficient shared-mesh transport architectures would be ruled out. It is therefore worth reviewing what is thought to be the real basis of the famous 50 ms requirement.

One viewpoint is that the figure arose as a requirement simply because that was what was technically feasible to achieve in early automatic protection switching systems (APS) (10 ms for fault detection, 10 ms for signaling, 10 ms for head-end and tail-end relays, and 10 ms margin). Then, as sometimes happens, the *capability* of one generation of equipment became a *requirement* of the next generation. In that way the figure becomes entrenched, even though not based on any actual service-related effects analysis. Another possibility is based on observing the numerical similarity between 50 ms and the time a DS1 reframe takes. Where a voice circuit is carried in a DS1, which is carried in a DS3, or over SONET directly, a failure of the transmission system will be protected by an APS function. After the transmission path is restored, the NE providing the DS3 sink will have to re-acquire framing on the DS3, and the NE terminating the DS1 will have to re-acquire framing on the DS1 signal. The re-framing process at the SONET level takes only microseconds, but the average time for the maximal-length re-framing hunt at DS3 and DS1 levels are about 2 ms and 50 ms, respectively. Thus, the DS1 maximum average reframe time of 50 ms is a worst-case minimum duration for any disruption that causes a reframing process to be triggered. The speculation is therefore that in the early development of specifications for protection switching times, 50 ms was stipulated in the absence of other criteria on the basis that it is of the same order as, and no worse than, the minimum interruption time implied by reframing considerations alone.

A more substantive theory relates the 50 ms figure to early values of the Carrier Group Alarm (CGA) time. The CGA is the duration of time above which, if a DS1 terminal bank has a loss of signal or loss of frame, it raises an alarm that would cause the associated switch to drop all the related connections in progress. In the early days of digital telephony, the CGA time was set at the stringent level of 230 ms. In this theory, it was then simply a budgeting consideration in that after allowing for fault detection time and worst-case reframes at DS1, 2, and 3 levels, plus some margin, there was only 50 msec left for protection switching to operate to "beat the CGA threshold." If this last theory is the true origin of the 50 ms specification, then it does stand on a weak footing because the CGA-related "call-dropping" threshold was subsequently raised to a minimum of two seconds. Voice circuits are quite robust to outages of less than a second. The subscriber will perceive such outages as a "pop" or "click" on the line. During longer outages, however, there is a danger that prior to the failure, the DS1 signal was carrying signaling information that should not be left unchanged for a long duration. (For example, if the signaling had turned on ringing for a subscriber's line, the subscriber would not want to have the phone ring constantly for the duration of a long outage.) Consequently, when outages persist for longer than 2 seconds (±0.5 s), the telephone network equipment will freeze the signaling states for the failed channels and begin dropping the telephone calls. But the fact that so many in the industry wonder where 50 ms comes from (but are unsure of deviating from it) is reflective that there is actually no service-based requirement for anything like 50 ms recovery speeds. This is confirmed by a comprehensive survey by Sosnosky [Sosn94] and by a series of experimental trials with virtually all types of services undergoing deliberately injected interruptions ranging from 50 ms up to near a second by Schallenburg [Scha01].[143] Schallenburg concluded that no service needed 50 ms restoration time to avoid undo effects and that all services tested could easily withstand at least 200 ms, with others tolerating considerably longer.

8.6.1.1 Understanding that Service Availability Does *Not* Significantly Depend on the Speed of Restoration

A further, important realization for planners and technology decision makers is that in network design to withstand any *single* failure, *service availability does not depend on the speed of restoration.* This may seem quite counter-intuitive. After all 50 ms is so much faster than say 500 msec or a second, that it seems reasonable at first glance to think faster switching means higher availability. However, virtually all restoration switching times are completely trivial sources of outage in determining overall service availability. To appreciate this, consider the two main things that interrupt service in a network that is already survivable against single-failures. They are: (i) switching times in response to restorable single failures (as intended by design), and (ii) outages arising from no restoration or incomplete restoration in the event of a second failure while in the first-failure restored state. Considering these two contributors to service outage, a simple numerical example shows that *availability depends on the probability of restoration in the face of <u>dual</u> failures, not on the speed of restoration to single failures.*

To illustrate this let us assume (as an extreme worst-case for argument's sake) that a service path through a restorable network is affected by four

---

[143] Yet another sometimes-encountered argument is that restoration times must be 50 ms "because the propagation delays in a voice-connection should not exceed 50 ms from an echo-loss-delay engineering standpoint." However, the linkage of these two considerations is a misunderstanding based on nothing more than numerical coincidence. Echo-delay in voice switching *in normal operation* has nothing to do with the restoration time needed upon failure. In the former, voice quality suffers from delay nearing 50 ms of *propagation* delay. But the latter pertains only to the length of the "click" or silence gap that would be heard in an affected voice connection arising from the restoration *time*. They are completely different issues, each with their own requirements. Restoration times makes no difference to the normal quality of the voice signal from an echo-delay standpoint.

restorable failures per year. In one case restoration happens in 50 ms, in the other we assume the slowest restoration time ever usually considered, of 2 seconds. The four restorable failures per year then result in path availability of 0.999999994 with 50 ms restoration and 0.999999746 with 2 seconds restoration. Whether 50 ms or 2 sec is assumed availability does not degrade below a "6 nines" level. With four restorable failures per year this high level of availability will go on indefinitely. With a more realistic number of failures per year affecting any one service and restoration times above 50 ms but below 2 seconds, the difference in availability due to restoration switching alone becomes almost vanishing small.

But now consider what does really impact availability—the probability of being partly exposed to a physical failure repair time. In a "restorable" network as usually defined this can only happen from dual or higher order failures. If two failures fall on spans of the same ring or 1+1 APS system, then on average half of the physical MTTR will be experienced as outage. This can easily be 6 hours. The resultant availability in the year that this happens then becomes 0.999316 or just "three nines"—generally unacceptable under any SLA. Thus, comparing the availability values, we see that restoration times are basically irrelevant from an availability standpoint. As long as restoration works, availabilities are well over 6 nines. But what *does* strongly impact availability is the ability to restore service in circumstances of more than one failure. This is an aspect in which it has recently been found that mesh networks excel over rings in the ability to come up with at least fractional restoration levels in the face of dual failures, and to allocate them to priority paths, giving them ultra-high service availability. See [ClGr00] [ClGr02][Grov03b], pp. 527-530. In summary, it is a red-herring for people to argue that restoration times must be 50 ms in a survivable network for high availability. What really dictates service path availability in such networks is the frequency of dual failures and whether there is an adaptive restoration response present to minimize the probability of a high availability service path being exposed to an outage on the time scale of the physical repair process itself. If the criteria of relevance in a premium service SLA is high availability, then the service is far better realized in a mesh network with 1 second restoration times and an adaptive dual-failure response than with 1+1 dedicated APS (or rings) even if the latter does offer 50 ms restoration switching.

8.6.2 "Future-Proof" Design: Planning for Uncertainty

For decades past in the telecommunications industry, forecasting the coming year's requirements for switching and transport were relatively easy and accurate because the dominant source of traffic was circuit-switched voice. Increasingly now, however, data traffic exceeds voice and is enormously volatile on all time scales. Thus, the growth of ever more Internet applications, wireless mobility, and economic variability all contribute to the difficulty of forecasting the demand for which an optical transport network should be planned. This is leading to some fundamentally new dimensions in network planning and design. Part of the problem is even to give technical meaning to how to measure demand forecast accuracy, and to develop theory for survivable networks that somehow by design can claim to be "future-proofed." Work in [GeAn01] and [LeGr02] are two early efforts to try to address these questions. They define quantitative measures of the extent to which a forecast demand pattern is different from the actual future pattern and defining measures such as "servability" to quantify the fraction of actual demands that can be both *routed and protected* within a network whose capacity was initially optimized for an earlier forecast that was ultimately inaccurate.

The most straightforward approach to dealing with uncertainty has been the use of "safety margins" of added capacity. This is, however, rather brute-force and expensive. The thinking is that with suitable theory one must be able to achieve more truly "future-proofed" architectures. One approach recently studied is to incorporate demand uncertainty in survivable network design as a two-part optimization problem [LeGr04]. The first part considers a budget to be invested at present and the second part represents the expected value of the corrective or "recourse" actions that would be required depending on a range of possible futures. The future costs are mainly for equipping and commissioning new fiber systems and/or additional channels on those systems needed for either protection or working capacity. In this context, economy of scale effects can be important in future-proof planning: to minimize present costs, small capacity modules may be preferred, e.g. OC-48s and/or single wavelengths. But if the demand over a certain span increases unexpectedly, the cost of adding more capacity in small modules in the future may exceed the cost of having simply invested once at the start in large-capacity modules, say OC-192s and/or whole multi-wavelength waveband equipment.

Also related to the issue of future proof planning, is work now in its beginning directed at fundamentally understanding how transport networks designed with different survivability principles inherently cope with mis-forecast demand patterns. It is reasonable to expect that as demand for communication services becomes ever harder to predict, and the industry becomes even more competitive, that this could be a significant contributor to business success.

8.6.3 Translucency, Transparency, and Opacity

In the jargon of optical networking, a network is referred to as *transparent* if all the nodes on the network are all-optical cross connects (OXCs) without any electronic regeneration function. Each lightpath must be routed from its source to its destination without any electronic processing at intermediate nodes en route. In practice today this requires a wavelength assignment that must be uniquely reserved for the path on each fiber en route. But in principle a transparent network will allow a lightpath to occupy different wavelengths on its links when all-optical wavelength conversion (WC) is available. The opposite of a transparent network is a so-called *opaque* network in which lightwave channels are detected at each node, and their payload is then electronically regenerated and switched and can be reassigned to any new outgoing wavelength.

A transparent optical network possesses the advantages of signal transparency, bit-rate transparency, and protocol transparency. Limitations and issues in network control and management, however, make fully transparent networks difficult to implement and manage. In addition, payloads may require electronic regeneration en route to retain required digital signal-to-noise ratios, and a fully transparent network does not provide regeneration en route. On the other hand, an opaque network requires electronic switching nodes with a large number of O–E (optical-to-electronic) and E–O transponders at each node. This is costly and involved and prevents this kind of electronic-core optical switch from being widely installed on the network.

A practical approach in future optical transport networks is therefore to strike a balance between transparency and opaqueness in terms of either transparent islands or a *translucent* optical network. Two basic architectures may be considered: In the former, a large-scale optical network is divided into several islands of optical transparency. In the same domain, a lightpath can transparently reach any node without any intermediate signal regeneration. But for communications between different domains, electronic switches are used at the domain boundaries. These switches perform electronic regeneration of payloads and provide wavelength conversion as needed to forward the lightpath on into the next transparent domain. A translucent network is a somewhat more general option based on the use of a relatively small number of strategically chosen opaque nodes where all other nodes are transparent switch nodes. A cost minimum is approached by finding the sparsest placement of the opaque (i.e., electronic) switches in the network that supports efficient (low-blocking) routing of lightpaths between all network nodes. Rather than being dedicated to routing lightpaths only in and out of transparent islands, these switches are shared by all paths of the network as a whole. The design and evolution of such translucent networks is expected to be an important focus in future, especially when both routability and survivability considerations are jointly addressed in the decisions about where and when to invest in an additional opaque switch node as a network grows. The optimum architecture will also continue to be a moving target as the transparent optical reach of long-haul DWDM systems increases. An important implication of the maturity of translucent networks and knowledge about their optimum design is that survivability schemes will not be limited to end-to-end methods based on the difficulties of fault detection and electronic access to payloads for signaling. With strategically placed O-E-O nodes amidst lower cost O-O-O nodes it should be possible to minimize overall network cost while also using more localized protection and restoration schemes with mesh efficiency such as p-cycles and/or adaptive span restoration protocols.

8.6.4   Automated Provisioning of Dynamic Demands: SBPP and PWCE Concepts

Another of the important ways in which next generation transport networks will differ from traditional solutions is in regard to dynamic random arrivals and departures of demand. The automatic handling of dynamic demands is a key feature of the SBPP approach. In fact, until recently it seemed that SBPP was the only approach available to accommodate incremental random arrivals and departures of service paths that required pre-planned protection arrangements as part of the service definition. Hundreds of papers have been written that approach dynamic demand through the basic SBPP paradigm. It is hard to find a paper that does not just assume SBPP as the paradigm for dynamic automated provisioning of protected services in either the lightpath or MPLS layers. However, a basic assumption of SBPP is that an up-to-date database is available in every service-provisioning node that includes complete topology and capacity information and all spare-channel sharing relationships in the network. This state data is both global in extent and is updated on the same time scale as the connection changes themselves. There seems reason to thus be concerned about the scalability and robustness of this architecture with network size and with frequency of connection requests and releases. These considerations have led recently to proposal of the Protected Working Capacity Envelope Concept (PWCE) as an alternate paradigm for dynamic provisioning of protected services.

Under a protected working capacity envelope, one provisions service paths over inherently protected capacity, as opposed to explicitly provisioning protection for every service. If you route the service through the available channels of a PWCE (and you designate it as a protected service) then it is inherently protected. Provisioning protected services through the envelope looks the same as point-to-point routing over a non-protected network. One does not have to make any explicit arrangements for protection of every individual path or globally update network state for every individual path setup (or takedown). A PWCE is created through restoration or protection mechanisms and off-line (or very slowly acting) spare-capacity planning processes which protect a defined set or pool of provisionable working channels on each span. If node failure protection is also required, node-encircling p-cycles can also be part of the strategy.

A key idea is that through the relatively simple theory of span-restorable networks (see Chapter 5 in [Grov03b] or [HeBy95]), once the graph and the vector of spare channel quantities on the network spans are given, there is a unique maximum number of protected working channels available on each span. Thus, a given distribution of spare capacity on a graph creates a uniquely determinable envelope of protected working capacity on each span. Within this operational envelope a vast number of simultaneous service path combinations are feasible—and all are inherently protected. Provisioning of a new protected service path is then only a matter of routing over the shortest path through the envelope using only spans which currently have one or more protected working channels available. Thus, in a network where the spare capacity distribution creates a PWCE that is well matched and dimensioned for the current average point-to-point demand intensities, there is no state-update signaling whatsoever, regardless of how fast individual demands are randomly coming and going through this envelope. The only signaling involved is for source-routed establishment of the new working paths (or release of any unneeded paths). Only if the random pattern of dynamic demand evolves—in a way such that a span approaches the envelope—is any updated network state dissemination required. A single LSA can then either withdraw the highly utilized span from further routing availability or issue an updated routing cost for OSPF-type routing over that span. Nodes operating under PWCE thus need only participate in a simple OSPF-type of topology discovery to support distributed end-node provisioning. At this level of transport, however, the basic topology is almost never changing so there are almost no state updates. As long as one or more working channels is still available for provisioning on each span, all such spans remain available for routing.

An important property of the PWCE concept is that actions of any type related to ensuring protection occur only on the time-scale of the statistical evolution of the demand pattern itself, not on the time-scale of individual connections. It takes a suitably large shift in the statistics of the demand pattern to require a logical change in the working envelope. Importantly, however, such actions also occur on a time scale where traffic behavior exhibits correlated daily trends that can be taken into account in capacity-configuration planning. Several envelope configurations can be pre-computed within the installed total capacities, each of which is known to suit the characteristic time of day to minimize any blocking. In contrast, SBPP works at the call-by-call time-scale where individual departures and arrivals are essentially random, and routing is individually controlled by end-users. This is an environment of inherently incremental reaction to the next arrival, not involving any opportunity for collectively optimizing capacity use or routing strategies to enhance performance.

Thus, the protected envelope is at most slowly changing or even static over long periods of time, even in the most frenetically dynamic network. No matter how rapidly individual lightpath demands come and go at random, the envelope requirement will not change at all if the demand process is at statistical equilibrium. The envelope is only sensitive to non-stationary drift in the underlying pattern of random arrival/departure processes. This is far simpler from the end-users viewpoint and is more scalable in the face of increasingly rapid provisioning changes than making globally coordinated protection arrangements individually for every connection. More on the detailed operational considerations for PWCEs can be found in [Grov04] and Chapter 5 of [Grov03b] and on the theory for capacity-design and demand-adaptive reconfiguration of PWCE capacities in [ShGr04] [ShGr05]. In detailed simulation studies undertaken by G. Shen total network signaling volumes have been found to be 40x lower for PWCE facing the exact same sequence of arrivals and departures on several test networks, and to require at least ten times less total network memory state maintenance.

8.6.5   Evolution of Rings to Next Generation Architectures

Another aspect that seems clear when considering NGN transport networks is the growing appreciation of the capacity efficiency and flexibility of mesh-based survivable architectures. Many network operators are now convinced of the long-term benefits of establishing mesh-based MPLS over DWDM transport architectures going forward, but currently have extensive ring-based transport deployed on the ground. A traditional approach when faced with a technology shift as apparently complete as from rings to mesh is to freeze investment in the existing technology and start planning for its eventual removal as maintenance costs rise and opportunities arise to roll traffic onto facilities in the new technology. This is the "cap and grow" approach to the technology discontinuity. Eventually with continued growth, the network will be almost all mesh-based and demands served in the residual rings could be rolled into the mesh at a suitable date in the future. An interesting and potentially beneficial alternate strategy also exists however, first pointed out in [ClGr01], where it was coined the "ring mining" approach to network evolution. Ring mining can be considered for evolution of an existing set of ring transport systems to any mesh-type target architecture or to *p*-cycles.

In ring-mining the installed capacity of rings is viewed as a sunken investment to be exploited in new ways. The enticing prospect is that for certain operators ring mining could significantly defer and partly eliminate new capital expenditures for incremental transport equipment. Ring mining involves logical disassembly and reuse of ring transmission capacities in a new target architecture. As a first study in this area, [ClGr01] looked at the potential for serving ongoing demand growth without any additional capacity by reclaiming the inefficiently used protection and working capacity resources in existing rings. Additional service-bearing capacity is extracted through routing and restoration re-design using mesh principles within the existing ring capacities.

[ClGr01] found that there can be dramatic potential for ring mining in some cases. In a few test cases a total growth to 290% over the initial demand served by the ring network was sustainable simply through conversion to mesh with no added capacity. In a related strategy of minimum-cost evolution with selective new capacity additions, complete deferral of new expense was also seen for up to 50 - 200% service growth followed by an approximately linear cost for continued growth at no higher expense rate than the slope of the cost curve for the cap and grow strategy. Analysis shows that ring mining strategies work so effectively for three main reasons: (i) accessing the 100% ring protection capacity, (ii) unlocking stranded ring working capacity, and (iii) shortening of the working path routes.

More recently, the perhaps even more obvious target architecture for ring mining strategies was considered: *p*-cycles. The cyclical nature of *p*-cycles makes one ask if existing rings can somehow be advantageously converted into *p*-cycles. This was studied in [KoGr03]. One possible mechanism to convert a BLSR ring ADM to a *p*-cycle node uses the "extra traffic" feature that most BLSR ring systems have. "Extra traffic" is normally a feature that allows the network operator to transport any other lower-priority traffic over the ring's protection channel. Extra traffic will be bumped off if the ring switches to protect its own working channels. Thus the missing functionality of a ring, related to the straddling spans of a *p*-cycle can be added, possibly allowing reuse of the ring ADM as part of a *p*-cycle node. If direct re-use of the ADM is not possible, the option still exists to freeze the ADM in the add-drop configuration and use cross-connect machines to implement the *p*-cycles. Figure 8-22 shows a generic ADM as part of a ring configuration to which a new device called a Straddling Span Interface Unit (SSIU), is coupled to support *p*-cycle straddling span access to the ring protection capacity.

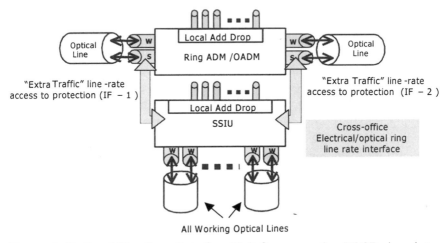

Figure 8-22. Straddling Span Interface Unit for converting BLSR rings into *p*-cycles.

The existing ring does not strictly need to even know that the straddling span unit is anything other than an apparent source/sink of some form of low priority traffic at its site. The SSIU can draw needed ring state information from the SONET K1-K2 line overhead bytes. Alternatively it could be given authority to source/sink protection protocol sequences as needed. Because the protection path continuity is now through the SSIU, not the ADM, the SSIU can completely observe the status of the protection channel, observe ring switches, and effectively block out or deny these ring switches without affecting them, when needed due to a prior SSIU switch. Further specific functions of the *p*-cycle SSIU device are defined in [GrSt00] and discussed in Chapter 10 of [Grov03b].

An alternative to use of SSIUs is a straight migration to a cross-connect based operation, or to develop custom *p*-cycle ADM-like nodal equipment, but it seems attractive to reuse the existing ADMs if possible, as SSIUs can also be an ADM with a specialized firmware upgrade. Note also that SSIUs are added only where the resulting *p*-cycles support straddling spans. All other *p*-cycle nodes require only their existing ADMs.

8.6.6   Multiple Quality of Protection (QoP) Classes

Not only as above is the 50 ms "requirement" increasingly under question but in the modern business environment there is also considerable interest in being able to differentiate between product offerings and, ideally, to charge a fair price for providing various degrees of service protection. This is a main idea of what we call multi-QoP design and operation of a survivable transport network—somewhat analogies to multiple QoS service offering in service layer applications. Lightpaths used as inter-router IP pipes may not need to be automatically protected within the optical networking layer because the IP service layer relies on routing re-convergence or other means (MPLS for instance) within the router-peer layer for fault recovery. Other customers or operators may, however, choose to refer survivability into the transport layer for faster response and/or simpler service-layer administration and software management. Ring-based transport provides protection inherently for all services.[144] This was always reasonable when services were almost entirely voice or circuit-switched 64 kbs services. In next-generation transport, however, it is going to be highly advantageous, business-wise, for carriers to be able to offer a flexible range of protection classes. The term Quality of Protection (QoP) has been coined in reference to this goal, analogous to more traditional QoS considerations. When it comes to simultaneously supporting different QoP classes in a transport network, mesh and cross-connected based p-cycle architectures turn out to fit the bill rather naturally.

One operational concept and related design theory for providing a range of QoP services on demand is based on an extension of span restoration techniques [GrCl02] (See also Chapter 5 in [Grov03b].) From that work it is fairly simple to see how corresponding multi-QoP p-cycle networks could be developed, or perhaps multi-QoP schemes based on SBPP as well. Each new service path is labeled (in its overhead bytes) as to its service class, when provisioned. The cross-connects adjacent to each span use these designations to classify the working capacity on the span. By being present in the path overheads themselves, this information is self-updating at every cross-connect. If the service class changes, or the routing of a path is changed at any time, the custodial cross-connects for a span see the updated working capacity priority composition of the span essentially the instant the new path or status change is made. There can then be up to five differentiated service offerings, based on efficient mesh capacity planning formulations and simple extensions to the real time restoration protocols to manage failed working, spare, and preemptible channels according to the following multi-QoP scheme:

- **Platinum** (assured *dual* failure restorability): Channels used to form platinum paths receive the benefit of any restoration paths

---

[144] "Extra traffic" on protection excepted. This statement applies to all services routed through the normal working capacity of the rings.

that are available following a dual-failure scenario. This service class is based on recent findings [ClGr00] that even in a minimal-capacity span-restorable mesh network for single failures, there is essentially always some level of dual-failure restorability. Up to the limits of this ability, platinum service offerings can be created, yielding ultra-high availability (and presumably revenues too) for premium services [ClGr02].

- **Gold** (assured single failure restoration): Channels currently in use for gold service paths enjoy assured single-failure restorability.
- **Silver** (best efforts restoration): These are working channels used in paths which should be restored when possible within existing spare capacity, following restoration of gold-class service capacity.
- **Bronze** (non-protected service): The bronze service class is comprised of working channels that are completely ignored from a restoration viewpoint, but are also not interrupted if they do not themselves experience a failure.
- **Economy** (preemptible services): Channels currently in use for economy services are not protected and moreover may be seized (interrupting their service paths) and used as spare channels if needed to satisfy gold or platinum-class restoration requirements.

Working within this general multi-QoP framework, there are many possible business strategies for NGN transport. An interesting example from [GrCl02] (and its extended version [GrCl04]) is the possibility of demand mixes where there is literally no "spare" capacity: the routing of gold and economy services can be worked out so that gold restorability is almost or completely provided through preemption of economy-class services. At the same time, this is not as bad as it might sound for the economy-class services, because in the multi-QoP context, there may only be say 20 to 30% of services in the gold class. Under most scenarios, high, even though un-assured, restorability for silver class demands makes it attractive for clients needing good transport layer availability like ISPs but that are not critical services like 911 call centers.

In more recent work looking at actual profitability of various multi-QoP service offering and pricing strategies [LeGr05] it was found that a mix of preemptible and protected service offerings can make as much profit as an offering only of protected services. In this strategy we can introduce preemptible (economy) services at 40% discount while simultaneously reducing gold service prices. Even with a pricing decrease in both services we achieve the same profit goal because of the strong network synergy and efficiency of this particular multi-QoP combination. The reason it works is that we still earn more for protected services but we don't have to incur the usual full cost of conventional spare channels for protection of those services. In effect the economy services pay for the gold service protection capacity. This is just one example of the more sophisticated and flexible transport network business strategies we expect to see in future. The key to such innovative and competitive service offerings and business strategies is the ever more intimate combining of network science with business strategy.

8.7. CONCLUSIONS

The transport network is a basic element of infrastructure for a modern society. It is as crucial to our ongoing activities, wealth, well-being and progress as are other traditional infrastructures such as water, power and transportation. As such, it is extremely important that the transport network be of ultra-high availability. To achieve this, it needs not only to have very high internal reliability in components and system design, but the whole network architecture needs to be conceived in ways that make if also survivable against external disruptions, such as a fiber optic cable cut. The fact that modern transport networking is essentially based on optical fiber technology means that active means of reconfigurability and rerouting are essential to achieve anything more than about "three nines" availability. This is because despite all its other advantages, cables of any type, are quite vulnerable in the real world. In the past, and sometimes recently as well, new engineers come to the field of network planning and design with a presumption that restoration should be a relatively easy consideration to address as an afterthought, following basic network design. But this is wrong, and costly. Efficient survivable network design is a ground-up set of considerations, basic to the entire choice of network architecture, technology, and operating strategy, and involving increasingly sophisticated mechanisms, theory, and concepts. When serious cable-related survivability issues with fiber optic networking first were realized in the 1980s, linear APS system concepts were evolved into the physically diverse BLSR and UPSR ring concepts. Especially in metropolitan access and transport, ring technology is so cost-optimized, and design methods well established, that rings will continue to be a mainstay in that arena. For longer distance regional or national-scale transport backbones, however, the high capacity efficiency and flexibility for both growth and reconfiguration of mesh architectures is far more attractive. The key idea of the mesh family of approaches is to maximize the sharing of protection capacity over multiple failure scenarios that do not occur simultaneously. This leads to networks that can be 100% restorable to any single failure, and restorable against any or all dual failures for high priority services, but use far less than 100% redundant capacity. Such networks can be centrally controlled, or controlled using GMPLS in an Internet-like way, but in either case comprehensive databases of current network state must be kept in the controlling nodes. There is growing recognition of the vision of truly self-organizing solutions for demand routing and restoration, such as the SHN protocol which was explained in detail. The SHN protocol is the origin of the concept of "the network is the database" as a principle to completely avoid the need for up-to-date databases of network state in every node. In the latter part of the chapter, we chose a selection of topics designed not only to inform and update readers, but also to excite those new to the field with the surprising originality and new challenges and concepts in the field of survivable transport networking. *p*-Cycles for instance, as clean and simple as the concept is, are a significant break-through because they provide ring-like speed without giving up mesh-like efficiency. This is a goal that was not achieved in over ten years of relatively intense ring versus mesh competition. Fully exploiting the *p*-cycle concepts is an area of vigorous ongoing study. Other important topics increasingly being addressed in the field include strategies for handling uncertainty and dynamic shifting in traffic patterns, ways of efficiently providing for multiple different grades of service protection, and strategies for evolution of current in-service architectures to new future target designs. Our hopes are that the selection of material for the chapter has been to varying degrees informative, useful, interesting and stimulating. Much of the corresponding design theory and further details of operation for the topics discussed can be found in the references provided.

REFERENCES

[AmCw00]   K. Ambs et al., "Optimizing restoration capacity in the AT&T network," *Interfaces*, vol. 30, no. 1, Jan./Feb. 2000, pp. 26-44.

[Ash98]   G. R. Ash, *Dynamic Routing in Telecommunications Networks*. McGraw-Hill, 1998.

[Bhan99]   R. Bhandari, *Survivable Networks: Algorithms for Diverse Routing*. Norwell, Massachusetts, USA: Kluwer Academic, 1999.

[BiKe03]   G. Birkan, J. Kennington, E. Olinick, A. Ortynski, G. Spiride, "Making a case for using integer programming to design DWDM networks," *Optical Networks Magazine*, vol. 4, no. 6, Nov./Dec. 2003, pp. 107-120.

[Bowm04]   L. Bowman, M. Broersma, ZDNet News, (10 Jun. 1998) [Online], "Severed MCI cable cripples the Net," http://zdnet.com.com/2100-11-510740.html, accessed 20 Jan. 2004.

[CDS95]   S. Cosares, D. N. Deutsch, I. Saniee, O. J. Wasem, "SONET toolkit: A decision support system for designing robust and cost-effective fiber-optic networks," *Interfaces*, vol. 25, no. 1, Jan./Feb. 1995, pp. 20-40.

[Chao94]   C. W. Chao, G. Fuoco, D. Kropfl, "FASTAR platform gives the network a competitive edge," *AT&T Technical Journal*, vol. 73, no. 4, Jul./Aug. 1994, pp. 69-81.

[ChDo91] C. W. Chao, P. M. Dollard, J. E. Weythman, L. T. Nguyen, H. Eslambolchi, "FASTAR: a robust system for fast DS-3 restoration," *Proc. IEEE Global Telecommunications Conference (GLOBECOM) '91*, Phoenix, Arizona, USA, 3-5 Dec. 1991, pp. 39.1.1-39.1.5.

[ChS97] S. Chamberland, B. Sansó, "Heuristics for ring network design when several types of switches are available," *Proc. IEEE International Conf. Communications (ICC) '97*, Montreal, Canada, 1997, pp. 570-574.

[ClGr00] M. Clouqueur, W. D. Grover, "Computational and design studies on the unavailability of mesh-restorable networks," *Proc. Second International Workshop on the Design of Reliable Communication Networks (DRCN 2000)*, Munich, Germany, 9-12 Apr. 2000, pp. 181-186.

[ClGr01] M. Clouqueur, W. D. Grover, D. Leung, O. Shai, "Mining the rings: Strategies for ring-to-mesh evolution," *Proc. Third International Workshop on the Design of Reliable Communication Networks (DRCN 2001)*, Budapest, Hungary, 7-10 Oct. 2001, pp. 113-120.

[ClGr02] M. Clouqueur, W. D. Grover, "Mesh-restorable networks with complete dual failure restorability and with selectively enhanced dual-failure restorability properties," *Proc. Third International Conference on Optical Networking and Communications (OptiComm 2002)*, Boston, Massachusetts, USA, 29 Jul.-2 Aug. 2002, pp. 1-12.

[CSW92] S. Cosares, I. Saniee, O. Wasem, "Network planning with the SONET toolkit," *Bellcore EXCHANGE*, Sep./Oct. 1992, pp. 8-13.

[DDH95] B. T. Doshi, S. Dravida, P. Harshavardhana, "Overview of INDT - A new tool for next generation network design," *Proc. IEEE GLOBECOM '95*, Singapore, 13-17 Nov. 1995, pp. 1942-1946.

[DDH97] B. T. Doshi, S. Dravida, P. Harshavardhana, P. K. Johri, R. Nagarajan, "Dual (SONET) ring interworking: High penalty cases and how to avoid them," *Proc. 15th International Teletraffic Congress (ITC 15)*, Washington, DC, USA, 23-27 Jun. 1997, pp. 361-370.

[DoGr00] J. Doucette, W. D. Grover, "Influence of modularity and economy-of-scale effects on design of mesh-restorable DWDM networks," *IEEE J. Selected Areas in Communications*, vol. 18, no. 10, Oct. 2000, pp. 1912-1923.

[DuGr94] D. A. Dunn, W. D. Grover, M. H. MacGregor, "A comparison of *k*-shortest paths and maximum flow routing for network facility restoration," *IEEE J. Selected Areas in Communications*, vol. 12, no. 1, Jan. 1994, pp. 88-99.

[Flan90] T. Flanagan, "Fiber network survivability," *IEEE Communications Magazine*, vol. 28, no. 6, Jun. 1990, pp. 46-53.

[Frye01] C. Frye, "Self-healing systems," *Application Development Trends*, 1 Sep. 2003.

[GeAn01] N. Geary, A. Antonopoulos, E. Drakopoulos, J. O'Reilly, J. Mitchell, "A framework for optical network planning under traffic uncertainty," *Proc. Third International Workshop on the Design of Reliable Communication Networks (DRCN 2001)*, Budapest, Hungary, 7-10 Oct. 2001.

[GMP01] L. Berger, P. Ashwood-Smith (Eds.), "Generalized MPLS - signaling functional description," draft-ietf-mpls-generalized-signaling-02.txt, Internet draft, work in progress, Mar. 2001.

[GrCl02] W. D. Grover, M. Clouqueur, "Span-restorable mesh network design to support multiple quality of protection (QoP) service-classes," *Proc. First International Conference on Optical Communications and Networks*

(ICOCN 2002), Singapore, 11-14 Nov. 2002, pp. 321-323.

[GrCl04]   W. D. Grover, M. Clouqueur, "Span-restorable mesh networks with multiple quality of protection (QoP) service classes," *Photonic Network Communications*, vol.9, Issue 1, 2005, pp. 19-34.

[GrDo02]   W. D. Grover, J. Doucette, "Advances in optical network design with *p*-cycles: Joint optimization and pre-selection of candidate *p*-cycles," *Proc. IEEE LEOS Summer Topical Meetings 2002*, Mont Tremblant, Quebec, Canada, 15-17 Jul. 2002, pp. 49-50.

[Grov02]   W. D. Grover, "Understanding *p*-cycles, enhanced rings, and oriented cycle covers," *Proc. First International Conference on Optical Communications and Networks (ICOCN 2002)*, Singapore, 11-14 Nov. 2002, pp. 305-308.

[Grov03]   W. Grover, "*p*-Cycles, ring-mesh hybrids and 'ring-mining:' Options for new and evolving optical networks," *Proc. Optical Fiber Communication Conference (OFC) 2003*, Atlanta, USA, 24-27 Mar. 2003, pp. 201-203.

[Grov03b]   W. D. Grover, *Mesh-Based Survivable Networks: Options and Strategies for Optical, MPLS, SONET, and ATM Networking*. Upper Saddle River, New Jersey, USA, Prentice Hall PTR, 2003.

[Grov04]   W. D. Grover, "The protected working capacity envelope concept: An alternate paradigm for automated service provisioning," *IEEE Communications Magazine*, vol. 42, no. 1, Jan. 2004, pp. 62-69.

[Grov87]   W. D. Grover, "The selfhealing network: A fast distributed restoration technique for networks using digital crossconnect machines," *Proc. IEEE GLOBECOM '87*, Tokyo, Japan, Nov. 1987, pp. 1090-1095.

[Grov94]   W. D. Grover, "Distributed restoration of the transport network," in *Telecommunications Network Management into the 21st Century: Techniques, Standards, Technologies, and Applications*, S. Aidarous, T. Plevyak (Eds.), New York City: Wiley-IEEE Press, 1995, pp. 337-417.

[Grov97]   W. D. Grover, "Self-organizing broadband transport networks," *Proceedings of the IEEE*, vol. 85, no. 10, Oct. 1997, pp. 1582-1611.

[Grov99]   W. D. Grover, "High availability path design in ring-based optical networks," *IEEE/ACM Transactions on Networking*, vol. 7, no. 4, Aug. 1999, pp. 558-574.

[Grov99b]   W. D. Grover, "Resource management for fault tolerant paths in SONET ring networks," *Journal of Network and Systems Management*, vol. 7, no. 4, Dec. 1999, pp. 373-394.

[GrSl95]   W. D. Grover, J. B. Slevinsky, M. H. MacGregor, "Optimized design of ring-based survivable networks," *Canadian Journal of Electrical and Computer Engineering*, vol. 20, no. 3, Aug. 1995, pp. 138-149.

[GrSt00]   W. D. Grover, D. Stamatelakis, "Bridging the ring-mesh dichotomy with *p*-cycles," *Proc. 2nd Int. Workshop on Design of Reliable Communication Networks (DRCN 2000)*, Munich, Germany, 9-12 Apr. 2000, pp. 92-104.

[GrSt98]   W. D. Grover, D. Stamatelakis, "Cycle-oriented distributed preconfiguration: Ring-like speed with mesh-like capacity for self-planning network restoration," *Proc. IEEE International Conference on Communications (ICC) '98*, Atlanta, Georgia, USA, 7-11 Jun. 1998, pp. 537-543.

[GrVe90]   W. D. Grover, B. D. Venables, J. H. Sandham, A. F. Milne, "Performance studies of a selfhealing network protocol in Telecom Canada long haul networks," *Proc. IEEE Global Telecommunications Conference (GLOBECOM) '90*, San Diego, USA, 5-7 Dec. 1990, pp. 452-458.

[HeBy95]   M. Herzberg, S. J. Bye, A. Utano, "The hop-limit approach for spare-capacity assignment in survivable networks," *IEEE/ACM Transactions on Networking*, vol. 3, no. 6, Dec. 1995, pp. 775-784.

[IrGr00]   R. R. Iraschko, W. D. Grover, "A highly efficient path-restoration protocol for management of optical network transport integrity," *IEEE J. Selected Areas in Communications*, vol. 18, no. 5, May 2000, pp. 779- 794.

[IrGr98]   R. R. Iraschko, M. H. MacGregor, W. D. Grover, "Optimal capacity placement for path restoration in STM or ATM mesh-survivable networks," *IEEE Trans. Networking*, v. 6, no. 3, Jun. 1998, pp. 325-336.

[ITU01]   ITU-T Recommendation G.7042/Y.1305 (2001), "Link capacity adjustment scheme for virtual concatenated signals."

[ITU03]   ITU-T Recommendation G.808.1 (2003), "Generic protection switching - Linear, trail, and subnetwork protection."

[ITU96]   ITU-T Recommendation G.707 (1996), "Synchronous digital hierarchy bit rates."

[ITU98]   ITU-T Recommendation G.841 (Oct. 1998), "Types and characteristics of SDH network protection architectures."

[KaGr03]   G. V. Kaigala, W. D. Grover, "On the efficacy of GMPLS auto-reprovisioning as a mesh-network restoration mechanism," *Proc. IEEE GLOBECOM 2003*, San Francisco, USA, 1-5 Dec. 2003, pp. 3797-3801.

[KaSa94]   R. Kawamura, K. Sato, I. Tokizawa, "Self-healing ATM networks based on virtual path concept," *IEEE Journal on Selected Areas in Communications*, vol. 12, no. 1, Jan. 1994, pp. 120-127.

[KiKo00]   S. Kini, M. Kodialam, T. V. Laksham, C. Villamizar, "Shared backup label switched path restoration," draft-kini-restoration-shared-backup-00.txt, Internet draft, work in progress, Nov. 2000.

[KoAc96]   M. Kovacevic, A. S. Acampora, "Electronic wavelength translation in optical networks," *IEEE Journal on Lightwave Technology*, vol. 4, no. 6, Jun. 1996, pp. 1161-1169.

[KoGr03]   A. Kodian, W. D. Grover, J. Slevinsky, D. Moore, "Ring-mining to *p*-cycles as a target architecture: Riding demand growth into network efficiency," *Proc. 19th National Fiber Optic Engineers Conference (NFOEC 2003)*, Orlando, Florida, USA, 7-11 Sep. 2003, pp. 1543-1552.

[KoLa00]   M. Kodialam, T. V. Lakshman, "Dynamic routing of bandwidth guaranteed tunnels with restoration," *Proc. 19th Annual Joint Conference of the IEEE Computer and Communications Societies (INFOCOM 2000)*, Tel-Aviv, Israel, 26-30 Mar. 2000, pp. 902-911.

[KoRe00]   K. Kompella et al., "OSPF extensions in support of generalized MPLS," draft-kompella-ospf-extensions-00.txt, Internet draft, work in progress, Jul. 2000.

[KoRe03a]   K. Kompella, Y. Rekhter (Eds.), "OSPF extensions in support of generalized multi-protocol label switching," draft-ietf-ccamp-ospf-gmpls-extensions-12.txt, Internet draft, work in progress, Oct. 2003.

[KoRe03b]   K. Kompella, Y. Rekhter (Eds.), "Routing extensions in support of generalized multi-protocol label switching," draft-ietf-ccamp-gmpls-routing-09.txt, Internet draft, work in progress, Oct. 2003.

[Lang03]   J. Lang (Ed.), "Link management protocol (LMP)," draft-ietf-ccamp-lmp-10.txt, Internet draft, work in progress, Oct. 2003.

[LeGr02]   D. Leung, W. D. Grover, "Comparative ability of span restorable and path protected network designs to withstand uncertainty in the demand forecast," *Proc. 18th National Fiber Optic Engineers Conference (NFOEC 2002)*, Dallas, Texas, USA, 15-19 Sep. 2002, pp. 1450-1461.

[LeGr04]   D. Leung, W. D. Grover, "Restorable mesh network design under demand uncertainty: Toward 'future proof' transport investments," *Proc. Optical Fiber Communications Conf. (OFC) 2004*, Los Angeles, Feb. 2004.

[LeGr05]   D. K. Leung, W.D. Grover, "Maximum-profit model for study of multi-QoP wavelength service offerings in survivable mesh networks," to appear in Proc. Combined NFOEC/OFC Conferences, Anaheim, March 6-11, 2005.

[LiWa02]   G. Li, D. Wang, C. Kalmanek, R. Doverspike, "Efficient distributed path selection for shared restoration connections," *Proc. 21st Annual Joint Conference of the IEEE Computer and Communications Societies (INFOCOM 2002)*, New York City, USA, 23-27 Jun. 2002, pp. 140-149.

[LiYa02]   G. Li, J. Yates, D. Wang, C. Kalmanek, "Control plane design for reliable optical networks," *IEEE Communications Magazine*, vol. 40, no. 2, Feb. 2002, pp. 90-96.

[Man03]   E. Mannie (Ed.), "Generalized multi-protocol label switching architecture," draft-ietf-ccamp-gmpls-architecture-07.txt, Internet draft, work in progress, May 2003.

[McAu04]   W. McAuliffe, ZDNet UK, (3 Aug. 2001) [Online], "Train crash could be to blame for Internet derailment," http://news.zdnet.co.uk/business/ 0,39020645,2092503,00.htm, accessed 20 Jan. 2004.

[MoGr01]   G. D. Morley, W. D. Grover, "Tabu search optimization of optical ring transport networks," *Proc. IEEE Global Telecommunications Conference (GLOBECOM) 2001*, San Antonio, USA, Nov. 2001, pp. 2160-2164.

[MoGr98]   D. Morley, W. Grover, "A comparative survey of methods for automated design of ring-based transport networks," TRLabs, Edmonton, Alberta, Canada, Technical Report TR-97-04, Issue 1.0, 28 Jan. 1998, available at: http://www.ece.ualberta.ca/~grover/publications (Tech Reports).

[Mok00]   S. Mokbel, "Canada's optical research and education network: CA*net3," *Proc. 2nd Intl. Workshop on Design of Reliable Communication Networks (DRCN 2000)*, Munich, Germany, 9-12 Apr. 2000, pp. 10-32 J.

[Nor96]   Nortel (Northern Telecom), (1996), "Introduction to SONET Networking"

[OkNo00]   E. Oki, N. Matsuura, K. Shiomoto, N. Yamanaka, "A disjoint path selection scheme with shared risk link groups in GMPLS networks," *IEEE Communications Letters*, vol. 6, no. 9, Sep. 2002, pp. 406-408.

[OrWe03]   S. Orlowski, R. Wessaly, "Comparing restoration concepts using optimal configurations with integrated hardware and routing decisions," *Proc. 4th Intl. Workshop on the Design of Reliable Communication Networks (DRCN 2003)*, Banff, Alberta, Canada, 19-22 Oct. 2003, pp. 15-22.

[QSIN04]   Quebec Scientific Information Network, (29 Jan. 2003) [Online], "Forestville-Rimouski underwater cable repaired," http://www.risq.qc.ca/nouvelles/nouvelle_item.php?LANG=EN&ART=123 1, accessed 20 Jan. 2004.

[Raja03]   B. Rajagopalan, J. Luciani, D. O. Awduche, "IP over optical networks: A framework," draft-ietf-ipo-framework-05.txt, Internet draft, work in progress, Sep. 2003.

[RPR04]   IEEE 802.17-2004, "Information technology – Telecommunications and information exchange between systems – Local and metropolitan area networks, Specific requirements – Part 17: Resilient Packet Ring Access Method & Physical Layer Specifications (RPR)."

[Sack04]   A. Sack, "The *p*-cycles home page," http://tomato.edm.trlabs.ca/p-cycles/. (If this URL changes in the future, an updated link will be available at http://www.ece.ualberta.ca/~grover/)

[SaGr04]   A. Sack, W. D. Grover, "Hamiltonian *p*-cycles for fiber-level protection in homogeneous and semi-homogeneous optical networks," *IEEE Network*, vol. 18, no. 2, Mar./Apr. 2004, pp. 49-56.

[ScGr02]   D. A. Schupke, C. G. Gruber, A. Autenrieth, "Optimal configuration of *p*-cycles in WDM networks," *Proc. IEEE Intl. Conf. on Communications (ICC) 2002*, New York City, USA, 28 Apr.-2 May 2002, pp. 2761-2765.

[ScGr04]   D.A. Schupke, W.D. Grover, M. Clouqueur, "Strategies for enhanced dual failure restorability with static or reconfigurable *p*-cycle networks," Proc. 2004 Intl.Conf. Communications (ICC 2004), Paris, France, June 2004.

[Scha01]   J. Schallenburg, "Is 50 ms restoration necessary?," slides presented at IEEE Bandwidth Management Workshop IX, Montebello, Quebec, Canada, Jun. 2001. Available: http://www.ece.ualberta.ca/~grover/ WDG_50ms.htm.

[ScSc03]   D. A. Schupke, M. C. Scheffel, W. D. Grover, "Configuration of *p*-cycles in WDM networks with partial wavelength conversion," *Photonic Network Communications*, vol. 6, no. 3, Nov. 2003, pp. 239-252.

[ShGr03]   G. Shen, W. D. Grover, "Extending the *p*-cycle concept to path-segment protection," *Proc. IEEE International Conference on Communications (ICC) 2003*, Anchorage, Alaska, USA, 11-15 May 2003, pp. 1314-1319.

[ShGr03b]   G. Shen, W. D. Grover, "Extending the *p*-cycle concept to path segment protection for span and node failure recovery," *IEEE Journal on Selected Areas in Communications*, vol. 21, no. 8, Oct. 2003, pp. 1306-1319.

[ShGr04]   G. Shen, W. D. Grover, "Design and performance of protected working capacity envelopes based on *p*-cycles: A fast, simple, and scalable framework for dynamic provisioning of survivable services," *Proc. Asia-Pacific Optical and Wireless Commun. Conf. (APOC) 2004*, Beijing, 7-11 Nov. 2004, vol. 5626.

[ShGr05]   G. Shen, W. D. Grover, "Design of protected working capacity envelopes based on *p*-cycles: An alternative framework for survivable automated lightpath provisioning," to appear in *Performance Evaluation and Planning Methods for the Next Generation Internet*, A. Girard, B. Sansò, F. Vazquez-Abad (Eds.), Kluwer Academic.

[Snow00]   A. P. Snow, M. W. Thayer, "Defeating telecommunication system fault-tolerant designs," *Proc. Third Information Survivability Workshop (ISW 2000)*, Boston, Massachusetts, USA, 24-26 Oct. 2000.

[Snow01]   A. P. Snow, "Network reliability: The concurrent challenges of innovation, competition, and complexity," *IEEE Transactions on Reliability*, vol. 50, no. 1, Mar. 2001, pp. 38-40.

[SONET00]   American National Standards Institute, "Synchronous optical network (SONET) - Automatic protection," ANSI T1.105.01-2000, Mar. 2000.

[SONET01]   American National Standards Institute, "Synchronous optical network (SONET) - Basic description including multiplex structure, rates, and formats," ANSI Standard T1.105-2001, May 2001.

[Sosn94]   J. Sosnosky, "Service applications for SONET DCS distributed restoration," *IEEE J. Selected Areas in Commun.*, vol. 12, no. 1, Jan. 1994, pp. 59-68.

[StGr00]   D. Stamatelakis, W. D. Grover, "IP layer restoration and network planning based on virtual protection cycles," *IEEE Journal on Selected Areas in Communications*, vol. 18, no. 10, Oct. 2000, pp. 1938-1949.

[Tel1400]   Telcordia GR-1400 (1999) SONET Dual-Fed Unidirectional Path Switched Ring (UPSR) Equipment Generic Criteria

[Tham04]   I. Tham, ZDNet UK, (20 Sep. 2001) [Online], "Anchor-draggers cut Asia's
            Internet   pipe,"   http://news.zdnet.co.uk/internet/
            0,39020369,2095715,00.htm, accessed 20 Jan. 2004.

[ToNe94]   M. To, P. Neusy, "Unavailability analysis of long-haul networks," *IEEE J.
            Selected Areas in Commun.*, vol. 12, no. 1, Jan. 1994, pp. 100-109.

[Wear04]   G. Wearden, ZDNet UK, (26 Nov. 2003) [Online], "Cable failure hits UK
            Internet   traffic,"   http://news.zdnet.co.uk/   communications/
            networks/0,39020345, 39118125,00.htm, accessed 20 Jan. 2004.

[Wu92]   T.-H. Wu, *Fiber Network Service Survivability.* Norwood, Massachusetts,
            USA: Artech House, 1992.

Chapter 9

CONTROL PLANE
Protocols and Architectures for Dynamic Service Setup

9.1. INTRODUCTION

In the TMN chapter, we presented telecom networks as consisting of service, management and control planes. As explained in that chapter, the purpose of the service plane is to provide transport services to transport customer traffic or carrier internal traffic. The purpose of the management plane is to manage the service plane and typically, that includes provisioning and monitoring of equipment and services in the service plane. The last plane, the focus of this chapter, is the control or signaling plane which serves the purpose of dynamic service setup with or without explicit help from the management plane.

While setting up services (i.e. provisioning) using the management plane represents a top-down approach, setting up services using the control plane represents a bottom-up approach. Recent interest in bottom-up approach to service creation is influenced by carriers desire to reduce operational expenses (i.e., the Opex) as well as the need to offer bandwidth on-demand services. The later feature enables service providers, such as ISPs, to request for bandwidth from transport network using signaling. Such flexibility will enable ISPs to alter their network topology to meet anticipated traffic patterns during webcasts or following some significant news events that draws more customers to one or few websites. While Opex savings was the initial driver and application of CPlane, switched connection might become an important application in future. Should that happen, the role of CPlane in optical transport network would become very important, probably more than the management plane and more important than the Opex savings.

Given the current drivers and importance attached to signaling, this chapter will discuss existing as well as emerging signaling standards for telecommunications networks and particularly those associated with optical networks.

The chapter will start with an introduction to Signaling System 7 (called #7), which is widely deployed in telecommunications networks followed by introduction to ATM/PNNI signaling. A good understanding of #7 is necessary to appreciate ATM signaling variants since they are in some ways derivatives of #7. Finally, we will provide an overview of GMPLS based signaling standards (ITU, OIF and IETF) as part of optical control plane discussion

9.2.   WHY SIGNALING?

Before jumping into details of specific signaling protocols, this section will try to present a theoretical perspective of signaling with a view to developing requirements. Some of the issues and concepts introduced here are addressed in each of the sections dealing with existing and emerging signaling standards.

A hypothetical network, shown in Figure 9-1, will be used to guide the discussion in this section.

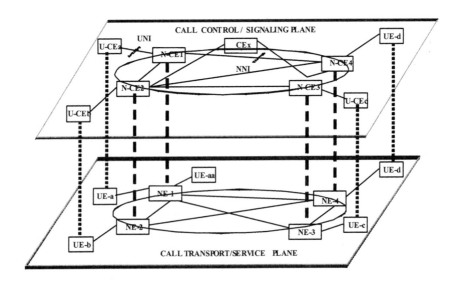

Figure 9-1. network and user entities with their associated control entities

The network shown has network entities (NE)[145] and user entities (UE). An UE is a manually operated telephone, computer, router, DSLAM, NGDLC, ADM, etc. An NE is a switch, router, or a transport network element to which one or more UEs are attached. Each NE will have switching or multiplexing hardware to cross-connect UEs attached to the NE. Each NE can also cross-connect any of its attached UEs with UEs attached to some other NE elsewhere in the network. For example, NE-1 can connect UE-a with UE-aa (both UEs are attached to NE-1). It can also connect any of its UEs with UE-c (attached to NE-3).

A UE (the *calling* UE), wanting to connect to another UE (the *called* UE), provides (i.e. signals) information about the called UE along with connection details. Using that information the network is expected to connect the *calling* UE with the *called* UE. UE requests for connection and subsequent interactions between the NEs to setup the requested connection is essentially what signaling is about. The interactions between the NEs may be direct or indirect depending on whether the connection request is handled by OSS or by the NEs themselves.

Though the term NE refers to network entity that interfaces to one or more UEs, usually there is a control entity (CE) in each network entity and hence we use N-CE to refer to the CE associated with the NE. Similarly, we use U-CE to refer to the CE associated with an UE. Given that there are CEs (U-CEs, N-CEs), signaling association exists between the various CEs. Signaling association between U-CE and N-CE is commonly called the user-to-network interface (UNI) while the association between two N-CEs is called the Network-to-Network Interface (NNI)[146].

It is possible to conceive of separation of CEs from their respective entities. Such stand alone CEs are typically called proxy controllers and they can operate independently[147]. Such separation also leads to the possibility, and as will become apparent later, that a CE in the control plane doesn't have an associated NE/UE in the service plane. Separation of U-CEs from their associated UEs doesn't influence the architecture and design of control plane. However, the same can't be said with respect to N-CEs because separation of N-CEs, from their associated NEs, allows the N-CE-to-N-CE connectivity (i.e. control plane topology) to be independent of the NE-to-NE connectivity (i.e. service plane topology). Obviously, control plane controls

---

[145] For all purposes a network entity (NE) is same as network element (NE)

[146] NNI is also defined as node- to- node interface. A degenerate case of a network is a node (NE) and therefore network-to-network interface is a more generic form of node-to-node interface.

[147] Across UNI, proxy or third party signaling is often discussed and is based on seperation of CE and UE as discussed here.

the service plane and hence it might be difficult to accept that there is no correspondence between these two planes. The beauty of such separation can be appreciated in the SS #7 signaling section.

Figure 9-1 captures the above discussion relating to separation of CEs from their associated service plane entities. Here, reader can notice that the topology of service plane is different from that of control plane. There is also a control entity, CEx, which isn't associated with any NE in the service plane.

From the above discussion, and based on the example network, the following questions may be asked:

- Does an NE need to know where a particular UE is attached to the network (i.e. know the identity of the NE to which a UE is attached). If yes then, how that information is made available.
- Does an UE need to know about the identity of its NE or of other NEs?
- Can two UEs, a UE-x and a UE-y, exchange user-to-user signaling information before connection is setup. If the answer is yes then, what is the support required from network to do so?

Let us discuss the issue of a NE knowing about each UE's identity as well as identity of the NE to which a given UE is attached. It is possible, in a small network, to advertise a UE's attachment information so that every NE knows where a particular UE is attached to the network. For example, if UE-x is attached to NE-n1 then it is possible to disseminate such information (called the reachability information). As a result, every NE in the network will know that UE-x is connected to NE-n1 (i.e. UE-x is reachable via NE-n1). Now, let us say there is a connection request at some NE-n2 to connect its attached UE-y with UE-x. Since NE-n2 knows where UE-x is attached it can proceed with required connection setup steps. However, the above scheme suffers from scalability limitations. A local switching center trying to store information for all phones out there in the world is next to impossible unless there is a means of greatly summarizing reachability. Besides, the structure, interpretation, and management of address of UEs and NEs is complicated due to non-homogeneous nature of telecom networks. The non-homogeneity exists due to:

- Multiple network operators or providers
- Network consisting of equipment from different vendors
- Multiple technologies used in an operator's network to support multiple services

Ignoring the UE and NE identity resolution issue for the moment, two further questions need to be asked:

1. Can there be some structure to UE's address so that every NE can automatically derive information about the NE to which a given UE is attached?
2. Is it possible that NEs are not required to translate UE address to find the NE to which a UE is attached?

If yes, which is true to some extent in the case of telephone network numbering, for a given UE-x, any NE can calculate or lookup the identity of NE (or at least region) to which UE-x is attached (belongs). Some addressing schemes used to identify UEs include NSAP, E.164, IP etc. Such structured address can encode direct or indirect information about the NE to which an UE is attached. For instance, we have globally unique UE identities in the case of PSTN; each telephone (a UE) is identified using CC-AC-PN address structure where CC is the country code, AC is the area code and PN is the phone number of the UE[148].

The second major issue is related to each UE knowing about the identity of the NE to which it is attached as well as identity of other NEs. In PSTN, it is not necessary that UEs know anything about NEs. For instance, we don't know anything about our local telephone switch or about the ones associated with our friends. We don't know that point codes (PC, explained later) are used for identification of switches in PSTN. But, we can decipher geographic zone based on CC or AC or both for a given telephone number represented by CC-AC-PN structure mentioned earlier. However, we still don't know anything about the NE itself. Are there applications that need to know about this? The answer is yes and we will discuss some later.

No matter how UE initiates connection request and how the network interprets and executes the request, there is a broad agreement on what functions signaling should provide. Some of them are listed here:

- Set up a service or circuit from point A to point Z
- Modify an existing service
- Delete an existing service
- Query for information such as VPN group or 800 number translation
- Equipment to equipment communication in a multi-vendor and heterogeneous service environment

---

[148] Using CC-AC-PN telephone number structure inferring zone/NE may not be possible when local number portability (LNP) is used.

Whether it is PSTN, ATM, or optical transport network, the above set of services or applications form a core part of signaling and will be the focus of the following discussion.

9.3.   PSTN CONTROL PLANE (SS7)

Signaling protocols are used in voice networks to setup voice or ISDN calls. There are many different classes of signaling and the two primary ones are: channel associated signaling (CAS), and common channel signaling (CCS). CAS is tightly tied to the voice channel in space, time or frequency (i.e. each traffic channel is associated with a dedicated signaling channel). As a result, the voice channel must be reserved before signaling can be done, and the information that can be carried is very limited. This makes CAS inflexible and slow. However, before standardization of CCS, telephone companies in Europe[149], South America, Africa, Australia, used CAS and it may still be in use in some of these places.

In CCS, a common channel carries signaling for circuits -called trunks[150] (as many as 1000) - between two switches. Figure 9-2 illustrates a simple application of CCS – to signal voice calls between two exchanges using a dedicated signaling link.

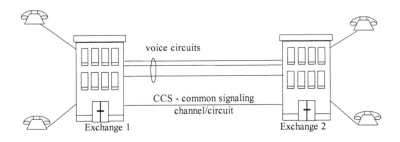

Figure 9-2. Simple CCS Example

The first CCITT CCS recommendation, referred to as CCITT No.6 (SS6) was introduced for international signaling in 1972. In 1976, AT&T introduced, based on CCITT No.6, Common Channel Interoffice Signaling System (CCIS). CCITT published the first set of standards on SS7 in 1980

---

[149] R2, a CAS type of signaling system is also used in Eastern Europe and Russia.
[150] A trunk is also commonly used to mean a multiplexed stream such as 2 Mbps PCM or 1.5 T1 link.

(Yellow Books), with modifications and enhancements in 1984 (Red Books), and in 1988 (Blue Books).

SS6 was used up until mid 80's before SS7 took over all signaling in PSTN. The following points are mentioned in literature in relation to SS6:

– It was designed for low speed links where each signaling link provided signaling for over 200 trunks.
– It was not a layered protocol and not flexible.
– It used fixed signaling messages compared to SS7, which allowed variable length messages.

Currently, SS7 is an integral part of global PSTN. It is used for many applications – voice call setup, 800 number translation, database lookup, value added call features (call forwarding, caller –id, etc). It is used not only in PSTN but also in wireless and VoIP networks.

As illustrated in Figure 9-3, an SS7 network consists of Signaling Points (SP[151]) and Signaling Transfer Points (STP).

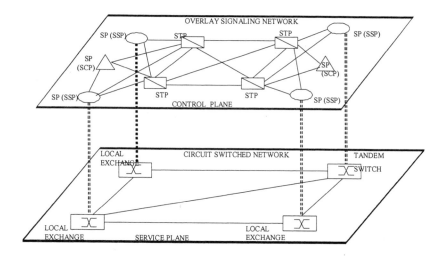

Figure 9-3. SS7 Signaling Network components

The SPs include Signaling Switching points (SSP) and Signaling Control Points (SCP). The SSPs are usually the local telephone exchanges (switching

[151] Some times we SP to refer to all – SSP, SCP, and STP

offices). SCPs are usually the database servers that provide such services as 800 number translation and others. SPs either originate or terminate signaling messages while STPs perform transfer or routing (layer three) of signaling messages and are analogous to IP routers in IP networks.

Figure 9-4 shows the protocol functional components of SS7; Message Transfer Part (MTP), ISDN User Part (ISUP), Telephony User Part (TUP), Signaling Connection Control Part (SCCP), and Transaction Capabilities Application Part (TCAP).

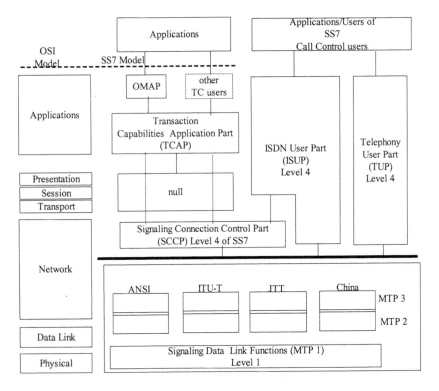

Figure 9-4. SS7 Protocol Components[152]

STPs, besides SS7 message routing functions, can also provide layer four functions, such as those implemented by SCCP and ISUP (discussed below), and such STPs are said to have *integrated STP* functionality. When an SP

---

[152] JIT – Japanese version of MTP. MTP as used in different countries may differ. However, we will highlight some of the differences between ITU-T (Europe) and ANSI (North American) variants only.

provides just STP or STP + SCCP capabilities, it is commonly called a *stand-alone STP*.

The users or clients to SS7 include Call Control, Transaction Control (TC), and other applications. Although the work in CCITT for the definition of the SS No.7 protocol started prior to the availability of the OSI reference model defined by International Standards Organization (ISO), significant efforts were made to bring the OSI structure into the SS7 protocol, particularly the ones published in Red (1984) and Blue (1988) Books. Figure 9-4 also shows OSI layering of SS7 to be discussed later.

Figure 9-5 illustrates user-to-network interface (UNI) and node-to-node interface (NNI) reference points of SS7/ISDN where a node refers to a SP, which can be an SSP, SCP, or STP.

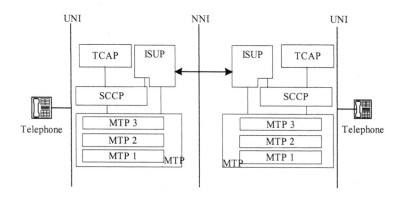

Figure 9-5. POTS/SS7 Signaling

9.3.1   SS7 Signaling Modes

'Signaling mode' is used to classify association between two signaling points involved in a signaling transaction for setting up a call or for some other purpose. There can be many definitions of signaling modes (or methods) used in voice networks. Some commonly used ones are:
- In-Band (IB)
- Out-Of-Band (OOB)
- Channel Associated (CAS)
- Common Channel (CCS)
- Facility-Associated (FAS)
- Non-Facility-Associated (NFAS)
- Associated
- Non-Associated
- Quasi-associated

In PSTN, "In-Band" can mean:

- Voice-frequencies used to signal address (digits) and signaling (off/on-hook)
- Voice-frequencies used to signal address (digits)
- Non-voice-band frequencies for signaling (off/on-hook)

An example of the "In Band" signaling method is SF[153] (2600 Hz) for signaling on/off hook condition and, MF for sending digits, as in the CCITT R1 and old AT&T analog carrier systems. An example of the OOB signaling method is E1/T1 digital carrier with signaling in dedicated channel or using "robbed bits". Many versions of MF were used for addressing (sending the called party digits). Signaling using non-voice-band frequencies can be technically called out-of-band since it uses extra bandwidth available on the channel/link. Both voice and non-voice frequency based signaling methods can be referred to as facility associated since signaling traffic travels over the same path as the voice traffic.

Today, "Out of Band" means no part of the signaling message travels over the chosen voice path. SS7 and ISDN D-channel signaling are modern examples. However, each of these can operate in both "facility associated" and "non-facility associated" modes. ISDN D-channels for BRI/BRA[154] are always "facility-associated" and PRI/PRA[155] is usually seen this way -- a single facility carries both the B- (user traffic) and D- (signaling traffic) channels. However, "non facility associated" PRI/PRA is also possible.

The SS7 signaling network itself is separate from the voice network. Therefore, signaling mode classification, such as in-band or out-of-band etc, used for early signaling implementations (before CCS) is no longer valid. Signaling modes used in SS7 are shown in Figure 9-6.

---

[153] SF and MF stand for Single Frequency and Multi Frequency respectively.

[154] BRI/BRA stands for Basic Rate Interface/Basic Rate Access and refers to standard 2B+D ISDN interface.

[155] PRI/PRA stands of Primary Rate Interface and Primary Rate Access respectively. This is commonly used to refer to ISDN service over 2 Mbps (Europe) or 1.5 T1 lines. The signaling is performed using one of the time slots of the trunk and the remaining slots are used as B channels.

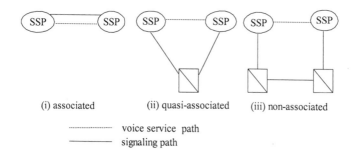

(i) associated   (ii) quasi-associated   (iii) non-associated

- - - - - - - - - - voice service path
_____ signaling path

Figure 9-6. POTS/SS7 Modes

When the two signaling entities involved in a signaling transaction have a direct signaling link between them, the signaling mode is defined as '*associated*'. A '*quasi-associated*' mode exists if the two signaling entities are connected to a common STP. A '*non-associated*' mode is defined if the two signaling entities involved depend on signaling packet network to transfer/route signaling messages between them. In other words, in '*non-associated*' mode there is no fixed path between the two signaling entities.

SS7 can work in all the above modes. Whatever may be the mode deployed all the signaling entities are part of the signaling packet network. The network required to support SS7 signaling messages can use spare bandwidth available in the circuit switching network or use a parallel or overlay network. According to [Jab91] it is more economical to use overlay networks for signaling as shown in Figure 9-3.

Though SS7 started with PSTN, it is now used in other type of networks, notable among them being the cellular/mobile networks. Cellular networks use many features requiring switching equipment to communicate with each other over a data communications network. Seamless roaming is one such feature of the cellular network that relies on the SS7 protocol to access subscriber information from Home Location Register (HLR) database of one provider by another provider. Similarly, in PSTN, besides 800 number database access, SS7 protocol provides the means for communication between switching equipment to support valued added features. . For example, if a caller dials a number that is busy, the caller may elect to invoke a feature such as automatic callback. When the called party becomes available, the network will ring the caller's phone. When the caller answers, the called party phone is then rung. This feature relies on the capabilities of SS7 to send messages from one switch to another, allowing the two systems to invoke features within each switch without setting up a circuit between the two systems.

Today, in North America, all operators (RBOCs, independent telephone companies, IXCs etc as well as cellular operators) have all deployed SS7 in their voice networks. All these carriers interconnect their SS7 signaling networks with each other so that calls can be setup across two operators' voice networks. This makes the SS7 signaling network the world's largest data communications network. At global level, SS7 network interconnects thousands of telephone company providers all over the world into one common signaling network linking separate signaling networks of local telephone companies, cellular service providers, and long distance carriers together into one large information-sharing network. Overall, the global SS7 network is a complex packet switching network that is in place for more than two decades, well before the birth of IP packet networks of today. In terms of the number of nodes and the number of links global SS7 packet switching network represents the most complex network of its kind. It operates with high reliability and performance to support millions of calls a day.

9.3.2   MTP

The Message Transfer Part (MTP) of SS7 is the basic building block for reliable connectionless transport of signaling information within a SS7 signaling network. The MTP consists of three layers, very similar to layers 1-3 of standard OSI. The description of the three layers follows.

9.3.2.1 MTP Level 1: Signaling Data Link Functions (Layer 1)

SS7 signaling data link is a bi-directional transmission link. The bi-directional nature of SS7 signaling links provides link status monitoring capabilities. The transmission link is standardized at 56 Kbps (ANSI), 64 Kbps (CCITT) though   link speeds of 4.8 Kbps[156] as well as higher speeds, such as 1.544 Mbps in North America, and 2.048 elsewhere are reported.

Essentially MTP-1 layer deals with physical, electrical, and functional characteristics of the signaling link. The link may be constructed using a dedicated time slot of an E1/T1 PCM trunk facility. As an example, on 2Mbps PCM trunks, there are 32 time slots and one of the time slots (slot 16) is commonly used for SS7 signaling. Remaining channels of E1/T1 carry voice/data circuits. Failure of the facility takes down both the signaling link and the voice circuits. Alternatively, the channel may be constructed using some reserved bits (also called robbed bits) in PCM/TDM frame.

---

[156] CCIT   specifies a minimum bit rate of   4.8 Kbps for call-control purposes

With VoIP, the underlying signaling link is no longer a dedicated bit-pipe. Various User-Adaptation components are standardized by IETF as part of SIGTRAN. All SS7 signaling transport over IP networks requires a basic protocol module called SCTP, shown in Figure 9-7.

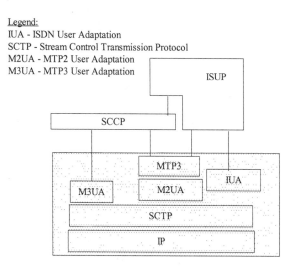

Figure 9-7. SS7/IP- VoIP SS7

One can consider SCTP as a lightweight TCP designed to provide reliability, sequence control, and record based delivery (as opposed to byte-stream delivery supported by TCP).

9.3.2.2 MTP Level 2: Signaling Link Functions (Layer 2)

The second layer of MTP provides signaling link functions consisting of *error correction, error monitoring* and *flow control*. It provides connection-less mode transport services. What this means is that clients to MTP don't need to have an existing channel or path or an end-to-end association before sending information to a destination. To put in perspective, in ATM two parties communicating must have virtual channel established between them. Similarly, TCP/IP applications communicating must establish a end-to-end TCP association before data can be transferred[157]. The MTP-2 transport model is very similar to datagram model of IETF's IP protocol. The SS7 signaling messages are transferred over the signaling link in variable length messages called "signal units" and there are three types of them - Link

---

[157] One can also equate MTP-2 to UDP's datagram model.

Status Signal Unit (LSSU), Fill In Signal Unit (FISU) and Message Signal Unit (MSU). Figure 9-8 illustrates the format of the three signal units

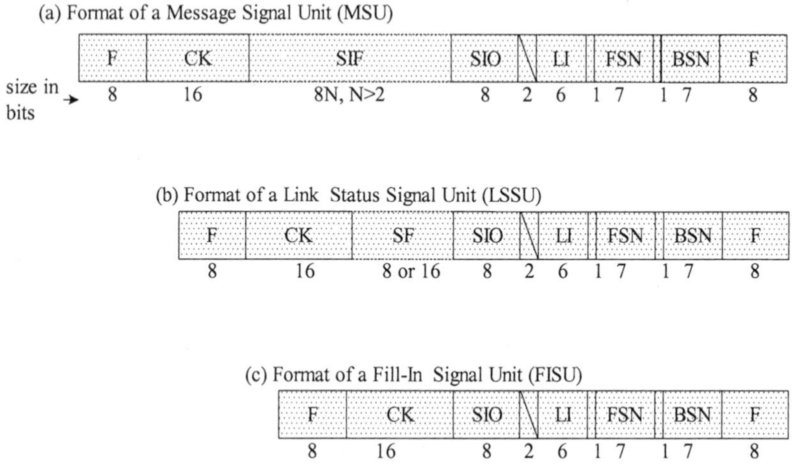

(a) Format of a Message Signal Unit (MSU)

F	CK	SIF	SIO	\	LI	FSN	BSN	F

size in bits → 8   16   8N, N>2   8   2   6   1 7   1 7   8

(b) Format of a Link Status Signal Unit (LSSU)

F	CK	SF	SIO	\	LI	FSN	BSN	F

8   16   8 or 16   8   2   6   1 7   1 7   8

(c) Format of a Fill-In Signal Unit (FISU)

F	CK	SIO	\	LI	FSN	BSN	F

8   16   8   2   6   1 7   1 7   8

BIB: Backward indicator bit   LI: Length indicator
BSN: Bankward Sequence Number   N: Number of octets in the SIF
CK: Check bits   SIF: Signal Information Field
F: Flag   SF: Status Field
FIB: Forward indicator bit   SIO: Service information Octet
FSN: Forward Sequence Number

Figure 9-8. SS7 Packet/Signal Unit Formats

The basic frame structure is similar to other data network bit-oriented link protocols (e.g., HDLC, LAP-B). SS7 uses standard flag (0111 1110 = 0x7E) to start a new signaling unit. The flag provides frame delineation. Error detection is supported by 16-bit CRC. Generally, bit-oriented data link protocols either transmit idle signals, which are all'1's or empty frames consisting of continuous sequence of back-to-back flags. In contrast, SS7 requires that FISUs be sent during idle times. The Signaling Information Field (SIF) and Signaling Information Octet (SIO) constitute the payload of SS7 layer two frames. The length of MTP2 frame can't be more 272 bytes including all the layer two support fields (CK, LI, FSN, BSN).

Data link protocols such as Ethernet don't have a two-way error correction mechanism. However, HDLC and MTP2 do support two-way handshake for error correction[158]. The two-way handshake consists of

[158] The rationale for having two-way handshake for error-control as well as flow-control at layer two as opposed to at layer three or at higher layers could be due to early views and

forward/backward sequence numbers and retransmission of errored/missed frames. MTP layer two is specified to use two types of retransmission: basic and preventive cyclic retransmission (PCR). In either retransmission method, a copy of the signal unit being transmitted is maintained in a retransmission buffer at the sending end until a positive acknowledgement for that signal is received from the remote end.

The basic method of error correction is based on the well-known "go-back-n" technique for retransmission. It is a noncompelled positive/negative acknowledgement (ACK/NAK) retransmission error correction system. Essentially the transmitter keeps sending unless a NAK is received from the remote station with the error-frame sequence number. The transmitter stops sending new frames and rolls-back to the sequence number received in the NAK. On the other hand, a received ACK indicates that the remote end has received frames up to and including the sequence number included in the ACK response. The local transmitter can discard any buffered frames with sequence number less than the ACKed sequence number. For PCR, which is primarily used with satellite-links or other long-delay signaling links, reader can refer to [Jab91],[Mod90]. Similarly, for information on *flow-control* and other functions of MTP2, reader can refer to [Jab92].

9.3.2.3 MTP Level 3: Signaling Network Functions (Layer 3)

The functions of MTP layer three consists of *message discrimination*, *message routing* and *message distribution*. Besides these, it contains functions for routing table management, signaling link management and signaling traffic management functions. Figure 9-9 shows the components of MTP layer three.

---

models available for data communications. With next generation (after 1975) of protocols such as IP, retransmission and flow-control are handled at application protocols such as TCP (layer 4). In other words, even at layer three functions as provided in MPT2 are not supported.

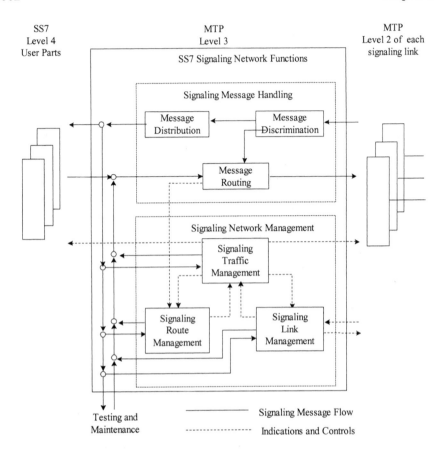

SS7	MTP	MTP
Level 4	Level 3	Level 2 of each
User Parts		signaling link

Figure 9-9. MTP Level 3 Functional Blocks

9.3.2.4 MTP Message Routing Function

SS7 uses point codes to identify SPs (an STP or an SCP or an SSP). The SPs are the NE control entities (N-CE) we talked about in the beginning. The point code consists of 32 bits (ITU) or 48 bits (ANSI). Figure 9-10 shows MSU with MTP3 specific details. MTP3 provides message-handling functions based on the "routing label" consisting of OPC and DPC where the OPC contains originating SP's point code while the DPC contains the destination point code.

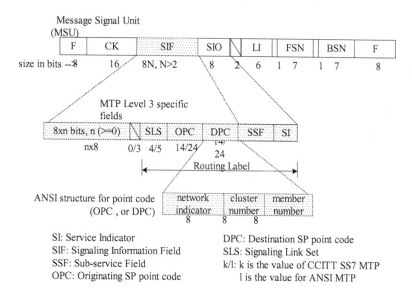

Figure 9-10. MTP Level 3 Routing Label

MTP3 routing function supports connectionless routing and SS7 messages are routed based on the DPC alone. The MTP3 routing function is used to route and transmit messages from a local level three user (ISUP, SCCP etc) as well as to route incoming messages from MTP2. With local messages from level 3 users, the choice of which outgoing signaling link to use is made by *message routing* function of MTP3. With incoming messages (received from MTP2), the *message discrimination* function is activated which determines, using DPC, if the message is addressed to the local SP. If not, and if the receiving SP has the transfer capability (i.e., the SP supports functions of STP) then the *message routing* function is activated. Message routing is based on the DPC and SLS in most cases. In some circumstances, the SIO, or parts of it may need to be used.0. If a message received from MTP2 is addressed to the local SP, then the *message distribution* function is activated which delivers the message to the appropriate local user or a MTP level three function based on the SI, a sub-field of SIO.

MTP layer three also supports what is called signaling link selection (SLS) function[159]. With SLS, MTP can control the signaling network path for a sequence of messages that are related. For instance, all signaling messages that are part of some application transaction can be constrained to take the same network path so that there is no sequence mis-ordering at the receiver.

---

[159] A comparable function is not available in IP protocol suite

Another important purpose of SLS is to provide capability to load balance-signaling traffic on multiple outgoing signaling links. The combination of DPC and SLS will completely determine the outgoing signaling link. The originating SP allocates the SLS code and it will be preserved except possibly for changes necessary for load sharing at subsequent nodes. All messages, in a given direction, associated with a given transaction will be assigned the same SLS code by MTP3. For more details on MTP routing and congestion control, reader can refer to [3].

The MTP routing tables contains multiple routing entries. Each entry describes, for a given DPC, which outgoing signaling link to use to forward messages destined for that DPC. MTP routing entries are much like the next hop routing entries in IP routing tables. The MTP routing entries are installed using network management systems. In other words, the routing entries are static and not updated automatically. There is some support in MTP for making sure the installed entries are valid. Next, we describe signaling network management functions that are used in validating static route entries.

9.3.2.5 MTP Signaling Network Management

The signaling network management consists of *signaling traffic management*, *signaling link management*, and *signaling route management*. These functions perform the internal network management of the signaling network and traffic such as re-routing of signaling traffic in case of failure or congestion. Failure may involve the SPs (particularly failure of STPs), or links. Specific procedures have been defined to change the network configuration or restore normal configuration upon availability of the faulty links or SPs.

For detailed description of functions and procedures associated with *signaling network management* reader can refer 0. Briefly, *signaling traffic management* part can be equated to similar function performed at higher layer such as TCP in IP protocol suite. The *signaling link* function can be thought of as similar to *neighbor link management* in OSPF using bi-directional *Hello* protocol.

It is important to understand that MTP provides static routing i.e. the routing tables are provisioned using OSS interfaces. For instance, if there are 100 SPs' in the network and a new SP is added to the network then each of the 100 SP's routing tables is manually added by OSS to contain the point code of the new SP. In contrast, dynamic routing protocols such as OSPF or

IS-IS are used to maintain routing tables in large IP networks. Also note that MTP, IP and OSI are all destination-based routing/forwarding protocols[160].

9.3.3 SCCP

SCCP provides transport services that are essentially an enhanced version of those offered by MTP3. The various transport classes supported:
- Basic connectionless
- Sequenced connectionless
- Basic connection-oriented
- Flow controlled connection-oriented
-

MTP3 is limited to delivering messages to a node based on the DPC and, to distributing messages to users within the node using a small four-bit service indicator (a sub-field of the SIO). Obviously, the number local users supported by MTP3 is limited. SCCP enhances this by providing an enhanced addressing capability that uses DCP plus Subsystem Numbers (SSNs). An SCCP sub-system is an SCCP user such as TCAP or ISUP. SSN is an eight-bit field and used identify SCCP users at a node. Figure 9-11 shows calling or called party address structure used in SCCP messages containing SSN. It also shows some pre-assigned SSN numbers (similar to well known or pre-assigned TCP/UDP port numbers in IP protocol standard). We will discuss this structure shortly.

SCCP also provides a management function, which controls the availability of the "sub-systems", and broadcasts this information to other nodes in the network, which have a need to know the status of the "sub-system". The combination of the MTP and the SCCP is called the "network service part" (NSP). The NSP should be seen as the network layer of the OSI Reference Model.

---

[160] IP protocol does support source routing. However, it rarely used since it takes more bandwidth. In addition, source routing s not useful if network configuration changes that invalidate computed source route resulting in dropped packets or routing loops.

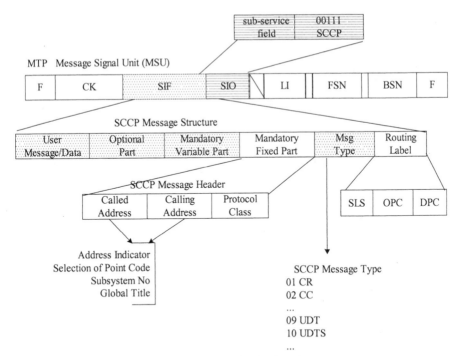

Figure 9-11. SCCP Message

Besides supporting more local users, SCCP provides a routing function, which allows signaling messages to be routed to a signaling point based on, for example, dialed digits [ITU00]. This capability involves a translation function that translates the global title (e.g. dialed digits) into a signaling point code (i.e. into SS7 network address) and possibly a SSN. As part of routing function, congestion control measures are provided to reduce traffic in the event of MTP, SCCP congestion or node (the SCCP node) congestion or to redirect traffic in case of a failure.

9.3.3.1 SCCP Routing function

SCCP allows users to specify a logical address, rather than a point code, to refer to a remote user. In such a case, Global Title Translation derives an SS7 network address (e.g. the physical point code of the remote user). If the association between the logical and the final physical address is not locally possible then, SCCP delivers the message to the next SCCP node to provide the association. This propagation is done until a translation in some SCCP node along the path delivers the final association between this address and a local SCCP user. The process of translating a logical address into a network address is similar to DNS service in IP networks.

Table 9-1. SCCP Address Structure

Calling and Called Party Address in SCCP messages				
8	7	6 5 4	3 2	1
National Use	RI Routing Indicator	GT Indicator	SSN Indicator	Point Code Indicator

Signaling Point Code (1..8)

00   Signaling Point Code (9..14)

Subsystem Number (SSN)

Global Title Translation Type (GTT)

Numbering Plan (NP)   GT encoding

Address Type (NA)

Address Information (digits)

...

	Note – SSN assignment
0	SSN not known/not used
1	SCCP management
2	TUP
3	ISUP
4	OMAP – SS7 Operation and Maintenance AP
5	MAP – Mobile AP
11	ISUP/SS ISUP supplementary services

.

.

.

The various pieces of information shown in Table 9-1 and used by SCCP routing function are:

1. **Signalling Point Code (SPC)** – This uniquely identifies a node in an MTP SS7 network. This PC identities the next SCCP node.
2. **Global Title (GT)** – This is an address used by the SCCP, comprising dialled digits or another form of address that will not be recognised by the SS7 network layer. Therefore, translation of this information to an SS7 Network Address is necessary.
3. **Sub-System Number (SSN)** – This identifies a sub-system accessed via the SCCP within a node and maybe a User Part (e.g. SCCP Management or an Application Entity containing the TCAP layer).

–

The node processing the SCCP address uses routing indicator bit to know whether it should use SPC (optionally SSN in the final node) or GT. If route-on-SPC then MTP will do the routing. If route-on-GT then the SCCP global title translation function will be invoked to determine the next node of the message. The node itself may have the capability to do the translation. If not it will need to forward to another SP (the address of which may be provisioned manually just like DNS server entries) that implements GTT. A node implementing GTT will examine TTN, NP, NA and address digits and derive   SS7 point code + network indicator. It is possible that SSN is computed as part of translation. After the translation, the address is an MTP routable and is forwarded to the node representing that address.   The following figure captures the above discussion with an example.

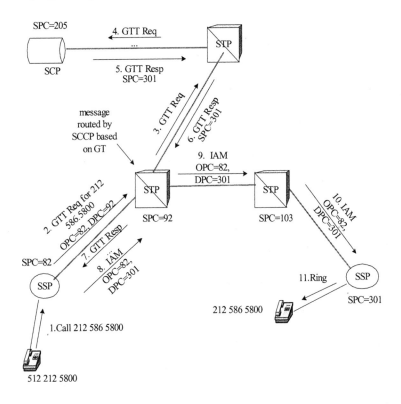

Figure 9-12. Global Title Translation Example

GTT is an interesting function supported by SS #7. It is possible to conceive similar function for ASON/ASTN. In OIF UNI 1.[OIF1], each user entity is provided or assigned Transport Network Address (TNA). Accordingly when a callee UE, identified as TNA-A, sends a connection request to connect to a remote UE, identified as TNA-Z, the ingress NE to which TNA-A is attached can query a central database using GTT mechanism and obtain identity of the NE to which the remote UE is attached. Subsequently the ingress NE can start signalling to set up the connection to the egress NE. This kind of mechanism provides separation of addressing schemes and hence flexibility. It is very easy to control the database through scripts or provisioning systems. Such a feature would give providers needed flexibility and enable faster deployment of optical control plane.

9.3.4   ISDN-UP (ISUP)

ISUP is point-to-point NNI (node-to-node) signaling protocol of SS7 and the main user of MTP.   It performs dual roles: call control (CallC) and connection control (CC). As part of connection control, it is responsible for setting up the switches along the path. The internal interface between ISUP and switch control software to setup connection is not part of ISUP and hence switch vendor specific.   We show the ISUP message structure in Figure 9-13. CIC, the circuit identification code, identifies the trunk (T1/E1) as well as time-slot with in the trunk for a call. What this means is that two SSPs connected with up to 128 trunks can use a single signaling link to setup voice calls over any time-slot of those trunks. If there are more than 128 trunks then a new signaling link must be established between the two SSPs. The following figure illustrates a simple call setup using ISDN-UP.

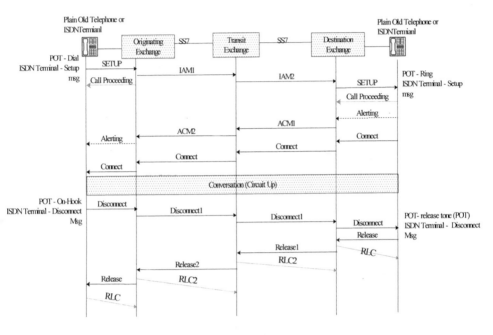

Figure 9-13. SS7/ISDN Call Setup/Clearing Sequence

9.3.5 OMAP

The Operations, Maintenance and Administration Part (OMAP) [GOL90] is a set of operations and maintenance audit procedures for maintaining the signaling network. For example, an audit procedure called MRVT – 'MTP Routing Verification Test' is defined to verify the entries in the MTP routing table to detect the following anomalies:

– Loops in MTP routing
– Routes with excessive length
– Unknown destinations
– Erroneous bidirectional signaling relations (A can reach B, but B can't reach A)

Some of the anomalies can be detected in IP routing as well. For instance, when a router receives a packet and the router can't forward that packet further because the destination address in the packet is not in the routing table, the router generates an ICMP error packet indicating 'destination unreachable' towards the sender of such packet. Similarly, both excessive route length and loops can be detected using time-to-live (TTL) field of IP packet header. However, all these are reactive methods of detecting routing issues. The auditing procedure defined for MTP can is a pro-active mechanism for detecting such anomalies. Refer to [Xin98] for more details on OMAP.

9.3.6 Signaling Network Structures

The components of SS7 networks (STPs-stand-alone or integrated, SCPs, SSPs, and signaling links) can be arranged in many different ways to form a signaling network. The signaling protocols of SS7 are independent of the resulting network structure (i.e. topology) as would be expected. However, a given choice of signaling network structure must meet unavailability objectives specified in SS7 standards. Route redundancies, backup for links and signaling points are important considerations.

A popular SS7 signaling network structure is the mesh structure, also known as the quad structure [Gol98], illustrated in Figure 9-14 (a).

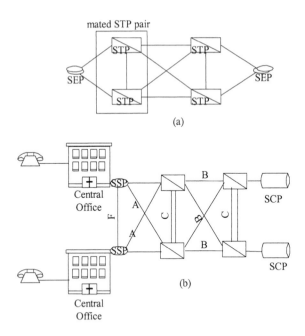

Figure 9-14. (a) SS7 Signaling Network structure (b) Link Types

In this structure, STPs are mated on a pair-wise basis. The quad structure is used in North America [Gol98]. The quad mesh network structure has 100% redundancy and as a result following any signaling link or SP failure, signaling traffic can be diverted to alternate path. In addition, the alternate path doesn't increase the number of transfer points involved. This is very important for voice networks where the call setup latency must be deterministic and bounded. The side effect of such redundancy is that each link and node must be properly engineered. For instance, assume that under normal conditions an STP handles a load of L. When a link fails somewhere in the network then the same STP must be capable of handling 2xL load.

As illustrated in Figure 9-14 (b) links connecting SPs are classified into the following types:

- A links (access links): Used between STP and SSP or STPs and SCPs. Provide access into network and to databases through STP.
- B links (bridge links): Used to connect mated pairs of STPs at the same hierarchical level. B-links are deployed in quad structure.
- C links (cross links): Used to connect two-mated STPs. Deployed in pairs. These links are used for signaling traffic only in the case of congestion and unavailability of other routes.

- D links (diagonal links): Used to connect mated STP pairs at different hierarchical levels. They are used only when different STPs are deployed at different levels. D-links are deployed in quad structure. An example will be shown later.
- E links (extended links): Used between SSPs and remote STPs for signaling route diversity. An example will be shown later.
- F links (fully associated links): Used between SSPs when signaling traffic volume justifies associated signaling between SSPs.

9.3.7 SS7 Signaling Network - Administrator Interconnections

According to [Gol98], the interconnection of one SS7 signaling network with another can take on many forms and decided based on such factors as robustness, reliability, flexibility, geographic, and jurisdictional needs. Both single and multi-STP gateway interconnection architectures are possible as illustrated in Figure 9-15.

The first one is a single gateway architecture where a mated pair of STPs in one network is connected to a mated pair of STPs in the other. The one shown in (c) uses two distinct pairs of STP gateways in each network; one pair in networks III and another in network I. Any other types of interconnections can be reduced to these two basic types.

An alternative, shown in (b) is to connect an SEP in one network directly to an STP in another network. This method is useful for connecting SEPs in smaller networks (networks without STPs) to SEPs in larger networks that will have STPs. This type of interconnection architecture, also called dual-or multi-homing, is increasingly being talked about for UE to NE IP services [BCR02].

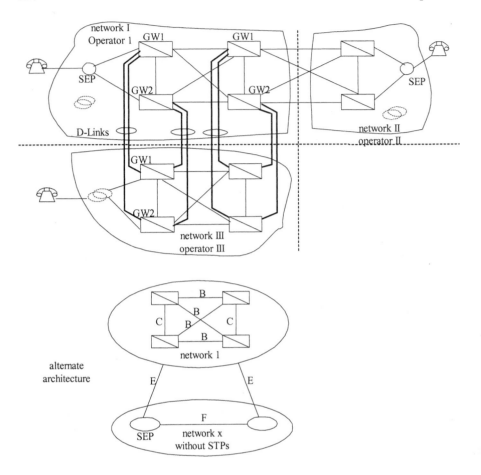

Figure 9-15. Network interconnection architectures

As explained earlier, ANSI standard for SS7 uses 24-bit point codes to identify each SP (an N-CE) with each point code constructed from three 8-bit fields: network identifier (NID), network cluster number (CLU) and cluster member number (MEM). An STP that is designated to route signaling traffic for a cluster can check the cluster number only and, if the cluster number is a foreign cluster then it can forward the message to a pre-designated STP gateway. Such function is also part of IP routers where a packet whose destination address is not in the routing table of a router is forwarded to pre-designated router (also called the default gateway). For instance, in the above illustration, all SPs within network I will have same NIDs and may be same CLUs. Similarly, all SPs in network III will have same NIDs and may be same CLUs.

A routing complexity arises when networks from different providers are interconnected. Figure 9-16 (a) is an example where the source and the destination network are directly interconnected using one of the redundant gateway interconnection schemes mentioned earlier. For the signaling traffic going from source network to the destination network, there is no alternative; the primary gateway can be used in each network to route messages destined for some switching point (SP) in the other network. However, in the case of (b) and (c), there will be multiple signaling paths, between the source to destination network. No MTP mechanism exists to dynamically determine alternate paths. Besides, no MPT policy exists to select the best alternative.

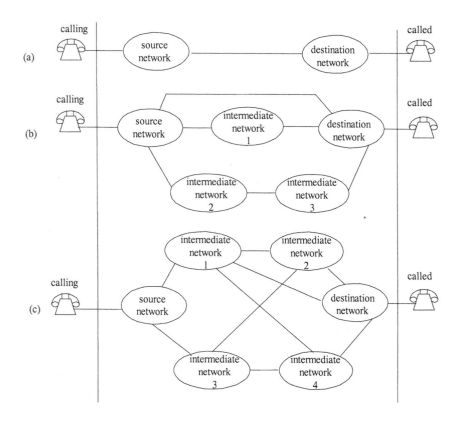

Figure 9-16. SS7 network interconnections with multiple transit networks

To provide some hypothetical case, consider configuration shown in (c). The source network may have a pair of GWs towards transit network one and, another pair towards transit network three. All these gateways can run inter-domain routing protocol, such as BGP [Rek95], between them and advertise all or a range of point codes of SPs belonging to their networks. For instance, the GW in transit network three can advertise to its counter part GW in the source network about all SPC of network three as well as foreign SPCs that can be reachable via network three. For each foreign SPC there can be multiple exit points (GW) in a network. Internal policies can select one of the gateways as exit point to reach a given foreign SPC. However, at the time SS7 work was done, such advanced features probably were not even discussed or deliberately the choice was made for manually configured solution. Given that SS7 supports worldwide call volume in excess of millions of calls per day and being in operation for over two decades, these issues are either minor or don't represent significant operational expense for carriers.

9.3.8 Gateway Screening

Carriers interconnect their SS7 networks using interconnection agreements (ICA)[161] such as the one in [BAC97]. Even if ICAs are in place, it is important to ensure that unexpected or erroneous signaling traffic is filtered [BAC00] by STPs. The screening performed at such interconnection points is called gateway screening and is performed by STPs. Without such capability, one or more malfunctioning SS7 nodes can storm messages through interconnection points potentially bringing down whole PSTN.

Gateway Screening [GR-802] involves examining incoming SS7 messages and determining whether the message should be accepted or rejected based on rules entered into the STP by an administrator. The rules can be constructed based on OPC, DPC, or other parameters of MTP3 or SCCP messages. For instance, a simple rule could be 'if DPC=201, then don't transfer'. The following are general screening rules:

- Allowed OPC
- Blocked OPC
- Allowed DPC
- Blocked DPC
- Allowed SIO
- ...

---

[161] IP network providers (the ISPs) interconnect their networks using peering agreements.

Besides, MTP screening, STP GWs can support screening based on fields of SCCP messages and the following are examples:

- Calling party address (point code/SSN, SCCP message type, routing indicator)
- GTT number and the typical screening

The screening is similar to IP filtering functionality to filter or drop packets based on source or destination IP address, UDP/TCP port numbers etc.

9.4. ATM CONTROL PLANE

ATM signaling is a derivative of ISDN signaling with additions and modifications to support QoS aspects of service creation and management. ATM Signaling is well defined, deployed and a mature standard. ITU-T has ratified ATM based signaling as one option (G.7713.1) for use in optical and transport networks. ATM technology was the holy grail of packet switching for a decade or more. Today packet switching and QoS guarantees offered by ATM technology are available with IP QoS as well. The industry is at a turning point with regards to using ATM as a data backbone technology. Currently, ATM carries significant amount of data traffic in the backbone and is part of RBOC and IXC networks for many years. In line with its maturity, applicability, and potential, this section will focus on ATM signaling.

9.4.1 ATM Signaling Reference Points

Figure 9-17 and Figure 9-18 illustrate reference points used in ATM signaling as well as components involved in both ITU and ATM Forum's signaling variants.

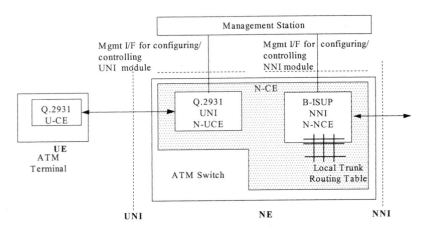

Figure 9-17. ITU ATM Signaling Reference Points

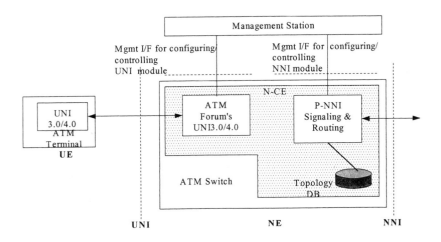

Figure 9-18. ATM Forums' Signaling Components

Table 9-2 lists some high level differences between these two variants and some of these differences will be explained in subsequent sections. The terminology N-UCE and N-NCE is used to refer to UNI and NNI parts respectively of an ATM NE.

Table 9-2. ATM Signaling Variants and differences

	ATM Forum	ITU	Remark
UNI	UNI 3.1/4.0	Q.2931	Both are supposed to be similar with some differences
NNI	P-NNI Applicable over private networks	B-ISUP Applicable over public networks	P-NNI is based on UNI for signaling but with additions to support routing of signaling messages. B-ISUP has no routing mechanism. Its layer 3 (MTP3) is empty in most implementations.
UE Addressing	NSAP NSAP	E.164 E.164	
NE Addressing			

9.4.2 ITU-T ATM Signaling: UNI (Q.2931), NNI (B-ISUP)

Figure 9-19 illustrates another view of ITU-T ATM signaling interface reference points. In this figure, Q.2931 protocol element supports call control functions. B-ISUP protocol element supports node to node call control signaling. The signaling message transport layer is represented by SAAL, Signaling ATM Adaptation Layer. Figure 9-20 illustrates a possible realization of ATM signaling infrastructure in a SONET/SDH NE using IB (embedded operations channels or dedicated time slots) or OOB (LAN or dialup connections). We show SAAL-IP, instead of regular SAAL, since IP protocol is becoming a common transport protocol of choice to transport signaling and management messages. This kind of realization may be used by vendors who already have products with Q.2931 and want a simple migration path.

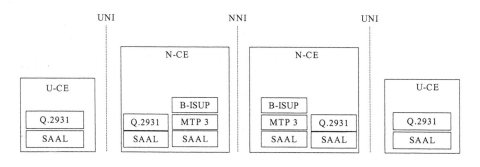

Figure 9-19. B-ISDN/ATM Signaling Stack

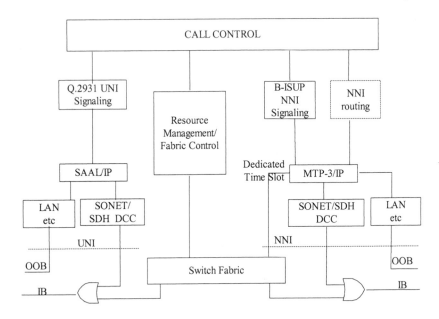

Figure 9-20. Possible implementation in a SONET/SDH NE

9.4.3   PNNI Routing

PNNI is the switch-to-switch (i.e. node to node) protocol that includes routing and signaling. It is standardized by ATM forum for use in private ATM networks, which include the data backbones operated by carriers. PNNI is a link state routing protocol, similar to OSPF and in fact, many aspects of PNNI were derived from OSPF.

9.4.3.1 PNNI Hierarchical Network

PNNI allows for scalable routing protocol deployment using partitioning– horizontally[162] using peer groups and vertically using levels (as many as 104 levels). The physical network consists of nodes (switches) and links connecting them. Figure 9-21   illustrates a sample PNNI network [Sha99] with three levels. Nodes at the highest-level (level L) are physical

---

[162] The horizontal partitioning in PNNI would be similar to area concept in OSPF. Such partitioning allows for containment of advertisements and hence scalability of advertisement based routing protocols such as OSPF and PNNI

while nodes at the lower hierarchies (level L-1, L-2) are logical. Level zero, the top level, contains the whole network.

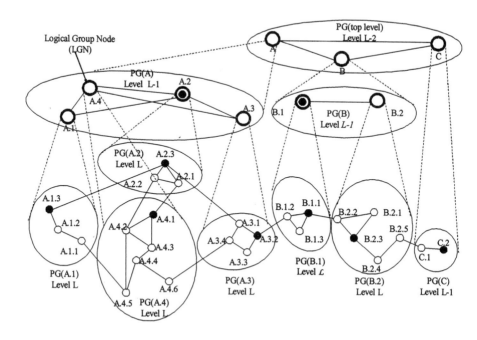

Figure 9-21. Sample ATM PNNI network

Nodes at each level are grouped into peer-groups. Within a peer-group, one of the nodes is elected as peer-group leader (PGL) to represent the group as an abstract logical node (LGN) in the next hierarchy. For instance, in the above illustration, A.1.3 is the PGL for PG (A.1). The concept of PGL is similar to Designate Router (DR)[163] concept in OSPF though the purpose of PGL is different. Both, OSPF and PNNI, provide mechanisms to configure and control the leader election process using priorities. In the case of OSF, router priority is used to help in the DR election. Similarly, in the case of PNNI, peer-group leader priority is defined and is used to elect a PGL for a

---

[163] In OSPF, two routers, a primary (DR) and a backup (BDR), on a given LAN are designated as DB keepers with the DR advertising the LAN into rest of OSPF area. Other routers on the LAN synchronize their topology DB with DR and BDR only. This scheme avoids the problem of *n-square* database synchronization between 'n' routers on the same LAN. As a result, non-DR and non-BDR routers can advertise the interface attached to the LAN as eligible for routing as soon DB synchronization is complete with DR and BDR routers only.

peer-group. When more than two nodes have same priority node ID (PNNI) is used to resolve the contention (in the case of OSPF, router ID (OSPF) is used resolve contention during DR election). Both OSPF and PNNI allow dynamic choice. When the elected DR or PGL fails, the neighbors detect that and all the nodes get involved again in the election of a new leader. Except for the fact that a PGL houses a LGN there is no difference between a PGL and another node in the same peer group. In addition, LGNs are not involved in signaling to setup circuit between user terminals attached to the network.

9.4.3.2 Routing Control Channel (RCC)

The channel used for PNNI route information exchange is called the Routing Control Channel (RCC). The RCC is constructed in three different ways:
– VPI/VCI 0/18 channel on each physical link to a neighbor node
– All virtual path connections for which the node is an end point
– Switched Virtual Circuit (SVC)
Logical nodes i.e. LGNs, which exist at lower hierarchies don't have direct connectivity and so use Switched Virtual Channel (SVC) to realize RCC. In the following section, we will see how LGNs setup RCC between them.

9.4.3.3 Border Nodes, Uplinks and Horizontal Links

A node in a peer-group with a link to a node in another peer group is called border node[164]. A link connecting border nodes of two different peer-groups is called an outside link. Links between nodes within a peer group are called inside links as illustrated in Figure 9-22.

---

[164] PNNI border nodes would be similar to area-border routers (ABR) in OSPF

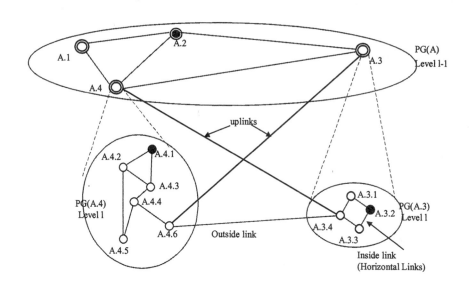

Figure 9-22. PNNI Uplinks and Outside links

Each border node in a peer group executes Hello protocol on each of its links and, determines if a link is an outside or an inside link. In addition, border nodes include their ancestor peer group Ids in Hello packets sent on each of their outside links. For instance, node A3.4, when it sends Hello packets on its outside link to A.4.6, will include its ancestors (assume that at this point of time A3.4 knows only about its immediate ancestor PG which is PG (A)). Similarly, A.4.6, in its Hello packets on its outside link to A.3.4, will include its ancestors (again assume that A.4.6 knows only about its immediate ancestor PG which is PG (A)). Since the two border nodes have found one common ancestor, each one will transition the state of the outside link to common-outside, meaning they found a common ancestor peer group. Subsequently, border node A.3.4 advertises, within its peer group PG (A.3), that it is connected to LGN A.4. The connection is called an uplink. Once nodes in PG (A.3) receive such advertisement, they will know that the border node, A.3.4, can be used to reach the LGN A.4. Same process occurs on A.4.6 and it will advertise an uplink to A.3 within its peer group. It is also interesting to note that there are multiple *uplinks* here: B.1.1—B.2; B.2.2—B.1; A.1.1—A.4; A.4.5—A.1; A.2.2—A.4; A.2—A.4; A.4.6—A.3; A.3.4—A.4; A.3.2—B.1; C.1—B; B.2—C with the last one called as *induced uplink,* derived from uplink B.2.5—C.

In the above illustration, logical nodes A.4 and A.3 are at the same hierarchical level and hence they can become routing neighbors. However, this requires an RCC between them. Since A.4 and A.3 are housed in physical nodes A.3.2 and A.4.1 respectively, these physical nodes need to know which border nodes in their respective peer groups to use to establish RCC. The information required would be available because of *uplink* advertisements described in the previous paragraph. For instance, A.3.2 will know that border node A.3.4 can be used to reach PG (3). Similarly, A.4.1 knows that it needs to use A.4.6 to reach PG (3). One of the nodes (i.e. A.3.4 or A.4.1) will initiate setting up of RCC by using SVC mechanism available in ATM signaling and shown for this particular example in Figure 9-23. The internal path for the SVC based RCC[165] with in each peer group is computed solely from internal topology information of that group.

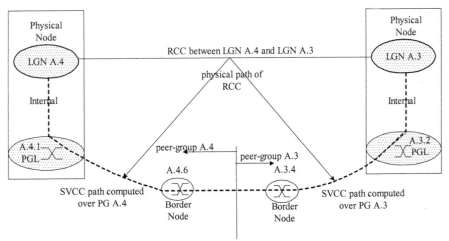

Figure 9-23. SVCC for RCC between two LGNs

9.4.3.4 Topology Aggregation

A unique feature of PNNI is its mechanisms for aggregating topology. In PNNI, and in general in any dynamic routing implementation, each node keeps a database store containing   network information. The database store

---

[165]One can draw parallel between the RCC/SVC and virtual link in OSPF. In OSPF, virtual link is used to carry OSPF protocol messages from an area that is not directly connected to backbone area-0. A router in such area is connected to one or more routers in the backbone area   through another non-back bone area and such connectivity is referred to as virtual link.

is assembled and updated from advertisements and updates received from its neighbors through flooding. A large network requires:

- Higher bandwidth for RCCs for exchanging large network information
- More powerful CPU for processing frequent and large routing information updates
- Large memory for storing large network information

Usually the same communication link is shared by management and control plane and hence bandwidth required to exchange large network information could have impact on performance of management applications. Similarly, time spent by CPU (NE's CPU) on processing routing protocol messages can impact other system related activities such as OAM&P. In NEs with dedicated routing processor this may not be an issue. However, in most practical implementations single CPU performs OAM&P and routing protocol jobs. Minimization of memory required using aggregation is important, as physical memory available in an NE can be limited.

Besides, security is another primary concern for providers who don't want to expose their internal topology to customers.

Typical solution to the above or any other scalability issues is to use network-partitioning technique and confining flooding of network information. With partitioning, the memory required to store information is less. Not only has that, partitioning reduces the amount of routing protocol traffic thereby freeing node's CPU for other activities. Therefore, scalability is achieved with partitioning and peer-groups of PNNI and areas in OSPF represent model this technique. Carriers can control routing protocol deployments to allow them to keep one network view as much as possible without stretching the limitations of equipment in terms of the above three parameters. To allow controlled deployment routing protocols used must provide configurable parameters such as peer-group id, summarization, timers etc.

Obviously, nodes in a partitioned network know only about the part of the network to which they belong and this presents another problem. How does a node know about nodes and reachable addresses that are outside of their networks? If we tell everything about nodes and reachable addresses in other networks then we are not solving any scalability related problems. A work around involves telling in some compact way about nodes and reachable addresses in other parts of network. One common compacting technique is aggregation - of topology and reachability information. Both OSPF and PNNI support aggregation. We will describe PNNI topology aggregation first and postpone reachability aggregation for later discussion until after we discuss ATM addressing schemes. For a description of OSPF aggregation mechanisms reader can refer to [Moy98]

Topology aggregation is a process of summarizing and compressing topology information at each hierarchical level to determine the topology information to be advertised at the level above [Lee]. To install a circuit it is necessary to compute a route or path through the network for the circuit. This requires that the node performing the path computation has the latest information about the network at its disposal. The information, obviously, consists of topology and reachability. The result of path computation, based on such information, will be accurate if the topology information maintained by each node is accurate and up to date. The computation will be detailed if the topology information is detailed. The path computation result can be sub-optimal or not accurate due to:

- Network information used by the node, performing path computation, is 'out of sync' with the actual network
- Aggregation or compacting of topology information resulting in loss of detail; routing computation with partial information can produce a path that may be not viable.

For the first problem, flooding ensures that the information is accurate or up to date. There may be small window during which a node's view of network is different from the real network. However, generally this problem is less of a concern than the second one relating to loss of topology information because of aggregation. Therefore, a desirable property of aggregation is that it must adequately represent the topology of a given network for efficient routing and network resource allocation. Reader should note that installing of a computed path using signaling or other means could fail if the path computation produced a wrong path because of the above reasons[166].

In general, aggregation results in loss of information as it hides the real topology and, instead, summarizes an abstract topology. In addition, aggregation is irreversible i.e. from aggregated information about a network we can't deduce the real topology of the network.

In a hierarchical network, a group of nodes many be abstractly represented by a single node commonly called the logical node. Applying

---

[166] Crank-back signaling mechanism is designed to take care of path computation errors during path provisioning phase. Resources in the network can change after a path has been computed making a segment of computed path invalid. The nodes at the end of such segments participate in crank-back mechanism.

this principle to PNNI, a peer-group can be abstractly represented and we already know this; PGL represents a peer-group as a logical node in the next level of hierarchy However, to abstract a peer-group and represent an aggregated view of the peer-group into higher level, we need only consider the connectivity of the peer-group to other peer-groups. Therefore, a logical node need only model the border nodes that connects the peer-group to neighbor peer-groups as illustrated in Figure 9-24 (a), containing a nine node peer-group (internally consisting of five border nodes).

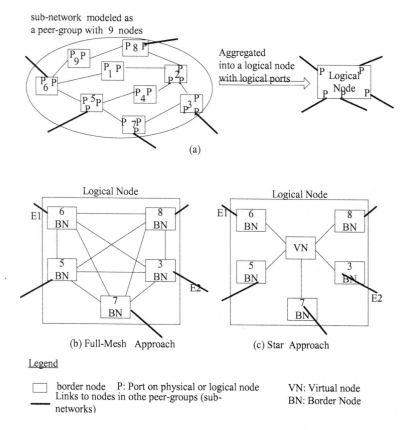

Figure 9-24. Topology aggregation (a) General Logical Node (b) Approaches (c) PNNI Logical Node - Internal connectivity representation

To provide a measure for aggregation, consider a PG with N nodes of which M ($\leq$N) are border nodes, interconnected by P physical links. Topology aggregation can map this PG into a logical node/network with M border nodes and L logical links where logical link refers to internal link between a pair of border nodes of the logical node. The objective of aggregation is to keep L small but give enough information so that a path computed using such aggregate information is as close as possible to the one computed using uncompressed topology information. In this context, there are two methods for topology aggregation 0: Symmetric-Node and Full-Mesh as illustrated in Figure 9-24 (b) taking the earlier example of 9-node peer-group. Any other method can be reduced to these two basic methods with modifications. Each such special method might represent some trade-off between accuracy and compactness, as we will see.

The Full-Mesh approach uses a logical link between each pair of border nodes to represent the internal connectivity of the logical node. Obviously, this approach requires MxM (i.e. L=MxM, worst case L=P) logical links. The cost of traversing a logical node (and hence a PG) is calculated based on the border nodes involved and the corresponding logical link cost between the border nodes. For instance, the cost of traversing, from E1 to E2 across the peer-group shown in Figure 9-24, is calculated based on the cost of logical link between border nodes 6 and 3. Using this approach, since each border node has a direct link to every other border node, the path across a logical node cost is based on just one link.

If the cost of all logical links is same then logical node's (constructed using Full-Mesh) internal connectivity models Symmetric-Node approach. A compromise between the two approaches is the Star approach illustrated in Figure 9-24 (c); logical node's internal connectivity is modeled using a virtual node with the each border node connected to the virtual node using logical links. The total number of logical links is only M (i.e. L=M and L<=P) as opposed to MxM in Full-Mesh approach. Further one can assume that all logical links have equal cost. However, real networks are hardly symmetric; some links may have higher bandwidths (implying they are more preferable), some border nodes may have direct links to one or more other border nodes in the same PG. Besides, operator may desire particular routes within a PG (i.e. administrative preferences) and such needs must be taken into account while finding a path across a PG. Taking all these factors into account PNNI modeled PG as a logical node with internal connectivity consisting of spokes, exceptions and express connections as illustrated in Figure 9-24 (c). Exception links represent pair-wise connectivity of a PG. For instance, one can compute, using shortest path algorithm, path to other border nodes in the same PG (i.e. a spanning tree). Once computed, a path from a border node to all other border nodes is modeled as bypass or express link and we show this for one border node in Figure 9-25.

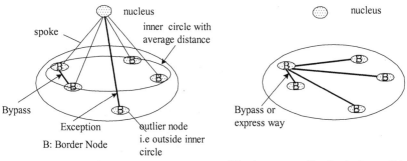

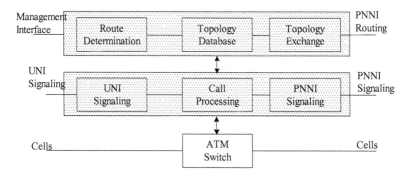

(i) PNNI Logical Node internal connectivity general representation

(ii) extreme case with pair-wise bypass links - by pass links shown only for one border node.

Figure 9-25. PNNI Logical Node - Internal connectivity representation

In PNNI, network operators can configure and control the process of aggregation. Such control provides flexibility and the operator can always by-pass automatic aggregation mechanisms built into PNNI if one is not comfortable with them. We will see later how the abstract, aggregated internal connectivity discussed thus far is advertised to all the nodes at the same hierarchical level and below.

9.4.3.5 PNNI Protocol Elements

Figure 9-26 illustrates PNNI node reference model.

Figure 9-26. PNNI Reference Model

PNNI, and in general any dynamic routing protocol, can be discussed in terms of:

- UE as well as NE Addressing Schemes
- Information Units – type codes, constants etc. (Information Groups)
- Packet formats for Hellos, Link state advertisements
- Neighbor discovery, Link and adjacency management
- Topology Database Synchronization
- Flooding
- Topology and Reachability Summarization/Aggregation

The following sections will elaborate some of the above topics.

9.4.3.6 ATM addressing scheme

In ATM, the term end system, a carry over from OSI/IS-IS, is used to refer to either ATM terminal (the UE) or ATM switch in the network (the NE). Regardless, each end system is assigned an address called ATM End System Address (AESA) and there are two formats for this:

- ITU-T: E.164 (for use in public ATM networks)
- ATM Forum's NSAP[167]

E.164 address format, illustrated inFigure 9-27 (a), consist of up to 15 digits with the first three digits allocated for country code and the remaining allocated for the subscriber number. The subscriber number itself can be structured to contain nation code + subscriber number as is common in many countries with the national code being usually called the city or area code.

ATM Forums' format, illustrated in Figure 9-27 (b), is designed for use within private ATM networks and is based on the OSI Network Service Access Point (NSAP) address format specified in ISO 8348 and ITU-T Recommendations.

---

[167] ATM Forum also allows for having group address.

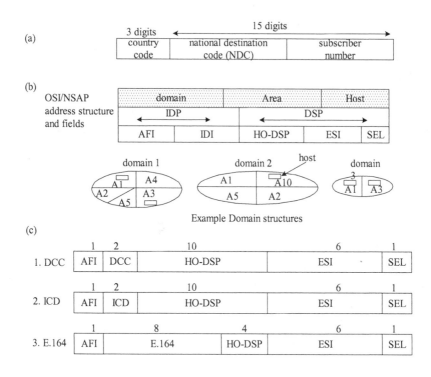

Figure 9-27. PNNI Address and Identity structures

A point to note is that ATM Forum's addresses are fixed in length to 20 octets while in OSI the length can vary from 7 to 20 octets. General NSAP format consists of network domain ID part (IDP) and domain specific part (DSP). Authority and Format Identifier (AFI) identifies the network addressing authority responsible for allocating IDI and the syntax of DSP. ATM Forum defines three AFIs, illustrated in Figure 9-27 (c): the data country code (DCC), international code designator (ICD), and E.164. The British Standards Institute administers the ICD codes and registrations, globally. DCC codes are assigned to countries; the ISO member organization in each country administers the assignment of organization specific codes in that country. The network administrator defines the specifics of high order DSP part. Based on the three AFIs, ATM defines three address formats: ICD, DCC, and E.164. The forum, while not preferring one format or the other, does recommend that the organization specific field of the private ATM address format should be constructed in such a way as to allow hierarchical routing and efficient use of resources. That is, the sub-allocation of fields within the organization specific part of the address should be assigned with topological significance[168].

In PNNI, UEs (terminals) and NEs (the nodes) are assigned end systems address/identifier (ESA/ESI) using one of the NSAP formats as explained earlier. Each UE or NE may have multiple sub or logical entities such as the LGNs. The selector (SEL) field available in all address formats may be used for identifying such entities. However, in all such cases, ESI field will be same as the one used with the physical node hosting the logical node.

9.4.3.6.1 PNNI Identifiers

Nodes and peer-groups occur at various hierarchical levels. As a result, both peer-groups and nodes must be assigned IDs that encode the level besides there assigned address/ID. Accordingly, a PNNI node ID is defined using a 22-byte structure and consists of two parts: a 1-byte level indicator and a 21-byte value that is same as node's assigned AESA. Similarly, PG ID is defined using 14-byte structure and consists of two parts – a byte level indicator followed by 13-byte bit-string that contains the peer-group's ESA. The 1-byte peer-group level indicator specifies the exact number of significant bits used for the peer group ID. The bit string can range from 0 to 104 (i.e. 13 octets), since PNNI supports from 0 up to 104 hierarchical levels. Figure 9-28 illustrates structure of peer-group ID, node ID as well as LGN ID.

---

168 There is always a hidden desire in conveying topological information using address. There was even a project to map IP addresses to zip codes

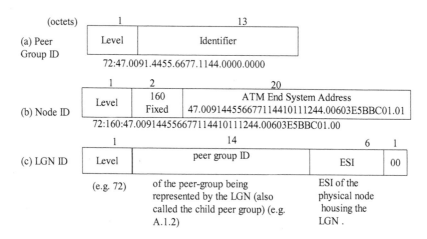

Figure 9-28. PNNI Address and Identity structures

9.4.3.7 Information Groups

In PNNI routing information is exchanged using information group (IG, a type-length-value (TLV)) units. Figure 9-29 illustrates the structure of IG.

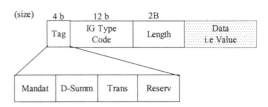

IG Name	IG Type Code
Nodal State IG	96
Horizontal Link IG	288
Uplink IG	289
IRA	224
ERA	256
PTSE	64
Requested PTSE Header	513
PTSE Ack	384
PTSE Summaries	512

Some common IG type codes

- **Mandat** - Mandatory bit. If this bit is set and a reciver doesn't understand the IG then it should not use information contained in the IG for path compuation.

- **D-Summ** - Don't Summarise bit. If this bit is set and a reciver doesn't understand the IG then the IG must not be summarized and advertised by the reciver.

- **Trans** - Trasitive bit. If this bit is set and a receiving PGL doesn't understand the IG then the IG must be preserved and not used in aggregation .

- **Reserv** - reserved amd set to 0.

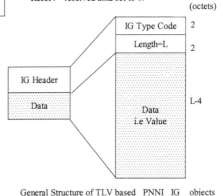

General Structure of TLV based PNNI IG objects

Figure 9-29. PNNI Information Group (IG) structure

In addition, an IG can contain other IGs (i.e. nest). IGs can be divided into three classes:

- Reachability information Group

  ATM End-System addresses (i.e. UE as well as NE identities) are advertised using Reachability IGs; their structure illustrated in Figure 9-30. A network with thousands of UEs and hundreds of NEs may result in large number of reachability IGs in the network causing too much protocol related traffic. To alleviate this problem, PNNI provides address summarization, to be explained later, allowing nodes to compress reachability information and thereby reducing the number of reachability IGs. Addresses associated with end systems are summarized in the reachable address prefixes advertised by the switching system to which the end systems are connected. Connections to those addresses are routed to the switch, which then forwards them to appropriate attached end system.

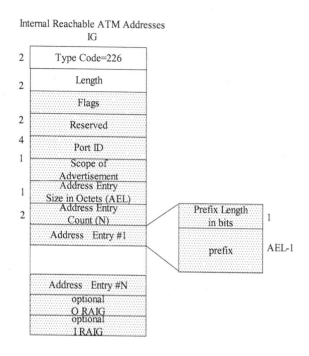

Figure 9-30. Reachable ATM Address IGs

- Nodal information class

  This class consists of two IGs - Node IG and Nodal State IGs. The Nodal IG, structure shown in Figure 9-31, is used to advertise information about a node such as its address, node ID, capabilities, PGL priority etc. If the node is a PGL then, the IG also contains information about the LGN by nesting Next Higher Level Binding IG. The information includes LGN node ID in the higher-level peer group, the higher-level peer-group ID and higher-level PGL ID if known at the time the Nodal IG is created. In general, Nodal IG includes all information about the node except reachability and resource information, which is advertised separately.

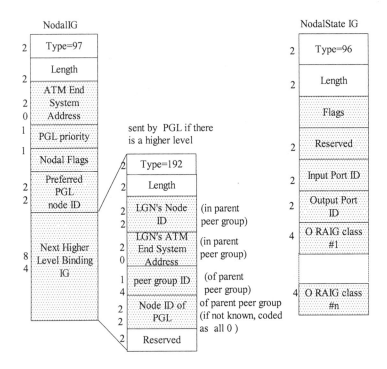

Figure 9-31. Nodal IGs

The *Nodal State IG* is used by LGNs to capture complex node representation explained in topology aggregation section. For each pair of logical port pairs the metric and attributes are captured in *Nodal State IG*. Multiple Nodal State IGs can be originated based on the number of logical ports modeled in the LGN.

- Topology State Information Class

IGs for describing the topological state of links and nodes belong to this group. Two important IGs in this class are used to attach cost (also called metric) and administrative attributes to nodes, links and reachable addresses. The first one is the Outgoing Resource Availability IG (O-RAIG) and the second one is the Incoming Resource Availability IG (I-RAIG). Both are similar, and Figure 9-32   shows the general structure of RAIG.

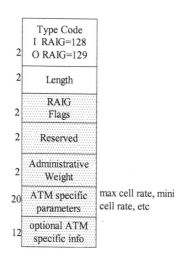

Figure 9-32. Resource Information Group (RAIG)

Some of the information of this IG is more specific to ATM networks and hence we show only some aspects of it. The administrative weight attribute is used by network operators to indicate if the metric associated with the IG is additive or not. By default it is additive.

The *Horizontal IG* is used to describe all outgoing links from a physical node. The information contained in *Horizontal IG* includes remote node ID, local and remote port IDs and one or more RAIGs that describe for each service class supported on the link the administrative cost. For using PNNI for non-ATM applications one can imagine using just one RAIG to advertise the metric for the link.

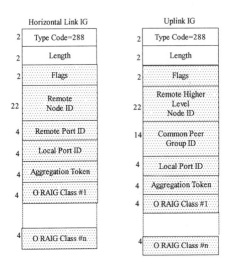

Figure 9-33. Horizontal and Uplink IGs

The *Uplink* IG is used by border nodes to advertise, into their own peer groups, about neighbor peer groups. For instance, border node, A.1.1.2, in peer-group, PG A.1.1 will advertise uplink to A.2.1. The process of forming uplinks has been explained earlier.

Aggregation token is yet another way of controlling what is aggregated. It is an administratively configured attribute with each link. All links between two peer groups (i.e. outside links) with the same aggregation token are aggregated into a single link. Reader may recall that outside links are not advertised; instead, an uplink to an upnode is advertised by border nodes. When there are multiple uplinks from a peer-group to same upnode then, they are aggregated into a single link if the aggregate token of associated outside links are the same. When there are multiple hierarchies, there can be uplinks between LGN in one level to an upnode in the next level. Such uplinks are called induced-uplinks but they are still supported over the same outside link.

From routing protocol perspective, contents of IGs are opaque and any of the above IGs can be included in any PNNI packet. Some, such as Hello, packets require mandatory inclusion of some IGs. We will discuss various packet formats, but before that, we need to introduce PTSE. Where applications exist, it is possible to advertise information that is not explicitly used by PNNI itself. Such information is commonly referred to as opaque information. Ability to advertise opaque information is also supported in OSPF using opaque LSAs.

9.4.3.8 PTSE

PNNI Topology State Elements or PTSEs are used to bundle different IGs covering a certain aspect of a topology. They are the units of flooding and retransmission. While IGs only carry values that describe topology or other information described earlier, PTSEs contain administrative information like remaining lifetime. However, PTSEs do not carry any information about the originating node and such information is part of PTSP packet, to be explained shortly.

A PTSE essentially serves as an envelope for a bundle of IGs. Only one type of IGs are allowed in a PTSE. As a result, there are many different PTSEs -address PTSEs, link state PTSEs etc. Figure 9-34 illustrates generic structure of PTSE. It is not necessary that all IGs of one type be bundled in one PTSE.

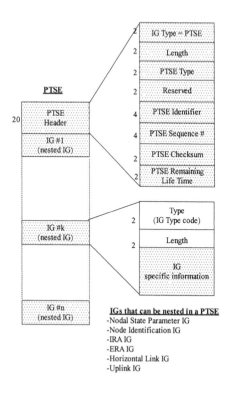

Figure 9-34. PSTE format

9.4.3.9 PNNI Packet Formats

PNNI routing protocol defines Hello, PTSP, DB Summary, PTSE Request, and PTSE Ack packets, exchanged in different phases of protocol operation. Each packet contains one or more IGs discussed above. The general structure of PNNI packet is shown in Figure 9-35.

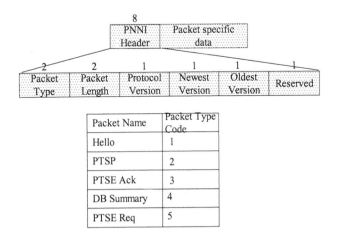

Packet Name	Packet Type Code
Hello	1
PTSP	2
PTSE Ack	3
DB Summary	4
PTSE Req	5

Figure 9-35. PNNI's (a) general packet format

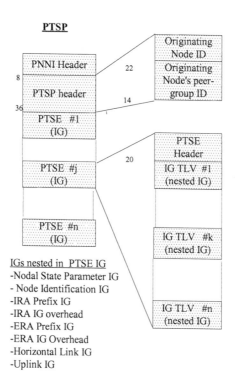

Figure 9-36. PTSP Packet

9.4.3.10 PTSP

Each node advertises information about itself (node identity, link and port information etc) using PTSEs. However, PTSEs don't contain information about the originating node; they need some kind of envelope when being sent to a neighbor. The PNNI Topology State Packet (PTSP), structure illustrated in Figure 9-36, is such an envelope that transports PTSEs.

While it is recommended to transmit as many PTSEs in one PTSP as possible, the size of the PTSP must not exceed the maximum PNNI packet size. If a node has lot of IGs of one type then they may not fit into a single PTSE without violating PNNI packet size limit. In such cases, multiple PTSEs may be originated with sets of PTSEs bundled into different PTSPs. A node receiving a PTSP from a neighbor need not flood the contained set of PTSE as a group together. The receiver can take out the individual PTSEs and group one or more of them with PTSEs received in some other PTSPs from the same neighbor and originate a new PTSP. The unit of flooding is a PTSE and not PTSP.

9.4.3.11 PNNI Hello Protocol

PNNI nodes transmit Hello packets, format shown in Figure 9-37, on RCCs. Figure 9-38 shows an example Hello exchange.

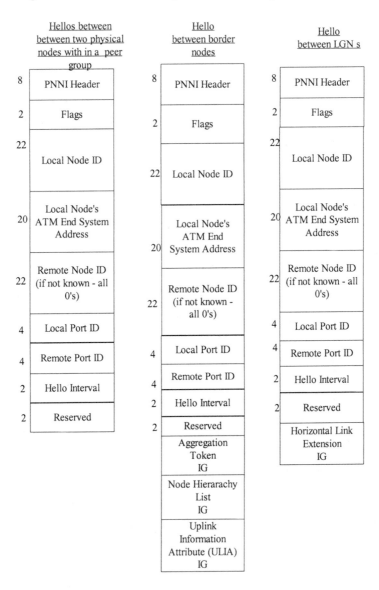

Figure 9-37. Hello Packet

Hellos are sent periodically every *HelloInterval* seconds or sent following an event, such as one of several hello protocol state machine changes. If no Hello is received with in some time   (= inactivity factor x hello interval), then the link is assumed to have failed. A node runs Hello protocol and a state machine, very similar to OSPF hello protocol, on each of its links (using RCCs).

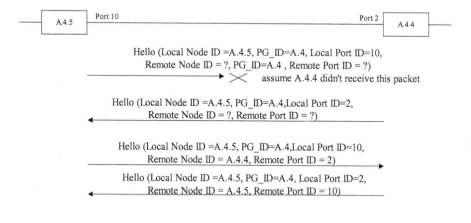

Figure 9-38. Sample Hello Protocol Exchange

On *outside links,* PTSE packets are not flooded over the well known 0/18 RCC, as done on *horizontal links*. Instead, the only way to exchange link parameters on an *outside link* is via the hello protocol.

9.4.3.12 PNNI Address Summarization

Address summarization reduces the number of individual addresses that need to be advertised. This results in reduced routing channel traffic; reduced CPU time spent on routing activity and finally reduced storage. We have already listed these as basic qualities of a scalable routing protocol. We will explain PNNI's summarization using Figure 9-39, which contains an example from PNNI specification.

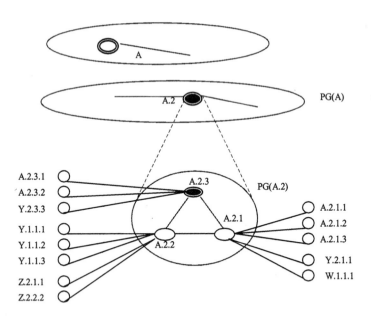

Figure 9-39. PNNI Address Summarization Example

We assume summarization is enabled on all nodes and further summarization prefixes are configured at each node. shown in Table 9-3. Accordingly, A.2.1 will advertise P<Y.2> and not the individual address Y.2.1.1. Similarly, A.2.2 is configured to summarize all Z.2.x.x into Z.2. Therefore it will summarize the two individual addresses Z.2.1.1 and Z.2.2.2 and advertises prefix P<Z.2>. For those addresses that don't have a configured prefix, they are advertised in their long form. For instance, P<W.1.1> is advertised all the way upward in its long form since there is no summarization prefix rule for that address

Table 9-3. PNNI Address Summarization Example

Summary Addresses at A.2.1	Summary Addresses at A.2.2	Summary Addresses at A.2.3
P<A.2.1> (configured) P<Y.2> (configured)	P<Y.1> (configured) P<Z.2> (configured)	P (A.2.3) (configured)
Reachable Address prefixes flooded by node A.2.1	Reachable Address prefixes flooded by node A.2.2	Reachable Address prefixes flooded by node A.2.3
P<A.2.1> P<Y.2> P<W.1.1>	P<A.2.2> P<Y.1>	P<A.2.3>
Summary Addresses at LGN A.2 Configured prefixes: P<A.2) (default, since it is the PG ID) P<Y> (configured)		
Reachability Information Advertised by LGN A.2 P<A.2> (PG ID A.2 should be automatically advertised by LGN)		
P<Y>		
P<Z.2>		
P<W.1.1>		

9.4.3.13 PNNI Database Synchronization

A new node (or an old node but after reboot/restart) will not have topology information for the network. Such   nodes can assemble topology information on their own from PTSE refreshes169 or updates received from neighbors. In a static network, only refresh PTSEs will be flooded If a new node or (a restarting old node) is to assemble its topology database from such refresh PTSEs alone then it can take quite long time, in the worst case as long as the   Refresh Interval which is in the range of 30 minutes. To avoid this problem and enable a new or restarting node to have global network information as quickly as possible, both PNNI and OSPF provide a mechanism called Database Synchronization. The method is executed on all links (i.e. on all RCCs) when a node starts. Besides, whenever a link goes down and comes up again, synchronization mechanism is re-executed. The mechanism consists of each node, at either   end of the link, exchanging a

---

[169] Link State routing protocols such as OSPF, PNNI require originator of link state advertisement packets   (LSA packets in OSPF and PTSE packets in PNNI) to refresh that advertisement periodically – typically 30 minutes if in between no triggered update occurs. An update following changes to link information is called triggered update.

summary of their DB contents. From the summary received, each node would know what pieces of database it doesn't have compared to the node at the other end of the link. Each node can then request for the missing pieces from the other node.

9.4.3.14 PNNI Flooding mechanism

PNNI, like OSPF, uses flooding to distribute topology information required for path computation. We already explained that topology information consist of nodal, link and Reachability information. The flooding mechanism ensures that all nodes in the network have identical information and hence each node can compute a route or path for a new service. Nodes within a single peer-group contain identical network information through flooding/database synchronization and we refer to such situation as convergence. If they compute routes using such database then the computed path will be correct. In the case of PNNI, since it uses source routing, it is not necessary that the convergence is a necessary condition. However, in hop-hop by routing networks such as IP, it is necessary that there be convergence in routing database otherwise there can be routing loops.

As a side note, we would like to point out that flooding, though a reliable technique to distribute information, leads to information redundancy [Joc]. For instance, can receive same topology updates from each of its connected neighbors. In such case a node need to discard all but one. Since the information is redundant, it results in wasted bandwidth, wasted CPU processing power.

When a PNNI node receives a new PTSE, it floods this (on the RCC) to all its neighbors, except the one the PTSE was received from. This is independent of the fact that some neighbors might already have flooded the same PTSE. When a PTSE is received, the following are possible:

- Discard the PTSE, if it is already installed in the database.
- Install it in the database, acknowledge to the sender of the PTSE, separately forward the TPSE via flooding
- There are two major reasons, why a node floods a PTSE to its neighbors:
- *Triggered Update*: Triggered flooding happens if a node originates a completely new PTSE or if there is a significant change in an IG with an existing PTSE (e.g. new end system addresses are added, the available bandwidth changed beyond a threshold etc.).
- *Aging:* Aging causes flooding (a) if the remaining lifetime of a PTSE reaches zero or (b) if the remaining lifetime of the PTSE reached a certain threshold in its originating node. A PTSE originated by a node

need to be refreshed by the same node so that other nodes in the network don't delete that PTSE just because its remaining lifetime has reached zero. The originating node needs to refresh i.e. flood an existing PTSE, even if the contents of the PTSE have not changed. It should refresh few seconds or minutes before the expiry so that it reaches every other node before their own timers expire.

In PNNI, to provide scalability, information flows across, up and down the hierarchy as illustrated in Figure 9-40. As a result, flooding is confined to each peer-group at each level (flooding across).

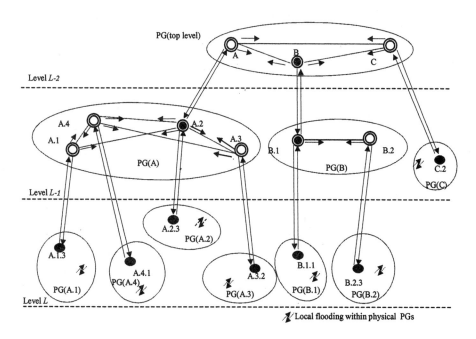

Figure 9-40. Flooding Across, Up and Down the Hierarchy

The PGL in each group takes responsibility to aggregate the peer-group information and advertise that information into the next level (flooding up). Similarly, nodes in a lower hierarchy take responsibility to send the information down (flooding down) into their respective child peer-groups. For instance, A.1 takes responsibility to send information down into PG (A.1) so that nodes in that peer group would know about rest of the network. Because the information coming down the hierarchy is in summarized form, as expected, a node in a child group need not have huge database store. However, because of such summarization, nodes in child peer groups will not have detailed information about internal structure of other peer-groups. As we explained already this is a compromise between scalability and the need for detailed information.

9.4.4   PNNI Signaling

In previous sections, we discussed PNNI routing used for distributing nodal, link and Reachability information. With that information, each node will have knowledge about the location of nodes and ATM terminals of the whole network. In the following paragraphs, we will describe how signaling works to setup a call170/connection between two ATM terminals attached to private or public ATM network.

---

[170] The term call is not appropriate for transport networks though it can be used in a generic sense.

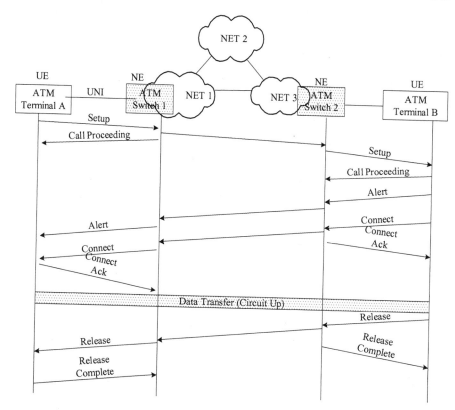

Figure 9-41. UNI Signaling Sequence

Table 9-4. ATM Signaling Message Structure

Bits								Octets
8	7	6	5	4	3	2	1	
Protocol Discriminator =0000 1001 for Q.2931								1
0	0	0	0	Length of Call Reference Value				2
Flag	Call reference value							3
Call reference value (continued)								4
								5
Message Type								6
								7
Message Length								8
								9
Variable Length Information Elements (IE) as required								etc

An ATM terminal wishing to connect to some other ATM terminal informs the switch to which it is attached using a SETUP signaling message across the UNI signaling interface. As mentioned earlier, ATM UNI comes in two flavors – public and private UNI. Both are similar and we will simply refer to them as UNI. Figure 9-41 shows the sequence involved.

The structure and contents of SETUP as well as other signaling messages (across both UNI and NNI) are standardized, shown in Table 9-4.

Some of the messages used in UNI (Q.2931 point-to-point, Q.2971 point-to-multi-point) are shown in Table 9-5.

Table 9-5. ATM UNI Messages[171]

Call Establishment	Call Control	Call Termination
Setup	Status Enquiry	Release
Call Proceeding	Status	Release Complete
Alerting	Notify	Restart
Connect	Information	Restart Ack
Connect Ack		Drop Party
Progress		Drop Party Ack
Add Party		
Party Alerting		
Add Party Ack		
Add Party Reject		

The switch that receives the SETUP message across its UNI interface is called the originating or source or ingress switch. The UE, which sent the SETUP message, is called the source or calling UE. The SETUP message contains information about the called UE as well as information about the connection such as required bandwidth, QoS etc. The switch node to which the called UE is attached is called the destination or egress or target switch.

The ingress switch node is responsible for setting up the end-to-end path. It can do that in two ways:

– Pass the information received in the SETUP message to higher management systems such as NMS/SMS, which take care of provisioning the end-to-end path.
– Use NNI signaling to setup the path.

The second method is of interest to us and we will learn the sequence of steps involved. The ingress switch, using the network information assembled using PNNI routing protocol, knows the advertised capabilities and

---

[171] ATM Forum's UNI 4.0 doesn't support notify, progress, and information messages. In addition, the forum defines new information elements (IEs) as well as services across the UNI interface. These services include point-to-multipoint connections, additional traffic parameters, private networking, proxy signaling, virtual UNI etc.

resources of other switches and links in the network. On this information, it can run some algorithm-based search, such as Dijkstra, to find a path that best supports the connection. The ingress node then requests the path to be setup using signaling. To ensure that the selected path is used, the ingress node includes explicit route information (a list of nodes) in the call SETUP message. The method consisting of (a) ingress node computing (or otherwise) the complete end-to-end path of a packet and (b) ingress node including explicit route information inside the packet is called *source routing*. One of the benefits of source routing is that loops in the path can be easily avoided thus increasing routing stability [Gre]. In PNNI, designated transit list (DTL) is used to specify explicit route information. A DTL contains a list of nodes that the call setup request needs to visit at a given hierarchical level. The destination pointer inside the DTL specifies which element in the list is the next node to be visited at that level. Figure 9-42 shows an example with the call setup message containing a DTL. When the end of a DTL is reached, the call setup message has visited all nodes and reached the egress switch node.

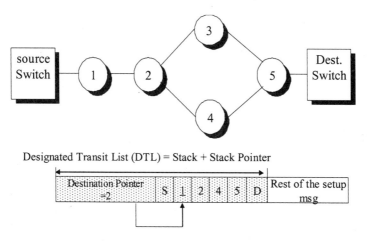

Designated Transit List (DTL) = Stack + Stack Pointer

Figure 9-42. Source Routing using Designated Transit List (DTL)

The simple source routing explained above works fine if the *'calling UE'* and *'called UE'* are attached to switches with in the same peer-group. In such a case, the source switch, since it has complete topology information for the peer-group, can compute the best path to the destination switch and prepare the DTL. However, if the *calling* and *called UEs are* attached to switches in two different peer-groups then, the ingress switch computes a *hierarchically complete* source route and prepares multiple DTLs to be included in the setup message. To understand what is meant by *hierarchically complete* source route, we revisit the hierarchical network example used earlier and place two ATM terminals (the UEs) with switches that are in different peer-groups as shown in Figure 9-43. We removed those parts of network that are unknown to the ingress switch. Also omitted is the prefix part of node IDs (to form full address for a node, the PG prefix shown with each peer-group need to be used). The lines in dark color indicate the path taken by the message as it travels through the network.

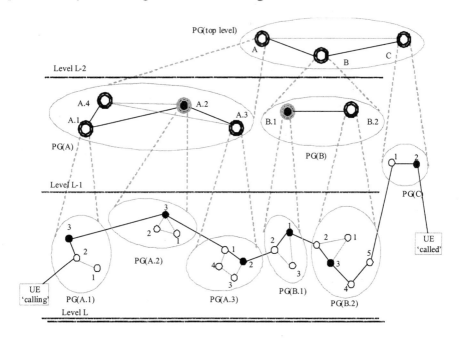

Figure 9-43. Source Routed Path

As shown in the figure, the ingress node, A.1.2 has detailed topology information about PG (A.1) and only summarized/aggregated information about other peer-groups. Using such information, assume A.1.2 computes the following path - {A.1.2, A.1.3, A.1, A.2, A.3, A, B, C} with C being the required last node since the 'called UE' is attached to some node inside PG (C). The ingress switch (A.1.2) prepares a set of DTL stacks, and appends them to a NNI call setup message and sends the message to the first next hop node (A.1.3) in the path. Finally, when the message arrives at the border node (A.2.3), since the path through PG (A.2) is not determined, A.2.3 will compute a detail path through PG (A.2). Since A.2.3 has, a direct link to PG (A.3), the path through this PG is simple and results in A.2.3 delivering the message to PG (A.3). The border node, A.3.1 computes detailed path through its PG and pushes a new DTL. The new top DTL is processed by nodes in PG (A.3) and finally resulting in message being delivered to PG

(A.4). Figure 9-44 shows DTL stack as the call setup message exits each node.

With the explanation of path computation and DTL processing, we revisit the call setup sequence across UNI shown in Figure 9-43. Next, we add signaling across NNI and show the resulting end-to-end call setup in Figure 9-45. The arrows with light color imply messages that are of local significance. Lines with dashed lines indicate optional messages that may or may not be implemented. Many of the messages defined across UNI are re-used across NNI with additional informational elements. One of the information elements (IE) carries DTL.

This concludes a brief introduction to ATM/PNNI signaling. Before jumping into ASON/Optical control plane, it is worth drawing some parallel between ATM and SONET/SDH circuits. ATM is a cell/packet-based technology with statistical multiplexing; circuits share bandwidth with other circuits and hence each circuit needs QoS guarantees. On the other hand, SONET/SDH circuits are dedicated circuits and the QoS is specified using branded service offerings such as silver, bronze, gold, and platinum. These branded offerings may translate to typical SONET/SDH service types – pre-emptible unprotected traffic (also called extra traffic), non-preemptible un-protected traffic (NUT), as well as regular protected or un-protected traffic. Where the underlying network is mesh and can't offer sub 50-ms restoration capability then the protected services may be further classified. For instance, real time -protected (<50 ms), near-real time protected (<200 ms), non-real time protected (>500 msec). Besides, QoS related to protection; other parameters can be associated with each service. Such additional parameters can be derived from Service Level Agreements (SLA). Typical parameters include downtime and BER. Besides differences in ATM and SONET/SDH circuits, there is also difference in the nature of calls. ATM supports point to point, point to multi point and others. Some of them may not be applicable to SONET/SDH. However, the basic point-to-point virtual circuit of ATM is similar to private line point-to-point SONET/SDH circuit and hence signaling functions and procedures of ATM/PNNI can be used to realize control plane for SONET/SDH transport networks as well. This is reflected in acceptance of PNNI (obviously with suitable extensions and modifications) by ITU as one of the signaling protocols for use in optical control plane, the topic of the next section.

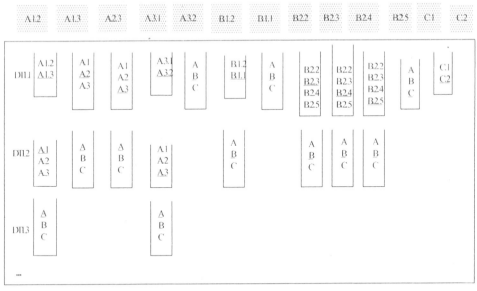

Legend: x.y.z - Destination Pointer points to x.y.z in the DTL with x.y.z being the next-hop node (physical or logical)

Figure 9-44. DTL Stack as the call setup exits each node

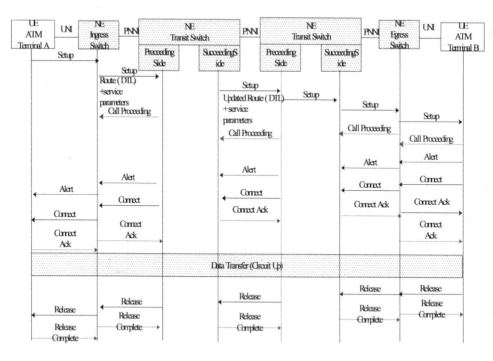

Figure 9-45. UNI + NNI Sequences

9.5.   OPTICAL CONTROL PLANE

The[172] purpose of the optical control plane, as with PSTN and ATM control planes, is to automate connection setup in an optical network. However, other control plane functions may be equally valuable to carriers, such as neighbor[173] and topology discovery.

Table 9-6. Optical Control Plane Glossary

Term	Comment
ASON/ASTN	Automatic Switched Optical /Transport Network
OIF	Optical Internetworking Forum
GMPLS	Generalized Multiprotocol Label Switching
UNI	User Network Interface
NNI	Network-to-Network Interface
MPLS	Multiprotocol Label Switching
OTN	Optical Transport Network (G.709)

Though ITU-T is the primary owner, others, such as IETF and OIF, are also involved in the standardization of optical control plane. Generally, the ITU effort is referred to as G.ASON, IETF's as GMPLS, and OIF's as the UNI/NNI. Table 9-6 provides glossary of frequently used terms (some may have been introduced already) to help the ensuing discussion

Before getting into details, there are three aspects worth highlighting with respect to standards activity related to optical control plane.

First, experience gained from routing and signaling protocols developed in the past decade or so is helping in providing richer set of abstractions than before. Standards now are moving more in the direction of protocol neutral functional specification, followed by specification of protocol specific bits and byte details that build on existing protocols rather than creating completely new protocols.

Second, clear separation is emerging between signaling and routing in optical control plane standards. Such separation exists already in the IP world, but in telephony these have been typically bound together (e.g., PNNI signaling and routing). In PNNI, a peer-group can implement its own routing policies but not its own routing protocol. G.ASON is generalizing this to say that each peer-group can implement its own signaling and routing protocols, interworking via the NNI.

---

[172] With help from Lyndon Ong. standards director, Ciena Corporation. Lyndon is very involved in standards discussion at ITU, IETF, OIF .

Useful for detecting miswiring and inventory mismatch [173]

Third is cooperation of IETF and ITU. Historically, ITU standards for optical/transport networks have been at layer 1 while IETF has been at layer 3. However, with GMPLS expanding the role of IP as a control plane into the optical domain, an overlap between ITU and IETF roles has developed that has required the two bodies to cooperate. Some of the following discussion will expand on this.

9.5.1 ITU-T ASON/ASTN Control Plane

Figure 9-46 illustrates ASON control plane reference points in the global ASTN. The UNI reference point is the point of demarcation between a transport network and its clients (a similar demarcation point is defined in IETF as GMPLS Overlay). In general, clients to a transport network are not aware of its internal topology (i.e. the UEs and NEs are not peers). The transport network itself can contain multiple smaller networks (subnetworks) and each such smaller network is called a control domain (CD). Policies, procedures, and protocols used inside a CD are hidden from clients as well as from other CDs. Each CD interacts with other CDs across an Exterior-NNI (E-NNI) interface. Across this interface, at the minimum, reachability of end users is exchanged. Exporting of complete or abstract topology across E-NNI is also possible. The above reference model meets security requirements discussed earlier; internal details (topology and reachability) of CDs and transport network are hidden from clients. Provider-to-provider network interfaces (similar to B-ICI) and associated protocols are also within the scope of ASON although complete requirements are still under study.

The role of the ASTN is intended in some ways to be equivalent of global PSTN; end users will be able to setup transport connection to other end users much like in PSTN today, with a policy/security boundary provided by the UNI.

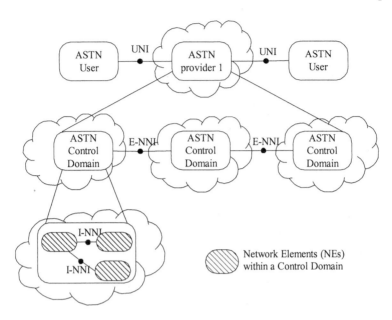

Figure 9-46. G.ASTN/G.ASON Signaling reference points

To avoid repetition, Table 9-7 provides a glossary of some of the commonly used terms from G.ASON, MPLS, GMPLS, and OIF. The terms will be explained in relevant sections.

Table 9-7. Optical Control Plane Glossary

Term	Comment
CC	Connection Control
NCC	Network Call Controller
CD	Control Domain
DCN	Data Communication Network (G.7712)
LRM	Link Resource Manager
PC	Protocol Controller - specific to each function
RA	Routing Area
RC	Routing Controller
RCD	Routing Control Domain , abstract representation of a lower level RA as a node and as seen by an RA at the next level
RP	Routing Performer
VC	Virtual Concatenation
TAP	Termination and Adaptation performer

Figure 9-47 illustrates the relation of ASON control plane to data and management planes.

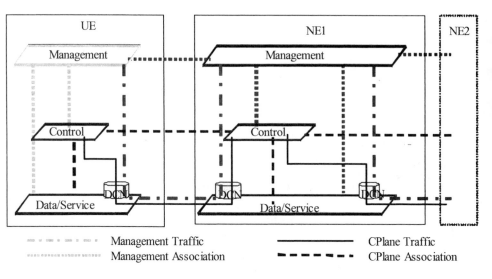

Figure 9-47. ASTN Control Plane relation to other planes

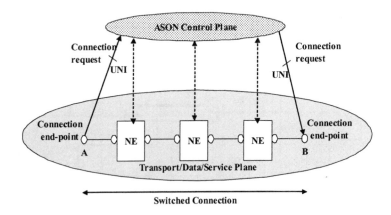

Figure 9-48. Switched Connection

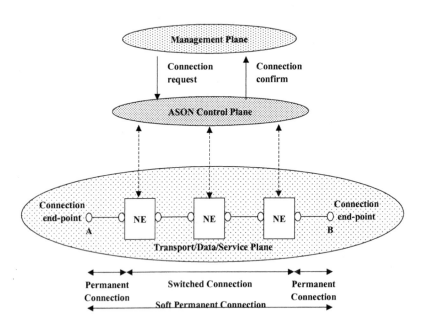

Figure 9-49. Soft Permanent Connection –SPC

Figures   9-48 and 9-49 illustrate the two types of connections to be supported by the control plane; switched connection (SC) and soft-permanent connection (SPC). A SC connection is setup by end user signaling to the ASTN using UNI signaling procedures while SPC is initiated by the management plane but setup using the control plane.

Figure 9-50 shows G.ASON standards published so far and their relationship to each other.

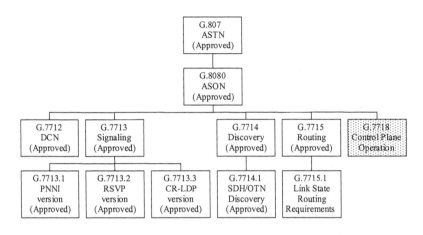

Figure 9-50. G-ASON Standards

G.807 provides requirements while G.8080, considered the root of ITU optical control plane, provides an architectural framework consisting of *routing control, call control* and *connection control* functional components without reference to any particular protocol. We start with a discussion of routing control (with some forward referencing to call control since both are related), followed by description of distributed connection management (DCM) that contains brief description of both call and connection control.

9.5.1.1 Routing

The purpose of routing function in optical control plane is to distribute information about optical links, nodes (physical or logical), and reachability (of nodes and UEs). Such information is used in determining network topology and finding a path for a connection. G.7715 details architecture and requirements for routing based on requirements outlined in G.8080, and G.7715.1 defines detailed requirements for link state routing protocol instantiation of ASON routing.

Routing Area (RA) is a collection of nodes and links that define a bounded network. For instance, a SONET/SDH ring (or a set of stacked rings) in a metropolitan area can be considered as a RA. Alternatively, one can split a large mesh network into multiple subnetworks with well-defined boundaries. RA exists at a particular level in a hierarchical routing. The lowest level of hierarchy contains RAs with physical nodes and links. The RAs at higher levels contain abstract nodes and links (similar concepts were discussed earlier in the PNNI section). Using the topology database for a bounded network i.e. the RA, an RC associated with that RA can perform route computation, such route being used to install an optical circuit.

The Routing Controller (RC) in G.8080 computes routes across a routing area (RA). The RC can be realized using centralized or distributed approach. With the latter approach, every node in the RA is a potential RC. The first approach can be used with legacy or existing network controlled by a subnetwork controller (SNC) – typically an EMS (refer to Chapter 5 for explanation of EMS).

G.8080 identifies three types of routing:

– Hierarchical
– Source routing
– Step-by-step

In all cases, it is assumed that a layer network is decomposed into a hierarchy of subnetworks. For the case of Hierarchical Routing, connections are routed in a top-down manner, first at the top of the hierarchy and then component RAs within that level of the hierarchy. Each RA has its own dynamic connection control that has knowledge of the topology of its subnetwork but has no knowledge of the topology of RAs above or below itself in the hierarchy (or other RAs at the same level in the hierarchy).

In Source or Step-by-Step routing, connection control is implemented by the federation of distributed connection and routing controllers and signaling flows from the originating to the destination ends of the connection. This model is consistent with the use of Distributed Connection Management as specified in G.7713.

When a hierarchy of RAs is established, RCs must be configured at the minimum with information that allows them to determine which RAs they belong to. The question arises, particularly from the operational context, as to how much information can flow automatically between different levels in a routing hierarchy. This is still an area under study at ITU.

9.5.1.2 Distributed Connection Management (DCM)

In DCM, described in G.7713, call control (CallCC) and connection control (CC) functional entities interact to control the establishment and release of call/connections across the network. .

Figure 9-51 shows the components of G.8080 control plane. The messages sent/received as part of the interaction between components are actually transported by protocol controllers (PC) as illustrated in the figure.

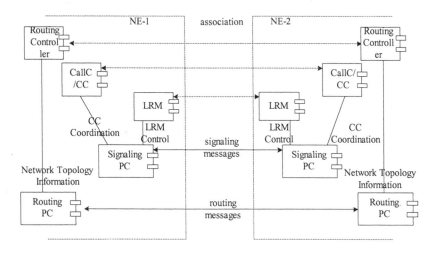

Figure 9-51. G.ASON's Architectural/Functional Components and associations

Briefly, a calling/originating UE *call controller* (OPCC) sends a call request to the ingress NE. The ingress NE's network call controller takes care of validation of incoming requests including any billing related parameters. If the network call controller is part of the ingress NE then, it may optionally contain a call admission control component that implements policy and administrative specific admission control. Once a call is accepted, the ingress NE passes the call to the next call controller in the series. Finally, the call is presented to the called/destination party call controller (DPCC). If the called party accepts the call then, a series of call confirmed messages are sent in the backward direction. When the ingress NE receives call confirmed message, it will contact the first connection controller to initiate setting up of switched connection. Finally, when the ingress NE receives connection confirmation, it will present call confirmation to the OPCC. Figure 9-52 shows an example scenario consisting of two network call controllers and three connection controllers. This also illustrates separation of call and connection controllers.

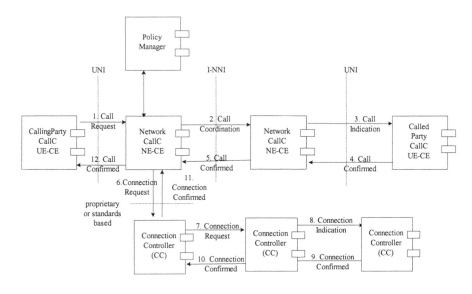

Figure 9-52. G.ASON Call Setup Example, Figure 20/G.8080

The Link Resource Manager (LRM) manages link resources associated with a link (physical or logical) between two neighbors. Resources can be discovered automatically or manually. The LRM components are responsible for the management of links including allocation and deallocation of link connections (c.f. Chapter 6). There will be LRMs on either side of a logical link connecting two subnetworks. One of the LRMs is called LRMA and other is called LRMZ. In case the link is bi-directional then there can be two instances of LRMA and two instances of LRMZ. The decision as to which side of the link will be LRMA is decided during LRM to LRM association or adjacency forming process.

The LRM should inform the routing controller about the resources so that the resources/links are advertised throughout the routing area (CD or peer-group or area). Through abstraction, G.ASON refers to each resource as a Subnetwork Point (SNP). A group of SNPs that is equivalent for routing purposes is referred to as SNP Pool (SNPP). Each SNP is identified by an SNP Id and so is each SNPP. An SNP is the control plane (G.ASON) representation of a transport plane (G.805) reference point while the CTP represents the management plane (M.3100) view of the same.

The abstractions ITU is defining in G.8080 allow for the possibility of each CD choosing its own routing protocol or method. To provide for flexibility, G.8080 allows both hierarchical as well non-hierarchical routing methods to co-exist as long as they fit in the routing model of G.8080. For instance, PNNI, OSPF-TE, IS-IS may   be deployed in three different domains of a transport network with the three domains interacting across E-NNI interface. Obviously, the domain with PNNI supports hierarchical routing while the domain with OSPF-TE doesn't. However, as long as the domains abstract and export topology across E-NNI, the internal topology representation is not visible outside.

The protocol neutral approach followed in defining the architectural as well as functional view of call control, connection control and routing control provides some advantages and disadvantages. The advantages include flexibility for operators to choose different protocols that blend with their current deployment and operational experience. In the case of vendors, it lets them package components or modules in a way that is appropriate to their product platforms as well as in re-use of their existing software base. Such flexibility will enable them to deliver solutions with less cost and compete with others on a level playing field. The disadvantages of protocol neutral approach include performance related issues associated with too much abstraction, conformance testing, too many test suites, increased number of combinations etc. As ITU control plane standards are in the early implementation and deployment stage, it is expected that they will converge on some mid-way solution and agree on some kind of protocol profiles to allow vendors to focus their resources on implementing some of the functionalities in a more cost effective way.

9.5.2   GMPLS

9.5.2.1 Label Specification

A TDM signal contains multiplexed traffic belonging to multiple O-D[174] pairs and in that sense is similar to a packet link carrying multiplexed IP or other packet/cell oriented data traffic of multiple O-D pairs. However, unlike data traffic, individual signals inside TDM signals don't contain explicit <O, D> information or label (such as MPLS label) that maps to a pre-designated <O, D> tuple. Instead, traffic in each time slot is implicitly understood to belong to a particular O-D pair. Not only that, intermediate NEs need not be aware of the identity of the source and destination end points. All they need to ensure is that incoming traffic on a particular interface and time slot is switched to an agreed output interface and time slot. Such agreement is

---

[174] Origination-Destination

established by a pair-wise label request/response negotiation process during connection setup. The SONET/SDH label format used for such pair-wise request/response negotiation is based on the SUKLM label structure illustrated in Table 9-9 along with   applicability and value range of S, U, K, L, M for SONET and SDH respectively.

Table 9-8. GMPLS Label structure for SONET/SDH

S				U	K	L	M
16b				4b	4b	4b	4b

S - indicates a specific AUG-1/STS-3 group inside an STM-N/STS-N multiplex

U - indicates a specific VC or STS-1 inside a given AUG-1/STS-3

K - only significant for SDH VC-4 (ignored for HO VC-3), indicates a specific branch of a VC-4.

L - indicates a specific branch of a TUG-3, VC-3 or STS-1 SPE (not significant for unstructured VC-4 or STS-1 SPE)

M - indicates a specific branch of a TUG-2/VT Group (not significant for unstructured VC-4, TUG-3, VC-3 or STS-1 SPE (M=0))

Figure 9-53 will be used to explain pair-wise request/response process and SUKLM usage.

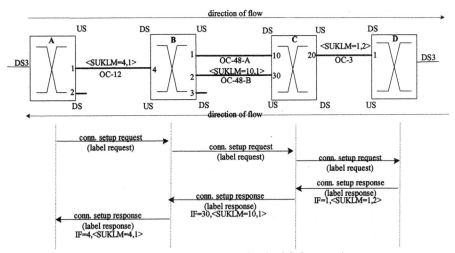

Figure 9-53. Connection Setup using pair-wise label request/response

Consider that a DS3 circuit needs to be setup between A-D. Further, assume that a craft person triggers node A, using CLI or some management interface, to initiate circuit setup. Following the trigger, A will send a connection setup request message (label request message) to B. Now, B will process the message (possibly allocating the label on the segment between

A-B). However, instead of responding to A, it will first request a label for the next segment between B-C and will do so by forwarding the connection setup message to C. Finally, the connection setup message arrives at D, which results in D responding with a label for the D-C segment. Now, C will respond to B for the B-C segment. Finally, B responds to A with the label for the A-B segment. Now all segments have labels and all NEs would have provisioned necessary cross-connects to support the DS3 between A and D. Reader may notice that each node includes an interface reference (ifIndex) in the label response. Somehow, each NE should know of the interface reference used by the remote NE to help them correlate remote interface index to their own. For instance, when B receives label, <SUKLM=10,1, 0,0,0> and ifIndex=30, from C, it should correlate that to its own ifIndex of two and hence to link OC-48-B. Such correlation is trivial and can be learned using simple neighbor discovery protocols (LMP for instance) if it is not initially configured in the device.

The pair-wise label request/response process just explained is formally called *downstream label allocation;* the label is allocated by downstream node on request from an upstream node. It is also possible for an upstream node to suggest a label and expect that the label response from downstream node will contain the same label. Given this, the upstream node can initiate cross-connect setup immediately (after or before sending/forwarding the connection setup request message) instead of waiting for a label response from downstream node. For instance, A can suggest <SUKLM=4,1,0,0,0> and ifIndex=1 to B in the setup/request message. Subsequently, A can go ahead and create the necessary cross-connection. Later when the response from B comes back, A can make sure that the received label is the same as the one it has suggested. If different, A can tear down the earlier cross-connection and setup a new one. As one can see, when the suggested label is honored, it can reduce the time to setup path.

9.5.2.2 Traffic Parameter Specification

To allocate a label to a connection, downstream nodes need bandwidth and other specifics for the connection. This information is specified using the traffic specification or TSpec object. The object is sent in the request message to downstream nodes. Table 9-9 lists SONET/SDH specific encoding of this object, with some of the codes for Signal Type (ST) enumerated in Table 9-10.

Table 9-9. GMPLS Traffic Specification for SONET/SDH

0 1 2 3 4 5 6 7	1 8 9 0 1 2 3 4 5	2 3 6 7 8 9 0 1 2 3 4 5 6 7 8 9 0 1
Signal type	RCC (vector of flags)	NCC
NVC		Multiplier (MT)
Transparency (T) (vector of flags)		
Profile (P)		

Signal Type
- SDH – LOVC/HOVC elementary signals
- SONET – STS/VT elementary signals

Requested Contiguous Concatenation
- Flag –1 Standard Contiguous Concatenation
- Others flags not defined

Number of Components (time slots)
- NCC - Contiguous Concatenation
- NVC – Virtual Concatenation

Multiplier
- Number of identical traffic connections

Transparency
- Flag –1 RS/SOH
- Flag –2 MS/LOH
- Other flags – 0 and reserved for now

Profile
- No profile defined yet

Table 9-10. GMPLS SONET/SDH Signal Type encoding

Signal Type Field	Elementary Signal	comments
1	VT 1.5 SPE/VC-11	
2	VT 2 SPE /VC-12	
3	VT 3 SPE	
4	VT6 SPE/VC-2	
5	STS-1 SPE/VC-3	
6	STS-3c SPE /VC-4	
7	STS-1 (OC-1) /STM-0	Only when requesting transparency
8	STS-3 (OC-3)/STM-1	' '
9	STS-12 (OC-12)/STM-4	' '
10	STS-48 (OC-48)/STM-16	' '
11	STS-192 (OC-192)/STM-64	' '
12	STS-768 (OC-192)/STM-256	' '

If a higher order path (such as STS-1, AU-3 or AU-4) is provisioned to carry one or more lower order paths (such as T1s, E1s) then the said higher order path may be modeled as a virtual interface on each of the NEs supporting the higher order path. In setting up a lower order path over such higher order paths, either long form   <S, U, K, L, M> or short form <0, 0, 0,L, M> can be used. In setting up higher order path the short form <S, U, K, 0, 0> is used since L and M don't make sense.

Transparency is a vector of flags that indicates the type of transparency desired. Not all combinations are valid. Transparency is applicable to fields in the SONET/SDH frame overheads. Transparency is only applicable when requesting connections of STM-N type in the SDH case and STS-3*N type in the SONET case (N=0,1,4,16,,...). SONET section/SDH regenerator section layer transparency requires the entire frame to be delivered unmodified and implies intact pointer fields. SONET line/SDH multiplex section layer transparency means that the LOH MSOH is delivered unmodified.

Examples in Table 9-11 explain how to set values for the various fields of SONET/SDH traffic specification object. Figure 9-54 illustrates applicability and use of the parameters discussed above.

Table 9-11. Examples to illustrate encoding of GMPLS SONET/SDH Connection parameters

Traffic		Mapping SONET/SDH	ST	RCC	NCC	NVC	MT	T
DS3		STS-1 /VC-3	5	0	1	0	1	0
Fast Ethernet (100 Mbps)		STS-3c/VC-4	6	0	1	0	1	0
	VC	STS-1-2v/VC-3-2v	5	0	0	0	1	0
GE (1000 Gbps) or 1G Fiber Channel	CC	STS-24c/VC-4-16c	6	0	8/16	0	1	0
	VC	STS-1-21v[175] /VC-4-7v	5	0	0	21/7	1	0
Fiber Channel or ESCON (200 Mbps)	CC	STS-12c/VC-4-4c	6	0	4/4	0	1	0
	VC	STS-1-4v/VC-3-4v	6	0	0	4/4	1	0
2x Fast Ethernet	CC	STS-3c/VC-4	6	0	1	0	2	0
	VC	STS-1-2v/VC-3-2v	5	0	0	2/2	2	0
Generic concatenated Signal (with possible exceptions) Y=1,4,16 STS-24c, no equivalent in SDH	CC	STS-3*Yc /VC-4*Yc	6	0	Y/Y	0	1	0
	VC	STS-3c-3*Yv/VC-4-Yv	6	0	0	Y/Y	1	0
OC-12 with Transparent SOH		STS-12	9	0	0	0	1	1
OC-48 with Transparent LOH		STS-48	10	0	0	0	1	2

---

[175] For GigE, another possible mapping is STS-3c-7v

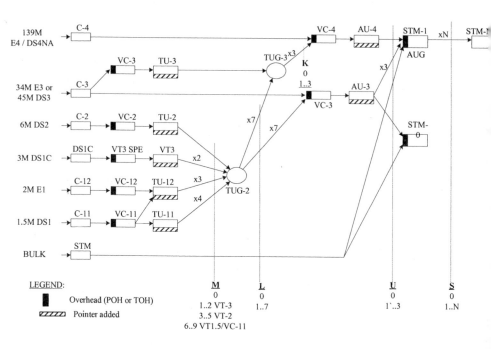

Figure 9-54. GMPLS Label Usage for SDH

9.5.3   RSVP-TE

Resource re**S**er**V**ation **P**rotocol (RSVP) [RFC 2205] was originally intended to do resource reservation in networks running the IP protocol to support Integrated Services (IntServ). IntServ architecture (together with RSVP) provides for per-flow QoS, signaling, and traffic management support. The per-flow view is a perfect match with point-to-point connections in TDM/optical networks, although in its original version, RSVP did not do routing of the packet flow, leaving this to the normal IP packet forwarding process. A set of extensions to support traffic engineering were added to the basic RSVP and published as RSVP-TE [RFC 3209]. The extensions include explicit route specification, setup and holding priority, path preemption, rerouting, and path reoptimization. The path along which resources are setup is called a label switched path (LSP). RSVP-TE supports setting up both strict and loose LSPs. In the case of loose LSPs, the path of the LSP is the path taken by the PATH message as it travels through the IP network towards the destination. Strict LSP is one whose path is pre-designed with the transit/intermediate node information encoded in the PATH message. Generalized MPLS (GMPLS) signaling in RFC 3473 extends basic MPLS TE signaling procedures and abstract messages to cover different types of switching applications such as TDM, wavelength switching, etc. A further set of extensions for SONET/SDH is currently in draft documents. G.7713.2 extends the basic GMPLS RSVP-TE to support requirements as specified in G.7713.

Following network topology changes, some of the existing connections may fail and need to be restored. While RSVP participates in restoring a failed connection, network planning in terms of spare capacity required etc is outside the scope RSVP-TE. Restoration in general is a complex topic and discussed in details in Chapter 9. Suffice to say that extensions to RSVP-TE are being discussed to support restoration. Among its other features, is opaque transport of traffic and policy control objects.

Connection setup starts with the originating node sending a PATH message. When the destination node receives an incoming PATH message, it originates a RESV message towards the originating node. As the RESV message travels back to the originating node, the resources are reserved. As such, RSVP is a two-phase protocol with the resources reserved in the second phase. To speed up, RSVP-TE does allow reservation of resources during the first phase i.e. during propagation of PATH message itself. Besides PATH and RESV, there are other messages used for tear down, notification etc. Table 9-12 lists some of the RSVP-TE messages and their mapping to G.ASON functional messages.

Table 9-12. RSVP-TE Signaling Messages

GG7713. DCM Call/Connection Messages	RSVP-TE Message
Setup Request	Path
Setup Indication	Resv, PathErr
Setup Confirm	ResvConf
Release Request	Path or Resv (W/ D&R bit) or
	Path or Resv (w/A&R bit)
Release Indication	PathErr (Path_State_Removed flag) or Path Tear
QueryRequest	Path (implicit via periodic refreshes)
QueryIndication	Resv (implicit in RSVP-TE periodic refreshes
Notification	Notify, ResvErr

The Path and Resv messages must be sent periodically to refresh the state maintained in all nodes along the path; this model of maintaining reservations based on refreshes is called *soft-state*. RSVP is often criticized for its soft-state approach. However, it is possible to set aside this limitation by implementation agreements (such as in OIF UNI 1.0) that don't delete connection due to refresh time outs.

Each GMPLS RSVP message consists of one or more objects. Some of the objects contain sub-objects. Table 9-14 lists some of the objects while Table 9-13 provides some of the attributes of G.ASON Call/Control Messages and their mapping to RSVP objects/sub objects. The details of many of the objects applicable to GMPLS RSVP are in RFC 3471 and 3473.

Original RSVP resource reservation is based on receiver specified per-flow QoS. With GMPLS, some of the models and procedures of original RSVP are irrelevant. A case in point, the TSPEC (S$\rightarrow$R) and FLOWSPEC (S$\leftarrow$R) are expected to be identical in G.ASON/OIF. A node receiving a RESV message with a FLOW SPEC that is different from the TSPEC for the same session should not send RESV upstream. Instead, it should send ResvErr message downstream.

In TDM networks, the resource is nothing but a time slot of particular type (STS-1, 3c etc). In the case of bi-directional TDM connection setup, identical time slots in the forward and backward direction can be signaled by including Upstream Label and Label Set objects in the PATH message.

Table 9-13. RSVP Objects (across RSVP, RSVP-TE, GMPLS, G.ASON)

RSVP Object Name (direction) (Class #, C-Type) sub objects if any	RSVP-TE	GMPLS, *OIF UNI, ITU G.ASON*	Description
SESSION (S→ R) 1, 1	LSP_TUNNEL_IPv4 (I,7)	*UNI_IPv4_Session (I, 11)* *ENNI_IPv4_Session (I, 15)*	Contains destination IP address + tunnel ID. It may also include ingress IP address (Extended tunnel ID) to make ingress-egress pair unique.
RSVP HOP (S→R)		IF_ID_RSVP_HOP	Carries link interface identifier
Error Spec (← →) (6,)			Specifies error information in path-error or resv-error messages
Flow Spec (R → S) 9		SONET/SDH FlowSpec (9,TBA)	Desired bandwidth and QoS/CoS
Filter Spec (10,)	LSP_TUNNEL_IPv4 FilterSpec 10,7		Identical to SenderTemplate
Sender Template (S→R) (11,)	LSP_TUNNEL_IPv4 Sender Template 11,7		Contains IP address and LSP ID to identify the source
Sender TSPEC (S→R) (12,)		SONET/SDH Sender TSPEC (12, TBA)	Desired QoS/CoS
ADSPEC (13,)			Not currently used
Resv Confirm (15,)			IP address of the node that requested explicit confirmation
	Admin Status		For administrative control of LSP such as lock/unlock, delete
	Label S←R (16,1)	Suggested Label	Used to suggest a specific timeslot or lambda

Table 9-13. RSVP Objects Cont.. (across RSVP, RSVP-TE, GMPLS, G.ASON) -

RSVP Object Name (direction) (Class #, C-Type) sub objects if any	RSVP-TE	GMPLS, *OIF UNI, ITU G.ASON*	Description
	Label Request S→R (19,x) 1 2 3	Gen. Label Request	
	Explicit Route (20,x) 1 IPv4 Address 2 IPv6 Address 3 Label		Specifies the route to be followed by the connection. Can be coded strict or loose
	Record Route (S← → R) (21,) 1 IPv4 Address 3 Label		The subobjects IPv4Address includes flags indicating whether the downstream link is protected or not. If protected, also indicates status of the LSP – on the W or P link.
	HELLO (22,)		Optionally used to monitor RSVP control plane connectivity
	Session Attribute (S→ R) (207,xx)		Optionally provides additional session-related information
		Call ID 230,x	Uniquely identifies the call
		Protection Label Set	Identifies the level of protection to be provided
		Generalized UNI 229,[1,2,3,4]	Provides UNI information such as the client-level address, service class, etc.

Table 9-14. Attributes used in G.ASON DCM Call/Connection Messages

	UNI/G.ASON Attributes	RSVP Object	Scope
Identity Attributes	A-end user name/Source TNA	GUA/Source TNA Address	End-to-End
	Z-end user name/Destination TNA	GUA/Destination TNA	End-to-End
	OPCC/CallC name	Sender Template/ Filter Spec	Local
	DPCC/CallC name	Session	Local
	Call ID	Call ID	End-To-End
	Local Connection ID	(UNI_IPv4_Session, LSP_TUNNEL_IPv4_Sender_Template) or (UNI_IPv4_Session, LSP_TUNNEL_IPv4_FilterSpec)	
Service Attributes	Source Port ID/ SNP ID	IPv4_IF_ID_RSVP_HOP, Gen. Label, Upstream_Label, Egress_Label,	Local
	Destination Port ID/SNP ID	Suggested_Label, GUA/SPC_Label	
	SONET/SDH Traffic Spec	SONET/SDH_Sender_TSpec, SONET/SDH_FlowSpec	End-to-End
	Directionality	Implied by Upstream_Label	Local
Policy attributes	CoS	GUA/Diversity, GUA/Service_Leve, PolicyData	End-To-End
	Explicit Resource list	ERO, RRO	End-to-End
	Recovery	Protection	Local
Protocol Specific Attributes			
	Path monitoring/control	ADMIN_STATUS	End-to-End
	Protocol robustness	HELLO_REQUEST, HELLOC_ACK	Local
	For status/error codes	ERROR_SPEC	End-to-End
	For protocol robustness	MESSAGE_ID, MESSAGE_ID_ACK etc	Local
	Node Failures	RESTART_CAP, RECOVERY_LABEL	Local
	Protocol timers	TIME_VALUES	Local

The main differences between IETF and ITU-T/OIF versions of RSVP have to do with the use of addressing fields and the support of multiple domains and associated boundary interfaces.

The *Sender_Template* and *Session objects* in IETF RSVP are sufficient to uniquely identify a connection across a GMPLS network. In ITU-T/OIF RSVP, a client access address (the TNA) is added, the supported address types being IPv4, IPv6, and NSAP. The combination of the Source and Destination TNA Addresses are required to uniquely identify a connection across an ASON network. The GENERALIZED_UNI (GUA) object defined in OIF UNI 1.0 is used to encapsulate A- and Z-end names, as well as service specific information in the setup request at the UNI interface. The

GUA object is of end-to-end significance- network may use it but should opaquely transport it from source UNI-C to destination UNI-C without modifying it.

At a higher level, a logical call model is added where a call is associated with zero or more connections. For instance, a call may have both primary and backup connections. To support the call model, a Call ID object is added plus some semantics of PATH/RESV and other messages are modified.

To take into account multiple domains in a connection, in the ITU/OIF model, there are multiple RSVP sessions; one between source UNI-C and ingress UNI-N; another between egress UNI-N and destination UNI-C; others as needed within each domain crossed and at each domain-to-domain boundary. The SESSION object contains IPv4 address of the destination of the session. Therefore, the session object in the PATH message leaving the UNI-C, since it is destined to the ingress UNI-N, contains the address of ingress UNI-N. Similarly, at the egress UNI-N, the outgoing path message contains the address of destination UNI-C. In a logical sense, the end-to-end connection consists of multiple segments "stitched" together.

For SPC connections, the A- and Z-end names/TNAs can be included in the PATH message, if available. Alternatively, the A and Z end network node IDs can be used. In both cases, the connection type is signaled by including SPC_LABEL sub-object in the GENERALIZED_UNI object. Figure 9-55 illustrate GUA object scope, while Figure 9-56 and 9-57 show general GMPLS implementation model and message flows respectively.

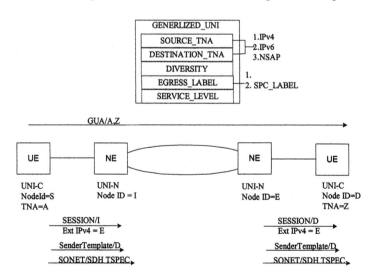

Figure 9-55. G.ASON/OIF GUA and scope

Both OIF and ITU efforts incorporate GMPLS architecture and extended versions of GMPLS protocols (RSVP-TE, OSPF-TE, IS-IS, CR-LDP) as illustrated in Table 9-15.

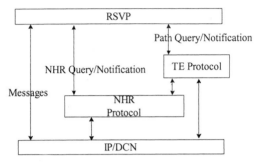

GMPLS/RSVP Implementation Model

Figure 9-56. GMLS/RSVP Implementation Model

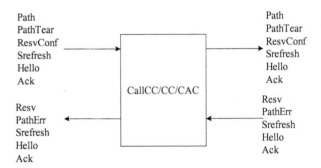

Figure 9-57. GMPLS Signaling Message Flows

Finally, an end-to-end setup and teardown sequence is shown in Figure 9-58 showing the use of RSVP messaging across multiple domains, which in ASON model can be using their own internal signaling protocols.

Table 9-15. Optical Control Plane standards

	Signaling	Routing	Discovery
IETF/GMPLS	RSVP-TE	OSPF-TE	LMP
	CR-LDP	IS-IS	
G.ASON			
- Functional specification	G.7713	G. 7715	G. 7714
- Protocol binding	G.7713.1 (PNNI)	G.7715.1	G.7714.1
	G.7713.2 (RSVP-TE)		
	G.7713.3 (CR-LDP)		
OIF	UNI 1.0/1.1		UNI 1.0/1.1
	- RSVP & CR-LDP		- LMP subset
	ENNI Signaling 1.0	E-NNI Routing	
	- PNNI, RSVP, CR-LDP	- OSPF	

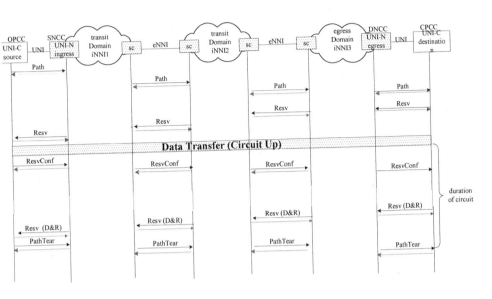

Figure 9-58. Multi Domain Circuit Setup using RSVP-TE

9.6. PSTN, ATM, AND OPTICAL CONTROL PLANE SIGNALING FEATURES

Table 9-16 lists some of the main elements of signaling standards discussed thus far to provide a summary as well as a comparison chart

Table 9-16. Summary of ATM, SS7, OIF/G.ASON Standards

	ATM	SS7	Optical	Remarks
Addressing of CE (control entity, U-CE or N-CE)	NSAP/E.164	DPC:OPC: SLS	IP/E.164 address in GMPLS. ASON/OIF also allows NSAP.	
Signaling/Control channel	VPI=0,VCI=5 Dedicated in-band signaling link using SAAL/Q.2934 signaling protocol	Dedicated signaling link using MTP data link protocol.	In-fiber out-of-band or Out-of-fiber out-of-band	Out-of-fiber could be Ethernet or IP network with proper QoS and security.
Routing Control Channel	VPI=0/VCI=18 dedicated in-band link using PNNI routing protocol	-	In-fiber out-of-band or Out-of-fiber out-of-band	Out-of-fiber could be Ethernet or IP network with proper QoS and security.
Signaling mode	associative	associative or quasi-associative or non-associative	associative or quasi-associative or non-associative[176]	

---

[176] There is no explicit and formal definition of signaling model in OIF. Rather it specifies different ways of realizing IP control channel; in-fiber, out-of-fiber, in-band, out-of-band. We tried to map these different realizations to signaling modes mentioned earlier.

Table 9-16. Summary of ATM, SS7, OIF/G.ASON Standards (Continued)

	ATM	SS7	Optical	Remarks
Neighbor Discovery	-	-	LMP or G.7714.1	
Third-party or Proxy Signaling	√	√	√	
Resource Allocation Collision avoidance	One-sided allocation in P-NNI to avoid collision	Collision resolution based on PC	Collision resolution based on node ID	
Carrier-to-carrier Interworking	B-ICI (ITU)	Supported but no explicit protocols required. Refer to SS7 administrator Interworking section.	Requirements are evolving with carriers contributions to OIF and ITU	
Point2Point	√	√	√	
Point2MultiPoint	√	√		In progress
MultiPoint2MultiPoint Tunneling	Using VPI	Not applicable	Not applicable	
Group address, anycast	√			
Multiple Signaling channels	√	√	√	
Supplementary Services	√	√	Limited	
Explicit Routing	Using designated-transit-list (DTL)	Only hop-by-hop routing	Uses ERO. No routing at UNI interface	
Signaling Deployment Scaling	√ using PNNI peer-groups and hierarchies	√	using RA hierarchy	

9.7. CONCLUSIONS

This chapter tried to present control plane issues associated with various networks; PSTN, ATM, and G.ASON/ASTN. First, a theoretical model was discussed to provide functional requirements for signaling applicable for any type of network. Following that, SS7 signaling, used in PSTN, was discussed. As part of this, many aspects of SS7, including signaling network interconnection architectures were discussed. Next, a description of ATM signaling and PNNI routing is provided. Finally, architecture of G.ASON/ASTN control plane was presented. OIF, IETF, and ITU discussions are still in progress on various aspects of signaling for optical networks and hence the description here is rather brief. Overall, the requirements and functionalities required have been mentioned.

Through this chapter, reader can see contrasting approaches to signaling. In SS7, a pragmatic approach was taken related to routing. SS7 avoided dynamic routing protocols by using static routing. Despite using static routing, SS7 signaling network supports millions of calls per day and takes care of complex provider interconnection scenarios. In the case of ATM, ITU continued with its legacy approach and standardized B-ISUP without dynamic routing protocol. On the other hand, ATM Forum took a bold step and standardized a dynamic routing protocol. Not only that, they went a step further by not using any existing routing protocols such as OSPF or IS-IS but by standardizing a new routing protocol, the PNNI, based on hierarchical routing model. Given the widespread deployment of OSPF, IS-IS, and PNNI, G.ASON tried to accommodate all variants or even none at all by abstracting networks into control domains and allowing each control domain the freedom in terms of routing implementation.

Telecommunication transport networks are at cross roads today in terms of automating service provisioning and network inventory management. The providers have multiple and complex requirements. When digital switching systems were deployed through 70's and 80's and as telephone traffic grew, it was necessary for providers to deploy the signaling to setup voice calls. Operator assisted long-distance call setup was not scalable, costly and slow and forced carriers to deploy signaling based solutions as quickly as they could. However, the same drivers are not present today with regards to ASON/ASTN transport networks. It is not expected that call setup volume will be huge in ASON/ASTN unlike in PSTN where millions of calls are setup and torn down every day. Typically, transport network connections stay in place for months and even years. In the future with the popularity of on-line music-on-demand and video-on-demand services , transport networks may need to start supporting small to moderate number of call volumes. Coupled with such future requirements, providers want the transport network control plane to reduce their operational expenses by automating service creation and inventory management. They want to reduce OSS system expenses by making the network more intelligent. Whether the push, from future broadband call volumes or carries Opex reduction measures is strong enough remains to be seen and will ultimately determine transport network control plane deployment.

REFERENCES AND FURTHER READING

[ATF]   PNNI Specification Version 1.1, ATM Forum

[Awd01]   RSVP-TE: Extensions to RSVP for LSP Tunnels, D. Awduche et al, RFC
            3209, IETF, Dec 2001

[BAC00]   Bell Atlantic Supplement Common Channel Signaling (CCS) Network
            Interface Specification, TPG-00-001, March 2000

[BAC97]   Interconnection Agreement between Bell Atlantic and Hyperion
            Telecommunications of Virginia, 1997

[BCR02]   "Optimizing Multi-Homed Connections," Business Communications
            Review, Jan 2002

[Ber03a]   Generalized Multi-Protocol Label Switching (GMPLS) Signaling
            Functional Description, L. Berger, RFC 3471, IETF, Jan 2003

[Ber03b]   Generalized Multi-Protocol Label Switching (GMPLS) Signaling
            Resource ReserVation Protocol-Traffic Engineering (RSVP-TE)
            Extensions, Lou Berger (Editor), RFC 3473, IETF, Jan 2003

[Bra97]   Resource ReSerVation Protocol-- Version 1 Functional Specification, R.
            Braden et al, RFC 2205, IETF, Sept 1997

[Gol90]   *"Common Channel Signaling Interface for Local Exchange Carrier to
            Interexchange Carrier Interconnection,"* Richard Goldberg, David C.
            Shrader, IEEE Comm. Magazine, July, 1990

[GR-802]   Telecordia GR-802, Gateway Screening

[Gre]   Performance Evaluation of the PNNI Routing Protocol Using an
            Emulation Tool, Ulrich Gremmelmaier et al

[Hui00]   Routing in the Internet, Christen Huitema, Prentice Hall PTR, 2nd Ed,
            2000

[IBM97]   IBM PNNI Control Point, IBM, 1997

[ITU01a]   Requirements for the Automatic Switched Transport Network (ASTN),
            ITU-T Rec. G.807, 2001

[ITU01b]   Architecture for Automatically Switched Optical Network (ASON),
            G.8080/Y.1304, ITU-T, 2001

[ITU01d]   Distributed Call and Connection management (DCM), G.7713/Y.1704,
            ITU-T, 2001

[ITU00]   http://www.itu.int/ITU-D/study_groups/SGP_1998-
            2002/SG2/Documents/2000/176E5.doc

[Jab91]   *"Common Channel Signalling System Number 7 for ISDN and Intelligent Networks,"* Bijan Jabbari, Proc. of IEEE, Feb 1991.

[Jab92]   *"Routing and Congestion Control in Common Channel Signalling System No. 7"*, Bijan Jabbari, Proc. of IEEE, April 1992.

[Joc]   "The Flooding Mechanism of the PNNI Routing Protocol," P.Jocher et al.

[Katz02]   Traffic Engineering Extensions to OSPF Version 2, Draft, D. Katz et al, IETF, Oct 2002

[Lang03]   Link Management Protocol (LMP), J. Lang, Editor, draft, IETF, June 2003

[Lee]   "Topology Aggregation for Hierarchical Routing in ATM networks," W.C.Lee, ACM SIGCOM

[Man02]   Generalized Multi-Protocol Label Switching Extensions for SONET and SDH Control, E. Mannie, D. Papadimitriou (Editors), Internet Draft, IETF, Oct 2002

[Mod90]   Telecommunications Networks, M. Schwartz, Addison-Wesley, 1988 *"Signaling System No.7: A Tutorial,"* A. Moderassi, R.A. Skoog, IEEE Comm. Magazine, July, 1990

[Moy98]   OSPF Version 2, J.Moy, Internet RFC 2328, April 1998

[OIF01]   User Network Interface (UNI) 1.0 Signaling Specification, Optical Interworking Forum Oct, 2001

[OIF02a]   Bandwidth Encodings for SONET/SDH Networks, OIF2002.543.000, Optical Interworking Forum, Nov, 2002

[OIF02b]   Non-Disruptive Bandwidth Modification for TDM Networks using GMPLS LBM –UNI 2.0 project, OIF2002.067.3, Optical Interworking Forum, Oct 2002

[Raz01]   Assessment of PNNI to Support ASON, R. Razdan, L. Ong, T1X1.5 Contribution, April 2001.

[Rek95]   A Border Gateway Protocol 4 (BGP-4), Y. Rekther, T.Li, RFC 1771, IETF March 1995

[Rus98]   Signaling System #7, Travis Russell, McGraw-Hill Telecommunications, Second edition, 1998.

[Sha99]   PNNI, NLANR/Internet 2 Techs Workshop, Kannan Shah, MCI/vBNS, 1999

[Swa03]   GMPLS RSVP Support for the Overlay Model, George Swallow, Draft, IETF, Feb 2003

Glossary

2-D	Two Dimensional
2R	Re-amplification, Reshaping
3R	Re-amplification, Reshaping and Retiming
8B/10B	Block line code in which 8 data bits are mapped into a 10-bit symbol
64B/65B	Block line code in which 64 data bits are mapped into a 65-bit symbol (GFP-T)

ADM	Add-Drop Multiplexer
AIS	Alarm Indication Signal
ANSI	American National Standards Institute
API	Applications Program Interface
APS	Automatic Protection Switching
ASON	Automatic Switched Optical Network
ARC	Alarm Report Control
ATIS	Alliance for Telecommunications Industry Solutions
ATM	Asynchronous Transfer Mode
AU	Administrative Unit (SDH)
AUG	Administrative Unit Group (SDH)

BCH	Bose-Chaudhuri-Hocquenghem FEC
BDI	Backward Defect Indicator (OTN)
BEI	Backward Error Indicator (OTN)
BER	Bit Error Rate
BIAE	Backward Incoming Alignment Error indicator (OTN)
BIP-n	n-bit Bit-Interleaved Parity

BLSR	Bidirectional Line-Switched Ring
BPON	Broadband Passive Optical Network (from FSAN)
Cap-Ex	Capital Expense
CBR	Constant Bit Rate
CBS	Committed Burst Size
CCITT	International Telegraph and Telephone Consultative Committee
CIR	Committed Information Rate
CIT	Craft Interface Terminal
CLEC	Competitive Local Exchange Carrier
CLI	Command Line Interface
Clos	Clos multi-stage switching network
CM	Connection Management
CMF	Client Management Frame (GFP)
CMIP	Common Management Information Protocol
CMIS	Common Management Information Service
CMISE	Common Management Information Service Element
CMTS	Cable Modem Termination System
CO	(Telephone company) Central Office
CO-PS	Connection-Oriented Packet Switched network
CORBA	Common Object Request Broker
CRC	Cyclic Redundancy Check
CRM	Customer Relation Management System
Cross-bar	A matrix of switching points to provide any-to-any non-blocking switching
CSF	Client Signal Fail (GFP)
CSU/DSU	Channel Service and Data Service Unit
CTRL	Control code (LCAS)
CV	Coding Violation
CWDM	Coarse Wave Division Multiplexing
DAF	Directory Access or Client Functions
dB	decibel
DCC	Data Communications Channel Also called ECC (Embedded Communication Channel) used for transport management traffic from OSS to NEs in the field)
DCN	Data Communications Network
DCS	Digital Cross connect System
DLC	Digital Loop Carrier system
DN	Distinguished Name
DNU	Do Not Use control code (LCAS)

DOCSIS™ Data Over Cable System Interface Specifications
DRA Dynamic Restoration Algorithms
DSF Directory System or Server Functions
DSL Digital Subscriber Loop
DSLAM Digital Subscriber Loop Access Multiplexer
DSn Digital Signal of the North American asynchronous hierarchy level n
DVB Digital Video Broadcast
DWDM Dense Wave Division Multiplexing

En Digital Signal of the ETSI PDH level n
EBS Excess Burst Size
EC-n Electrical Carrier for a SONET STS-n
EDFA Erbium-Doped Fiber Amplifier
EIR Excess Information Rate
EMS Element Management System used to manage single or set of Network Elements
EOC Embedded Operations Channel
EOS End of Sequence control code (LCAS)
EPON Ethernet-based Passive Optical Network
ERDI Enhanced Remote Defect Indicator (SONET)
ESCON Enterprise Systems Connection protocol
ETSI European Telecommunications Standards Institute
ES Errored Seconds

FCAPS Fault, Configuration, Accounting, Performance and Security
FCC (U.S.) Federal Communications Commission
FCS Frame Check Sequence
FDM Frequency Division Multiplexing
FE Fairness Eligible (RPR)
FEC Forward Error Correction
FICON Fiber Connection protocol
FN Fiber Node
FR Frame Relay
FSAN Full Service Access Network consortium
FTFL Fault Type and Fault Location
FTTx Fiber To The X (e.g., X = Home, Curb, Cabinet, Building)

G.ASON Automatic Switched Optical Network. Generic name sometimes given to the ITU-T architecture, framework, and suite of standards related to setting up dynamic end-to-end

KSP K-Shortest path First algorithm

L2VPN Layer 2 Virtual Private Network
LAN Local Area Network
LAPD Link Access Protocol for the D-channel
LAPS Link Access Protocol – SDH/SONET
LCAS Link Capacity Adjustment Scheme
LEC Local Exchange Carrier
LLA Logical Layered Architecture
LLC Logical Link Control
LMDS Local Multipoint Distribution System
LOVC Low-Order Virtual Container
LRTT Loop Round Trip Time (RPR)
LSB Least Significant Bit
LSR Label Switched Router (MPLS)
LTE Line Terminating Equipment/Element (SONET)

MAC Medium Access Control protocol
MAF Management Application Function
MAN Metropolitan Area Network
MCF Message Communication Function
MDA Model Driven Architecture
MEF Metro Ethernet Forum
MEMS Micro-Electrical-Mechanical Switches
MF Mediation Function (sometimes Multi-Frame)
MFAS Multiframe Alignment Signal
MFI Multi-Frame Indicator
MIB Management Information Base
MMDS Multi-channel Multipoint Distribution Systems
MMS Multi-Media Standards
MOC Managed Object Class
MPLS Multi-Protocol Label Switching
MS Multiplex Section (SDH) or Manual Switch
MSB Most Significant Bit
MSO Multi-service Operator
MST Member Status (LCAS)

NE Network Element – an element in the network that provides functions such as switching and transport of electrical or optical transport signals
NEF Network Element Function

NJO	Negative Justification Opportunity
NM	Network Management
NMS	Network Management Systems
NNI	Network-to-Network Interface
NRZ	Non-Return to Zero line code

OA	Optical Alignment bytes (OTN)
OADM	Optical ADM
OAM	Operations, Administration, and Maintenance
OAM&P	Operations, Administration, Maintenance & Provisioning
OC-n	Optical Carrier for a SONET STS-n
OCC	Optical Carrier Channel
OCh	Optical Channel (OTN)
ODI	Outgoing Defect Indication
OEI	Outgoing Error Indication
ODU	Optical Data Unit (OTN)
OH	Overhead
OIF	Optical Industry Forum
OLT	Optical Line Terminal (for PON)
OMG	Object Management Group
OMS	Optical Multiplex Section (OTN)
ONU	Optical Network Unit (for PON)
Op-Ex	Operations Expense
OPTX	Optical Transmission and Synchronization Committee (formerly T1X1)
OPU	Optical Payload Unit (OTN)
OSMINE	Telcordia's Operations Systems Modification for the Integration of Network Elements
OSPF	Open Shortest Path Routing protocol used widely in the Internet and considered for G.ASON as a candidate to implement E-NNI routing
OTN	Optical Transport Network
OTS	Optical Transport Section (OTN)
OTU	Optical Transport Unit (OTN)
OVH	Overhead
OVTG	Optical Virtual Tributary Group
OXC	Optical DCS

PDH	Plesiochronous Digital Hierarchy
PDI	Payload Defect Indicator (SONET)
PDU	Protocol Data Unit
pFCS	payload FCS (GFP)

PJO	Positive Justification Opportunity
PLI	Payload Length Indicator (GFP)
PLM	Payload Label Mismatch (SONET/SDH)
PM	Performance Monitoring
POH	Path Overhead
PON	Passive Optical Network
POP	Point of Presence
PoS	Packet over SONET/SDH
PPP	Point-to-Point Protocol (IETF)
PRBS	Pseudo-Random Bit Sequence
PSTN	Public Switched Telephone Network
PTE	Path Terminating Equipment/Element (SONET/SDH)
PTI	Payload Type Indicator (GFP)
PTQ	Primary Transit Queue
PTT	Postal Telephone and Telegraph administration
QoS	Quality of Service
RDI	Remote Defect Indicator (SONET/SDH)
RDT	Remote Digital Terminal
REI	Remote Error Indicator (SONET/SDH)
RFI	Remote Failure Indication (SONET VT1.5)
RM	Remote Multiplexer
RPR	Resilient Packet Ring
RS	Regenerator Section (SDH) or Reconciliation Sublayer (RPR)
RS	Reed-Solomon FEC
RS-ACK	Re-sequence Acknowledgement (LCAS)
SAN	Storage Area Network
SD	Signal Degrade
SDCC	Section Data Communications Channel (SONET)
SDH	Synchronous Digital Hierarchy
SES	Severely Errored Seconds
SF	Signal Fail (sometimes, Security Functions)
SHN	Self Healing network
SLA	Service Level Agreement
SM	Service Management
SMC	SONET Minimum Clock
SNC	Subnetwork Connection
SNCP	Sub-Network Connection Protection
SNMP	Simple Network Management Protocol

SOAP	Simple Object Access Protocol
SOH	Section Overhead (SONET)
SONET	Synchronous Optical NETwork
SPE	Synchronous Payload Envelope (SONET)
SPR	Span Restoration
SQ	Sequence number (LCAS)
SS7	Signaling System No 7 (also called CCS #7)
STE	Section Terminating Equipment/Element
STM	Synchronous Transport Module (SDH)
STQ	Secondary Transit Queue (RPR)
STS	Synchronous Transport Signal (SONET)

TC	Tandem Connection
TCM	Tandem Connection Monitoring
TCN	Telecommunications Network
TDM	Time Division Multiplexing
TDMA	Time Division Multiple Access
TF	Transformation Function
TIRKS	Trunk Inventory Record Keeping System
TL1	Transaction Language 1
TMF	Telemanagement Forum
TMN	Telecommunication Management Network
TMOC	Transmission Management and Operations Committee (formerly T1M1)
TOH	Transport Overhead
TTI	Trail Trace Identifier (SDH and OTN)
TTL	Time To Live
TU	Tributary Unit (SDH)
TUG	Tributary Unit Group (SDH)

UML	Unified Modeling Language
UNI	User-to-Network Interface
UPI	User Payload Indicator (GFP)
UPSR	Unidirectional Path-Switched Ring
UTRAD	Unified TMN Requirements Analysis and Design

VC	Virtual Container (SDH)
VCAT	Virtual Concatenation
VCG	Virtually Concatenated Group
VDSL	Very high-speed Digital Subscriber Loop
VoIP	Voice over Internet Protocol
VT	Virtual Tributary (SONET)

VTG	Virtual Tributary Group (SONET)
WAN	Wide Area Network
WDM	Wave Division Multiplexing
WSF	Workstation Function
WSSF	Workstation Support Function
WTR	Wait-To-Restore
XML	Extensible Markup Language
XOR	Exclusive OR logic operation

Author Biographies

Manohar Naidu Ellanti is a software project management consultant. He received his M.Sc (Chemistry) and B.E (EEE) from Birla Institute of Tech. and Science, Pilani, India in 1990. He has over 14 years of experience in hardware and software development. He designed hardware for network synchronization, cellular base station system. He was involved in development of FTTx/ATM based multimedia interactive system for VOD/Gaming, in design and development of software for network management system related to SONET/SDH mesh-networks composed of Digital Cross Connects; in architecting MPLS IP/QoS control software for an access device. Also associated with development of OSPF infrastructure for DCN, management transport protocols, control plane and OIF UNI. He holds one patent and two pending.

Steven Scott Gorshe is a Principal Engineer with PMC-Sierra's Product Research Group. He received his B.S.E.E. (University of Idaho), and M.S.E.E. and Ph.D. (Oregon State University) in 1979, 1982, and 2002, respectively. He has been involved in applied research and the development of transmission and access system architectures and ASICs since 1982, including over five years at GTE and over 12 years with NEC America where he became Chief Architect for NEC Eluminant Technologies. His current work at PMC-Sierra involves technology development for applications specific standard product ICs, including those for Ethernet WAN transport over telecommunications networks. Dr. Gorshe is a Senior Member of the IEEE and Co-Editor for the regular Broadband Access series and guest editor for multiple Feature Topics in the *IEEE Communications Magazine*. He has also been involved in telecommunications network standards continuously since 1985 and serves as Senior Editor for OPTSX (formerly T1X1, responsible for North American

SONET and optical network interface standards); technical editor for multiple standards within the SONET series; and a technical editor for multiple ITU-T Recommendations including G.7041 (GFP), G.8011.1 (Ethernet Private Line Service), and G.7043 (Virtual Concatenation of PDH Signals). Areas in which he has made key contributions include architectures for multiple generations of SONET/SDH equipment and much of the transparent GFP protocol. He is a recipient of the Committee T1 Alvin Lai Outstanding Achievement Award for his standards contributions. He has 26 patents issued or pending and multiple published papers.

Lakshmi G. Raman is a Senior Director of Carrier Platform Group in Radisys Corporation. She received her Ph.D in Physics (IIT Madras) and a second masters in Computer Engineering (U Mass Amherst) in 1974 and 1986 respectively. She has been responsible for developing network management standards in Bellcore (now Telcordia), ITU, and T1. In these forums she has been a technical editor for several information models, served as working party chair and rapporteur for over 18 years. In addition to standards development, she has been responsible for engineering products in HFC Access Platforms, Optical Cross Connect System, Collaborative visualization transport system. She has published a book and authored a chapter in the IEEE Network Management Series. She received the first T1 Alvin Lai Outstanding Achievement Award for her standards contributions. She has 1 patent pending and published papers in Physics and Network Management.

Wayne D. Grover- holds a B.Eng. (EE) with High Distinction from Carleton University, Ottawa, an M.Sc.(EE Science) from the University of Essex, England, and a Ph.D. (EE) from the University of Alberta (1989). He was a *Commonwealth Scholar* to the United Kingdom in 1980-81. Dr. Grover has produced issued patents on 26 topics, 58 journal publications, three book chapters and over 100 technical reports and conference papers. In Fall 2003 his book *Mesh-based Survivable Networks: Options and Strategies for Optical, MPLS, SONET and ATM Networking* was published by Prentice-Hall (841 pages plus web-based appendices). Previously he had 10 years experience as scientific staff and management at BNR (now Nortel Networks) on fiber optics, switching systems, digital radio and network planning areas before joining TRLabs as its founding Technical VP in 1986. In this position he was responsible for the development of the TRLabs research program and the TRLabs sponsorship base from start-up to over the 100-person level. He now functions as Chief Scientist-Network Systems at TRLabs and as Professor, Electrical and Computer Engineering, at the University of Alberta. In 1999 he received the *IEEE Baker Prize Paper Award* for his paper "Self-organizing broadband transport networks" in the Oct. 1997 *IEEE Proceedings*. Other research contributions are in the areas of SONET format design, high-speed synchronization, precise time transfer, wireless traffic analysis, rate-adaptive subscriber loops, radio-location in wireless systems and availability analysis of transport networks. Dr. Grover was an *NSERC E.W.R. Steacie Fellow* for 2001-2002. Previously he was the *McCalla Professor in Engineering* and recipient of the *Martha Cook-Piper Research Prize*, the *"Smart City" Award* (City of Edmonton) and a *Technology Commercialization Award* from TR*Labs* (1997) for the licensing of technology to industry. In 2002 he was cited as an *IEEE Fellow* "for contributions to survivable and self-organizing broadband transport networks," and *Fellow of the Engineering Institute of Canada*. In 1999 he was the Canadian *Outstanding Electrical Engineer* of the year. He recently served as General Chair for the 4*th* *International Workshop on Design of Reliable Networks (DRCN 2003)*, Banff, Alberta, October 19-22, 2003.

Index